Lecture Notes in Computer Science 11194

Commenced Publication in 1973
Founding and Former Series Editors:
Gerhard Goos, Juris Hartmanis, and Jan van Leeuwen

More information about this series at http://www.springer.com/series/7407

Jules Desharnais · Walter Guttmann
Stef Joosten (Eds.)

Relational and Algebraic Methods in Computer Science

17th International Conference, RAMiCS 2018
Groningen, The Netherlands, October 29 – November 1, 2018
Proceedings

Springer

Editors
Jules Desharnais
Université Laval
Québec, QC
Canada

Walter Guttmann
University of Canterbury
Christchurch
New Zealand

Stef Joosten
Open Universiteit
Heerlen
The Netherlands

ISSN 0302-9743 ISSN 1611-3349 (electronic)
Lecture Notes in Computer Science
ISBN 978-3-030-02148-1 ISBN 978-3-030-02149-8 (eBook)
https://doi.org/10.1007/978-3-030-02149-8

Library of Congress Control Number: 2015949476

LNCS Sublibrary: SL1 – Theoretical Computer Science and General Issues

This Springer imprint is published by the registered company Springer Nature Switzerland AG
The registered company address is: Gewerbestrasse 11, 6330 Cham, Switzerland

Preface

This volume contains the proceedings of the 17th International Conference on Relational and Algebraic Methods in Computer Science (RAMiCS 2018), which was held in Groningen, The Netherlands, from October 29 to November 1, 2018.

The plan to initiate this series of conferences was put in place during the 38th Banach Semester on Algebraic Methods in Logic and Their Computer Science Application in Warsaw, Poland, in September and October 1991. The first numbered occurrence was a Dagstuhl seminar on Relational Methods in Computer Science (RelMiCS 1), held in Germany in 1994. From then on and until 2009, at intervals of about one year and a half, there was a RelMiCS conference. Starting in 2003, RelMiCS conferences were held jointly with Applications of Kleene Algebras (AKA) conferences. At RelMiCS 11/AKA 6 in Doha, Qatar, it was decided to use the current name for the series, that is, Relational and Algebraic Methods in Computer Science (RAMiCS).

Recurrent topics of RAMiCS conferences include semiring- and lattice-based structures such as relation algebras and Kleene algebras, their connections with program logics and other logics, their use in theories of computing, their formalization with theorem provers, and their application to modeling and reasoning about computing systems and processes.

In total, 30 papers were submitted to RAMiCS 2018 and the Program Committee selected 21 of them for presentation at the conference. In these proceedings the selected papers are grouped under three headings: "Theoretical Foundations," "Reasoning About Computations and Programs," and "Applications and Tools." Each submission was evaluated by at least four independent reviewers, and further discussed electronically during two weeks. The chairs are very grateful to all Program Committee members and to the external reviewers for their time and expertise.

Roland Backhouse, Manuel Bodirsky, and Philippa Gardner kindly accepted our invitation to present their research at the conference. The abstracts of Roland Backhouse's talk, "The Importance of Factorisation in Algorithm Design," and Philippa Gardner's talk, "Scalable Reasoning About Concurrent Programs," are included in the proceedings. So is the full paper related to Manuel Bodirsky's talk, "Finite Relation Algebras with Normal Representations." This conference featured a tutorial on motivating students for relation algebra presented by one of us (Stef Joosten). The abstract of this tutorial is also included in this volume. An experimental session of this conference was a meeting with software engineers who work in practice, most of whom have little contact with the theories discussed in our conference. This meeting was planned as a way to foster applications of relational and algebraic methods in computer science.

We thank the RAMiCS Steering Committee for their support and the Open Universiteit Nederland for organizing this conference. We gratefully acknowledge the financial support of Ordina Nederland BV for their sponsorship and their help with organizing this conference. We also thank the Groningen Congres Bureau and the Rijksuniversiteit Groningen for the warm welcome we received in the Academiegebouw and the city itself. A special thank you is due to Sebastiaan J. C. Joosten, author in previous editions and this time a very active, talented publicity chair.

We also appreciate the excellent facilities offered by the EasyChair conference administration system, and Alfred Hofmann and Anna Kramer's help in publishing this volume with Springer. Finally, we owe much to all authors and participants for their support of this RAMiCS conference.

October 2018

Jules Desharnais
Walter Guttmann
Stef Joosten

Organization

Organizing Committee

Conference Chair

Stef Joosten	Open Universiteit, The Netherlands

PC Chairs

Jules Desharnais	Université Laval, Canada
Walter Guttmann	University of Canterbury, New Zealand

Publicity Chair

Sebastiaan Joosten	Universiteit Twente, The Netherlands

Program Committee

Luca Aceto	Reykjavík University, Iceland Gran Sasso Science Institute, Italy
Rudolf Berghammer	Christian-Albrechts-Universität zu Kiel, Germany
Jules Desharnais	Université Laval, Canada
Uli Fahrenberg	École Polytechnique, France
Hitoshi Furusawa	Kagoshima University, Japan
Walter Guttmann	University of Canterbury, New Zealand
Robin Hirsch	University College London, UK
Peter Höfner	Data61, CSIRO, Australia
Marcel Jackson	La Trobe University, Australia
Jean-Baptiste Jeannin	University of Michigan, USA
Peter Jipsen	Chapman University, USA
Stef Joosten	Open Universiteit, The Netherlands
Wolfram Kahl	McMaster University, Canada
Barbara König	Universität Duisburg-Essen, Germany
Dexter Kozen	Cornell University, USA
Agi Kurucz	King's College London, UK
Tadeusz Litak	Friedrich-Alexander Universität Erlangen-Nürnberg, Germany
Roger Maddux	Iowa State University, USA
Annabelle McIver	Macquarie University, Australia
Szabolcs Mikulás	Birkbeck, University of London, UK
Ali Mili	New Jersey Institute of Technology, USA
Bernhard Möller	Universität Augsburg, Germany
José N. Oliveira	Universidade do Minho, Portugal
Alessandra Palmigiano	Technische Universiteit Delft, The Netherlands

Damien Pous	CNRS, France
Mehrnoosh Sadrzadeh	Queen Mary University of London, UK
John Stell	University of Leeds, UK
Georg Struth	University of Sheffield, UK
Michael Winter	Brock University, Canada

Steering Committee

Rudolf Berghammer	Christian-Albrechts-Universität zu Kiel, Germany
Jules Desharnais	Université Laval, Canada
Ali Jaoua	Qatar University, Qatar
Peter Jipsen	Chapman University, USA
Bernhard Möller	Universität Augsburg, Germany
José N. Oliveira	Universidade do Minho, Portugal
Ewa Orłowska	National Institute of Telecommunications, Poland
Gunther Schmidt	Universität der Bundeswehr München, Germany
Michael Winter (Chair)	Brock University, Canada

Additional Reviewers

Musa Al-Hassy
Benjamin Cabrera
Christian Doczkal
Jérémy Dubut
Sabine Frittella
Ilias Garnier
Roland Glück
Giuseppe Greco
Lucy Ham
Sebastiaan Joosten
Fei Liang
Konstantinos Mamouras
Koki Nishizawa
Filip Sieczkowski
Jorge Sousa Pinto
Norihiro Tsumagari
Thorsten Wißmann

Sponsors

Ordina Nederland BV (Main Sponsor)
Groningen Congres Bureau

Local Organizers

Chrisja Muris	Open Universiteit
Heleen Bakker	Open Universiteit
Hillebrand Meijer	Ordina
Erick Koster	Ordina
Astrid Patz	Ordina
Margot Spee	Groningen Congres Bureau
Kelly Scholtens	Groningen Congres Bureau

Abstracts of Invited Talks

Scalable Reasoning About Concurrent Programs

Philippa Gardner

Imperial College London, UK
p.gardner@imperial.ac.uk

Scalable reasoning about complex concurrent programs interacting with shared memory is a fundamental, open research problem. Developers manage the complexity of concurrent software systems by designing software components that are *compositional* and *modular*. With compositionality, a developer designs local subcomponents with well-understood *interfaces* that connect to the rest of the system. With modularity, a developer designs reusable subcomponents with *abstract* software interfaces that can hide the complexity of the subcomponents from the rest of the system. The challenge is to develop compositional, modular reasoning of concurrent programs, which follows the intuitions of the developer in how to structure their software components with precisely defined specifications of software interfaces. These specifications should not leak implementation details and should be expressed at the level of abstraction of the client.

I will describe the work done by my group and others on compositional and modular reasoning about concurrent programs using modern concurrent separation logics. I will present work on reasoning about *safety properties*, highlighting the CAP logic [ECOOP'10] which introduced logical abstraction (the fiction of separation) to concurrent separation logics and the TaDA logic [ECOOP'14] which introduced abstract atomicity (the fiction of atomicity). I will also present new work on *progress properties*, introducing the TaDA-Live logic for reasoning about the termination of blocking programs. I will demonstrate the subtlety of the reasoning using a simple lock module, and also compare this work with linearizability, contextual refinement and other concurrent separation logics.

Papers to read:

O'Hearn, P.W.: Resources, concurrency, and local reasoning. Theor. Comput. Sci. **375** (1–3), 271–307 (2007)
http://www0.cs.ucl.ac.uk/staff/p.ohearn/papers/concurrency.pdf

Dinsdale-Young, T., da Rocha Pinto, P., Gardner, P.: A perspective on specifying and verifying concurrent modules. J. Logical Algebraic Methods Program. **98**, 1–25 (2018)
https://www.doc.ic.ac.uk/~pg/publications/Dinsdale-Young2018perspective.pdf

The Importance of Factorisation in Algorithm Design

Roland Backhouse

School of Computer Science, University of Nottingham
rcb@cs.nott.ac.uk

In 1971, J.H. Conway [Con71] published a slim volume entitled "Regular Algebra and Finite Machines" which was to have great influence on my own work (eg. [Bac06]). I was particularly impressed by the chapter on factor theory and its subsequent application in the construction of biregulators. Although some elements of Conway's book are now well cited, this part of the book still appears to be much less well known. The goal of this talk is to explain why factor theory is important in the design of algorithms.

We introduce Conway's factor matrix and then show how the (unique) reflexive-transitive-reduction of the factor matrix, dubbed the "factor graph" [Bac75], is the basis of the well-known Knuth-Morris-Pratt pattern-matching algorithm [KMP77, BL77]. This serves as an appetiser for a review of fixed-point theory and Galois connections, focusing particularly on the relevance of the theory in the design of algorithms.

We then return to factor theory and how it forms the basis of practical applications in program analysis [SdML04]. We conclude with some speculation on how a greater focus on factorisation might help us to better understand the complexity of algorithms.

References

[Bac75] Backhouse, R.C.: Closure algorithms and the star-height problem of regular languages. Ph.D. thesis, University of London (1975). Scanned-in copy of the chapters on factor theory available from www.cs.nott.ac.uk/~psarb2/MPC/FactorGraphs.pdf

[Bac06] Backhouse, R.: Regular algebra applied to language problems. J. Log. Algebraic Program. **66**, 71–111 (2006)

[BL77] Backhouse, R.C., Lutz, R.K.: Factor graphs, failure functions and bi-trees. In: Salomaa, A., Steinby, M. (eds.) ICALP 1977. LNCS, vol. 52, pp. 61–75. Springer, Heidelberg (1977)

[Con71] Conway, J.H.: Regular Algebra and Finite Machines. Chapman and Hall, London (1971)

[KMP77] Knuth, D.E., Morris, J.H., Pratt, V.R.: Fast pattern matching in strings. SIAM J. Comput. **6**, 325–350 (1977)

[SdML04] Sittampalam, G., de Moor, O., Larssen, K.F.: Incremental execution of transformation specifications. In: Proceedings of the 31st SIGPLAN-SIGACT Symposium on Principles of Programming Languages, POPL 2004, vol. 39. ACM SIGPLAN Notices, pp. 26–38, January 2004

Tutorial: Relation Algebra in the Classroom with Ampersand

Stef Joosten

Open University of the Netherlands
stef.joosten@ou.nl

This tutorial explores a way to motivate students for relation algebras by applying it in software engineering. Participants will get hands-on experience with Ampersand [1], which is a compiler that transforms a specification in relation algebra in working software. They will be directed to the documentation site [2] for more precise information about the language and tools.

This tutorial starts with a motivation: Why might working with Ampersand motivate students for relation algebra? Then the participants will create an information system online, just like students do in the course Rule Based Design [3]. For this purpose we ask participants to bring their laptops. The presentation proceeds by pointing out which learning points are relevant in the tutorial. It finalizes by giving an overview in the available materials and tools. All materials are freely available in open source, so participants can take it to their own classrooms.

Professors who want to use these materials are cordially invited to partake in the further development.

Background

Ampersand was originally intended as a means to specify requirements [4] in heterogeneous relation algebra [5]. The toolset evolved into a tool for students [6]. The Ampersand toolset has been used since 2013 [7] at the Open University in two courses and numerous research assignments.

The novel feature of Ampersand is that a theory in relation algebra is being used as a database program. It specifies persistence (i.e. the database) and user interfaces. This is achieved by using one interpretation of the algebra: a relation is interpreted as a finite set of pairs.

The user gets a programming language that is declarative, strongly typed, and easily subjected to proofs of correctness [8]. The benefits are fast development (because the Ampersand compiler generates working software), maintainability (because software is easily divided into independent chunks), and adaptability (because generating an application and deploying it is automated).

References

1. Joosten, S.: RAP3 (2009–2018). rap.cs.ou.nl/RAP3
2. Joosten, S., Hageraats, E.: Documentation of Ampersand (2016–2018). ampersandtarski.gitbook.io/documentation/
3. Rutledge, L., Wetering, R.v.d., Joosten, S.: Course IM0103 Rule Based Design for CS (2009–2018). www.ou.nl/-/IM0403_Rule-Based-Design
4. Dijkman, R.M., Ferreira Pires, L., Joosten, S.M.M.: Calculating with concepts: a technique for the development of business process support. In: Evans, A. (ed.) Proceedings of the UML 2001 Workshop on Practical UML-Based Rigorous Development Methods: Countering or Integrating the eXstremists, Lecture Notes in Informatics, vol. 7. FIZ, Karlsruhe (2001)
5. van der Woude, J., Joosten, S.: Relational heterogeneity relaxed by subtyping. In: de Swart, H. (ed.) RAMICS 2011. LNCS, vol. 6663, pp. 347–361. Springer, Heidelberg (2011)
6. Michels, G., Joosten, S., van der Woude, J., Joosten, S.: Ampersand: applying relation algebra in practice. In: Proceedings of the 12th Conference on Relational and Algebraic Methods in Computer Science RAMICS 2011. LNCS, vol. 6663, pp. 280–293. Springer-Verlag, Berlin (2011)
7. Michels, G., Joosten, S.: Progressive development and teaching with RAP. In: Proceedings of the 3rd Computer Science Education Research Conference on Computer Science Education Research, CSERC 2013, pp. 3:33–3:43. Open University, Heerlen, The Netherlands (2013)
8. Joosten, S.: Relation algebra as programming language using the ampersand compiler. J. Log. Algebraic Methods Program. **100**, 113–129 (2018)

Contents

Reasoning About Computations and Programs

Applications and Tools

Invited Paper

Finite Relation Algebras with Normal Representations

Manuel Bodirsky(✉)

Institut für Algebra, TU Dresden, 01062 Dresden, Germany
manuel.bodirsky@tu-dresden.de

Abstract. One of the traditional applications of relation algebras is to provide a setting for infinite-domain constraint satisfaction problems. Complexity classification for these computational problems has been one of the major open research challenges of this application field. The past decade has brought significant progress on the theory of constraint satisfaction, both over finite and infinite domains. This progress has been achieved independently from the relation algebra approach. The present article translates the recent findings into the traditional relation algebra setting, and points out a series of open problems at the interface between model theory and the theory of relation algebras.

1 Introduction

One of the fundamental computational problems for a relation algebra **A** is the *network satisfaction problem for* **A**, which is to determine for a given **A**-network N whether it is satisfiable in some representation of **A** (for definitions, see Sects. 2 and 3). Robin Hirsch named in 1995 the *Really Big Complexity Problem (RBCP)* for relation algebras, which is to *'clearly map out which (finite) relation algebras are tractable and which are intractable'* [Hir96]. For example, for the Point Algebra the network satisfaction problem is in P and for Allen's Interval Algebra it is NP-hard. One of the standard methods to show that the network satisfaction problem for a finite relation algebra is in P is via establishing local consistency. The question whether the network satisfaction problem for **A** can be solved by local consistency methods is another question that has been studied intensively for finite relation algebras **A** (see [BJ17] for a survey on the second question).

If **A** has a fully universal square representation (we follow the terminology of Hirsch [Hir96]) then the network satisfaction problem for **A** can be formulated as a constraint satisfaction problem (CSP) for a countably infinite structure. The complexity of constraint satisfaction is a research direction that has seen quite some progress in the past years. The *dichotomy conjecture* of Feder and

Manuel Bodirsky—The author has received funding from the European Research Council under the European Community's Seventh Framework Programme (FP7/2007-2013 Grant Agreement no. 257039, CSP-Infinity).

J. Desharnais et al. (Eds.): RAMiCS 2018, LNCS 11194, pp. 3–17, 2018.
https://doi.org/10.1007/978-3-030-02149-8_1

Vardi from 1993 states that every CSP for a finite structure is in P or NP-hard; the *tractability conjecture* [BKJ05] is a stronger conjecture that predicts precisely which CSPs are in P and which are NP-hard. Two independent proofs of these conjectures appeared in 2017 [Bul17,Zhu17], based on concepts and tools from universal algebra. An earlier result of Barto and Kozik [BK09] gives an exact characterisation of those finite-domain CSPs that can be solved by local consistency methods.

Usually, the network satisfaction problem for a finite relation algebra **A** cannot be formulated as a CSP for a finite structure. However, suprisingly often it can be formulated as a CSP for a countably infinite ω-*categorical* structure $\mathfrak{B}$. For an important subclass of ω-categorical structures we have a tractability conjecture, too. The condition that supposedly characterises containment in P can be formulated in many non-trivially equivalent ways [BKO+17,BP16,BOP17] and has been confirmed in numerous special cases, see for instance the articles [BK08,BMPP16,KP17,BJP17,BMM18,BM16] and the references therein.

In the light of the recent advances in constraint satisfaction, both over finite and infinite domains, we revisit the RBCP and discuss the current state of the art. In particular, we observe that if **A** has a *normal representation* (again, we follow the terminology of Hirsch [Hir96]), then the network satisfaction problem for **A** falls into the scope of the infinite-domain tractability conjecture. We also show that there is an algorithm that decides for a given finite relation algebra **A** with a fully universal square representation whether **A** has a normal representation. (In other words, there is an algorithm that decides for a given **A** whether the class of atomic **A**-networks has the amalgamation property.) The scope of the tractability conjecture is larger, though. We describe an example of a finite relation algebra which has an ω-categorical fully universal square representation (and a polynomial-time tractable network satisfaction problem) which is not normal, but which does fall into the scope of the conjecture.

Whether the infinite-domain tractability conjecture might contribute to the resolution of the RBCP in general remains open; we present several questions in Sect. 7 whose answer would shed some light on this question. These questions concern the existence of ω-categorical fully universal square representations and are of independent interest, and in my view they are central to the theory of representable finite relation algebras.

2 Relation Algebras

A *proper relation algebra* is a set B together with a set $\mathcal{R}$ of binary relations over B such that

1. $\mathrm{Id} := \{(x,x) \mid x \in B\} \in \mathcal{R}$;
2. If R_1 and R_2 are from $\mathcal{R}$, then $R_1 \vee R_2 := R_1 \cup R_2 \in \mathcal{R}$;
3. $1 := \bigcup_{R \in \mathcal{R}} R \in \mathcal{R}$;
4. $0 := \emptyset \in \mathcal{R}$;
5. If $R \in \mathcal{R}$, then $-R := 1 \setminus R \in \mathcal{R}$;
6. If $R \in \mathcal{R}$, then $R^{\smile} := \{(x,y) \mid (y,x) \in R\} \in \mathcal{R}$;

7. If R_1 and R_2 are from $\mathcal{R}$, then $R_1 \circ R_2 \in \mathcal{R}$; where

$$R_1 \circ R_2 := \{(x,z) \mid \exists y((x,y) \in R_1 \wedge (y,z) \in R_2)\} .$$

We want to point out that in this standard definition of proper relation algebras it is *not* required that 1 denotes B^2. However, in most examples, 1 indeed denotes B^2; in this case we say that the proper relation algebra is *square.* The inclusion-wise minimal non-empty elements of $\mathcal{R}$ are called the *basic relations* of the proper relation algebra.

Example 1 (The Point Algebra). Let $B = \mathbb{Q}$ be the set of rational numbers, and consider

$$\mathcal{R} = \{\emptyset, =, <, >, \leq, \geq, \neq, \mathbb{Q}^2\} .$$

Those relations form a proper relation algebra (with the basic relations $<, >, =$, and where 1 denotes $\mathbb{Q}^2$) which is known under the name *point algebra.* □

The *relation algebra associated to* $(B, \mathcal{R})$ is the algebra $\mathbf{A}$ with the domain $A := \mathcal{R}$ and the signature $\tau := \{\vee, -, 0, 1, \circ, ^\smile, \mathrm{Id}\}$ obtained from $(B, \mathcal{R})$ in the obvious way. An *abstract relation algebra* is a τ-algebra that satisfies some of the laws that hold for the respective operators in a proper relation algebra. We do not need the precise definition of an abstract relation algebra in this article since we deal exclusively with *representable* relation algebras: a *representation* of an abstract relation algebra $\mathbf{A}$ is a relational structure $\mathfrak{B}$ whose signature is A; that is, the elements of the relation algebra are the relation symbols of $\mathfrak{B}$. Each relation symbol $a \in A$ is associated to a binary relation $a^{\mathfrak{B}}$ over B such that the set of relations of $\mathfrak{B}$ induces a proper relation algebra, and the map $a \mapsto a^{\mathfrak{B}}$ is an isomorphism with respect to the operations (and constants) $\{\vee, -, 0, 1, \circ, ^\smile, \mathrm{Id}\}$. In this case, we also say that $\mathbf{A}$ is the *abstract relation algebra of* $\mathfrak{B}$. An abstract relation algebra that has a representation is called *representable.* For $x, y \in A$, we write $x \leq y$ as a shortcut for the partial order defined by $x \vee y = y$. The minimal elements of $A \setminus \{0\}$ with respect to $\leq$ are called the *atoms* of $\mathbf{A}$. In every representation of $\mathbf{A}$, the atoms denote the basic relations of the representation. We mention that there are abstract finite relation algebras that are not representable [Lyn50], and that the question whether a finite relation algebra is representable is undecidable [HH01].

Example 2. The (abstract) point algebra is a relation algebra with 8 elements and 3 atoms, $=$, $<$, and $>$, and can be described as follows. The values of the composition operator for the atoms of the point algebra are shown in the table of Fig. 1. Note that this table determines the full composition table. The inverse $(<)^\smile$ of $<$ is $>$, and Id denotes $=$ which is its own inverse. This fully determines the relation algebra. The proper relation algebra with domain $\mathbb{Q}$ presented in Example 1 is a representation of the point algebra. □

$\circ$	$=$	$<$	$>$
$=$	$=$	$<$	$>$
$<$	$<$	$<$	1
$>$	$>$	1	$>$

Fig. 1. The composition table for the basic relations in the point algebra.

3 The Network Satisfaction Problem

Let **A** be a finite relation algebra with domain A. An **A**-*network* $N = (V; f)$ consists of a finite set of nodes V and a function $f \colon V \times V \to A$.

A network N is called

- *atomic* if the image of f only contains atoms of **A** and if

$$f(a,c) \leq f(a,b) \circ f(b,c) \text{ for all } a,b,c \in V \tag{1}$$

  (here we follow again the definitions in [Hir96]);
- *satisfiable in* $\mathfrak{B}$, for a representation $\mathfrak{B}$ of **A**, if there exists a map $s \colon V \to B$ (where B denotes the domain of $\mathfrak{B}$) such that for all $x, y \in V$

$$(s(x), s(y)) \in f(x,y)^{\mathfrak{B}};$$

- *satisfiable* if N is satisfiable in some representation $\mathfrak{B}$ of **A**.

The *(general) network satisfaction problem for a finite relation algebra* **A** is the computational problem to decide whether a given **A**-network is satisfiable. There are finite relation algebras **A** where this problem is undecidable [Hir99]. A representation $\mathfrak{B}$ of **A** is called

- *fully universal* if every atomic **A**-network is satisfiable in $\mathfrak{B}$;
- *square* if its relations form a proper relation algebra that is square.

The point algebra is an example of a relation algebra with a fully universal square representation. Note that if **A** has a fully universal representation, then the network satisfaction problem for **A** is decidable in NP: for a given network (V, f), simply select for each $x \in V^2$ an atom $a \in A$ with $a \leq f(x)$, replace $f(x)$ by a, and then exhaustively check condition (1). Also note that a finite relation algebra has a fully universal representation if and only if the so-called path-consistency procedure decides satisfiability of atomic **A**-networks (see, e.g., [BJ17, HLR13]).

However, not all finite relation algebras have a fully universal representation. An example of a relation algebra with 4 atoms which has a representation with seven elements but where path consistency of atomic networks does not imply consistency, called $\mathbf{B}_9$, has been given in [LKRL08]. A representation of $\mathbf{B}_9$ with domain $\{0, 1, \ldots, 6\}$ is given by the basic relations $\{R_0, R_1, R_2, R_3\}$ where $R_i = \{(x, y) \mid x + y = i \mod 7\}$, for $i \in \{0, 1, 2, 3\}$. In fact, every representation of $\mathbf{B}_9$ is isomorphic to this representation. Let N be the network (V, f) with $V = \{a, b, c, d\}$, $f(a, b) = f(c, d) = R_3$, $f(a, d) = f(b, c) = R_2$, $f(a, c) = f(b, d) = R_1$, $f(i, i) = R_0$ for all $i \in V$, and $f(i, j) = f(j, i)$ for all $i, j \in V$. Then N is atomic but not satisfiable.

4 Constraint Satisfaction Problems

Let $\mathfrak{B}$ be a structure with a (finite or infinite) domain B and a finite relational signature ρ. Then the *constraint satisfaction problem for* $\mathfrak{B}$ is the computational problem of deciding whether a finite ρ-structure $\mathfrak{C}$ homomorphically maps to $\mathfrak{B}$. Note that if $\mathfrak{B}$ is a square representation of $\mathbf{A}$, then the input $\mathfrak{C}$ can be viewed as an $\mathbf{A}$-network N. The nodes of N are the elements of $\mathfrak{C}$. To define $f(x,y)$ for variables x, y of the network, let $a_1, \ldots, a_k$ be a list of all elements $a \in A$ such that $(x,y) \in a^{\mathfrak{C}}$. Then define $f(x,y) = (a_1 \wedge \cdots \wedge a_k)$; if $k = 0$, then $f(x,y) = 1$. Observe that $\mathfrak{C}$ has a homomorphism to $\mathfrak{B}$ if and only if N is satisfiable in $\mathfrak{B}$ (here we use the assumption that $\mathfrak{B}$ is a square representation).

Conversely, when N is an $\mathbf{A}$-network, then we view N as the A-structure $\mathfrak{C}$ whose domain are the nodes of N, and where $(x,y) \in r^{\mathfrak{C}}$ if and only if $r = f(x,y)$. Again, $\mathfrak{C}$ has a homomorphism to $\mathfrak{B}$ if and only if N is satisfiable in $\mathfrak{B}$.

Proposition 1. *Let $\mathfrak{B}$ be a fully universal square representation of a finite relation algebra $\mathbf{A}$. Then the network satisfaction problem for $\mathbf{A}$ equals the constraint satisfaction problem for $\mathfrak{B}$ (up to the translation between $\mathbf{A}$-networks and finite A-structures presented above).*

Proof. We have to show that a network is satisfiable if and only if it has a homomorphism to $\mathfrak{B}$. Clearly, if N has a homomorphism to $\mathfrak{B}$ then it is satisfiable in $\mathfrak{B}$, and hence satisfiable. For the other direction, suppose that the $\mathbf{A}$-network $N = (V, f)$ is satisfiable in some representation of $\mathbf{A}$. Then there exists for each $x \in V^2$ an atomic $a \in A$ such that $a \leq f(x)$ and such that the network N' obtained from N by replacing $f(x)$ by a satisfies (1); hence, N' is atomic and satisfiable in $\mathfrak{B}$ since $\mathfrak{B}$ is fully universal. Hence, N is satisfiable in $\mathfrak{B}$, too.

For general infinite structures $\mathfrak{B}$ a systematic understanding of the computational complexity of CSP($\mathfrak{B}$) is a hopeless endeavour [BG08]. However, if $\mathfrak{B}$ is a *first-order reduct of a finitely bounded homogeneous structure* (the definitions can be found below), then the universal-algebraic tractability conjecture for finite-domain CSPs can be generalised. This condition is sufficiently general so that it includes fully universal square representations of almost all the concrete finite relation algebras studied in the literature, and the condition also captures the class of finite-domain CSPs. As we will see, the concepts of *finite boundedness* and *homogeneity* are conditions that have already been studied in the relation algebra literature.

4.1 Finite Boundedness

Let ρ be a relational signature, and let $\mathcal{F}$ be a set of ρ-structures. Then Forb($\mathcal{F}$) denotes the class of all finite ρ-structures $\mathfrak{A}$ such that no structure in $\mathcal{F}$ embeds into $\mathfrak{A}$. For a ρ-structure $\mathfrak{B}$ we write Age($\mathfrak{B}$) for the class of all finite ρ-structures that embed into $\mathfrak{B}$. We say that $\mathfrak{B}$ is *finitely bounded* if $\mathfrak{B}$ has a finite relational signature and there exists a finite set of finite τ-structures $\mathcal{F}$ such that Age($\mathfrak{B}$) = Forb($\mathcal{F}$). A simple example of a finitely bounded structure is $(\mathbb{Q}; <)$. It is easy to see that the constraint satisfaction problem of a finitely bounded structure $\mathfrak{B}$ is in NP.

Proposition 2. *Let* **A** *be a finite relation algebra with a fully universal square representation* $\mathfrak{B}$. *Then* $\mathfrak{B}$ *is finitely bounded.*

Proof (Proof sketch). Besides some bounds of size at most two that make sure that the atomic relations partition B^2, it suffices to include appropriate three-element structures into $\mathcal{F}$ that can be read off from the composition table of **A**.

4.2 Homogeneity

A relational structure $\mathfrak{B}$ is *homogeneous* (or *ultra-homogeneous* [Hod97]) if every isomorphism between finite substructures of $\mathfrak{B}$ can be extended to an automorphism of $\mathfrak{B}$. A simple example of a homogeneous structure is $(\mathbb{Q}; <)$.

A representation of a finite relation algebra **A** is called *normal* if it is square, fully universal, and homogeneous [Hir96]. The following is an immediate consequence of Propositions 1 and 2.

Corollary 1. *Let* **A** *be a finite relation algebra with a normal representation* $\mathfrak{B}$. *Then the network satisfaction problem for* **A** *equals the constraint satisfaction problem for a finitely bounded homogeneous structure.*

A versatile tool to construct homogeneous structures from classes of finite structures is *amalgamation* à la Fraïssé. We present it for the special case of *relational structures*; this is all that is needed here. An *embedding* of $\mathfrak{A}$ into $\mathfrak{B}$ is an isomorphism between $\mathfrak{A}$ and a substructure of $\mathfrak{B}$. An *amalgamation diagram* is a pair $(\mathfrak{B}_1, \mathfrak{B}_2)$ where $\mathfrak{B}_1, \mathfrak{B}_2$ are τ-structures such that there exists a substructure $\mathfrak{A}$ of both $\mathfrak{B}_1$ and $\mathfrak{B}_2$ such that all common elements of $\mathfrak{B}_1$ and $\mathfrak{B}_2$ are elements of $\mathfrak{A}$. We say that $(\mathfrak{B}_1, \mathfrak{B}_2)$ is a *2-point amalgamation diagram* if $|B_1 \setminus A| = |B_2 \setminus A| = 1$. A τ-structure $\mathfrak{C}$ is an *amalgam of* $(\mathfrak{B}_1, \mathfrak{B}_2)$ *over* $\mathfrak{A}$ if for $i = 1, 2$ there are embeddings f_i of $\mathfrak{B}_i$ to $\mathfrak{C}$ such that $f_1(a) = f_2(a)$ for all $a \in A$. In the context of relation algebras **A**, the amalgamation property can also be formulated with atomic **A**-networks, in which case it has been called the *patchwork property* [HLR13]; we stick with the model-theoretic terminology here since it is older and well-established.

Definition 1. *An isomorphism-closed class* $\mathcal{C}$ *of* τ*-structures has the* amalgamation property *if every amalgamation diagram of structures in* $\mathcal{C}$ *has an amalgam in* $\mathcal{C}$. *A class of finite* τ*-structures that contains at most countably many non-isomorphic structures, has the amalgamation property, and is closed under taking induced substructures and isomorphisms is called an* amalgamation class.

Note that since we only look at relational structures here (and since we allow structures to have an empty domain), the amalgamation property of $\mathcal{C}$ implies the *joint embedding property (JEP)* for $\mathcal{C}$, which says that for any two structures $\mathfrak{B}_1, \mathfrak{B}_2 \in \mathcal{C}$ there exists a structure $\mathfrak{C} \in \mathcal{C}$ that embeds both $\mathfrak{B}_1$ and $\mathfrak{B}_2$.

Theorem 1 (Fraïssé [Fra54, Fra86]; see [Hod97]). *Let C be an amalgamation class. Then there is a homogeneous and at most countable τ-structure $\mathfrak{C}$ whose age equals C. The structure $\mathfrak{C}$ is unique up to isomorphism, and called the* Fraïssé-limit *of C.*

The following is a well-known example of a finite relation algebra which has a fully universal square representation, but not a normal one.

Example 3. The *left linear point algebra* (see [Hir97, Dün05]) is a relation algebra with four atoms, denoted by $=$, $<$, $>$, and $|$. Here we imagine that '$x < y$' signifies that x is *earlier in time than* y. The idea is that at every point in time the past is linearly ordered; the future, however, is not yet determined and might branch into different worlds; incomparability of time points x and y is denoted by $x|y$. We might also think of $x < y$ as x *is to the left of* y if we draw points in the plane, and this motivates the name *left linear point algebra.* The composition operator on those four basic relations is given in Fig. 2. The inverse $(<)^{\smile}$ of $<$ is $>$, Id denotes $=$, and $|$ is its own inverse, and the relation algebra is uniquely given by this data. It is well known (for details, see [Bod04]) that the left linear point algebra has a fully universal square representation. On the other hand, the networks drawn in Fig. 3 show the failure of amalgamation.

$\circ$	$=$	$<$	$>$	$\|$
$=$	$=$	$<$	$>$	$\|$
$<$	$<$	$<$	$\{<,>\}$	$\{<,\|\}$
$>$	$>$	1	$>$	$\|$
$\|$	$\|$	$\|$	$\{>,\|\}$	1

Fig. 2. The composition table for the basic relations in the left linear point algebra.

An algorithm to test whether a finite relation algebra has a normal representation can be found in Sect. 6.

4.3 The Infinite-Domain Dichotomy Conjecture

The infinite-domain dichotomy conjecture applies to a class which is larger than the class of homogeneous finitely bounded structures. To introduce this class we need the concept of *first-order reducts.*

Suppose that two relational structures $\mathfrak{A}$ and $\mathfrak{B}$ have the same domain, that the signature of a structure $\mathfrak{A}$ is a subset of the signature of $\mathfrak{B}$, and that $R^{\mathfrak{A}} = R^{\mathfrak{B}}$ for all common relation symbols R. Then we call $\mathfrak{A}$ a *reduct of* $\mathfrak{B}$, and $\mathfrak{B}$ an *expansion of* $\mathfrak{A}$. In other words, $\mathfrak{A}$ is obtained from $\mathfrak{B}$ by dropping some of the relations. A *first-order reduct of* $\mathfrak{B}$ is a reduct of the expansion of $\mathfrak{B}$ by all relations that are first-order definable in $\mathfrak{B}$. The CSP for a first-order reduct of a finitely bounded homogeneous structure is in NP (see [Bod12]). An example of a structure which is not homogeneous, but a reduct of finitely

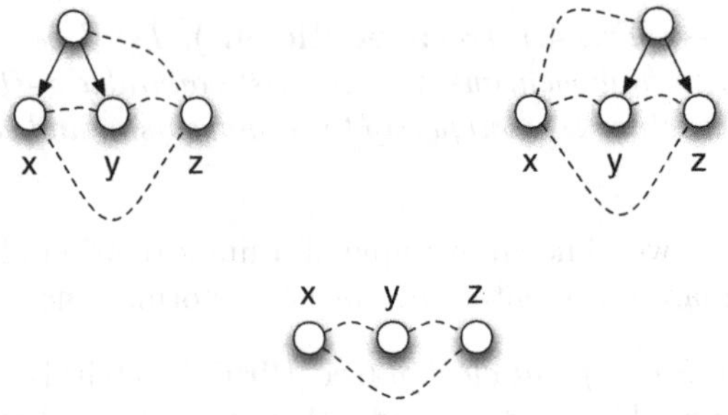

Fig. 3. Example showing that atomic networks for the left linear point algebra do not have the amalgamation property. A directed edge from x to y signifies $x < y$, and a dashed edge between x and y signifies $x|y$.

bounded homogeneous structure is the representation of the left-linear point algebra (Example 3) given in [Bod04].

Conjecture 1 (Infinite-domain dichotomy conjecture). Let $\mathfrak{B}$ be a first-order reduct of a finitely bounded homogeneous structure. Then CSP($\mathfrak{B}$) is either in P or NP-complete.

Hence, the infinite-domain dichotomy conjecture implies the RBCP for finite relation algebras with a normal representation. In Sect. 5 we will see a more specific conjecture that characterises the NP-complete cases and the cases that are in P.

5 The Infinite-Domain Tractability Conjecture

To state the infinite-domain tractability conjecture, we need a couple of concepts that are most naturally introduced for the class of all ω-categorical structures. A theory is called *ω-categorical* if all its countably infinite models are isomorphic. A structure is called *ω-categorical* if its first-order theory is ω-categorical. Note that finite structures are ω-categorical since their first-order theories do not have countably infinite models. Homogeneous structures $\mathfrak{B}$ with finite relational signature are ω-categorical. This follows from a very useful characterisation of ω-categoricity given by Engeler, Svenonius, and Ryll-Nardzewski (Theorem 2). The set of all automorphisms of $\mathfrak{B}$ is denoted by $\text{Aut}(\mathfrak{B})$. The *orbit* of a k-tuple $(t_1,\dots,t_n)$ under $\text{Aut}(\mathfrak{B})$ is the set $\{(a(t_1),\dots,a(t_n)) \mid a \in \text{Aut}(\mathfrak{B})\}$. Orbits of pairs (i.e., 2-tuples) are also called *orbitals.*

Theorem 2 (see [Hod97]). *A countable structure $\mathfrak{B}$ is ω-categorical if and only if $\text{Aut}(\mathfrak{B})$ has only finitely many orbits of n-tuples, for all $n \geq 1$.*

The following is an easy consequence of Theorem 2.

Proposition 3. *First-order reducts of ω-categorical structures are ω-categorical.*

First-order reducts of homogeneous structures, on the other hand, need not be homogeneous. An example of an ω-categorical structure which is not homogeneous is the ω-categorical representation of the left linear point algebra given in [Bod04] (see Example 3). Note that every ω-categorical structure $\mathfrak{B}$, and more generally every structure with finitely many orbitals, gives rise to a finite relation algebra, namely the relation algebra associated to the unions of orbitals of $\mathfrak{B}$ (see [BJ17]); we refer to this relation algebra as the *orbital relation algebra* of $\mathfrak{B}$.

We first present a condition that implies that an ω-categorical structure has an NP-hard constraint satisfaction problem (Sect. 5.1). The tractability conjecture says that every reduct of a finitely bounded homogeneous structure that does not satisfy this condition is NP-complete. We then present an equivalent characterisation of the condition due to Barto and Pinsker (Sect. 5.2), and then yet another condition due to Barto, Opršal, and Pinsker, which was later shown to be equivalent (Sect. 5.3).

5.1 The Original Formulation of the Conjecture

Let $\mathfrak{B}$ be an ω-categorical structure. Then $\mathfrak{B}$ is called

- a *core* if all endomorphisms of $\mathfrak{B}$ (i.e., homomorphisms from $\mathfrak{B}$ to $\mathfrak{B}$) are embeddings (i.e., are injective and also preserve the complement of each relation).
- *model complete* if all self-embeddings of $\mathfrak{B}$ are elementary, i.e., preserve all first-order formulas.

Clearly, if $\mathfrak{B}$ is a representation of a finite relation algebra $\mathbf{A}$, then $\mathfrak{B}$ is a core. However, not all representations of finite relation algebras are model complete. A simple example is the orbital relation algebra of the structure $(\mathbb{Q}_0^+; <)$ where $\mathbb{Q}_0^+$ denotes the non-negative rationals: its representation with domain $\mathbb{Q}_0^+$ has self-embeddings that do not preserve the orbital $\{(0,0)\}$.

Let τ be a relational signature. A τ-formula is called *primitive positive* if it is of the form $\exists x_1, \ldots, x_n(\psi_1 \wedge \cdots \wedge \psi_m)$ where ψ_i is of the form $y_1 = y_2$ or of the form $R(y_1, \ldots, y_k)$ for $R \in \tau$ of arity k. The variables $y_1, \ldots, y_k$ can be free or from $x_1, \ldots, x_n$. Clearly, primitive positive formulas are preserved by homomorphisms.

Theorem 3 ([Bod07,BHM10]). *Every ω-categorical structure is homomorphically equivalent to a model-complete core $\mathfrak{C}$, which is unique up to isomorphism, and again ω-categorical. All orbits of k-tuples are primitive positive definable in $\mathfrak{C}$.*

Let $\mathfrak{B}$ and $\mathfrak{A}$ be structures, let $D \subseteq B^n$, and let $I \colon D \to A$ be a surjection. Then I is called a *primitive positive interpretation* if the pre-image under I of A, of the equality relation $=_A$ on A, and of all relations of $\mathfrak{A}$ is primitive positive definable in $\mathfrak{A}$. In this case we also say that $\mathfrak{B}$ *interprets* $\mathfrak{A}$ *primitively positively.* The complete graph with three vertices (but without loops) is denoted by K_3.

Theorem 4 **([Bod08]).** *Let $\mathfrak{B}$ be an ω-categorical structure. If the model-complete core of $\mathfrak{B}$ has an expansion by finitely many constants so that the resulting structure interprets K_3 primitively positively, then* CSP($\mathfrak{B}$) *is NP-hard.*

We can now state the infinite-domain tractability conjecture.

Conjecture 2. Let $\mathfrak{B}$ be a first-order reduct of a finitely bounded homogeneous structure. If $\mathfrak{B}$ does not satisfy the condition from Theorem 4 then CSP($\mathfrak{B}$) is in P.

This conjecture has been verified in numerous special cases (see, for instance, the articles [BK08, BMPP16, KP17, BJP17, BMM18, BM16]), including the class of finite-structures [Bul17, Zhu17].

5.2 The Theorem of Barto and Pinsker

The tractability conjecture has a fundamentally different, but equivalent formulation: instead of the *non-existence* of a hardness-condition, we require the *existence* of a polymorphism satisfying a certain identity; the concept of polymorphisms is fundamental to the resolution of the Feder-Vardi conjecture in both [Bul17] and [Zhu17].

Definition 2. *A* polymorphism *of a structure $\mathfrak{B}$ is a homomorphism from $\mathfrak{B}^k$ to $\mathfrak{B}$, for some $k \in \mathbb{N}$. We write* Pol($\mathfrak{B}$) *for the set of all polymorphisms of $\mathfrak{B}$.*

An operation $f : B^6 \to B$ is called

- *Siggers* if it satisfies

$$f(x, y, x, z, y, z) = f(z, z, y, y, x, x)$$

  for all $x, y, z \in B$;
- *pseudo-Siggers modulo* $e_1, e_2 : B \to B$ if

$$e_1(f(x, y, x, z, y, z)) = e_2(f(z, z, y, y, x, x))$$

  for all $x, y, z \in B$.

Theorem 5 **([BP16]).** *Let $\mathfrak{B}$ be an ω-categorical model-complete core. Then either*

- *$\mathfrak{B}$ can be expanded by finitely many constants so that the resulting structure interprets K_3 primitively positively, or*
- *$\mathfrak{B}$ has a pseudo-Siggers polymorphism modulo endomorphisms of $\mathfrak{B}$.*

5.3 The Wonderland Conjecture

A weaker condition that implies that an ω-categorical structure has an NP-hard CSP has been presented in [BOP17]. For reducts of homogeneous structures with finite signature, however, the two conditions are equivalent [BKO+17]. Hence, we obtain another different but equivalent formulation of the tractability conjecture. The advantage of the new formulation is that it does not require that the structure is a model-complete core.

Let $\mathfrak{B}$ be a countable structure. A map $\mu\colon \operatorname{Pol}(\mathfrak{B}) \to \operatorname{Pol}(\mathfrak{A})$ is called *minor-preserving* if for every $f \in \operatorname{Pol}(\mathfrak{B})$ of arity k and all k-ary projections $\pi_1, \dots, \pi_k$ we have $\mu(f) \circ (\pi_1, \dots, \pi_k) = \mu(f \circ (\pi_1, \dots, \pi_k))$ where $\circ$ denotes composition of functions. The set $\operatorname{Pol}(\mathfrak{B})$ is equipped with a natural complete ultrametric d (see, e.g., [BS16]). To define d, suppose that $B = \mathbb{N}$. For $f, g \in \operatorname{Pol}(\mathfrak{B})$ we define $d(f,g) = 1$ if f and g have different arity; otherwise, if both f, g have arity $k \in \mathbb{N}$, then

$$d(f,g) := 2^{-\min\{n \in \mathbb{N} \mid \exists s \in \{1,\dots,n\}^k : f(s) \neq g(s)\}}.$$

Theorem 6 (of [BOP17]). *Let $\mathfrak{B}$ be ω-categorical. Suppose that* $\operatorname{Pol}(\mathfrak{B})$ *has a uniformly continuous minor-preserving map to* $\operatorname{Pol}(K_3)$. *Then* $\operatorname{CSP}(\mathfrak{B})$ *is NP-complete.*

We mention that there are ω-categorical structures where the condition from Theorem 6 applies, but not the condition from Theorem 4.

Theorem 7 (of [BKO+17]). *If $\mathfrak{B}$ is a reduct of a homogeneous structure with finite relational signature, then the conditions given in Theorem 4 and in Theorem 6 are equivalent.*

6 Testing the Existence of Normal Representations

In this section we present an algorithm that tests whether a given finite relation algebra has a normal representation. This follows from a model-theoretic result that seems to be folklore, namely that testing the amalgamation property for a class of structures that has the JEP and a signature of maximal arity two which is given by a finite set of forbidden substructures is decidable. We are not aware of a proof of this in the literature.

Theorem 8. *There is an algorithm that decides for a given finite relation algebra* **A** *which has a fully universal square representation whether* **A** *also has a normal representation.*

Proof. First observe that the class $\mathcal{C}$ of all atomic **A**-networks, viewed as A-structures, has the JEP: if N_1 and N_2 are atomic networks, then they are satisfiable in $\mathfrak{B}$ since $\mathfrak{B}$ is fully universal, and hence embed into $\mathfrak{B}$ when viewed as structures. Since $\mathfrak{B}$ is square the substructure of $\mathfrak{B}$ induced by the union of the images of N_1 and N_2 is an atomic network, too, and it embeds N_1 and N_2.

Let k be the number of atoms of $\mathbf{A}$. It clearly suffices to show the following claim, since the condition given there can be effectively checked exhaustively.

Claim. $\mathcal{C}$ has the AP if and only if all 2-point amalgamation diagrams of size at most $k+2$ amalgamate.

So suppose that $D = (\mathfrak{B}_1, \mathfrak{B}_2)$ is an amalgamation diagram without amalgam. Let $\mathfrak{B}'_1$ be a maximal substructure of $\mathfrak{B}_1$ that contains $B_1 \cap B_2$ such that $(\mathfrak{B}'_1, \mathfrak{B}_2)$ has an amalgam. Let $\mathfrak{B}'_2$ be a maximal substructure of $\mathfrak{B}_2$ that contains $B_1 \cap B_2$ such that $(\mathfrak{B}'_1, \mathfrak{B}'_2)$ has an amalgam. Then $B_i \neq B'_i$ for some $i \in \{1, 2\}$; let $\mathfrak{C}_1$ be a substructure of $\mathfrak{B}_i$ that extends $\mathfrak{B}'_i$ by one element, and let $\mathfrak{C}_2 := \mathfrak{B}'_{3-i}$. Then $(\mathfrak{C}_1, \mathfrak{C}_2)$ is a 2-point amalgamation diagram without an amalgam. Let $C_0 := C_1 \cap C_2$. Let $C_1 \setminus C_0 = \{p\}$ and $C_2 \setminus C_0 = \{q\}$. For each $a \in A$ there exists an element $r_a \in C_0$ such that the network $(\{r, p, q\}, f)$ with $f(p, q) = a$, $f(p, r) = f^{\mathfrak{B}_1}(p, r)$, $f(r, q) = f^{\mathfrak{B}_2}(r, q)$ fails the atomicity property (1). Let $\mathfrak{C}'_1$ be the substructure of $\mathfrak{C}_1$ induced by $\{p\} \cup \{r_a \mid a \in A\}$ and $\mathfrak{A}'_1$ be the substructure of $\mathfrak{C}_2$ induced by $\{q\} \cup \{r_a \mid a \in A\}$. Then the amalgamation diagram $(\mathfrak{C}'_1, \mathfrak{C}'_2)$ has no amalgam, and has size at most $k+2$.

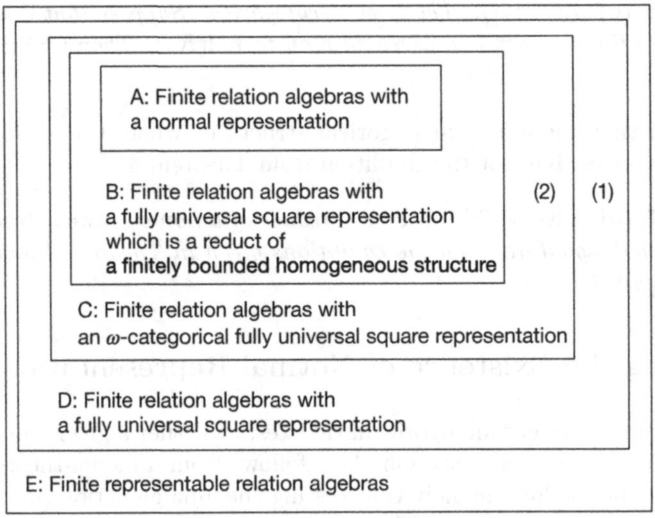

Fig. 4. Subclasses of finite representable relation algebras. Membership of relation algebras from D to the innermost box A is decidable (Theorem 8). Example 3 separates Box A and Box B. The finite relation algebra from [Hir99] separates Box D and E. Box B falls into the scope of the infinite-domain tractability conjecture. Boxes C and D might also fall into the scope of this conjecture (see Problem (1) and Problem (2)).

7 Conclusion and Open Problems

Hirsch's Really Big Complexity Problem (RBCP) for finite relation algebras remains really big. However, the network satisfaction problem of every finite relation algebra known to the author can be formulated as the CSP of a structure that falls into the scope of the infinite-domain tractability conjecture. Most of the classical examples even have a normal representation, and therefore the RBCP for those is implied by the infinite-domain tractability conjecture (Corollary 1). We presented an algorithm that tests whether a given finite relation algebra has a normal representation.

To better understand the RBCP in general, or at least for finite relation algebras with fully universal square representation, we need a better understanding of representations of finite relation algebras with good model-theoretic properties. We mention some concrete open questions; also see Fig. 4.

1. Is there a finite relation algebra with a fully universal square representation, but without an ω-categorical fully universal square representation?
2. Is there a finite relation algebra with an ω-categorical fully universal square representation but without a fully universal square representation which is not the reduct of a finitely bounded homogeneous structure?
3. Find a finite relation algebra **A** such that there is no ω-categorical structure $\mathfrak{B}$ such that the general network satisfaction problem for **A** equals the constraint satisfaction problem for $\mathfrak{B}$. (Note that we do not insist on $\mathfrak{B}$ being a representation of **A**.)
4. Find a finite relation algebra **A** with an ω-categorical fully universal square representation which is not the orbital relation algebra of an ω-categorical structure.

References

[BG08] Bodirsky, M., Grohe, M.: Non-dichotomies in constraint satisfaction complexity. In: Aceto, L., Damgård, I., Goldberg, L.A., Halldórsson, M.M., Ingólfsdóttir, A., Walukiewicz, I. (eds.) ICALP 2008. LNCS, vol. 5126, pp. 184–196. Springer, Heidelberg (2008). https://doi.org/10.1007/978-3-540-70583-3_16

[BHM10] Bodirsky, M., Hils, M., Martin, B.: On the scope of the universal-algebraic approach to constraint satisfaction. In: Proceedings of the Annual Symposium on Logic in Computer Science (LICS), pp. 90–99. IEEE Computer Society, July 2010

[BJ17] Bodirsky, M., Jonsson, P.: A model-theoretic view on qualitative constraint reasoning. J. Artif. Intell. Res. **58**, 339–385 (2017)

[BJP17] Bodirsky, M., Jonsson, P., Van Pham, T.: The complexity of phylogeny constraint satisfaction problems. ACM Trans. Comput. Logic (TOCL) **18**(3) (2017). An extended abstract appeared in the conference STACS 2016

[BK08] Bodirsky, M., Kára, J.: The complexity of temporal constraint satisfaction problems. In: Dwork, C. (ed.) Proceedings of the Annual Symposium on Theory of Computing (STOC), pp. 29–38. ACM, May 2008

[BK09] Barto, L., Kozik, M.: Constraint satisfaction problems of bounded width. In: Proceedings of the Annual Symposium on Foundations of Computer Science (FOCS), pp. 595–603 (2009)

[BKJ05] Bulatov, A.A., Krokhin, A.A., Jeavons, P.G.: Classifying the complexity of constraints using finite algebras. SIAM J. Comput. **34**, 720–742 (2005)

[BKO+17] Barto, L., Kompatscher, M., Olšák, M., Pinsker, M., Van Pham, T.: The equivalence of two dichotomy conjectures for infinite domain constraint satisfaction problems. In: Proceedings of the 32nd Annual ACM/IEEE Symposium on Logic in Computer Science - LICS 2017 (2017). Preprint arXiv:1612.07551

[BM16] Bodirsky, M., Mottet, A.: Reducts of finitely bounded homogeneous structures, and lifting tractability from finite-domain constraint satisfaction. In: Proceedings of the 31th Annual IEEE Symposium on Logic in Computer Science - LICS 2016, pp. 623–632 (2016). Preprint available at ArXiv:1601.04520

[BMM18] Bodirsky, M., Madelaine, F., Mottet, A.: A universal-algebraic proof of the complexity dichotomy for Monotone Monadic SNP. In: Proceedings of the Symposium on Logic in Computer Science - LICS 2018 (2018). Preprint available under ArXiv:1802.03255

[BMPP16] Bodirsky, M., Martin, B., Pinsker, M., Pongrácz, A.: Constraint satisfaction problems for reducts of homogeneous graphs. In: 43rd International Colloquium on Automata, Languages, and Programming, ICALP 2016, 11–15 July 2016, Rome, Italy, pp. 119:1–119:14 (2016)

[Bod04] Bodirsky, M.: Constraint satisfaction with infinite domains. Dissertation, Humboldt-Universität zu Berlin (2004)

[Bod07] Bodirsky, M.: Cores of countably categorical structures. Logical Methods Comput. Sci. **3**(1), 1–16 (2007)

[Bod08] Bodirsky, M.: Constraint satisfaction problems with infinite templates. In: Creignou, N., Kolaitis, P.G., Vollmer, H. (eds.) Complexity of Constraints. LNCS, vol. 5250, pp. 196–228. Springer, Heidelberg (2008). https://doi.org/10.1007/978-3-540-92800-3_8

[Bod12] Bodirsky, M.: Complexity classification in infinite-domain constraint satisfaction. Mémoire d'habilitation à diriger des recherches, Université Diderot - Paris 7 arXiv:1201.0856 (2012)

[BOP17] Barto, L., Opršal, J., Pinsker, M.: The wonderland of reflections. Isr. J. Math. (2017, to appear) Preprint arXiv:1510.04521

[BP16] Barto, L., Pinsker, M.: The algebraic dichotomy conjecture for infinite domain constraint satisfaction problems. In: Proceedings of the 31th Annual IEEE Symposium on Logic in Computer Science - LICS 2016, pp. 615–622 (2016). Preprint arXiv:1602.04353

[BS16] Bodirsky, M., Schneider, F.M.: A topological characterisation of endomorphism monoids of countable structures. Algebra Universalis **77**(3), 251–269 (2016). Preprint available at arXiv:1508.07404

[Bul17] Bulatov, A.A.: A dichotomy theorem for nonuniform CSPs. In: 58th IEEE Annual Symposium on Foundations of Computer Science, FOCS 2017, Berkeley, CA, USA, 15–17 October 2017, pp. 319–330 (2017)

[Dün05] Düntsch, I.: Relation algebras and their application in temporal and spatial reasoning. Artif. Intell. Rev. **23**, 315–357 (2005)

[Fra54] Fraïssé, R.: Sur l'extension aux relations de quelques propriétés des ordres. Annales Scientifiques de l'École Normale Supérieure **71**, 363–388 (1954)

[Fra86] Fraïssé, R.: Theory of Relations. Elsevier Science Ltd, North-Holland (1986)

[HH01] Hirsch, R., Hodkinson, I.: Representability is not decidable for finite relation algebras. Trans. Am. Math. Soc. **353**(4), 1387–1401 (2001)

[Hir96] Hirsch, R.: Relation algebras of intervals. Artif. Intell. J. **83**, 1–29 (1996)

[Hir97] Hirsch, R.: Expressive power and complexity in algebraic logic. J. Logic Comput. **7**(3), 309–351 (1997)

[Hir99] Hirsch, R.: A finite relation algebra with undecidable network satisfaction problem. Logic J. IGPL **7**(4), 547–554 (1999)

[HLR13] Huang, J., Li, J.J., Renz, J.: Decomposition and tractability in qualitative spatial and temporal reasoning. Artif. Intell. **195**, 140–164 (2013)

[Hod97] Hodges, W.: A Shorter Model Theory. Cambridge University Press, Cambridge (1997)

[KP17] Kompatscher, M., Van Pham, T.: A complexity dichotomy for poset constraint satisfaction. In: 34th Symposium on Theoretical Aspects of Computer Science (STACS 2017), volume 66 of Leibniz International Proceedings in Informatics (LIPIcs), pp. 47:1–47:12 (2017)

[LKRL08] Li, J.J., Kowalski, T., Renz, J., Li, S.: Combining binary constraint networks in qualitative reasoning. In: ECAI 2008–18th European Conference on Artificial Intelligence, Patras, Greece, 21–25 July 2008, Proceedings, pp. 515–519 (2008)

[Lyn50] Lyndon, R.: The representation of relational algebras. Ann. Math. **51**(3), 707–729 (1950)

[Zhu17] Zhuk, D.: A proof of CSP dichotomy conjecture. In: 58th IEEE Annual Symposium on Foundations of Computer Science, FOCS 2017, Berkeley, CA, USA, 15–17 October 2017, pp. 331–342 (2017)

Theoretical Foundations

C-Dioids and μ-Continuous Chomsky-Algebras

Hans Leiß[1(✉)] and Mark Hopkins[2]

[1] Munich, Germany
leiss@cis.uni-muenchen.de
[2] UW-Milwaukee (Alumnus), Milwaukee, USA
federation2005@netzero.net

Abstract. We prove that the categories of $\mathcal{C}$-dioids of Hopkins 2008 and of μ-continuous Chomsky-algebras of Grathwohl, Henglein and Kozen 2013 are the same.

1 Introduction

The equational theory of the class of context-free languages has been axiomatized in 2013 by Grathwohl, Henglein and Kozen [3], using μ-terms as notation system for context-free grammars. In order to be able to interpret μ as a least fixed point operator, they consider algebraically closed idempotent semirings, called "Chomsky algebras". An idempotent semiring $(M, +, \cdot, 0, 1)$ is algebraically closed if every finite system of inequations $x_1 \geq p_1(x_1, \ldots, x_n), \ldots, x_n \geq p_n(x_1, \ldots, x_n)$, where $p_1, \ldots, p_n$ are polynomials, has a least solution. The essential part of their axiomatization is the μ-continuity axiom, relating $\cdot$ with the least fixed point operator μ and a least upper bound operation $\sum$:

$$a \cdot \mu x t \cdot b = \sum \{\, a \cdot m x t \cdot b \mid m \in \mathbb{N} \,\},$$

where $a, b \in M$ and mxt is the m-fold iteration of the map $x \mapsto t$.

To give an algebraic generalization of the Chomsky hierarchy of language classes to classes of subsets of monoids, Hopkins [6] has introduced "monadic operators" $\mathcal{A}$ that assign to each monoid M a set $\mathcal{A}M \subseteq \mathcal{P}M$ of its subsets; $\mathcal{A}M$ is assumed to contain all finite subsets of M, so $(\mathcal{A}M, \cup, \cdot, \emptyset, \{1\})$ is an idempotent semiring or "dioid" in the terminology of [4,6]. An "$\mathcal{A}$-dioid" is a dioid $(M, +, \cdot, 0, 1)$ where (i) each $U \in \mathcal{A}M$ has a least upper bound $\sum U \in M$ and (ii) for each $U, V \in \mathcal{A}M$, $(\sum U) \cdot (\sum V) = \sum (U \cdot V)$. This can also be seen as a continuity assumption for $\cdot$, and, assuming (i), is equivalent to:

$$a(\sum U)b = \sum (aUb), \qquad \text{for all } U \in \mathcal{A}M, a, b \in M.$$

H. Leiß—Retired from: Centrum für Informations- und Sprachverarbeitung, Ludwig-Maximilians-Universität München, Oettingenstr. 67, 80539 München.

J. Desharnais et al. (Eds.): RAMiCS 2018, LNCS 11194, pp. 21–36, 2018.
https://doi.org/10.1007/978-3-030-02149-8_2

We here show that if $\mathcal{A}$ is specialized to the operator $\mathcal{C}$ that returns the context-free subsets of monoids, then the $\mathcal{C}$-dioids and the μ-continous Chomsky-algebras are the same. Since both approaches give a notion of context-freeness, this is to be expected, but had been left unresolved in Leiß [11].

Both definitions have their advantages and drawbacks. The main advantage of the μ-continuous Chomsky algebras is, of course, that they lead to an infinitary axiomatization of the equational theory of context-free languages. A disadvantage is that μ-terms are just formal solution terms and, when nested, intuitively incomprehensible; moreover, the context-free subsets are kept in the background. An advantage of the $\mathcal{C}$-dioids is that they bring the context-free subsets in front and separate a completeness property of the partial order from the sup-continuity of the product; this allows for algebraic constructions like coproduct, coequalizer and tensor product in fairly standard ways that extend to other classes of dioids (cf. Hopkins and Leiß [8]). A drawback of the notion of $\mathcal{C}$-dioid may be its universal second-order formulation, which hides an equivalent formulation by infinitary equational implications.

2 Monadic Operators and Language Classes

Let $\mathbb{M}$ be the category of monoids $(M,\cdot,1)$ and homomorphisms between monoids. A *partially ordered monoid* $(M,\cdot,1,\leq)$ is a monoid $(M,\cdot,1)$ with a partial order $\leq\,\subseteq M\times M$, with respect to which $\cdot$ is monotone in each argument.

A *semiring* $R=(R,+,0,\cdot,1)$ is a set R with two operations $+,\cdot : R\times R\to R$, such that $(R,+,0)$ and $(R,\cdot,1)$ are monoids, $+$ is commutative, and the zero and distributivity laws holds:

$$\forall a,b,c,d:\quad a0b=0,\quad a(b+c)d=abd+acd.$$

A *dioid* or *idempotent semiring* $D=(D,+,0,\cdot,1)$ is a semiring in which $+$ is idempotent. Each dioid D has a natural partial order $\leq$, defined by

$$a\leq b:\iff a+b=b.$$

Let $\mathbb{D}$ be the category of dioids and dioid-homomorphisms.

If $M=(M,\cdot^M,1^M)$ is a monoid, its power set $\mathcal{P}M$ is a partially ordered monoid $(\mathcal{P}M,\cdot,1,\subseteq)$, using

$$A\cdot B:=\{\,a\cdot^M b\mid a\in A, b\in B\,\}\quad\text{and}\quad 1:=\{1^M\},$$

and a diod $(\mathcal{P}M,+,\cdot,0,1)$, using $A+B:=A\cup B$ and $0:=\emptyset$.

We review some definitions of Hopkins [6]. A *monadic operator* $\mathcal{A}$ is a functor $\mathcal{A}:\mathbb{M}\to\mathbb{D}$ such that for each monoid M

(A_0) $\mathcal{A}M$ is a set of subsets of M,
(A_1) $\mathcal{A}M$ contains each finite subset of M,
(A_2) $\mathcal{A}M$ is closed under product (hence a monoid),

(A_3) $\mathcal{A}M$ is closed under union of sets from $\mathcal{AA}M$ (hence a dioid), and
(A_4) $\mathcal{A}$ preserves monoid-homomorphisms: if $f : M \to N$ is a homomorphism, so is $\mathcal{A}f : \mathcal{A}M \to \mathcal{A}N$, where for $U \subseteq M$

$$\mathcal{A}f(U) := \{ f(u) \mid u \in U \}.$$

For brevity, we often write $\widetilde{f}$ instead of $\mathcal{A}f$.

Remark 1. Each monadic operator $\mathcal{A}$ gives rise to a subcategory $\mathbb{D}\mathcal{A}$ of $\mathbb{D}$ and is the left adjoint of an adjunction $(\mathcal{A}, \widehat{\mathcal{A}}, \eta, \epsilon) : \mathbb{M} \to \mathbb{D}\mathcal{A}$ (cf. Mac Lane [13]), where $\widehat{\mathcal{A}} : \mathbb{D}\mathcal{A} \to \mathbb{M}$ is the forgetful functor and for $M \in \mathbb{M}$, $\eta_M(m) = \{m\} \in \mathcal{A}M$, and $\epsilon_M : \mathcal{A}M \to M$ maps $U \in \mathcal{A}M$ to its least upper bound $\sum U \in M$. This adjunction gives rise to a monad $T = (\widehat{\mathcal{A}} \circ \mathcal{A}, \eta, \mu)$, an endofunctor on $\mathbb{M}$, where the unit η is taken from the adjunction and the product μ maps $\mathcal{U} \in \mathcal{AA}M$ to $\bigcup \mathcal{U} \in \mathcal{A}M$. While $\widehat{\mathcal{A}}$ is called a *monadic functor* in category theory (see [13], p. 139), Hopkins [6] calls $\mathcal{A}$ a *monadic operator*, a term we also use here.

Example 1. The power set operator $\mathcal{P}$ is a monadic operator. The operator $\mathcal{F}$ that assigns to M the set $\mathcal{F}M$ of all finite subsets of M is a monadic operator.

Example 2. The operator $\mathcal{R}$ that assigns to M the closure of $\mathcal{F}M$ under $+$ (union), $\cdot$ (elementwise product) and * (iteration, Kleene's star), is monadic. Hopkins [6] defines monadic operators $\mathcal{C}$ and $\mathcal{T}$ that select the context-free subsets $\mathcal{C}M$ and the Turing/Thue-subsets $\mathcal{T}M$ of a monoid.

We give a simpler, but equivalent definition of $\mathcal{C}$ in Sect. 3 below. A correction of the definition of $\mathcal{T}$ in [6] is given in Hopkins and Leiß [8].

Let M be a partially ordered monoid. For $a \in M$ and $U \subseteq M$ let $U < a$ mean that a is an upper bound of U, i.e. for all $u \in U$, $u \leq a$. M is *$\mathcal{A}$-complete*, if each $U \in \mathcal{A}M$ has a least upper bound $\sum U \in M$. M is *$\mathcal{A}$-distributive*, if for all $U, V \in \mathcal{A}M$, $\sum(UV) = \sum U \cdot \sum V$.

$\mathcal{A}$-distributivity states $\sum$-continuity of $\cdot$ in both arguments simultaneously. One can as well state it in each argument separately, by $\sum(aU) = a(\sum U)$ and $\sum(Ub) = (\sum U)b$ for all $a, b \in M, U \in \mathcal{A}M$, or combined to $a(\sum U)b = \sum aUb$.

Proposition 1. *Let M be a partially ordered monoid and $U, V \in \mathcal{P}M$ such that least upper bounds $u := \sum U$ and $v := \sum V$ exist. Then (i) implies (ii), where*

(i) for all $a, b \in M$, $\sum aUb = a(\sum U)b$ and $\sum aVb = a(\sum V)b$,
(ii) $\sum(UV) = \sum U \cdot \sum V$.

Notice that the existence of $\sum aUb$ and $\sum aVb$ in (i) and of $\sum(UV)$ in (ii) is not assumed, but is part of the claims.

Proof. Clearly, $UV < uv$. To prove that uv is $\sum(UV)$, we take $c \in M$ with $UV < c$ and show $uv \leq c$. For each $a \in U$, by (i), $\sum aV1$ exists, and as $aV1 \subseteq UV < c$, using (i) we have

$$av = a(\sum V)1 = \sum aV1 \leq c.$$

Hence $Uv = 1Uv < c$. By (i), $\sum 1Uv$ exists, and $uv = 1(\sum U)v = \sum 1Uv \leq c$.

Corollary 1. *If the partially ordered monoid M is $\mathcal{A}$-complete, the following conditions are equivalent:*

(D_1) (weak $\mathcal{A}$-distributivity): for all $a, b \in M$ and $U \in \mathcal{A}M$, $\sum aUb = a(\sum U)b$.
(D_2) (strong $\mathcal{A}$-distributivity): for all $U, V \in \mathcal{A}M$, $\sum(UV) = \sum U \cdot \sum V$.

An $\mathcal{A}$-*dioid* D is a partially ordered monoid that is $\mathcal{A}$-complete and $\mathcal{A}$-distributive.

Every $\mathcal{A}$-dioid $(M, \cdot, 1, \leq)$ is a dioid, using 0 and + defined by $0 := \sum \emptyset$ and $a + b := \sum\{a, b\}$. The zero and distributivity laws follow from D_1.

The monadic operator $\mathcal{A}$ provides us with a notion of continuous homomorphisms between $\mathcal{A}$-dioids, as follows.

(D_3) A homomorphism $f : M \to M'$ is $\mathcal{A}$-*continuous*, if for all $U \in \mathcal{A}M$ and $y > \widetilde{f}(U)$ there is some $x > U$ with $y \geq f(x)$.

An $\mathcal{A}$-*morphism* is a $\leq$-preserving, $\mathcal{A}$-continuous monoid-homomorphism. We write $\mathbb{D}\mathcal{A}$ for the category of $\mathcal{A}$-dioids and $\mathcal{A}$-morphisms.

For order-preserving homomorphism $f : M \to M'$, $\mathcal{A}$-continuity of f reduces to:

$$f(\sum U) = \sum \widetilde{f}(U) \qquad \text{for all } U \in \mathcal{A}M.$$

Every $\mathcal{A}$-morphism is also a dioid-homomorphism.

Proposition 2. *If M is an $\mathcal{A}$-dioid, then for all $a, b \in M$ and $U, V \in \mathcal{A}M$:*

1. *$a(\sum U)b = \sum aUb$ and $\sum(UV) = \sum U \cdot \sum V$,*
2. *$a + (\sum U) = \sum(a + U)$ and $\sum(U + V) = \sum U + \sum V$.*

Proof. 2. Since $\{U, V\} \in \mathcal{F}\mathcal{A}M \subseteq \mathcal{A}\mathcal{A}M$, we have $U + V = \bigcup\{U, V\} \in \mathcal{A}M$, and so there is a least upper bound $\sum(U + V) \in M$. Hence

$$\sum U + \sum V \leq \sum(U + V) + \sum(U + V) = \sum(U + V).$$

Besides, $U + V < \sum U + \sum V$, so $\sum(U + V) \leq \sum U + \sum V$. Claim 1 follows from Corollary 1. □

The *free monoid* X^* generated by the set X consists of all finite sequences of elements from X, with concatenation as $\cdot$ and the empty sequence as 1. A *monomial* $m(x_1, \ldots, x_n)$ *in* $x_1, \ldots, x_n$ is a formal product of elements of X, and a *polynomial* $p(x_1, \ldots, x_n)$ a formal sum of monomials in $x_1, \ldots, x_n$. For elements $a_1, \ldots, a_n \in M$, we write $m^M(a_1, \ldots, a_n)$ for the value of the monomial in the monoid M and $p^M(a_1, \ldots, a_n)$ for the value of the polynomial p in the dioid M.

Corollary 2. *If M is an $\mathcal{A}$-dioid and $p(x_1, \ldots, x_n)$ a polynomial in $x_1, \ldots, x_n$, then $p^{\mathcal{A}M}(U_1, \ldots, U_n) \in \mathcal{A}M$ for all $U_1, \ldots, U_n \in \mathcal{A}M$, and*

$$\sum p^{\mathcal{A}M}(U_1, \ldots, U_n) = p^M(\sum U_1, \ldots, \sum U_n).$$

Proof. In an $\mathcal{A}$-dioid M, for $U, V \in \mathcal{A}M$ and $a, b \in M$ we have

$$(\sum U)(\sum V) = \sum(UV) \quad \text{and} \quad a(\sum U)b = \sum(aUb).$$

As $\mathcal{A}M \supseteq \mathcal{F}M$ is closed under product, we have $m^{\mathcal{A}M}(U_1, \ldots, U_n) \in \mathcal{A}M$ for each monomial $m(x_1, \ldots, x_n)$, and

$$m^M(\sum U_1, \ldots, \sum U_n) = \sum m^{\mathcal{A}M}(U_1, \ldots, U_n).$$

Since $\mathcal{A}M$ is closed under $\cup$ and $(\sum U) + (\sum V) = \sum(U \cup V)$, this extends to

$$p^M(\sum U_1, \ldots, \sum U_n) = \sum p^{\mathcal{A}M}(U_1, \ldots, U_n). \qquad \square$$

A polynomial in $x_1, \ldots, x_n$ *over* M or *with parameters from* M is a polynomial in $x_1, \ldots, x_n$ whose monomials may have additional factors taken from M. The corollary holds for polynomials with parameters as well.

3 $\mathcal{C}$-Dioids

Every monoid M gives rise to the idempotent semiring $(\mathcal{P}M, +, \cdot, \emptyset, \{1^M\})$. $\mathcal{P}M$ is *complete*: every $Y \subseteq \mathcal{P}M$ has a supremum, $\sum Y := \bigcup Y$, and the operations $+$ and $\cdot$ are $\sum$-*continuous*, i.e. for $Y, Z \subseteq \mathcal{P}M$ we have

$$\bigcup Y + \bigcup Z = \bigcup\{\, A + B \mid A \in Y, B \in Z\,\},$$
$$\bigcup Y \cdot \bigcup Z = \bigcup\{\, A \cdot B \mid A \in Y, B \in Z\,\}.$$

Any finite system $\boldsymbol{p}$ of polynomial inequations (with parameters from $\mathcal{P}M$)

$$x_1 \geq p_1(x_1, \ldots, x_n), \; \ldots, \; x_n \geq p_n(x_1, \ldots, x_n)$$

has a least solution, a least $\boldsymbol{A} \in (\mathcal{P}M)^n$ with $\boldsymbol{A} \geq \boldsymbol{p}^{\mathcal{P}M}(\boldsymbol{A})$ (componentwise), where

$$\boldsymbol{A} := \bigcup\{\, \boldsymbol{A}_k \mid k \in \mathbb{N}\,\} \quad \text{with } A_i = \bigcup\{\, A_{i,k} \mid k \in \mathbb{N}\,\}$$

and the sequence of tuples $\boldsymbol{A}_k \in (\mathcal{P}M)^n$, $k \in \mathbb{N}$ is defined by

$$\boldsymbol{A}_0 := \emptyset^n, \qquad \boldsymbol{A}_{k+1} := \boldsymbol{p}^{\mathcal{P}M}(\boldsymbol{A}_k) := (p_1^{\mathcal{P}M}(\boldsymbol{A}_k), \ldots, p_n^{\mathcal{P}M}(\boldsymbol{A}_k)).$$

For complete partial order $(P, \leq, 0)$ and $\sum$-continuous monotone $f : P \to P$ let μf be the least $a \in P$ with $f(a) \leq a$. A well-known fact about least solutions is:

Lemma 1. *Let $f : P \to P$ and $g : Q \to Q$ be $\sum$-continuous between the complete partial orders $(P, \leq^P, 0^P)$ and $(Q, \leq^Q, 0^Q)$. If $h : P \to Q$ is $\sum$-preserving with $h(0) = 0$ and $h \circ f = g \circ h$, then $h(\mu f) = \mu g$.*

We are henceforth considering idempotent semirings, like the $\mathcal{A}M$, which need not be complete: the existence of least upper bounds is restricted to $\mathcal{A}$-subsets.

For a monoid $M = (M, \cdot^M, 1^M)$, let $\mathcal{C}M$, the set of *context-free subsets of* M, be the closure of $\mathcal{F}M$ under (binary) union and under components of least solutions in $\mathcal{P}M$ of polynomial systems over $\mathcal{C}M$: the components $A_1, \ldots, A_n$ of the least solution in $\mathcal{P}M$ of any inequation system

$$x_1 \geq p_1(x_1, \ldots, x_n), \;\ldots,\; x_n \geq p_n(x_1, \ldots, x_n)$$

with polynomials p_i in $x_1, \ldots, x_n$ with parameters from $\mathcal{C}M$ belong to $\mathcal{C}M$.

This inductive definition of $\mathcal{C}M$ and $\mathcal{C}$ differs from the grammatical one in Hopkins [6]. We prove the equivalence of both definitions in Remark 2 below.

Example 3. For a monoid M with $a, b, c \in M$, the least solution of $x \geq \{a\} \cdot x \cdot \{b\} + \{c\}$ in $\mathcal{P}M$ is $L = \{\, a^n c b^n \mid n \in \mathbb{N}\}$, hence $L \in \mathcal{C}M$.

Theorem 1. *$\mathcal{C}$ is a monadic operator.*

Proof. Let M be a monoid M. By definition, $\mathcal{C}M$ satisfies A_0, A_1, as $\mathcal{F}M \subseteq \mathcal{C}M \subseteq \mathcal{P}M$, and it satisfies A_2 as it contains with $A, B \in \mathcal{C}M$ the least solution in $\mathcal{P}M$ of $x \geq Ay, y \geq B$, which is AB. By Theorem 9 of Hopkins [6], the conjunction of A_3 and A_4 is equivalent (over A_0, A_1, A_2) to the condition

(A_5) For monoids M, N, every "substitution" homomorphism $\sigma : M \to \mathcal{C}N$ extends to a homomorphism $\sigma^* : \mathcal{C}M \to \mathcal{C}N$ by

$$\sigma^*(U) := \bigcup \{\, \sigma(m) \mid m \in U\,\}.$$

To show (A_5), we prove $\sigma^*(U) \in \mathcal{C}N$ by induction on $U \in \mathcal{C}M$. If $U \in \mathcal{F}M$, then $\tilde{\sigma}(U) \in \mathcal{F}\mathcal{C}N$, hence $\sigma^*(U) = \bigcup \tilde{\sigma}(U) \in \mathcal{C}N$ since $\mathcal{C}N$ is closed under binary union. If $U = A \cup B$ with $A, B \in \mathcal{C}M$, then $\sigma^*(U) = \sigma^*(A) \cup \sigma^*(B) \in \mathcal{C}N$, by induction and closure of $\mathcal{C}N$ under binary union.

Finally, let U be the first component of the least solution in $\mathcal{P}M$ of the system $\boldsymbol{x} \geq \boldsymbol{p}(\boldsymbol{x})$ over $\mathcal{C}M$, where $\boldsymbol{p}(\boldsymbol{x}) = \boldsymbol{q}(\boldsymbol{x}, \boldsymbol{y})[\boldsymbol{y}/\boldsymbol{A}]$ and $\boldsymbol{A}$ are the parameters $A_j \in \mathcal{C}M$ in $\boldsymbol{p}(\boldsymbol{x})$. By induction, $\sigma^*(\boldsymbol{A}) \in \mathcal{C}N$. Let $\boldsymbol{x} \geq \boldsymbol{p}^{\sigma^*}(\boldsymbol{x})$ be the system over $\mathcal{C}N$, where $\boldsymbol{p}^{\sigma^*}(\boldsymbol{x}) := \boldsymbol{q}(\boldsymbol{x}, \boldsymbol{y})[\boldsymbol{y}/\sigma^*(\boldsymbol{A})]$ is obtained by replacing the parameters as shown. As $\sigma^* : \mathcal{P}M \to \mathcal{P}N$ is a homomorphism and preserves $\cup$, we have

$$\sigma^*(p_i^{\mathcal{P}M}(\boldsymbol{B})) = \sigma^*(q_i^{\mathcal{P}M}(\boldsymbol{B}, \boldsymbol{A})) = (q_i^{\mathcal{P}N}(\sigma^*(\boldsymbol{B}), \sigma^*(\boldsymbol{A})) = (p_i^{\sigma^*})^{\mathcal{P}N}(\sigma^*(\boldsymbol{B}))$$

for each i, so $\sigma^* \circ p_i = p_i^{\sigma^*} \circ \sigma^* : \mathcal{P}M \to \mathcal{P}N$. Since σ^* preserves $\bigcup$, it maps the least fixed-point of $\boldsymbol{p}$ in $(\mathcal{P}M)^n$ to the least fixed-point of $\boldsymbol{p}^{\sigma^*}$ in $(\mathcal{P}N)^n$, by Lemma 1. It follows that $\sigma^*(U)$ is the first component of the least solution of $\boldsymbol{p}^{\sigma^*}(\boldsymbol{x})$ in $\mathcal{P}N$, hence lies in $\mathcal{C}N$. □

Remark 2. Hopkins [6,7] defines $\mathcal{C}$ differently, as follows. First, for free monoids X^*, one puts

$$\mathcal{C}X^* := \{\, L(G) \mid G \text{ is a context-free grammar over } X\,\}.$$

Here, a context-free grammar $G = (Q, S, H)$ over X has a set Q disjoint from X, with distinguished element $S \in Q$, and a finite set $H \subseteq Q \times (Q \cup X)^*$ of "context-free" rules. The language $L(G) \subseteq X^*$ is defined as usual. The class of context-free languages is closed under homomorphisms: If $h : X^* \to Y^*$ is a homomorphism, so is $\widetilde{h} : \mathcal{C}X^* \to \mathcal{C}Y^*$: a context-free grammar $G = (Q, S, H)$ over X gives rise to a context-free grammar $G^h = (Q, S, H^h)$ over Y with $\widetilde{h}(L(G)) = L(G^h)$.

For arbitrary monoid M, one takes a generating subset $X \subseteq M$ and uses the canonical homomorphism $h_X : X^* \to M$, where $h_X(x) = x$ for $x \in X$, to put

$$\mathcal{C}M = \{\, \widetilde{h_X}(L(G)) \mid G \text{ a context-free grammar over } X \,\}.$$

This is independent of the choice of X, as for any generating set $Y \subseteq M$ there is a homomorphism $h : X^* \to Y^*$ with $h_X = h_Y \circ h$, so that $\widetilde{h_X}(L(G)) = \widetilde{h_Y}(L(G^h))$.

Let $\mathcal{C}_H$ be the operator defined by Hopkins. To show $\mathcal{C} = \mathcal{C}_H$, fix a monoid M and a generating subset $Y \subseteq M$.

Claim. $\mathcal{C}_H M \subseteq \mathcal{C}M$.

Proof. Let $U \in \mathcal{C}_H M$ and $G = (X, S, H)$ a context-free grammar over Y, with finite set $H \subseteq X \times (X \cup Y)^*$, such that $U = \widetilde{h_Y}(L(G))$. For each nonterminal $x_i \in X = \{x_1, \ldots, x_n\}$, let $p_i(\boldsymbol{x}, \boldsymbol{y}) = \sum\{\, m(\boldsymbol{x}, \boldsymbol{y}) \mid (x_i, m(\boldsymbol{x}, \boldsymbol{y})) \in H \,\}$. Then $\boldsymbol{x} \geq \boldsymbol{p}(\boldsymbol{x}, \boldsymbol{y})$ is a polynomial system over Y^*. Its least solution $\boldsymbol{A}$ in $\mathcal{P}Y^*$ consists of the languages $A_i = L_G(x_i) \subseteq Y^*$, where $L(G) = L_G(x_1)$, say. To show $\widetilde{h_Y}(\boldsymbol{A}) = (\mu \boldsymbol{x} \boldsymbol{p})^{\mathcal{P}M}(h_Y(\boldsymbol{y}))$ it is suffices by Lemma 1 that for any $\boldsymbol{B} \in \mathcal{P}Y^*$,

$$\widetilde{h_Y}(\boldsymbol{p}^{\mathcal{P}Y^*}(\boldsymbol{B}, \boldsymbol{y})) = \boldsymbol{p}^{\mathcal{P}M}(\widetilde{h_Y}(\boldsymbol{B}), h_Y(\boldsymbol{y})).$$

(Parameters y resp. $h_Y(y)$ are interpreted by the corresponding singleton sets.) But this is clear by the definition of h_Y. It follows that $U = \widetilde{h_Y}(L_G(x_1))$ is the first component of $\mu \boldsymbol{x} \boldsymbol{p}^{\mathcal{P}M}(h_Y(\boldsymbol{y}))$, and hence belongs to $\mathcal{C}M$. ◁

Claim. $\mathcal{C}M \subseteq \mathcal{C}_H M$.

Proof. We proceed by induction on $U \in \mathcal{C}M$. If $U = \{w_1, \ldots, w_m\} \in \mathcal{F}M$, there are $y_1, \ldots, y_m \in Y^*$ with $h_Y(y_1) = w_1, \ldots, h_y(y_m) = w_m$ and a context-free grammar $G = (X, S, H)$ over Y with $L(G) = \{y_1, \ldots, y_m\}$. Hence $U = \{h_Y(y_1), \ldots, h_Y(y_m)\} = \widetilde{h_Y}(L(G)) \in \mathcal{C}_H M$. The case $U = U_1 \cup U_2$ is left to the reader. Finally, let U be a component of the least solution $\boldsymbol{U}$ in $\mathcal{P}M$ of a polynomial system $\boldsymbol{x} \geq \boldsymbol{p}(\boldsymbol{x})$ over $\mathcal{C}M$, using $\boldsymbol{p}(\boldsymbol{x}) = \boldsymbol{q}(\boldsymbol{x}, \boldsymbol{B})$ to show the parameters $\boldsymbol{B}$ from $\mathcal{C}M$. By induction, for each B_j of $\boldsymbol{B}$ there is a context-free grammar $G_j = (X_j, S_j, H_j)$ over Y such that $B_j = \widetilde{h_Y}(L(G_j))$. Let X be the disjoint union of $\boldsymbol{x}$ with these X_j. Each inequation $x_i \geq q_i(\boldsymbol{x}, \boldsymbol{B})$ of $\boldsymbol{x} \geq \boldsymbol{p}(\boldsymbol{x})$ gives rise to context-free grammar rules $(x_i, m_{i,1}(\boldsymbol{x}, \boldsymbol{S})), \ldots, (x_i, m_{i,n_i}(\boldsymbol{x}, \boldsymbol{S})) \in X \times (X \cup Y)^*$, where $m_{i,k}(\boldsymbol{x}, \boldsymbol{B})$ are the monomials of $q_i(\boldsymbol{x}, \boldsymbol{B})$. Let $G(X, S, H)$ be the context-free grammar over Y where S is the variable from $\boldsymbol{x}$ corresponding to U and H is the union of the H_j with the rules obtained from the inequations $x_i \geq q_i(\boldsymbol{x}, \boldsymbol{S})$, $i = 1, \ldots, n$. Then $L_G(S_j) = L(G_j)$, hence $\widetilde{h_Y}(L_G(S_j)) = \widetilde{h_Y}(L(G_j)) = B_j$, and $\widetilde{h_Y}(L_G(x_i)) = U_i$. Therefore, $U = \widetilde{h_Y}(L(G))$. ◁

4 μ-Continuous Chomsky Algebras

Let X be an infinite set of variables. The set of *μ-terms over* X is defined by the grammar

$$t := x \mid 0 \mid 1 \mid (s \cdot t) \mid (s + t) \mid \mu x t.$$

A term not containing μ will be called (somewhat unprecisely) a *polynomial*. The free occurrences of variables in a term are defined as usual. By $\mathit{free}(t)$ we denote the set of variables having a free occurrence in t; in particular, $\mathit{free}(\mu x t) = \mathit{free}(t) \setminus \{x\}$. By $t(x_1, \ldots, x_n)$ we indicate $\mathit{free}(t) \subseteq \{x_1, \ldots, x_n\}$. In $\mu x t$ all free occurrences of x in t are *bound* by μx. By $t[x/s]$ we denote the result of substituting all free occurrences of x in t by s, renaming bound variables of t to avoid capture of free variables of s by bindings in t.

A *partially ordered μ-semiring* $(M, +, \cdot, 0, 1, \leq)$ is a semiring $(M, +, \cdot, 0, 1)$ with a partial order $\leq$ on M, where every term t defines a function $t^M : (X \to M) \to M$, so that for all variables $x \in X$ and terms s, t we have:

1. for all valuations $g : X \to M$,

$$\begin{aligned} 0^M(g) &= 0, & (s+t)^M(g) &= s^M(g) + t^M(g), \\ 1^M(g) &= 1, & (s \cdot t)^M(g) &= s^M(g) \cdot t^M(g), \\ x^M(g) &= g(x), & \text{if } s^M \leq t^M&, \text{ then } \mu x s^M \leq \mu x t^M, \end{aligned}$$

2. t^M is monotone with respect to the pointwise order on $X \to M$,
3. $t^M(g) = t^M(h)$, for all valuations $g, h : X \to M$ which agree on $\mathit{free}(t)$,
4. $t[x/s]^M(g) = t^M(g[x/s^M(g)])$, for all valuations $g : X \to M$.

When $\mathit{free}(t) \subseteq \{x_1, \ldots, x_n\}$ and $g(x_i) = a_i$ for $1 \leq i \leq n$, instead of $t^M(g)$ we often write $t^M[x_1/a_1, \ldots, x_n/a_n]$ or just $t^M(a_1, \ldots, a_n)$.

The final two conditions above are called the *coincidence* and *substitution properties*; in the latter, $g[x/a]$ denotes the valuation that agrees with g, except that it assigns a to x. Clearly, the substitution property extends to simultaneous substitutions $[x_1/s_1, \ldots, x_n/s_n]$.

Following Grathwohl et al. [3], an idempotent semiring $(M, +, \cdot, 0, 1)$ is *algebraically closed* or a *Chomsky algebra*, if every finite system of polynomial inequations

$$\begin{aligned} x_1 &\geq p_1(x_1, \ldots, x_n, y_1, \ldots, y_m), \\ &\vdots \\ x_n &\geq p_n(x_1, \ldots, x_n, y_1, \ldots, y_m), \qquad \text{abbreviated } \boldsymbol{x} \geq \boldsymbol{p}(\boldsymbol{x}, \boldsymbol{y}), \end{aligned} \tag{1}$$

has least solutions, i.e. for all $\boldsymbol{b} \in M^m$ there is a least $\boldsymbol{a} = a_1, \ldots, a_n \in M^n$ such that $a_i \geq p_i^M(\boldsymbol{a}, \boldsymbol{b})$ for $i = 1, \ldots, n$, where $\leq$ is the natural partial order on M defined by $a \leq b$ iff $a + b = b$. Of course, for each $\boldsymbol{b}$ the least solution $\boldsymbol{a}$ is unique.

Example 4. The set $\mathcal{C}X^*$ of *context-free languages* over X is the smallest set $\mathcal{L} \subseteq \mathcal{P}X^*$ such that (i) each finite subset of $X \cup \{\epsilon\}$ is in $\mathcal{L}$ and (ii) if $\boldsymbol{x} \geq \boldsymbol{p}(\boldsymbol{x}, \boldsymbol{y})$ is a polynomial system, and $\boldsymbol{B} = B_1, \ldots, B_m \in \mathcal{L}$, then the components A_i of the least $\boldsymbol{A} = A_1, \ldots, A_n \in \mathcal{P}X^*$ with $\boldsymbol{A} \supseteq \boldsymbol{p}^{\mathcal{P}X^*}(\boldsymbol{A}, \boldsymbol{B})$ belong to $\mathcal{L}$. With the operations inherited from $\mathcal{P}X^*$, $(\mathcal{C}X^*, +, \cdot, 0, 1)$ is a Chomsky algebra. For example, $\{\, a^n b^n \mid n \in \mathbb{N}\,\}$ is a context-free language over $X \supseteq \{a, b\}$; it is the least solution of $x \geq axb + 1$ relative to the valuation $g(a) = \{a\}, g(b) = \{b\}$.

The class of regular languages over X does not form a Chomsky algebra.

Lemma 2 *(Grathwohl et al. [3]). Every Chomsky algebra M is an idempotent, partially ordered μ-semiring, if for all μ-terms t, variables x and valuations $g : X \to M$ we take*

$$\mu x t^M(g) := \textit{the least } a \in M \textit{ such that } t^M(g[x/a]) \leq a. \tag{2}$$

Moreover, every inequation system $\boldsymbol{t}(\boldsymbol{x}, \boldsymbol{y}) \leq \boldsymbol{x}$ with μ-terms $\boldsymbol{t}(\boldsymbol{x}, \boldsymbol{y})$ has least solutions in M, i.e. for all parameters $\boldsymbol{b}$ from M there is a least tuple $\boldsymbol{a}$ in M such that $\boldsymbol{t}^M(\boldsymbol{a}, \boldsymbol{b}) \leq \boldsymbol{a}$.

Proof. See Lemma 2.1 in [3] or, in more detail, Lemma 8 in [11].

A Chomsky algebra M is *μ-continuous*, if for all $a, b \in M$, all μ-terms t, variables x and valuations $g : X \to M$ it satisfies the μ-continuity condition

$$a \cdot \mu x t^M(g) \cdot b = \sum\{\, a \cdot m x t^M(g) \cdot b \mid m \in \mathbb{N}\,\}, \tag{3}$$

where mxt is defined inductively by $0xt := 0$, $(m+1)xt := t[x/mxt]$. This condition of Grathwohl et al. [3] generalizes the *-continuity condition

$$a \cdot c^* \cdot b = \sum\{\, a \cdot c^m \cdot b \mid m \in \mathbb{N}\,\}$$

of Kozen [9]; every μ-continuous Chomsky algebra is a *-continuous Kleene algebra, if c^* is defined by $\mu x(cx + 1)$.

To prove that every μ-continuous Chomsky algebra is a $\mathcal{C}$-dioid, we will below need a vector-version of the μ-continuity condition. By a theorem of Bekić [1], deBakker and Scott [2], the n-ary least-fixed-point operator can be reduced to the unary one. The theorem applies to complete (or at least ω-complete) partial orders only, but $\mathcal{C}X^*$ is far from being ω-complete. So it needs to be checked that the suprema used in Bekić's reduction exist also in the incomplete partial orders of $\mathcal{C}M$. This can be done, leading to a definition of term vectors $\mu\boldsymbol{x}\boldsymbol{t}$ that embody Bekić's reduction (cf. Leiß and Ésik [12]).

For vectors $\boldsymbol{t} = t_1, \ldots, t_n$ of terms and $\boldsymbol{x} = x_1, \ldots, x_n$ of pairwise different variables, we define the term vector $\mu\boldsymbol{x}\boldsymbol{t}$ as follows. If $n = 1$, then $\mu\boldsymbol{x}\boldsymbol{t} := \mu x_1 t_1$. If $n > 1$, $\boldsymbol{x} = (\boldsymbol{y}, z)$ and $\boldsymbol{t} = (\boldsymbol{r}, s)$ with term vectors $\boldsymbol{r}, s$ of lengths $|\boldsymbol{y}|, |z| < n$, then $\mu\boldsymbol{x}\boldsymbol{t}$ is

$$\mu(\boldsymbol{y}, z)(\boldsymbol{r}, s) := (\mu\boldsymbol{y}.\boldsymbol{r}[z/\mu z s], \mu z.s[\boldsymbol{y}/\mu\boldsymbol{y}\boldsymbol{r}]). \tag{4}$$

It can be shown that the values of $\mu\boldsymbol{x}\boldsymbol{t}$ in Chomsky algebras do not depend on the choice of splitting $\boldsymbol{x}$ into $\boldsymbol{y}, z$ in the definition.

Lemma 3. *For any Chomsky algebra M and valuation $g : X \to M$, $\mu \boldsymbol{x} \boldsymbol{t}^M(g)$ is the least tuple $\boldsymbol{a}$ in M such that $\boldsymbol{t}^M(g[\boldsymbol{x}/\boldsymbol{a}]) \leq \boldsymbol{a}$.*

Proof. See Lemma 14 of Leiß [11].

It is essential that the unary version of μ-continuity implies the n-ary version:

Lemma 4. *Let M be a μ-continuous Chomsky algebra and $g : X \to M$. Then*

$$\boldsymbol{a} \cdot \mu \boldsymbol{x} \boldsymbol{t}^M(g) \cdot \boldsymbol{b} = \sum \{ \boldsymbol{a} \cdot m \boldsymbol{x} \boldsymbol{t}^M(g) \cdot \boldsymbol{b} \mid m \in \mathbb{N} \},$$

for any term vector $\boldsymbol{t}$ and vectors $\boldsymbol{a}, \boldsymbol{b}$ of elements of M of the same length as $\boldsymbol{t}$.

Proof. See Corollary 23 in Leiß [11].

5 $\mathcal{C}$-Dioids and μ-Continuous Chomsky Algebras

We now prove our result, that the categories of $\mathcal{C}$-dioids and μ-continuous Chomsky algebras are the same.

Theorem 2. *Every $\mathcal{C}$-dioid M is a μ-continuous Chomsky algebra.*

Proof. To simplify the notation, we talk of a "polynomial system $\boldsymbol{x} \geq \boldsymbol{p}(\boldsymbol{x})$ *over M*" or *with parameters from M*, meaning that $\boldsymbol{x} \geq \boldsymbol{p}(\boldsymbol{x}, \boldsymbol{y})$ is considered with a fixed valuation $g : X \to M$ that provides the values for the parameters $\boldsymbol{y}$. Hence we also write $\mu \boldsymbol{x} \boldsymbol{p}^M$ instead of $\mu \boldsymbol{x} \boldsymbol{p}^M(g)$.

(i) M is algebraically closed: Let $\boldsymbol{x} \geq \boldsymbol{p}(\boldsymbol{x})$ be a polynomial system with parameters b from M. In $\mathcal{P}M$ it has a least solution $\boldsymbol{A} = \mu \boldsymbol{x} \boldsymbol{p}^{\mathcal{C}M}$ with components $A_i \in \mathcal{C}M$ (where parameters $b \in M$ are interpreted by $\{b\} \in \mathcal{F}M \subseteq \mathcal{C}M$). Since M is a $\mathcal{C}$-dioid, the suprema $a_i := \sum A_i \in M$ $(1 \leq i \leq n)$ exist. We show that $\boldsymbol{a} = (a_1, \ldots, a_n) = \sum \boldsymbol{A}$ is the least solution of $\boldsymbol{x} \geq \boldsymbol{p}(\boldsymbol{x})$ in M, i.e.

$$\mu \boldsymbol{x} \boldsymbol{p}^M = \sum \mu \boldsymbol{x} \boldsymbol{p}^{\mathcal{C}M} = \sum \boldsymbol{A}. \tag{5}$$

By the distributivity properties of $\mathcal{C}$-dioids, $\boldsymbol{a}$ is a solution of $\boldsymbol{x} \geq \boldsymbol{p}(\boldsymbol{x})$ in M, using Corollary 2:

$$p_i^M(\boldsymbol{a}) = p_i^M(\sum A_1, \ldots, \sum A_n) = \sum p_i^{\mathcal{C}M}(A_1, \ldots, A_n) \leq \sum A_i = a_i.$$

To show that $\boldsymbol{a}$ is the least solution of $\boldsymbol{x} \geq \boldsymbol{p}(\boldsymbol{x})$ in M, let $\boldsymbol{c}$ be any solution in M, hence $c_i \geq p_i^M(\boldsymbol{c})$ for $1 \leq i \leq n$. It is sufficient to show $\boldsymbol{c} > \boldsymbol{A}$, because $\boldsymbol{a}$ is the least upper bound of $\boldsymbol{A}$. We know that

$$A_i = \bigcup \{ p_i^{\mathcal{P}M}(\boldsymbol{A}_m) \mid m \in \mathbb{N} \}$$

where $\boldsymbol{A}_0 := \boldsymbol{\emptyset}$, $\boldsymbol{A}_{m+1} := \boldsymbol{p}^{\mathcal{P}M}(\boldsymbol{A}_m)$. For $m = 0$, obviously $\boldsymbol{c} > \boldsymbol{A}_0$. Suppose $\boldsymbol{c} > \boldsymbol{A}_m$ for some m. By induction on p_i, $p_i^{\mathcal{C}M}(\boldsymbol{A}_m) < p_i^M(\boldsymbol{c})$ for each i, hence $\boldsymbol{A}_{m+1} < \boldsymbol{p}^M(\boldsymbol{c}) \leq \boldsymbol{c}$. Therefore, $\boldsymbol{A} < \boldsymbol{c}$.

(ii) M is μ-continuous: Any valuation $g : X \to M$ is the composition of a valuation $g' : X \to \mathcal{C}M$ with $\sum : \mathcal{C}M \to M$: if $g'(x) = \{g(x)\}$, $g(x) = \sum g'(x)$.

Claim. For every μ-term $t(x_1, \ldots, x_n)$ and sets $A_1, \ldots, A_n \in \mathcal{C}M$,

$$t^M(\sum A_1, \ldots, \sum A_n) = \sum t^{\mathcal{C}M}(A_1, \ldots, A_n). \tag{6}$$

Proof. By induction on t. We abbreviate $A_1, \ldots, A_n$ by $\boldsymbol{A}$. For atomic terms 0 and 1, $0^M = \sum \emptyset = \sum 0^{\mathcal{C}M}$ and $1^M = \sum \{1^M\} = \sum 1^{\mathcal{C}M}$. For variables,

$$x_i^M(\sum \boldsymbol{A}) = \sum A_i = \sum x_i^{\mathcal{C}M}(\boldsymbol{A}).$$

For term $(r + s)$,

$$\begin{aligned}(r+s)^M(\sum \boldsymbol{A}) &= r^M(\sum \boldsymbol{A}) +^M s^M(\sum \boldsymbol{A}) \\ &= \sum\{\sum r^{\mathcal{C}M}(\boldsymbol{A}), \sum s^{\mathcal{C}M}(\boldsymbol{A})\} \\ &= \sum(r^{\mathcal{C}M}(\boldsymbol{A}) \cup s^{\mathcal{C}M}(\boldsymbol{A})) \\ &= \sum(r+s)^{\mathcal{C}M}(\boldsymbol{A}).\end{aligned}$$

For term $(r \cdot s)$, we use the distributivity property of the $\mathcal{C}$-dioid M

$$\begin{aligned}(r \cdot s)^M(\sum \boldsymbol{A}) &= r^M(\sum \boldsymbol{A}) \cdot^M s^M(\sum \boldsymbol{A}) \\ &= (\sum r^{\mathcal{C}M}(\boldsymbol{A})) \cdot^M (\sum s^{\mathcal{C}M}(\boldsymbol{A})) \\ &= \sum(r^{\mathcal{C}M}(\boldsymbol{A}) \cdot^{\mathcal{C}M} s^{\mathcal{C}M}(\boldsymbol{A})) \\ &= \sum(r \cdot s)^{\mathcal{C}M}(\boldsymbol{A}).\end{aligned}$$

Finally, consider $\mu x r$. Let $f : \mathcal{C}M \to \mathcal{C}M$ be $f(B) = r^{\mathcal{C}M}(\boldsymbol{A}, B)$, $g : M \to M$ be $g(b) = r^M(\sum \boldsymbol{A}, b)$. By the induction hypothesis, we have $h \circ f = g \circ h$ for $h = \sum$, i.e. for each $B \in \mathcal{C}M$,

$$\sum r^{\mathcal{C}M}(\boldsymbol{A}, B) = r^M(\sum \boldsymbol{A}, \sum B).$$

We would like to use Lemma 1 to conclude that $\sum = h$ maps the least fixed-point of f, $\mu x r^{\mathcal{C}M}(\boldsymbol{A})$, to the least fixed point of g, $\mu x r^M(\sum \boldsymbol{A})$. The lemma does not apply literally, since the orders on $\mathcal{C}M$ and M are *not* complete. For $\mathcal{C}M$, this is not a problem, since we *do* have $\mu x r^{\mathcal{C}M}(\boldsymbol{A}) = \bigcup\{ m x r^{\mathcal{C}M}(\boldsymbol{A}) \mid m \in \mathbb{N} \}$, because $t^{\mathcal{C}M}(v) = t^{\mathcal{P}M}(v)$ for each valuation $v : X \to \mathcal{C}M$ (c.f. Lemma 10 in [11]), and the equation holds in $\mathcal{P}M$, which *is* complete. For M, by the argument below the least fixed point of g also is the $\sum$ of the finite iterations of g, i.e.

$$\mu x r^M(\sum \boldsymbol{A}) = \sum\{ m x r^M(\sum \boldsymbol{A}) \mid m \in \mathbb{N} \}.$$

Namely, by induction along the well-ordering $\prec$ on μ-terms of Kozen [10], in which $mxr \prec \mu xr$ for each $m \in \mathbb{N}$, we may also assume $mxr^M(\sum \boldsymbol{A}) = \sum mxr^{\mathcal{C}M}(\boldsymbol{A})$ for all $m \in \mathbb{N}$. Therefore,

$$\begin{aligned}\sum \mu xr^{\mathcal{C}M}(\boldsymbol{A}) &= \sum \bigcup \{ mxr^{\mathcal{C}M}(\boldsymbol{A}) \mid m \in \mathbb{N} \} \\ &= \sum \{ \sum mxr^{\mathcal{C}M}(\boldsymbol{A}) \mid m \in \mathbb{N} \} \\ &= \sum \{ mxr^M(\sum \boldsymbol{A}) \mid m \in \mathbb{N} \}.\end{aligned}$$

This implies $\sum \mu xr^{\mathcal{C}M}(\boldsymbol{A}) \le \mu xr^M(\sum \boldsymbol{A})$, since $mxr^M(\sum \boldsymbol{A}) \le \mu xr^M(\sum \boldsymbol{A})$ by induction on m. On the other hand, $\sum \mu xr^{\mathcal{C}M}(\boldsymbol{A})$ solves $r^M(\sum \boldsymbol{A}, x) \le x$ in M, by the induction hypothesis for r and a fixed-point property of $\mu xr^{\mathcal{C}M}(\boldsymbol{A})$:

$$r^M(\sum \boldsymbol{A}, \sum \mu xr^{\mathcal{C}M}(\boldsymbol{A})) = \sum r^{\mathcal{C}M}(\boldsymbol{A}, \mu xr^{\mathcal{C}M}(\boldsymbol{A})) \le \sum \mu xr^{\mathcal{C}M}(\boldsymbol{A}).$$

Hence, $\sum \mu xr^{\mathcal{C}M}(\boldsymbol{A})$ lies above the least solution of $r^M(\sum \boldsymbol{A}, x) \le x$, i.e. we have the reverse inequation $\mu xr^M(\sum \boldsymbol{A}) \le \sum \mu xr^{\mathcal{C}M}(\boldsymbol{A})$ also. ◁

We can now prove the μ-continuity condition (3) for valuations $g : X \to M$ of the form $g = \sum g'$ for some $g' : X \to \mathcal{C}M$.

Claim. For all μ-terms μxt with $t(x, \boldsymbol{x})$, all $\boldsymbol{A} = A_1, \ldots, A_n \in \mathcal{C}M$, and $a, b \in M$:

$$a \cdot \mu xt^M(\sum \boldsymbol{A}) \cdot b = \sum \{ a \cdot mxt^M(\sum \boldsymbol{A}) \cdot b \mid m \in \mathbb{N} \}. \tag{7}$$

Proof. Using the previous Claim (6) in the first and last step, we have

$$\begin{aligned}a \cdot \mu xt^M(\sum \boldsymbol{A}) \cdot b &= (\sum\{a\})(\sum \mu xt^{\mathcal{C}M}(\boldsymbol{A}))(\sum\{b\}) \\ &= \sum(\{a\} \cdot \mu xt^{\mathcal{C}M}(\boldsymbol{A}) \cdot \{b\}) \\ &= \sum(\{a\} \cdot \bigcup \{ mxt^{\mathcal{C}M}(\boldsymbol{A}) \mid m \in \mathbb{N} \} \cdot \{b\}) \\ &= \sum(\bigcup \{ \{a\} \cdot mxt^{\mathcal{C}M}(\boldsymbol{A}) \cdot \{b\} \mid m \in \mathbb{N} \}) \\ &= \sum \{ \sum(\{a\} \cdot mxt^{\mathcal{C}M}(\boldsymbol{A}) \cdot \{b\}) \mid m \in \mathbb{N} \} \\ &= \sum \{ (\sum\{a\}) \cdot (\sum mxt^{\mathcal{C}M}(\boldsymbol{A})) \cdot (\sum\{b\}) \mid m \in \mathbb{N} \} \\ &= \sum \{ a \cdot mxt^M(\sum \boldsymbol{A}) \cdot b \mid m \in \mathbb{N} \}. \quad ◁\end{aligned}$$

This completes the proof of the theorem. □

We now come to the reverse inclusion, that every μ-continuous Chomsky algebra is a $\mathcal{C}$-dioid, i.e. is $\mathcal{C}$-complete and $\mathcal{C}$-distributive.

The idea is, of course, that for any polynomial system $\boldsymbol{x} \ge \boldsymbol{p}(\boldsymbol{x}, \boldsymbol{y})$, if $U \in \mathcal{C}M$ is a component of the least solution $\boldsymbol{U}$ of $\boldsymbol{x} \ge \boldsymbol{p}^{\mathcal{C}M}(\boldsymbol{x}, \boldsymbol{A})$ with parameters $\boldsymbol{A}$, then all components of $\boldsymbol{U}$ have least upper bounds, namely the components of the least solution $\boldsymbol{u}$ of $\boldsymbol{x} \ge \boldsymbol{p}^M(\boldsymbol{x}, \sum \boldsymbol{A})$ in M. And since $\boldsymbol{U}$ is the union of finite iterations $\boldsymbol{U}_m = \boldsymbol{mxp}^{\mathcal{C}M}(\boldsymbol{A})$, these $\boldsymbol{U}_m$ should have least upper bounds $\boldsymbol{u}_m = \boldsymbol{mxp}^M(\sum \boldsymbol{A})$, which make up $\boldsymbol{u} = \sum\{\, \boldsymbol{u}_m \mid m \in \mathbb{N}\,\}$. Since the p_i contain products, to show $\boldsymbol{u}_m = \sum \boldsymbol{U}_m$ the induction must provide the distributivity property $\sum(UV) = (\sum U)(\sum V)$ for all U, V among $\boldsymbol{U}_m, \boldsymbol{A}$.

Theorem 3. *Every μ-continuous Chomsky algebra M is a $\mathcal{C}$-dioid.*

Proof. By induction on the construction of $\mathcal{C}M$, we show that for all $U, V \in \mathcal{C}M$

(a) U and V have least upper bounds, $\sum U$ resp. $\sum V \in M$, and
(b) UV has a least upper bound, and $\sum(UV) = (\sum U)(\sum V)$.

Then M is a $\mathcal{C}$-dioid. By Proposition 1, it is sufficient to show (a) and

(b') For all $a, b \in M$, $\sum(aUb) = a(\sum U)b$ and $\sum(aVb) = a(\sum V)b$.

If U, V belong to $\mathcal{F}M$, (a) and (b') are true since M is a dioid. Otherwise, there is a polynomial system $\boldsymbol{x} \geq \boldsymbol{p}(\boldsymbol{x}, \boldsymbol{y})$ and parameters $\boldsymbol{A} \in (\mathcal{C}M)^k$ such that U, V belong to the least solution $\boldsymbol{U}$ of $\boldsymbol{x} \geq \boldsymbol{p}^{\mathcal{P}M}(\boldsymbol{x}, \boldsymbol{A})$ in $\mathcal{P}M$. By induction, (a) and (b'), hence (b), hold for all $U, V \in \boldsymbol{A}$. We must show (a) and (b') for all $U, V \in \boldsymbol{U}$.

By induction on m, we first prove for $\boldsymbol{U}_m := \boldsymbol{mxp}^{\mathcal{C}M}(\boldsymbol{A})$, $\boldsymbol{u}_m := \boldsymbol{mxp}^M(\sum \boldsymbol{A})$

(i) $\sum \boldsymbol{U}_m$ exists (componentwise),
(ii) for all monomials $q(\boldsymbol{x}, \boldsymbol{y})$, $q^M(\sum \boldsymbol{U}_m, \sum \boldsymbol{A}) = \sum q^{\mathcal{C}M}(\boldsymbol{U}_m, \boldsymbol{A})$,
(iii) $\boldsymbol{u}_m = \sum \boldsymbol{U}_m$.

Clearly, (ii) extends to polynomials $q(\boldsymbol{x}, \boldsymbol{y})$, as $\sum(A \cup B) = \sum A + \sum B$.

For $m = 0$, (i) and (iii) are clear: $\boldsymbol{0} = \sum \boldsymbol{\emptyset}$. Therefore, (ii) follows from the hypothesis (a) and (b) for members of $\boldsymbol{A}$.

For $m + 1$, by induction $\sum \boldsymbol{U}_m$ exists by (i), and then

$$\begin{aligned}
\boldsymbol{u}_{m+1} &= \boldsymbol{p}^M(\boldsymbol{u}_m, \textstyle\sum \boldsymbol{A}) && \text{(by definition)}\\
&= \boldsymbol{p}^M(\textstyle\sum \boldsymbol{U}_m, \sum \boldsymbol{A}) && \text{(by (iii))}\\
&= \textstyle\sum \boldsymbol{p}^{\mathcal{C}M}(\boldsymbol{U}_m, \boldsymbol{A}) && \text{(by (ii))}\\
&= \textstyle\sum \boldsymbol{U}_{m+1}. && \text{(by definition)}
\end{aligned}$$

Hence, (i) $\sum \boldsymbol{U}_{m+1}$ exists, and (iii) $\boldsymbol{u}_{m+1} = \sum \boldsymbol{U}_{m+1}$. For (ii), let $q(\boldsymbol{x}, \boldsymbol{y})$ be a monomial in $\boldsymbol{x}, \boldsymbol{y}$, and $r(\boldsymbol{x}, \boldsymbol{y})$ the polynomial obtained by distribution from $q(\boldsymbol{x}, \boldsymbol{y})[\boldsymbol{x}/\boldsymbol{p}(\boldsymbol{x}, \boldsymbol{y})]$. Then

$$\begin{aligned}
q^M(\textstyle\sum \boldsymbol{U}_{m+1}, \sum \boldsymbol{A}) &= r^M(\textstyle\sum \boldsymbol{U}_m, \sum \boldsymbol{A})\\
&= \textstyle\sum r^{\mathcal{C}M}(\boldsymbol{U}_m, \boldsymbol{A}) \qquad \text{(by (ii) for } r)\\
&= \textstyle\sum q^{\mathcal{C}M}(\boldsymbol{U}_{m+1}, \boldsymbol{A}).
\end{aligned}$$

Since M is a Chomsky algebra, $\boldsymbol{x} \geq \boldsymbol{p}^M(\boldsymbol{x}, \sum \boldsymbol{A})$ has a least solution, $\boldsymbol{u} := \mu \boldsymbol{x}\boldsymbol{p}^M(\sum \boldsymbol{A})$. It follows that it is the least upper bound of $\boldsymbol{U} = \mu \boldsymbol{x}\boldsymbol{p}^{\mathcal{C}M}(\boldsymbol{A})$:

$$\begin{aligned}
\boldsymbol{u} &= \mu \boldsymbol{x}\boldsymbol{p}^M(\textstyle\sum \boldsymbol{A}) \\
&= \textstyle\sum\{\, m\boldsymbol{x}\boldsymbol{p}^M(\sum \boldsymbol{A}) \mid m \in \mathbb{N}\} && (M \text{ is } \mu\text{-continuous}) \\
&= \textstyle\sum\{\, \boldsymbol{u}_m \mid m \in \mathbb{N}\} \\
&= \textstyle\sum\{\sum \boldsymbol{U}_m \mid m \in \mathbb{N}\} && (\text{by (iii)}) \\
&= \textstyle\sum \bigcup\{\boldsymbol{U}_m \mid m \in \mathbb{N}\} \\
&= \textstyle\sum \boldsymbol{U} = \sum \mu \boldsymbol{x}\boldsymbol{p}^{\mathcal{C}M}(\boldsymbol{A}).
\end{aligned}$$

In particular, we have shown (a) for $U, V \in \boldsymbol{U}$. To show (b') extend a, b to some $\boldsymbol{a}, \boldsymbol{b} \in M^n$. Having $\boldsymbol{a}(\sum \boldsymbol{U}_m)\boldsymbol{b} = \sum \boldsymbol{a}\boldsymbol{U}_m\boldsymbol{b}$ inductively by (ii), we obtain

$$\begin{aligned}
\boldsymbol{a}(\textstyle\sum \boldsymbol{U})\boldsymbol{b} &= \boldsymbol{a} \cdot \boldsymbol{u} \cdot \boldsymbol{b} \\
&= \textstyle\sum\{\, \boldsymbol{a} \cdot \boldsymbol{u}_m \cdot \boldsymbol{b} \mid m \in \mathbb{N}\} && (M \text{ is } \mu\text{-continuous}) \\
&= \textstyle\sum\{\, \boldsymbol{a}(\sum \boldsymbol{U}_m)\boldsymbol{b} \mid m \in \mathbb{N}\} && (\text{by (iii)}) \\
&= \textstyle\sum\{\sum(\boldsymbol{a}\boldsymbol{U}_m\boldsymbol{b}) \mid m \in \mathbb{N}\} && (\text{by (ii)}) \\
&= \textstyle\sum \bigcup\{\, \boldsymbol{a}\boldsymbol{U}_m\boldsymbol{b} \mid m \in \mathbb{N}\} && (\textstyle\sum \text{ property}) \\
&= \textstyle\sum(\boldsymbol{a} \cdot \bigcup\{\boldsymbol{U}_m \mid m \in \mathbb{N}\} \cdot \boldsymbol{b}) && (\cdot^{\mathcal{C}M} \text{ is } \textstyle\bigcup\text{-continuous}) \\
&= \textstyle\sum(\boldsymbol{a}\boldsymbol{U}\boldsymbol{b}).
\end{aligned}$$

Hence, for $U \in \boldsymbol{U}$ we have (b') $a(\sum U)b = \sum aUb$ for all a, b. □

The morphisms in the category $\mathbb{D}\mathcal{A}$ of $\mathcal{C}$-dioids are the $\mathcal{C}$-morphisms. In the category of μ-continuous Chomsky algebras of Grathwohl et al. [3], the morphisms are the semiring homomorphisms that preserve least solutions of polynomial inequalities. It remains to be checked that these two types of morphisms are the same.

Proposition 3. *Let $f : M \to M'$ be a homomorphism between $\mathcal{C}$-dioids M and M'. Then f is a $\mathcal{C}$-morphism iff f is a semiring homomorphism that preserves least solutions of polynomial inequalities.*

Proof. If f is a dioid-homomorphism, we have $\widetilde{f}(m\boldsymbol{x}\boldsymbol{p}^{\mathcal{C}M}(\boldsymbol{A})) = m\boldsymbol{x}\boldsymbol{p}^{\mathcal{C}M'}(\widetilde{f}(\boldsymbol{A}))$ for all m, which implies $\widetilde{f}(\mu\boldsymbol{x}\boldsymbol{p}^{\mathcal{C}M}(\boldsymbol{A})) = \mu\boldsymbol{x}\boldsymbol{p}^{\mathcal{C}M'}(\widetilde{f}(\boldsymbol{A}))$.

⇒: Let f be a $\mathcal{C}$-morphism, $\boldsymbol{x} \geq \boldsymbol{p}(\boldsymbol{x}, \boldsymbol{y})$ a system of polynomial inequalities with $n = |\boldsymbol{x}|, k = |\boldsymbol{y}|$ and $\boldsymbol{a} \in M^k$. We have $\boldsymbol{a} = \sum \boldsymbol{A}$ for suitable sets $\boldsymbol{A} \in (\mathcal{C}M)^k$. Since f is a $\mathcal{C}$-morphism, hence a dioid-homomorphism, we obtain

$$\begin{aligned} f(\mu \boldsymbol{x} \boldsymbol{p}^M(\sum \boldsymbol{A})) &= f(\sum \mu \boldsymbol{x} \boldsymbol{p}^{\mathcal{C}M}(\boldsymbol{A})) = \sum \widetilde{f}(\mu \boldsymbol{x} \boldsymbol{p}^{\mathcal{C}M}(\boldsymbol{A})) \\ &= \sum (\mu \boldsymbol{x} \boldsymbol{p}^{\mathcal{C}M'}(\widetilde{f}(\boldsymbol{A}))) = \mu \boldsymbol{x} \boldsymbol{p}^{\mathcal{C}M'}(\sum \widetilde{f}(\boldsymbol{A})) \\ &= \mu \boldsymbol{x} \boldsymbol{p}^{\mathcal{C}M'}(f(\sum \boldsymbol{A})). \end{aligned}$$

$\Leftarrow$: Let f be a semiring homomorphism that preserves least solutions of polynomial inequations. We show $f(\sum U) = \sum \widetilde{f}(U)$, by induction on $U \in \mathcal{C}M$. It is clear for $U \in \mathcal{F}M$. Otherwise, U is a component of the least solution $\boldsymbol{U}$ of some polynomial system $\boldsymbol{x} \geq \boldsymbol{p}^{\mathcal{C}M}(\boldsymbol{x}, \boldsymbol{A})$, where $f(\sum \boldsymbol{A}) = \sum \widetilde{f}(\boldsymbol{A})$ for the parameters $\boldsymbol{A} \in \mathcal{C}M$ by induction. From the proof of Theorem 3 we know $\sum \boldsymbol{U} = \sum \mu \boldsymbol{x} \boldsymbol{p}^{\mathcal{C}M}(\boldsymbol{A}) = \mu \boldsymbol{x} \boldsymbol{p}^M(\sum \boldsymbol{A})$, so by the assumption on f

$$\begin{aligned} f(\sum \boldsymbol{U}) &= f(\mu \boldsymbol{x} \boldsymbol{p}^M(\sum \boldsymbol{A})) = \mu \boldsymbol{x} \boldsymbol{p}^{M'}(f(\sum \boldsymbol{A})) \\ &= \mu \boldsymbol{x} \boldsymbol{p}^{M'}(\sum \widetilde{f}(\boldsymbol{A})) = \sum \mu \boldsymbol{x} \boldsymbol{p}^{\mathcal{C}M'}(\widetilde{f}(\boldsymbol{A})) \\ &= \sum \widetilde{f}(\mu \boldsymbol{x} \boldsymbol{p}^{\mathcal{C}M}(\boldsymbol{A})) = \sum \widetilde{f}(\boldsymbol{U}). \end{aligned}$$

6 Conclusion

We have shown that the categories of C-dioids of Hopkins [6] and of μ-continuous Chomsky algebras of Grathwohl et al. [3] coincide. To do so, we have replaced the somewhat technical grammar-based definition of context-free subsets $\mathcal{C}M$ of a monoid M from Hopkins [6] by a more natural, but equivalent definition as the closure of the collection of finite subsets of M under least solutions of polynomial inequations with parameters from $\mathcal{C}M$, which avoids a detour through free monoids. Our proofs exhibit a direct correspondence between the stages of the construction of least solutions $\boldsymbol{U} \in (\mathcal{C}M)^n$ of polynomial inequations and their least upper bounds $\sum \boldsymbol{U} \in M^n$, as was to be expected.

If one is not interested in this correspondence, one can obtain Theorem 3, as pointed out by a reviewer, from Lemma 3.1 of Grathwohl et al. [3], which "asserts that the supremum of a context-free language over a μ-continuous Chomsky algebra K exists, interpreting strings over K as products in K. Moreover, multiplication is continuous with respect to context-free languages." And given that for a $\mathcal{C}$-dioid M the sets in $\mathcal{C}M$ are the context-free sets in the sense just indicated and have least upper bounds denoted by μ-terms, the $\mathcal{C}$-distributivity amounts to the μ-continuity property by Corollary 1, which roughly gives Theorem 2.

By the equivalence of $\mathcal{C}$-dioids and μ-continuous Chomsky algebras, we can transfer closure under coproducts, coequalizers and tensor products from the former to the latter, and transfer axiomatizability and completeness results from the latter to the former.

Let us close with an open question. A polynomial with parameters from a monoid M is *linear*, if each of its monomials has at most one variable factor. Let $\mathcal{L}M$, the *metalinear subsets* of M, be the closure of $\mathcal{F}M$ under (binary) union and finite products of components of least solutions of systems $\boldsymbol{x} \geq \boldsymbol{p}(\boldsymbol{x})$ of linear polynomials with parameters from M (cf. Harrison [5], p. 64, for the metalinear languages). Similar to the proof of Theorem 1, we can show that $\mathcal{L}$ is a monadic operator, thereby obtaining a category $\mathbb{D}\mathcal{L}$ of $\mathcal{L}$-dioids and $\mathcal{L}$-morphisms. On the other hand, call a dioid M *linear-algebraically closed* if each system of linear polynomial inequations has least solutions. Let the *linear-μ-terms* be those that arise as solution terms of linear polynomial inequations through Bekić's reduction. It seems that our proofs of Theorems 2 and 3 can be specialized to show that the $\mathcal{L}$-dioids and the linear-algebraically closed, linear-μ-continuous dioids are the same. Moreover, can this be continued by specializing the completeness theorem of Grathwohl et al. [3] to provide a complete axiomatization of the equational theory of metalinear languages, consisting of the dioid axioms and a linear-μ-continuity axiom?

References

1. Bekić, H.: Definable operations in general algebras, and the theory of automata and flowcharts. In: Jones, C.B. (ed.) Programming Languages and Their Definition. LNCS, vol. 177, pp. 30–55. Springer, Heidelberg (1984). https://doi.org/10.1007/BFb0048939
2. de Bakker, J., Scott, D.: A theory of programs. IBM Seminar, Vienna (1969)
3. Grathwohl, N.B.B., Henglein, F., Kozen, D.: Infinitary axiomatization of the equational theory of context-free languages. In: Baelde, D., Carayol, A. (eds.) Fixed Points in Computer Science (FICS 2013). EPTCS, vol. 126, pp. 44–55 (2013)
4. Gunawardena, J. (ed.): Idempotency. Publications of the Newton Institute, Cambridge University Press, Cambridge (1998)
5. Harrison, M.: Introduction to Formal Languages. Addison Wesley, Reading (1978)
6. Hopkins, M.: The algebraic approach I: the algebraization of the chomsky hierarchy. In: Berghammer, R., Möller, B., Struth, G. (eds.) RelMiCS 2008. LNCS, vol. 4988, pp. 155–172. Springer, Heidelberg (2008). https://doi.org/10.1007/978-3-540-78913-0_13
7. Hopkins, M.: The algebraic approach II: dioids, quantales and monads. In: Berghammer, R., Möller, B., Struth, G. (eds.) RelMiCS 2008. LNCS, vol. 4988, pp. 173–190. Springer, Heidelberg (2008). https://doi.org/10.1007/978-3-540-78913-0_14
8. Hopkins, M., Leiß, H.: Coequalizers and tensor products for continuous idempotent semirings. In: Desharnais, J., Guttmann, W., Joosten, S. (eds.) RAMiCS 2018. LNCS, vol. 11194, pp. 37–52. Springer, Cham (2018)
9. Kozen, D.: On induction vs. *-continuity. In: Kozen, D. (ed.) Logic of Programs 1981. LNCS, vol. 131, pp. 167–176. Springer, Heidelberg (1982). https://doi.org/10.1007/BFb0025782
10. Kozen, D.: Results on the propositional μ-calculus. Theor. Comput. Sci. **27**(3), 333–354 (1983)
11. Leiß, H.: The matrix ring of a μ-continuous Chomsky algebra is μ-continuous. In: Regnier, L., Talbot, J.-M. (eds.) 25th EACSL Annual Conference on Computer Science Logic (CSL 2016). Leibniz International Proceedings in Informatics, pp. 1–16. Leibniz-Zentrum für Informatik, Dagstuhl Publishing (2016)
12. Leiß, H., Ésik, Z.: Algebraically complete semirings and Greibach normal form. Ann. Pure Appl. Log. **133**, 173–203 (2005)
13. Mac Lane, S.: Categories for the Working Mathematician. Springer, New York (1971). https://doi.org/10.1007/978-1-4612-9839-7

Coequalizers and Tensor Products for Continuous Idempotent Semirings

Mark Hopkins[1] and Hans Leiß[2(✉)]

[1] UW-Milwaukee (Alumnus), Milwaukee, USA
federation2005@netzero.net
[2] Munich, Germany
leiss@cis.uni-muenchen.de

Abstract. We provide constructions of coproducts, free extensions, coequalizers and tensor products for classes of idempotent semirings in which certain subsets have least upper bounds and the operations are sup-continuous. Among these classes are the *-continuous Kleene algebras, the μ-continuous Chomsky-algebras, and the unital quantales.

1 Introduction

The theory of formal languages and automata has well-recognized connections to algebra, as shown by work of S. C. Kleene, H. Conway, S. Eilenberg, D. Kozen and many others. The core of the algebraic treatment of the field deals with the class of regular languages and finite automata/transducers. Here, right from the beginnings in the 1960s one finds "regular" operations + (union), · (elementwise concatenation), * (iteration, i.e. closure under 1 and ·) and equational reasoning, and eventually a consensus was reached that drew focus to so-called Kleene algebras to model regular languages and finite automata.

An early effort to expand the scope of algebraization to context-free languages was made by Chomsky and Schützenberger [3]. Somewhat later, around 1970, Gruska [5], McWhirter [19], Yntema [20] suggested to add a least-fixed-point operator μ to the regular operations. But, according to [5], "one can hardly expect to get a characterization of CFL's so elegant and simple as the one we have developed for regular expressions." Neither approach found widespread use.

After the appearance of a new axiomatization of Kleene algebras by Kozen [10] in 1990, a formalization of the theory of context-free languages within the algebra of idempotent semirings with a least-fixed-point operator was suggested in Leiß [13] and Ésik et al. [15], leading to the first complete axiomatization of the equational theory of context-free languages and the introduction of the "μ-continuous Chomsky-algebras" by Grathwohl et al. [4] in 2013. This formalism is cast in the same mould as that for the *-continuous Kleene algebra, with an infinitary identity related to the distributivity axiom below.

H. Leiß—Retired from: Centrum für Informations- und Sprachverarbeitung, Ludwig-Maximilians-Universität München, Oettingenstr. 67, 80539 München.

J. Desharnais et al. (Eds.): RAMiCS 2018, LNCS 11194, pp. 37–52, 2018.
https://doi.org/10.1007/978-3-030-02149-8_3

Already in 2008 (see Hopkins [7,8]), a second-order formalism for context-free languages emerged, in the guise of the category $\mathbb{D}\mathcal{C}$, as part of a larger endeavor to embody all of the Chomsky hierarchy by a family of categories $\mathbb{D}\mathcal{A}$ (discussed below). The primary models are idempotent semirings $\mathcal{A}M$ obtained by lifting the operations of monoids M to a family of $\mathcal{A}$-subsets of M, such that the lifted operations are continuous with respect to a supremum operator. The $\mathcal{A}M$ form the Kleisli subcategory of an Eilenberg-Moore category (cf. Mac Lane [18]) $\mathbb{D}\mathcal{A}$.

As shown in Hopkins [8], there is a complete lattice of suitable subfunctors $\mathcal{A}$ of the powerset functor $\mathcal{P}$; among them are $\mathcal{R}$ and $\mathcal{C}$, selecting the regular and context-free subsets of a monoid, for which $\mathbb{D}\mathcal{R}$ coincides with the *-continuous Kleene algebras and $\mathbb{D}\mathcal{C}$ with the μ-continuous Chomsky algebras. For any two functors $\mathcal{A} \leq \mathcal{B}$ in the lattice there is an adjunction $(Q^{\mathcal{B}}_{\mathcal{A}}, Q^{\mathcal{A}}_{\mathcal{B}}, \eta, \epsilon)$ where $Q^{\mathcal{A}}_{\mathcal{B}} : \mathbb{D}\mathcal{B} \to \mathbb{D}\mathcal{A}$ is the forgetful functor and $Q^{\mathcal{B}}_{\mathcal{A}} : \mathbb{D}\mathcal{A} \to \mathbb{D}\mathcal{B}$ extends $D \in \mathbb{D}\mathcal{A}$ to a certain ideal-completion $\overline{D} \in \mathbb{D}\mathcal{B}$ of D. An interesting open problem is whether and when the ideal-completion can be replaced by a more algebraic method. In particular, can we do so for $Q^{\mathcal{C}}_{\mathcal{R}} : \mathbb{D}\mathcal{R} \to \mathbb{D}\mathcal{C}$, the extension of *-continuous Kleene algebras $K \in \mathbb{D}\mathcal{R}$ to their μ-continuous completions $\overline{K} \in \mathbb{D}\mathcal{C}$? Below we provide basic algebraic and categorical constructions with this goal in mind.

By the classical theorem of Chomsky and Schützenberger [3], the context-free languages $\mathcal{C}X^*$ can be reduced to the regular languages $\mathcal{R}(X \cup Y)^*$ over an extended alphabet. The algebraic and category-theoretic constructions given below allow us to sharpen and generalize the Chomsky-Schützenberger result and construct $\mathcal{C}X^*$ from $\mathcal{R}X^*$, then $\mathcal{C}M$ from $\mathcal{R}M$ for arbitrary monoids M, and finally to provide an algebraic construction of $Q^{\mathcal{C}}_{\mathcal{R}}$ and analogous results for the Thue/Turing subsets $\mathcal{T}M$. However, this can only be indicated below; its development has to be deferred to a forthcoming publication.

Section 2 introduces the categories $\mathbb{D}\mathcal{A}$ of $\mathcal{A}$-dioids and mentions the main examples. Section 3 shows that $\mathbb{D}\mathcal{A}$ has coproducts and free extensions. Section 4 provides coequalizers for $\mathbb{D}\mathcal{A}$, shows how they relate to $\mathcal{A}$-congruences, and finally introduces a tensor product for $\mathbb{D}\mathcal{A}$. Section 5 sketches two applications, the construction of the matrix ring $D^{n\times n}$ of a $\mathcal{A}$-dioid D as a tensor product of D with the boolean matrices $\mathbb{B}^{n\times n}$ and the construction of the context-free languages $\mathcal{C}X^*$ as tensor product of $\mathcal{R}X^*$ with a regular "bracket-algebra" $C_2 \in \mathbb{D}\mathcal{R}$. Hence the context-free languages over X are the values of regular expressions in this particular (non-free) *-continuous Kleene algebra –achieving what "we can hardly expect to get" according to [5]–, which has implications for parsing theory to be worked out. Section 6 discusses potential generalizations.

2 The Category of $\mathcal{A}$-Dioids and $\mathcal{A}$-Morphisms

Let $\mathbb{M}$ be the category of monoids $(M, \cdot, 1)$ and homomorphisms between monoids. A *dioid* $(D, +, \cdot, 0, 1)$ is an idempotent semiring. Idempotency of $+$ provides a partial order $\leq$ on D, via $d \leq d'$ iff $d + d' = d'$, with 0 as least element. Distributivity makes $\cdot$ monotone with respect to $\leq$, and $+$ guarantees a least upper bound $\sum U = d_1 + \ldots + d_n$ for each finite subset $U = \{d_1, \ldots, d_n\}$

of D. Let $\mathbb{D}$ be the category of dioids with dioid homomorphisms. Following [7], a *monadic operator* is a functor $\mathcal{A} : \mathbb{M} \to \mathbb{D}$ where for all monoids M, N, (A_0) $\mathcal{A}M$ is a set of subsets of M, (A_1) $\mathcal{A}M$ contains all finite subsets of M, (A_2) $\mathcal{A}M$ is closed under products, hence $(\mathcal{A}M, \cdot, \{1\})$ with

$$A \cdot B := \{ a \cdot b \mid a \in A, b \in B \} \quad \text{for } A, B \in \mathcal{A}M$$

is itself a monoid, (A_3) $\mathcal{A}M$ is closed under unions of sets from $\mathcal{A}\mathcal{A}M$, which implies that $\mathcal{A}M$ is an idempotent semiring with

$$0 := \bigcup \emptyset, \qquad A + B := \bigcup \{A, B\}, \quad \text{for } A, B \in \mathcal{A}M,$$

and (A_4) $\mathcal{A}$ preserves homomorphisms: if $f : M \to N$ is a homomorphism, so is $\mathcal{A}f : \mathcal{A}M \to \mathcal{A}N$ –abbreviated as $\widetilde{f}$–, where for $U \subseteq M$,

$$(\mathcal{A}f)(U) := \{ f(u) \mid u \in U \} =: \widetilde{f}(U).$$

We write $\mathcal{A}M$ for both the set of subsets of M and the dioid $(\mathcal{A}M, +, \cdot, 0, 1)$.

An $\mathcal{A}$-*dioid* $(D, \cdot, 1, \leq)$ is a partially ordered monoid where each $U \in \mathcal{A}D$ has a least upper bound, $\sum U \in D$, and distributivity[1] holds:

$$(\sum U)(\sum V) = \sum (UV) \qquad \text{for all } U, V \in \mathcal{A}D.$$

An $\mathcal{A}$-*morphism* $f : D \to D'$ between $\mathcal{A}$-dioids D and D' is an order-preserving homomorphism such that $f(\sum U) = \sum \widetilde{f}(U)$ for all $U \in \mathcal{A}D$. Let $\mathbb{D}\mathcal{A}$ be the category of $\mathcal{A}$-dioids and $\mathcal{A}$-morphisms between $\mathcal{A}$-dioids.

For every monoid M, $\mathcal{A}M$ is an $\mathcal{A}$-dioid, by Theorem I.1[2]. Every $\mathcal{A}$-dioid becomes a dioid, using $a + b := \sum\{a, b\}$ and $0 := \sum \emptyset$. Every $\mathcal{A}$-morphism is a dioid-homomorphism. Hence we view $\mathbb{D}\mathcal{A}$ as a subcategory of $\mathbb{D}$.

In fact, $\mathcal{A} : \mathbb{M} \to \mathbb{D}\mathcal{A}$ and the forgetful functor $\widehat{\mathcal{A}} : \mathbb{D}\mathcal{A} \to \mathbb{M}$ form an adjunction, by Theorem II.16, and combine to a monad $T_{\mathcal{A}} = (\widehat{\mathcal{A}}\mathcal{A}, \eta, \mu) : \mathbb{M} \to \mathbb{M}$ with unit $\eta : m \in M \mapsto \{m\} \in \mathcal{A}M$ and product $\mu : \mathcal{U} \in \mathcal{A}\mathcal{A}M \mapsto \bigcup \mathcal{U} \in \mathcal{A}M$.

For a partial order D, the *down-closure* $U^{\leq}$ of $U \subseteq D$ is $\{ d \in D \mid d \leq u \text{ for some } u \in U \}$. We say $U, V \subseteq D$ are *cofinal* (in symbols: $U \simeq V$), if U and V have the same down-closure.

Example 1. If $\mathcal{F}$ assigns to each monoid M its finite subsets, then $\mathcal{F}$ is monadic and $\mathbb{D}\mathcal{F}$ is the category of dioids and dioid-homomorphisms. The power set operator $\mathcal{P}$ is monadic and $\mathbb{D}\mathcal{P}$ is the category of quantales with unit.

Example 2. For infinite cardinal κ, $\mathcal{P}_\kappa M = \{ X \mid X \subseteq M, |X| \leq \kappa \}$ is a monadic operator; $\mathbb{D}\mathcal{P}_{\aleph_0}$ is the category of closed semirings [9]. For regular cardinal κ, $\mathcal{F}_\kappa M = \{ X \mid X \subseteq M, |X| < \kappa \}$ is monadic; (A_3) corresponds to regularity.

---

[1] Distributivity is $\sum$-continuity of $\cdot$ and equivalent to $\sum(aUb) = a(\sum U)b$ for all $a, b \in D, U \in \mathcal{A}D$, an instance of which is *-continuity $\sum\{ ac^m b \mid m \in \mathbb{N} \} = ac^*b$.

[2] Theorem I.1 means Theorem 1 of [7], Theorem II.1 means Theorem 1 of [8], etc.

Example 3. Regular, context-free, and Turing/Thue-subsets $\mathcal{R}M$, $\mathcal{C}M$, and $\mathcal{T}M$ of a monoid M can be defined by generalizing the grammatical approach of doing so for free monoids $M = X^*$. In this case, for $\mathcal{A} \in \{\mathcal{R}, \mathcal{C}, \mathcal{T}\}$ one puts

$$\mathcal{A}X^* := \{\, L(G) \mid G \text{ is a grammar of type } \mathcal{A} \text{ over } X \,\}.$$

Here, a grammar $G = (Q, S, H)$ of type $\mathcal{A}$ over X has a set Q disjoint from X, with distinguished element $S \in Q$, and a finite subset H of "right-linear" rules $Q \times (XQ \cup X^*)$ in case $\mathcal{A} = \mathcal{R}$, of "context-free" rules $Q \times (Q \cup X)^*$ in case $\mathcal{A} = \mathcal{C}$, and of "contextual"[3] rules $Q^+ \times (Q \cup X)^*$ in case $\mathcal{A} = \mathcal{T}$, with $Q^+ = Q^* \setminus \{1\}$. The language defined by G is $L(G) = \{ w \in X^* \mid S \Rightarrow_G w \}$, where $\Rightarrow_G$ is the least reflexive, transitive relation above H that is compatible with the monoid operation on $(X \cup Q)^*$. The class of $\mathcal{A}$-languages is closed under homomorphisms:[4] If $h : X^* \to Y^*$ is a homomorphism, so is $\widetilde{h} : \mathcal{A}X^* \to \mathcal{A}Y^*$: a grammar $G = (Q, S, H)$ over X gives rise to a grammar $G^h = (Q, S, H^h)$ over Y, of the same type, with $\widetilde{h}(L(G)) = L(G^h)$.

For arbitrary monoid M, take a generating subset $X \subseteq M$ and use the canonical homomorphism $h_X : X^* \to M$, where $h_X(x) = x$ for $x \in X$, to put

$$\mathcal{A}M = \{\, \widetilde{h_X}(L(G)) \mid G \text{ a grammar of type } \mathcal{A} \text{ over } X \,\}.$$

This is independent of the choice of X, because for any generating set $Y \subseteq M$ there is a homomorphism $h : X^* \to Y^*$ with $h_X = h_Y \circ h$, so that $\widetilde{h_X}(L(G)) = \widetilde{h_Y}(L(G^h))$. We sketch why these $\mathcal{A}$ are monadic operators. Obviously, $\mathcal{A}M \subseteq \mathcal{P}M$, showing (A_0). For (A_1) and (A_2), note that any finite set $F \subseteq M$ is the language of a grammar G of type $\mathcal{A}$ over a set $X \supseteq F$ of generators, and that from grammars G_1, G_2 of type $\mathcal{A}$ one easily constructs a grammar of type $\mathcal{A}$ for $L(G_1)L(G_2)$. According to Theorem I.8, $(A_3) \wedge (A_4)$ are equivalent to (A_5): every "substitution" homomorphism $\sigma : M \to \mathcal{A}N$ lifts to a homomorphism $\sigma^* : \mathcal{A}M \to \mathcal{A}N$ by $\sigma^*(U) = \bigcup\{ \sigma(m) \mid m \in U \}$, for $U \in \mathcal{A}M$. To show (A_5), assume generating subsets $X \subseteq M$, $Y \subseteq N$, homomorphisms $h_X : X^* \to M$, $h_Y : Y^* \to N$, $\sigma : M \to \mathcal{A}N$ and $U \in \mathcal{A}M$. There is a grammar G of type $\mathcal{A}$ over X with $U = \widetilde{h_X}(L(G))$. For each $x \in X$, let G_x be a grammar of type $\mathcal{A}$ over Y with $\sigma(h_X(x)) = \widetilde{h_Y}(L(G_x)) \in \mathcal{A}N$. The map $x \mapsto L(G_x) \in \mathcal{A}Y^*$ extends to a homomorphism $\hat{\sigma} : X^* \to \mathcal{A}Y^*$, and $\sigma \circ h_X = \widetilde{h_Y} \circ \hat{\sigma}$. Then

$$\sigma^*(U) = \bigcup \widetilde{\sigma}(\widetilde{h_X}(L(G))) = \bigcup \widetilde{\widetilde{h_Y}}(\widetilde{\hat{\sigma}}(L(G))) = \widetilde{h_Y}(\bigcup \widetilde{\hat{\sigma}}(L(G))) = \widetilde{h_Y}(\hat{\sigma}^*(L(G))).$$

As the $\mathcal{A}$-languages are closed under substitutions (cf. [6], Theorems 3.4, 6.2, Exercise 9.11), we get $\hat{\sigma}^* : \mathcal{A}X^* \to \mathcal{A}Y^*$, hence $\hat{\sigma}^*(L(G)) \in \mathcal{A}Y^*$ and $\sigma^*(U) \in \mathcal{A}N$.

---

[3] A normal form where $H \subseteq Q^+ \times (X \cup Q)^*$ instead of $H \subseteq (X \cup Q)^+ \times (X \cup Q)^*$, making rule application effective. This corrects a mistaken definition of $\mathcal{T}X^*$ in [7].

[4] This is not true for the context-sensitive languages (cf. [6], Exercise 9.14), so we have to correct Corollary I.2: there is no monadic operator $\mathcal{S}$ of context-sensitive subsets.

Remark 1. Alternative definitions of $\mathcal{RM}$ resp. $\mathcal{CM}$ can be given as the closure of $\mathcal{FM}$ under binary union, elementwise product and iteration * resp. least solutions in $\mathcal{PM}$ of polynomial inequations $x_1 \geq p_1(x_1,\ldots,x_n),\ldots,x_n \geq p_n(x_1,\ldots,x_n)$ with parameters from $\mathcal{CM}$. $\mathbb{D}\mathcal{R}$ is the category of *-continuous Kleene algebras [9]; for a proof, see [7]. $\mathbb{D}\mathcal{C}$ is the category of μ-continuous Chomsky algebras [4]; a proof appears in [16] (in these proceedings).

Let ρ be a dioid-congruence on an $\mathcal{A}$-dioid D. The set D/ρ of congruence classes is a dioid under the operations defined by $(d/\rho)(d'/\rho) := (dd')/\rho$, $1 := 1/\rho$, $d/\rho + d'/\rho := (d+d')/\rho$, $0 := 0/\rho$. For $U \subseteq D$, $U/\rho := \{\, d/\rho \mid d \in U \,\}$.

Let $\leq$ be the partial order on D/ρ derived from $+$. An *$\mathcal{A}$-congruence* on D is a dioid-congruence ρ on D such that for all $U, U' \in \mathcal{A}D$, if $U/\rho \simeq U'/\rho$, then $(\sum U)/\rho = (\sum U')/\rho$.

For any $\rho_0 \subseteq D \times D$, there is a least $\mathcal{A}$-congruence on D above ρ_0, the intersection of all $\mathcal{A}$-congruences $\rho \supseteq \rho_0$ on D.

Lemma 1. *Let $q : D \to Q$ be an $\mathcal{A}$-morphism between $\mathcal{A}$-dioids D, Q and ρ the least $\mathcal{A}$-congruence on D above the relation $\rho_0 \subseteq D \times D$. If $q(a) = q(b)$ for all $(a,b) \in \rho_0$, then $q(a) = q(b)$ for all $(a,b) \in \rho$.*

Proof. Since q is an $\mathcal{A}$-morphism,

$$\rho_q := \{\, (a,b) \mid a,b \in D, q(a) = q(b) \,\}$$

is an $\mathcal{A}$-congruence. By assumption, $\rho_0 \subseteq \rho_q$, hence $\rho \subseteq \rho_q$. □

Proposition 1. *If D is an $\mathcal{A}$-dioid and ρ an $\mathcal{A}$-congruence on D, then D/ρ is an $\mathcal{A}$-dioid and the map $d \mapsto d/\rho$ is an $\mathcal{A}$-morphism.*

Proof. We first show that each $U' \in \mathcal{A}(D/\rho)$ has a least upper bound $\sum U'$. Since $\cdot/\rho$ is a surjective homomorphism, $U' = U/\rho$ for some $U \in \mathcal{A}D$, by Theorem I.9. If $U/\rho = \tilde{U}/\rho$ for some $\tilde{U} \in \mathcal{A}D$, then U/ρ and $\tilde{U}/\rho$ are cofinal, hence $(\sum U)/\rho = (\sum \tilde{U})/\rho$, so $(\sum U)/\rho$ depends on U' only. Clearly $d \mapsto d/\rho$ is monotone, so $(\sum U)/\rho$ is an upper bound of U'. To show that it is least, let e/ρ be any upper bound of U'. Since $\{U, \{e\}\} \in \mathcal{FAD} \subseteq \mathcal{AAD}$, we have $U \cup \{e\} = \bigcup\{U, \{e\}\} \in \mathcal{A}D$. By the choice of e, $(U \cup \{e\})/\rho \simeq \{e\}/\rho$ and so

$$e/\rho + (\textstyle\sum U)/\rho = (e + \sum U)/\rho = (\sum(U \cup \{e\}))/\rho = (\sum\{e\})/\rho = e/\rho.$$

This shows $(\sum U)/\rho \leq e/\rho$. Hence we can define $\sum U' := (\sum U)/\rho$. It follows that for $U \in \mathcal{A}D$, $(\sum U)/\rho = \sum\{\, d/\rho \mid d \in U \,\}$, showing that $d \mapsto d/\rho$ is an $\mathcal{A}$-morphism. Distributivity of $\sum$ can be reduced to distributivity of $\sum$ on D. □

3 Coproducts and Free Extensions

3.1 Coproducts

A *coproduct* of two objects M_1 and M_2 in a category is an object $M_1 \oplus M_2$ with two morphisms $\iota_1 : M_1 \to M_1 \oplus M_2$ and $\iota_2 : M_2 \to M_1 \oplus M_2$ such that for any two morphisms $f : M_1 \to M$ and $g : M_2 \to M$ there is a unique morphism $[f, g] : M_1 \oplus M_2 \to M$ with $f = [f, g] \circ \iota_1$ and $g = [f, g] \circ \iota_2$.

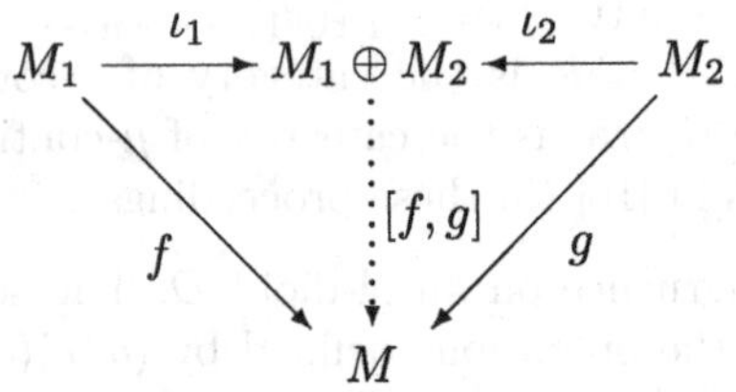

A similar definition may be applied dually with arrows reversed to yield the *product* $M_1 \times M_2$ of objects M_1 and M_2 with corresponding morphisms $\pi_1 : M_1 \times M_2 \to M_1$ and $\pi_2 : M_1 \times M_2 \to M_2$. By the universal property, the coproduct and product of M_1 and M_2 are unique up to isomorphism (cf. [12]).

Example 4. In the category $\mathbb{M}$, a coproduct $M \oplus M'$ of M and M' can be constructed from the interleaved sequences of elements from M and M', i.e. as $(M \times M')^+/E$, the non-empty finite sequences of pairs from $M \times M'$ modulo the congruence E on $(M \times M')^+$ generated by the equations

$$\begin{aligned} &\{\, (m_0, 1^{M'})(m_1, m') = (m_0 m_1, m') \mid m_0, m_1 \in M, m' \in M' \,\} \\ \cup &\{\, (m, m'_0)(1^M, m'_1) = (m, m'_0 m'_1) \mid m \in M, m'_0, m'_1 \in M' \,\}, \end{aligned}$$

with unit $1 := (1^M, 1^{M'})$, injections $\iota_1(m) := (m, 1^{M'})$ and $\iota_2(m') := (1^M, m')$, product $(u/E)(v/E) := (uv)/E$ and induced function

$$[f, g]((m_0, m'_0) \cdots (m_k, m'_k)/E) := f(m_0)g(m'_0) \cdots f(m_k)g(m'_k).$$

Of course, for free monoids Γ^*, Δ^* with Γ disjoint from Δ, $\Gamma^* \oplus \Delta^* \simeq (\Gamma \cup \Delta)^*$.

Theorem 1. *The category $\mathbb{D}\mathcal{A}$ has coproducts. A coproduct*

$$\iota_1 : D_1 \to D_1 \oplus_{\mathcal{A}} D_2 \leftarrow D_2 : \iota_2$$

of $D_1, D_2 \in \mathbb{D}\mathcal{A}$ can be constructed from $\hat{\iota}_1 : M_1 \to M_1 \oplus M_2 \leftarrow M_2 : \hat{\iota}_2$, the coproduct of the monoids M_1, M_2 underlying D_1, D_2, as follows: Let $\equiv$ be the least $\mathcal{A}$-congruence on $\mathcal{A}(M_1 \oplus M_2)$ containing[5]

$$(\{(\sum A, \sum B)\}, A \times B), \ \textit{for all } A \in \mathcal{A}D_1, B \in \mathcal{A}D_2,$$

π the canonical map $U \mapsto U/_\equiv$ and $\eta : M_1 \oplus M_2 \to \mathcal{A}(M_1 \oplus M_2)$ be $\alpha \mapsto \{\alpha\}$. Then let $D_1 \oplus_{\mathcal{A}} D_2$ be $\mathcal{A}(M_1 \oplus M_2)/_\equiv$ and $\iota_k = \pi \circ \eta \circ \hat{\iota}_k$, for $k = 1, 2$.

Proof. $M_1 \oplus M_2$ is a monoid by Example 4, so $\mathcal{A}(M_1 \oplus M_2)$ is an $\mathcal{A}$-dioid. By Proposition 1, its quotient $D_1 \oplus_{\mathcal{A}} D_2 = \mathcal{A}(M_1 \oplus M_2)/_\equiv$ is an $\mathcal{A}$-dioid and π is an $\mathcal{A}$-morphism. Clearly, ι_1 and ι_2 are homomorphisms, since $\hat{\iota}_1, \hat{\iota}_2, \eta$ and π are. The reader may check that they are $\mathcal{A}$-morphisms, hence order-preserving.

[5] Here we use (a, b) etc. for its equivalence class in $M_1 \oplus M_2 = (M_1 \times M_2)^+/E$.

Let $f{:}D_1 \to D$, $g : D_2 \to D$ be $\mathcal{A}$-morphisms and M the monoid underlying D. By the universal property of $M_1 \oplus M_2$, there is a unique homomorphism $[f,g] : M_1 \oplus M_2 \to M$ with $f = [f,g] \circ \hat{\iota}_1$ and $g = [f,g] \circ \hat{\iota}_2$. It extends uniquely to an $\mathcal{A}$-morphism $[f,g]^* : \mathcal{A}(M_1 \oplus M_2) \to D$, by Theorem I.4, with

$$[f,g]^*(U) = \sum \{\, [f,g](u) \mid u \in U \,\} \quad \text{for } U \in \mathcal{A}(M_1 \oplus M_2).$$

For the pairs $(\{(\sum A, \sum B)\}, A \times B)$ generating $\equiv$, we have

$$\begin{aligned}&[f,g]^*(\{(\textstyle\sum A, \sum B)\}) = [f,g](\sum A, \sum B) = f(\sum A) g(\sum B)\\ &= (\textstyle\sum \widetilde{f}(A))(\sum \widetilde{g}(B)) = \sum(\widetilde{f}(A)\widetilde{g}(B)) = \sum \widetilde{[f,g]}(A \times B) = [f,g]^*(A \times B).\end{aligned}$$

Hence by Lemma 1, $[f,g]^*$ is constant on congruence classes of $\equiv$, so we can define $[f,g]_{\mathcal{A}} : D_1 \oplus_{\mathcal{A}} D_2 \to D$ by

$$[f,g]_{\mathcal{A}}([U]) := [f,g]^*(U) \quad \text{for } U \in \mathcal{A}(M_1 \oplus M_2).$$

Then $f = [f,g]_{\mathcal{A}} \circ \iota_1$ and $g = [f,g]_{\mathcal{A}} \circ \iota_2$, and the uniqueness of $[f,g]_{\mathcal{A}}$ can be shown from the uniqueness of $[f,g]$. □

From the uniqueness of coproducts up to isomorphism, one can derive:

Proposition 2. *$\mathcal{A}M_1 \oplus_{\mathcal{A}} \mathcal{A}M_2 \simeq \mathcal{A}(M_1 \oplus M_2)$ for monoids M_1, M_2.*

Proof (Sketch). Let $f : (\mathcal{A}M_1 \times \mathcal{A}M_2)^+ \rightleftarrows (M_1 \times M_2)^+ : g$ map $\langle (A_i, B_i) \mid i \leq n\rangle$ to $\langle (\sum A_i, \sum B_i) \mid i \leq n\rangle$ and $\langle (a_i, b_i) \mid i \leq n\rangle$ to $\langle (\{a_i\}, \{b_i\}) \mid i \leq n\rangle$, respectively. Then $h : (\mathcal{A}M_1 \oplus_{\mathcal{A}} \mathcal{A}M_2) \rightleftarrows \mathcal{A}(M_1 \oplus M_2) : h^{-1}$, with $h([U]) := \widetilde{f}(U)$ for $U \in \mathcal{A}(\mathcal{A}M_1 \times \mathcal{A}M_2)^+$, and $h^{-1}(V) := [\widetilde{g}(V)]$ for $V \in \mathcal{A}(M_1 \oplus M_2)$, are $\mathcal{A}$-morphisms and inverse to each other. □

3.2 Freely Generated Objects and Free Extensions

An object C in a category *is freely generated by the set S* if there is a map $i : S \to C$ such that for all objects D and maps $s : S \to D$ there is a unique morphism $h_s : C \to D$ with $s = h_s \circ i$.

Example 5. In $\mathbb{M}$, the object freely generated by Q is the set Q^* of finite sequences of elements of Q with concatenation as product and the empty sequence as unit element. $i : Q \to Q^*$ maps q to the sequence of q's of length 1.

Proposition 3. *In the category $\mathbb{D}\mathcal{A}$, the object freely generated by the set Q is $\mathcal{A}Q^*$ with map $\eta \circ i : Q \to \mathcal{A}Q^*$, where $\eta : Q^* \to \mathcal{A}Q^*$ is $w \mapsto \{w\}$.*

Proof. Let $D \in \mathbb{D}\mathcal{A}$ and $s : Q \to D$ be a map. As D is a monoid, there is a unique homomorphism $h_s : Q^* \to D$ with $s = h_s \circ i$. By Theorems I.3 and I.2, $\widetilde{h}_s : \mathcal{A}Q^* \to \mathcal{A}D$ and $\sum : \mathcal{A}D \to D$ are $\mathcal{A}$-morphisms. Let $h_s^* : \mathcal{A}Q^* \to D$ be their composition $\sum \circ \widetilde{h}_s$. Then $h_s(w) = \sum \widetilde{h}_s(\{w\}) = h_s^*(\{w\})$ for all $w \in Q^*$, so $s = h_s \circ i = h_s^* \circ (\eta \circ i)$. The uniqueness of h_s^* follows from that of h_s. □

A *free extension* of an object M by a set Q is an object $M[Q]$ with a morphism $\iota : M \to M[Q]$ and a map $\sigma : Q \to M[Q]$ such that for each morphism $f : M \to M'$ and map $s : Q \to M'$ there is a unique morphism $[f,s] : M[Q] \to M'$ with $f = [f,s] \circ \iota$ and $s = [f,s] \circ \sigma$. Again, the free extension of M by Q is unique up to an isomorphism.

Example 6. For monoids M, $\iota = \iota_1 : M \to M \oplus Q^* \leftarrow Q : \iota_2 \circ i = \sigma$ form a free extension of M by Q, where $\iota_1 : M \to M \oplus Q^* \leftarrow Q^* : \iota_2$ is the coproduct of M and Q^* and $i : Q \to Q^*$ the canonical embedding. Hence, $M[Q] \simeq M \oplus Q^*$.

Proposition 4. *In $\mathbb{D}\mathcal{A}$, the free extension $\iota : D \to D[Q] \leftarrow Q : \sigma$ of D by Q consists of the coproduct $D[Q] := D \oplus_{\mathcal{A}} \mathcal{A}Q^*$ of D and $\mathcal{A}Q^*$, with the embedding $\iota_1 : D \to D \oplus_{\mathcal{A}} \mathcal{A}Q^*$ as ι and the composition $\iota_2 \circ \eta \circ i$ of the maps $i : Q \to Q^*$, $\eta : Q^* \to \mathcal{A}Q^*$ with the embedding $\iota_2 : \mathcal{A}Q^* \to D \oplus_{\mathcal{A}} \mathcal{A}Q^*$ as σ.*

Proof. For an $\mathcal{A}$-morphism $f : D \to D'$ and map $s : Q \to D'$, construct an $\mathcal{A}$-morphism $[f,s] : D[Q] \to D'$ with $f = [f,s] \circ \iota$ and $s = [f,s] \circ \sigma$ from the unique homomorphism $h_s : Q^* \to D'$ with $s = h_s \circ i$, its unique extension to an $\mathcal{A}$-morphism $h_s^* : \mathcal{A}Q^* \to D'$ with $h_s = h_s^* \circ \eta$, and the unique $\mathcal{A}$-morphism $[f, h_s^*] : D \oplus_{\mathcal{A}} \mathcal{A}Q^* \to D'$ provided by the coproduct:

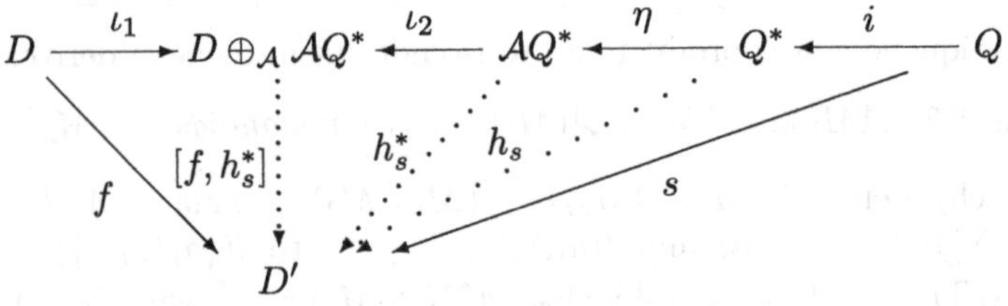

Putting $[f,s] := [f, h_s^*]$, $\iota := \iota_1$ and $\sigma := \iota_2 \circ \eta \circ i$, we get $f = [f,s] \circ \iota$ and $s = [f,s] \circ \sigma$. The uniqueness follows using Proposition 3 and Theorem 1. □

Corollary 1. $(\mathcal{A}M)[Q] = \mathcal{A}M \oplus_{\mathcal{A}} \mathcal{A}Q^* \simeq \mathcal{A}(M \oplus Q^*) = \mathcal{A}(M[Q])$.

4 Coequalizers and Tensor Products

4.1 Coequalizers and Quotients

A *coequalizer* of two morphisms $f, g : A \to B$ is an object Q with a morphism $q : B \to Q$ such that $q \circ f = q \circ g$ and for every morphism $q' : B \to Q'$ with $q' \circ f = q' \circ g$ there is a unique morphism $h_{q'} : Q \to Q'$ with $q' = h_{q'} \circ q$. By the universal property, coequalizers are unique up to isomorphism.

Example 7. In the category $\mathbb{M}$, a coequalizer of $f, g : N \to M$ consists of the quotient monoid $M/_{\equiv_{f,g}}$ with the canonical map $m \mapsto m/_{\equiv_{f,g}}$, where $\equiv_{f,g}$ is the least congruence E on M with $\{\, (f(n), g(n)) \mid n \in N \,\} \subseteq E$.

Conversely, if $E \subseteq M \times M$ is a congruence on M, the quotient monoid M/E with the canonical map $m \mapsto m/E$ is the coequalizer of the homomorphisms $f, g : N \to M$ where N is the submonoid of $M \times M$ with universe E and f, g are the restrictions of the projections $\pi_1, \pi_2 : M \times M \to M$ to N.

Likewise, in $\mathbb{D}\mathcal{A}$ coequalizers and quotients correspond to each other.

Theorem 2. *The category $\mathbb{D}\mathcal{A}$ has coequalizers.*

Proof. Let $f, g : A \to B$ be $\mathcal{A}$-morphisms between $\mathcal{A}$-dioids. Let ρ be the least $\mathcal{A}$-congruence on B above $\{ (f(a), g(a)) \mid a \in A \}$, $Q := B/\rho$ and $q : B \to Q$ the canonical map, $b \mapsto b/\rho$. By Proposition 1, Q is an $\mathcal{A}$-dioid and q an $\mathcal{A}$-morphism. Clearly, $q \circ f = q \circ g$. Concerning the universal property, let $q' : B \to Q'$ be an $\mathcal{A}$-morphism with $q' \circ f = q' \circ g$. To define $h : Q \to Q'$, we put $h(b/\rho) := q'(b)$; this is well-defined by Lemma 1, since if $(b, b') = (f(a), g(a))$ for some $a \in A$, $q'(b) = q'(b')$ holds. Clearly $q' = h \circ q$, and h is an $\mathcal{A}$-morphism, because q' is; in particular, for $U \in \mathcal{A}B$,

$$h(\sum(U/\rho)) = h((\sum U)/\rho) = q'(\sum U) = \sum\{ h(b/\rho) \mid b \in U \} = \sum \widetilde{h}(U/\rho).$$

As $q : B \to Q = B/\rho$ is surjective, the h with $q' = h \circ q$ is unique. □

Corollary 2. *The category $\mathbb{D}\mathcal{A}$ has colimits.*

This follows from the existence of coproducts and coequalizers, see [11], p. 24.

Proposition 5. *Let D be an $\mathcal{A}$-dioid and ρ an $\mathcal{A}$-congruence on D. There are an $\mathcal{A}$-dioid N and two $\mathcal{A}$-morphisms $f, g : N \to D$ such that ρ is the least $\mathcal{A}$-congruence on D above $\{ (f(n), g(n)) \mid n \in N \}$ and D/ρ with $d \mapsto d/\rho$ is a coequalizer of $f, g : N \to D$.*

Since we don't make use of this fact below, we omit the proof.

Proposition 6. *Suppose $\pi : M \to Q$ is a coequalizer of $f, g : N \to M$ in $\mathbb{M}$. Then $\mathcal{A}\pi : \mathcal{A}M \to \mathcal{A}Q$ is a coequalizer of $\mathcal{A}f, \mathcal{A}g : \mathcal{A}N \to \mathcal{A}M$ in $\mathbb{D}\mathcal{A}$.*

Proof. As coequalizers are unique up to isomorphism, we can assume Q is M/E and π is $m \mapsto m/E$, where E is the least congruence on M above $\{ (f(n), g(n)) \mid n \in N \}$. We return to our abbreviation $\widetilde{f}$ for $\mathcal{A}f$ etc. Clearly, $\widetilde{\pi} \circ \widetilde{f} = \widetilde{\pi} \circ \widetilde{g}$ follows from the assumption $\pi \circ f = \pi \circ g$.

To show the universal property, let $\pi' : \mathcal{A}M \to Q'$ be an $\mathcal{A}$-morphism with $\pi' \circ \widetilde{f} = \pi' \circ \widetilde{g}$. Since π is surjective, so is $\widetilde{\pi} : \mathcal{A}M \to \mathcal{A}Q$, by Theorem I.9, and there can be at most one $\mathcal{A}$-morphism $h : \mathcal{A}Q \to Q'$ with $\pi' = h \circ \widetilde{\pi}$. As E is the closure of $\{ (f(n), g(n)) \mid n \in N \}$ under reflexivity, symmetry, transitivity and monoid-congruence, one sees by induction that if $(m, m') \in E$, then $\pi'(\{m\}) = \pi'(\{m'\})$, using $\pi' \circ \widetilde{f} = \pi' \circ \widetilde{g}$ in the base case. Since $m \mapsto \{m\}$ is a homomorphism, $\{ \{m\} \mid m \in B \} \in \mathcal{A}\mathcal{A}M$, so

$$\pi'(B) = \pi'(\bigcup\{ \{m\} \mid m \in B \}) = \sum\{ \pi'(\{m\}) \mid m \in B \}.$$

On $U = B/E = \tilde{\pi}(B) \in \mathcal{A}Q$ with $B \in \mathcal{A}M$, put $h(B/E) := \pi'(B)$. This is well-defined, since $\pi'(\{m\}) = \pi'(\{m'\})$ for $m/E = m'/E$. Finally, for $\mathcal{U} \in \mathcal{A}\mathcal{A}(M/E)$ there is $\mathcal{V} \in \mathcal{A}\mathcal{A}M$ with $\mathcal{U} = \{ V/E \mid V \in \mathcal{V} \}$, so h is an $\mathcal{A}$-morphism:

$$\begin{aligned} h(\bigcup \mathcal{U}) &= h((\bigcup \mathcal{V})/E) = \pi'(\bigcup \mathcal{V}) = \sum \{ \pi'(V) \mid V \in \mathcal{V} \} \\ &= \sum \{ h(V/E) \mid V \in \mathcal{V} \} = \sum \tilde{h}(\mathcal{U}). \end{aligned}$$

□

Theorem 3. *Let E be a congruence on the monoid M, $\mathcal{A}E$ the least $\mathcal{A}$-congruence on $\mathcal{A}M$ above $\{ (\{m\}, \{m'\}) \mid (m, m') \in E \}$. Then $\mathcal{A}M/\mathcal{A}E \simeq \mathcal{A}(M/E)$.*

Proof. By Example 7, there are a monoid N and homomorphisms $f, g : N \to M$ such that $\pi : M \to M/E$ is the coequalizer of $f, g : N \to M$ and E is the least congruence on M above $\{ (f(n), g(n)) \mid n \in N \}$. We show that the canonical map $c : \mathcal{A}M \to \mathcal{A}M/\mathcal{A}E$ is a coequalizer of $\tilde{f}, \tilde{g} : \mathcal{A}N \to \mathcal{A}M$. Then, by the uniqueness of coequalizers and Proposition 6, $\mathcal{A}M/\mathcal{A}E \simeq \mathcal{A}(M/E)$. Write $[U]$ for the $\mathcal{A}E$-congruence class $c(U)$ of $U \in \mathcal{A}M$.

First, $c \circ \tilde{f} = c \circ \tilde{g}$: By Proposition 1, c is an $\mathcal{A}$-morphism, so for $A \in \mathcal{A}N$, from $\{ \{n\} \mid n \in A \} \in \mathcal{A}\mathcal{A}N$ we get

$$(c \circ \tilde{f})(A) = \bigcup \{ (c \circ \tilde{f})(\{n\}) \mid n \in A \} = \bigcup \{ [\{f(n)\}] \mid n \in A \}.$$

Therefore, it is sufficient to show $[\{f(n)\}] = [\{g(n)\}]$ for each $n \in N$. But since $(f(n), g(n)) \in E$, we have $(\{f(n)\}, \{g(n)\}) \in \mathcal{A}E$.

Second, $c : \mathcal{A}M \to \mathcal{A}M/\mathcal{A}E$ has the universal property for coequalizers of $\tilde{f}, \tilde{g}$: Let $q : \mathcal{A}M \to Q$ be an $\mathcal{A}$-morphism with $q \circ \tilde{f} = q \circ \tilde{g}$. We have to show that q uniquely factors through c. By Proposition 6, $\tilde{\pi} : \mathcal{A}M \to \mathcal{A}(M/E)$ is a coequalizer of $\tilde{f}, \tilde{g} : \mathcal{A}N \to \mathcal{A}M$. As $q \circ \tilde{f} = q \circ \tilde{g}$, there is a unique $\mathcal{A}$-morphism h_q with $q = h_q \circ \tilde{\pi}$. We show that $\tilde{\pi}$ and hence q are constant on congruence classes of $\mathcal{A}E$, so that by

$$h([U]) := q(U), \quad \text{for } U \in \mathcal{A}M,$$

$h : \mathcal{A}M/\mathcal{A}E \to Q$ is well-defined. By Lemma 1, $\tilde{\pi}$ is constant on $\mathcal{A}E$-congruence classes, if $\tilde{\pi}(U) = \tilde{\pi}(U')$ for all $(U, U') \in \{ (\{m\}, \{m'\}) \mid (m, m') \in E \}$. But in this case, $(m, m') \in E$ gives $\tilde{\pi}(U) = U/E = \{m/E\} = \{m'/E\} = U'/E = \tilde{\pi}(U')$.

Finally, as c is surjective, for every $\mathcal{V} \in \mathcal{A}(\mathcal{A}M/\mathcal{A}E)$ there is $\mathcal{U} \in \mathcal{A}\mathcal{A}M$ with $\mathcal{V} = \{ [U] \mid U \in \mathcal{U} \}$; therefore, h is an $\mathcal{A}$-morphism:

$$h(\sum \mathcal{V}) = h([\bigcup \mathcal{U}]) = q(\bigcup \mathcal{U}) = \sum \{ q(U) \mid U \in \mathcal{U} \} = \sum \tilde{h}(\mathcal{V}).$$

Since c is surjective, h is the unique $\mathcal{A}$-morphism with $q = h \circ c$. □

4.2 Tensor Products

Two monoid-homomorphisms $f : M_1 \to M$ and $g : M_2 \to M$ are *relatively commuting*, if for all $m_1 \in M_1$ and $m_2 \in M_2$, $f(m_1)g(m_2) = g(m_2)f(m_1)$.

In a category whose objects have a monoid structure, a *tensor product* of two objects M_1 and M_2 is an object $M_1 \otimes M_2$ with two relatively commuting morphisms $\top_1 : M_1 \to M_1 \otimes M_2$ and $\top_2 : M_2 \to M_1 \otimes M_2$ such that for any pair of relatively commuting morphisms $f : M_1 \to M$ and $g : M_2 \to M$ there is a unique morphism $h_{f,g} : M_1 \otimes M_2 \to M$ with $f = h_{f,g} \circ \top_1$ and $g = h_{f,g} \circ \top_2$:

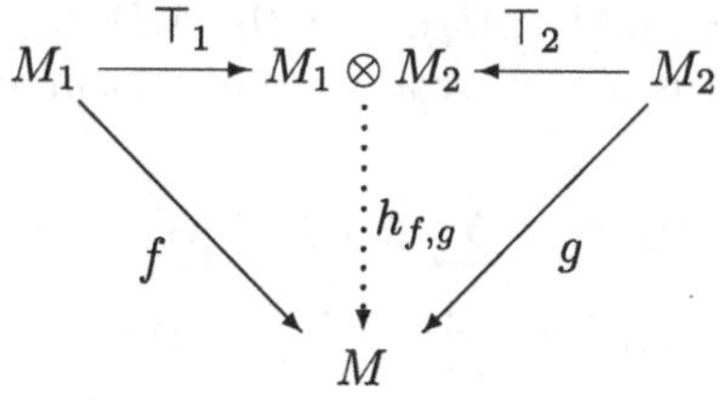

Intuitively, a tensor product is a free extension of both objects in which elements of one commute with elements of the other.

Example 8. A tensor product $\top_1 : M_1 \to M_1 \otimes M_2 \leftarrow M_2 : \top_2$ of two monoids M_1 and M_2 can be constructed as the coequalizer $(M_1 \oplus M_2)/_{\equiv_{a,b}}$ of the homomorphisms $a, b : M_1 \times M_2 \to M_1 \oplus M_2$ defined by

$$a(m_1, m_2) = (m_1, 1)(1, m_2), \qquad b(m_1, m_2) = (1, m_2)(m_1, 1),$$

with the embeddings $\top_1(m_1) = (m_1, 1)/_{\equiv_{a,b}}$, $\top_2(m_2) = (1, m_2)/_{\equiv_{a,b}}$.

Proof. By the universal property of the coproduct $\iota_1 : M_1 \to M_1 \oplus M_2 \leftarrow M_2 : \iota_2$, there is a unique homomorphism $[f, g] : M_1 \oplus M_2 \to M$ with $f = [f, g] \circ \iota_1$ and $g = [f, g] \circ \iota_2$. Since f and g are relatively commuting, $[f, g] \circ a = [f, g] \circ b$. Hence, by the universal property of the coequalizer $\cdot/_{\equiv_{a,b}} : M_1 \oplus M_2 \to Q$ of a, b there is a unique homomorphism $h_{[f,g]} : Q \to M$ such that $[f, g] = h_{[f,g]} \circ (\cdot/_{\equiv_{a,b}})$.

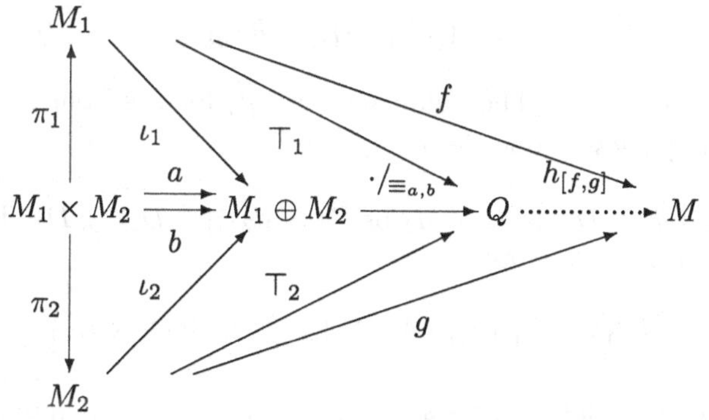

It follows that $f = [f, g] \circ \iota_1 = h_{[f,g]} \circ (\cdot/_{\equiv_{a,b}}) \circ \iota_1 = h_{[f,g]} \circ \top_1$, likewise $g = h_{[f,g]} \circ \top_2$. Thus, $h_{f,g} := h_{[f,g]}$ is the induced homomorphism for f, g. □

To obtain a tensor product for $\mathcal{A}$-dioids, we can lift this construction to $\mathbb{D}\mathcal{A}$.

Theorem 4. *A tensor product* $\top_1 : D_1 \to D_1 \otimes_{\mathcal{A}} D_2 \leftarrow D_2 : \top_2$ *of* $\mathcal{A}$*-dioids* D_1 *and* D_2 *can be obtained from the tensor product* $\hat{\top}_1 : M_1 \to M_1 \otimes M_2 \leftarrow M_2 : \hat{\top}_2$ *of the monoids* M_1, M_2 *underlying* D_1, D_2 *by taking* $D_1 \otimes_{\mathcal{A}} D_2 := \mathcal{A}(M_1 \otimes M_2)/_{\equiv}$ *and* $\top_k = \pi \circ \eta \circ \hat{\top}_k$, $k = 1, 2$, *where* π *is the canonical map* $U \mapsto U/_{\equiv}$ *and* $\equiv$ *is the least* $\mathcal{A}$*-congruence on* $\mathcal{A}(M_1 \otimes M_2)$ *containing*

$$(\{(\sum A, \sum B)\}, A \times B), \ \textit{for all } A \in \mathcal{A}D_1, B \in \mathcal{A}D_2.$$

Proof. The proof is analogous to the proof of Theorem 1. We write $[U]$ for $\pi(U)$, where $U \in \mathcal{A}(M_1 \otimes M_2) = \mathcal{A}(M_1 \times M_2)$. $\top_1 : D_1 \to D_1 \otimes_{\mathcal{A}} D_2$ is an $\mathcal{A}$-morphism, and so is $\top_2$, because $\top_1$ is obviously a homomorphism and for $A \in \mathcal{A}D_1$,

$$\begin{aligned}\top_1(\sum A) &= [\{\hat{\top}_1(\sum A)\}] = [\{(\sum A, 1)\}] = [A \times \{1\}] \\ &= \sum \{ [\{(a,1)\}] \mid a \in A \} = \sum \{ [\hat{\top}_1(a)] \mid a \in A \} = \sum \widetilde{\top_1}(A).\end{aligned}$$

Let $f : D_1 \to D$ and $g : D_2 \to D$ be relatively commuting $\mathcal{A}$-morphisms. By the universal property for the tensor product in $\mathbb{M}$, there is a unique homomorphism $\hat{h}_{f,g} : M_1 \otimes M_2 \to D$ such that $f = \hat{h}_{f,g} \circ \hat{\top}_1$ and $g = \hat{h}_{f,g} \circ \hat{\top}_2$. By Theorem I.4, $\hat{h}_{f,g}$ extends uniquely to an $\mathcal{A}$-morphism $\hat{h}^*_{f,g} : \mathcal{A}(M_1 \otimes M_2) \to D$ with $\hat{h}_{f,g} = \hat{h}^*_{f,g} \circ \eta$, where for $U \in \mathcal{A}(M_1 \otimes M_2)$,

$$\hat{h}^*_{f,g}(U) = \sum \widetilde{\hat{h}_{f,g}}(U) = \sum \{ \hat{h}_{f,g}(m_1, m_2) \mid (m_1, m_2) \in U \}.$$

Define $h_{f,g} : D_1 \otimes_{\mathcal{A}} D_2 \to D$ by

$$h_{f,g}([U]) := \hat{h}^*_{f,g}(U) = \sum \widetilde{\hat{h}_{f,g}}(U), \qquad \text{for } U \in \mathcal{A}(M_1 \otimes M_2).$$

As for $[f, g]^*$ in the proof of Theorem 1 one sees that $\hat{h}^*_{f,g}$ is constant on $\equiv$-classes, so $h_{f,g}$ is well-defined. Then for $d \in D_1$,

$$(h_{f,g} \circ \top_1)(d) = h_{f,g}([\{\hat{\top}_1(d)\}]) = \hat{h}_{f,g}(\hat{\top}_1(d)) = f(d),$$

and likewise $h_{f,g} \circ \top_2 = g$. The uniqueness of $h_{f,g}$ follows from the surjectivity of π and the uniqueness properties of $\hat{h}^*_{f,g}$ and $\hat{h}_{f,g}$. □

For $d_1 \in D_1$ and $d_2 \in D_2$, let $d_1 \otimes d_2$ be $[\{(d_1, d_2)\}] \in D_1 \otimes_{\mathcal{A}} D_2$. The elements of $D_1 \otimes_{\mathcal{A}} D_2$ can be written as

$$\sum \{ d_1 \otimes d_2 \mid (d_1, d_2) \in U \}, \quad U \in \mathcal{A}(D_1 \times D_2).$$

Remark 2. A slightly different tensor product for $\mathbb{D}\mathcal{P}$, the quantales with unit, has been constructed by Liang [17], admitting lattice operations and using bimorphisms rather than relatively commuting morphisms.

Proposition 7. *$\mathcal{A}M_1 \otimes_{\mathcal{A}} \mathcal{A}M_2 \simeq \mathcal{A}(M_1 \otimes M_2)$ for monoids M_1, M_2.*

Proof (Sketch). Define $h : (\mathcal{A}M_1 \otimes_{\mathcal{A}} \mathcal{A}M_2) \rightleftarrows \mathcal{A}(M_1 \otimes M_2) : h^{-1}$ by

$$[U] \mapsto \{ (\sum A, \sum B) \mid (A, B) \in U \}, \quad U \in \mathcal{A}(\mathcal{A}M_1 \times \mathcal{A}M_2)$$
$$V \mapsto [\{ (\{a\}, \{b\}) \mid (a, b) \in V \}], \quad V \in \mathcal{A}(M_1 \otimes M_2).$$
□

5 Applications

We note that if the dioid $D^{n\times n}$ of $n \times n$ square matrices of an $\mathcal{A}$-dioid D is an $\mathcal{A}$-dioid, it is isomorphic to the tensor product $D \otimes_{\mathcal{A}} \mathbb{B}^{n\times n}$.

Proposition 8. *If $D \in \mathbb{D}\mathcal{A}$ and $D^{n\times n} \in \mathbb{D}\mathcal{A}$, then $D^{n\times n} \simeq D \otimes_{\mathcal{A}} \mathbb{B}^{n\times n}$.*

Proof (Sketch). One shows that $I_D : D \to D^{n\times n} \leftarrow \mathbb{B}^{n\times n} : Id$ has the properties of a tensor product, where $I_D(d) := dI_n$ and $Id(B) = B$, for $d \in D$ and $B \in \mathbb{B}^{n\times n}$. If $f : D \to D' \leftarrow \mathbb{B}^{n\times n} : g$ are relatively commuting $\mathcal{A}$-morphisms,

$$h_{f,g}(A) := \sum \{ f(A_{i,j})g(E_{(i,j)}) \mid i, j < n \}, \quad A \in D^{n\times n},$$

is the induced $\mathcal{A}$-morphism with $f = h_{f,g} \circ I_D$ and $g = h_{f,g} \circ Id$. Here, $E_{(i,j)} \in \mathbb{B}^{n\times n}$ is the matrix with 1 only in line i, row j. □

In general, to have $D^{n\times n} \in \mathbb{D}\mathcal{A}$, every $U \in \mathcal{A}D^{n\times n}$ must have a supremum, with $(\sum U)_{i,j} := \sum U_{i,j}$, where $U_{i,j} = \{ A_{i,j} \mid A \in U \}$ for $i, j < n$. Thus, $\sum U$ exists if $U_{i,j} \in \mathcal{A}D$ for each $U \in \mathcal{A}D^{n\times n}$ and $i, j < n$.

Example 9. For the operators $\mathcal{A}$ of Example 1 ($\mathcal{F}$, $\mathcal{P}$) and Example 2 ($\mathcal{F}_\kappa$, $\mathcal{P}_\kappa$), $\mathbb{D}\mathcal{A}$ is closed under matrix ring formation, i.e. for each $D \in \mathbb{D}\mathcal{A}$ and $n \in \mathbb{N}$, $D^{n\times n} \in \mathbb{D}\mathcal{A}$. This also holds for operators $\mathcal{R}$ and $\mathcal{C}$ of Example 3 (cf. [10,14]).

Finally, we return to the earlier attempt by Chomsky and Schützenberger to expand the programme of algebraization beyond the regular languages.

Example 10. Let $X = \{u, v\}$, $Y = \{b, d, p, q\}$ and view b, d and p, q as two pairs of brackets. Let $D \subseteq (X \cup Y)^*$ be the Dyck-language of strings with balanced brackets and $h : (X \cup Y)^* \to X^*$ the bracket-erasing homomorphism. By a theorem of Chomsky/Schützenberger [3], $\mathcal{C}X^* = \{ h(R \cap D) \mid R \in \mathcal{R}(X \cup Y)^* \}$. The theorem admits a form relating $\mathcal{C}$ with $\otimes_{\mathcal{R}}$, which we indicate by an example.

Let $C = \mathcal{R}Y^*/\{bd = 1 = pq, bq = 0 = pd\}$ be the *polycyclic* $\mathcal{R}$-dioid, in which matching brackets reduce to 1 and bracket mismatches to 0; in the semiring equations $bq = 0$ etc., we use $y \in Y$ for $\{y\} \in \mathcal{R}Y^*$. Then the tensor product $\mathcal{R}X^* \otimes_{\mathcal{R}} C$ contains all context-free languages $L \in \mathcal{C}X^*$; for example, if $L = \{ u^n v^n \mid n \in \mathbb{N} \}$, then L belongs to $\mathcal{R}X^* \otimes_{\mathcal{R}} C$, since by *-continuity

$$b(up)^*(qv)^*d = b(\sum_n (up)^n)(\sum_m (qv)^m)d = \sum_{n,m} u^n v^m b p^n q^m d = \sum_n u^n v^n,$$

where b, d, p, q and u, v stand for their images $1 \otimes \{b\}$, $\{u\} \otimes 1$ etc. in $\mathcal{R}X^* \otimes_{\mathcal{R}} C$.

The reduction of $\mathcal{C}X^*$ to $\mathcal{R}(X \cup Y)^*$ by Chomsky and Schützenberger can be improved to a reduction of $\mathcal{C}M$ to $\mathcal{R}M$ for arbitrary monoids M as follows:

Theorem 5. *For any monoid M, $\mathcal{C}M \simeq Z_{C_2}(\mathcal{R}M \otimes_{\mathcal{R}} C_2)$, where*

$$C_2 = \mathcal{R}\{b, d, p, q\}^* / \{bd = 1 = pq, bq = 0 = pd, db + qp = 1\}$$

and $Z_{C_2}(\mathcal{R}M \otimes_{\mathcal{R}} C_2)$ is the centralizer of C_2 in $\mathcal{R}M \otimes_{\mathcal{R}} C_2$, the set of those elements that commute with C_2 in $\mathcal{R}M \otimes_{\mathcal{R}} C_2$.

This generalizes to a construction of $\mathbb{D}\mathcal{C}$ from $\mathbb{D}\mathcal{R}$. The proofs have to be deferred to a future publication.

To give an idea, in the classical theorem of [3], each $L \in \mathcal{C}X^*$ has a *rational kernel* $\hat{L} \in \mathcal{R}(X \cup Y)^*$ in the extended alphabet $X \cup Y$, where intuitively, $\hat{L} \cap D$ consists of the strings in L with begin- and end-markers of phrases according to a context-free grammar for L (or the push- and pop-actions of a push-down automaton for L) inserted. In the general case, $L \in \mathcal{C}M$ is elementwise embedded in $\mathcal{R}M \otimes_{\mathcal{R}} C_2$ and gives a copy $\sum\{\{m\} \otimes 1 \mid m \in L\} \in \mathcal{R}M \otimes_{\mathcal{R}} C_2$. It can be obtained from a regular kernel $\hat{L} \in \mathcal{R}(M \times Y^*) \simeq \mathcal{R}M \otimes_{\mathcal{R}} \mathcal{R}Y^*$ by performing in $\mathcal{R}Y^*$ analogs of the intersection with D and the bracket-erasure via calculations modulo the defining relations of C_2. The equation $db + qp = 1$ not present in C is needed to encode stack operations in C_2.

A similar reduction of $\mathcal{T}M$ to $\mathcal{R}M \otimes_{\mathcal{R}} C_2 \otimes_{\mathcal{R}} C_2$ can be made. To this end, a second copy of C_2 is used to enable computations of a 2-stack or Turing machine.

Open Problems

1. Can an operator $\mathcal{S}$ of context-sensitive subsets be provided by restricting to a subcategory of $\mathbb{M}$ in which erasing homomorphisms are excluded?
2. Are all categories $\mathbb{D}\mathcal{A}$ closed under matrix ring formation, with a uniform proof?
3. Concerning transductions between monoids M and M', is the image of a set $A \in \mathcal{A}M$ under a relation $T \in \mathcal{A}(M \times M')$ a set in $\mathcal{A}M'$?

6 Conclusion

We have studied functors $\mathcal{A} : \mathbb{M} \to \mathbb{D}$ between the finite-subset functor $\mathcal{F}$ and the power set functor $\mathcal{P}$ that give rise to subcategories $\mathbb{D}\mathcal{A}$ of $\mathbb{D}$ and are left adjoints of adjunctions $(\mathcal{A}, \widehat{\mathcal{A}}, \eta, \epsilon)$ between $\mathbb{M}$ and $\mathbb{D}\mathcal{A}$, where $\widehat{\mathcal{A}} : \mathbb{D}\mathcal{A} \to \mathbb{M}$ is the forgetful functor. Each $D \in \mathbb{D}\mathcal{A}$ has a subset $\mathcal{A}D$ between $\mathcal{F}D$ and $\mathcal{P}D$ whose elements $U \in \mathcal{A}D$ have a least upper bound $\sum U \in D$ satisfying the distributivity property $(\sum U)(\sum V) = \sum(UV)$. Based on a notion of $\mathcal{A}$-congruence we have provided constructions of coproduct, coequalizers, quotient and tensor product by lifting corresponding constructions from $\mathbb{M}$ to $\mathbb{D}\mathcal{A}$.

Do our results hold more generally or are they better expressed at a higher level of abstraction like universal algebra or category theory? A reviewer suggested that our $\mathbb{D}\mathcal{A}$'s are infinitary quasi-varieties (or *prevarieties* in

Bergman [2]) and, for example, the existence of coproducts follows from known results. As far as we see, many of our classes $\mathbb{D}\mathcal{A}$ indeed are infinitary quasi-varieties, but probably not all. An *infinitary quasi-variety* is a class of algebras M that can be axiomatized by a set of equational implications

$$\forall \boldsymbol{x} \in M^I(\bigwedge\{\, a_j(\boldsymbol{x}) = b_j(\boldsymbol{x}) \mid j \in J\,\} \to c(\boldsymbol{x}) = d(\boldsymbol{x})),$$

where I and J are sets and a_j, b_j, c, d are derived operations (resp. terms) of arity I. The idea is to to express the least-upper-bound property of $\sum U$ by

$$\bigwedge\{\, u \leq \sum U \mid u \in U\,\} \wedge \forall y \in M(\bigwedge\{\, u \leq y \mid u \in U\,\} \to \sum U \leq y)$$

and replace quantification over sets $U \in \mathcal{A}M$ by quantification over families $\boldsymbol{x} \in M^I$ of individuals, for a set I of size $|I| \geq |U|$. For an algebra signature with operations $\sum$ and projections π_i of arbitrary large arities I the least-upper-bound property of $\sum \boldsymbol{x}$ becomes an infinite equational implication. By quantifying over families $\boldsymbol{x}$ instead of subsets U, the classes of size-bounded subsets, $\mathbb{D}\mathcal{P}_\kappa$ and $\mathbb{D}\mathcal{F}_\kappa$ of Example 2, as well as $\mathbb{D}\mathcal{P}$ of Example 1 become infinitary quasi-varieties (in signatures of class size). Clearly, $\mathbb{D}\mathcal{F} = \mathbb{D}\mathcal{F}_{\aleph_0} = \mathbb{D}$, the idempotent semirings, form a variety (in the signature $+, \cdot, 0, 1$).

For $\mathbb{D}\mathcal{R}$ of Example 3, one can use the equational implications of Kozen's [9] axioms for Kleene algebra (in $0, 1, +, \cdot, ^*$), plus *-continuity of $\cdot$ in the form

$$\forall a, b, c, y[\bigwedge\{\, ab^n c \leq ab^* c \mid n \in \mathbb{N}\,\} \wedge (\bigwedge\{\, ab^n c \leq y \mid n \in \mathbb{N}\,\} \to ab^* c \leq y)].$$

Likewise, for $\mathbb{D}\mathcal{C}$ one can use the μ-continuity condition of [4] (say, with $|\boldsymbol{y}|$-ary Skolem functions $f_{\boldsymbol{x},\boldsymbol{p}}$ for each system of polynomial inequations $\boldsymbol{x} \geq \boldsymbol{p}(\boldsymbol{x}, \boldsymbol{y})$, and first-order terms $f_{\boldsymbol{p}(\boldsymbol{x},\boldsymbol{y})}(\boldsymbol{y})$ for $\mu \boldsymbol{x}\boldsymbol{p}$). Thus, $\mathbb{D}\mathcal{R}$ and $\mathbb{D}\mathcal{C}$ form infinitary quasi-varieties. For $\mathbb{D}\mathcal{T}$ we know of no such axiomatization.

For arbitrary $\mathcal{A}$, the $\mathcal{A}$-subsets $\mathcal{A}D$ on which $\sum$ is defined are not images under arbitrary maps of fixed sets I given by a signature. Hence it seems more appropriate to view the $\mathcal{A}$-dioids as two-sorted algebras $(D, +, \cdot, 0, 1\,;\, \mathcal{A}D, \sum_D)$ or as T-algebras $(D, \sum_D : TD \to D)$ of the Eilenberg-Moore category $\mathbb{M}^T$ of the monad $T = T_\mathcal{A}$ of the adjunction $(\mathcal{A}, \hat{\mathcal{A}}, \eta, \epsilon) : \mathbb{M} \to \mathbb{D}\mathcal{A}$. Embeddings that are T-algebra morphisms would give the right notion of $\mathcal{A}$-subdioids, but we don't know if there are Birkhoff-type theorems for T-algebras (cf. [1]) which show that the $\mathbb{D}\mathcal{A}$ are quasi-varieties in a generalized sense, and if so, whether this implies some of our results. Our lifting of closure properties of $\mathbb{M}$ to closure properties of the categories $\mathbb{D}\mathcal{A}$ is based on taking quotients of $\mathcal{A}$-dioids under $\mathcal{A}$-congruences. It may be possible to perform these liftings from a base category $\mathbb{C}$ with suitable notion of congruence to the category $\mathbb{C}^T$ under more general conditions.

Acknowledgement. We wish to thank the referees for their helpful and challenging comments. The first author wishes to acknowledge the support of his family and support and inspiration of Melanie and Lydia.

References

1. Barr, M.: HSP subcategories of Eilenberg-Moore algebras. Theory Appl. Categ. **18**, 461–468 (2002)
2. Bergman, G.M.: On coproducts in varieties, quasivarieties and prevarieties. Algebr. Number Theory **3**, 847–879 (2009)
3. Chomsky, N., Schuetzenberger, M.: The algebraic theory of context free languages. In: Braffort, P., Hirschberg, D. (eds.) Computer Programming and Formal Systems, pp. 118–161. North-Holland, Amsterdam (1963)
4. Grathwohl, N.B.B., Henglein, F., Kozen, D.: Infinitary axiomatization of the equational theory of context-free languages. In: Baelde, D., Carayol, A. (eds.) Fixed Points in Computer Science (FICS 2013). EPTCS, vol. 126, pp. 44–55 (2013)
5. Gruska, J.: A characterization of context-free languages. J. Comput. Syst. Sci. **5**, 353–364 (1971)
6. Hopcroft, J., Ullman, J.D.: Formal Languages and their Relation to Automata. Addison Wesley, Reading (1979)
7. Hopkins, M.: The algebraic approach I: the algebraization of the Chomsky hierarchy. In: Berghammer, R., Möller, B., Struth, G. (eds.) RelMiCS 2008. LNCS, vol. 4988, pp. 155–172. Springer, Heidelberg (2008). https://doi.org/10.1007/978-3-540-78913-0_13
8. Hopkins, M.: The algebraic approach II: dioids, quantales and monads. In: Berghammer, R., Möller, B., Struth, G. (eds.) RelMiCS 2008. LNCS, vol. 4988, pp. 173–190. Springer, Heidelberg (2008). https://doi.org/10.1007/978-3-540-78913-0_14
9. Kozen, D.: On Kleene algebras and closed semirings. In: Rovan, B. (ed.) MFCS 1990. LNCS, vol. 452, pp. 26–47. Springer, Heidelberg (1990). https://doi.org/10.1007/BFb0029594
10. Kozen, D.: A completeness theorem for Kleene algebras and the algebra of regular events. Inf. Comput. **110**(2), 366–390 (1994)
11. Lambek, J., Scott, P.: Introduction to Higher Order Categorical Logic. Cambridge University Press, New York (1986)
12. Lang, S.: Algebra. Addison-Wesley, Reading (1965)
13. Leiß, H.: Towards Kleene algebra with recursion. In: Börger, E., Jäger, G., Kleine Büning, H., Richter, M.M. (eds.) CSL 1991. LNCS, vol. 626, pp. 242–256. Springer, Heidelberg (1992). https://doi.org/10.1007/BFb0023771
14. Leiß, H.: The matrix ring of a μ-continuous Chomsky algebra is μ-continuous. In: Regnier, L., Talbot, J.-M. (eds.) 25th EACSL Annual Conference on Computer Science Logic (CSL 2016). Leibniz International Proceedings in Informatics, pp. 1–16. Leibniz-Zentrum für Informatik, Dagstuhl Publishing (2016)
15. Leiß, H., Ésik, Z.: Algebraically complete semirings and Greibach normal form. Ann. Pure Appl. Log. **133**, 173–203 (2005)
16. Leiß, H., Hopkins, M.: C-dioids and μ-continuous Chomsky algebras. In: Desharnais, J., et al. (eds.) RAMiCS 2018. LNCS, vol. 11194, pp. 21–36. Springer, Cham (2018)
17. Liang, S.: The tensor product of unital quantales. Stud. Math. Sci. **7**(1), 15–21 (2013)
18. Mac Lane, S.: Categories for the Working Mathematician. Springer, New York (1971). https://doi.org/10.1007/978-1-4612-9839-7
19. McWhirter, I.P.: Substitution expressions. J. Comput. Syst. Sci. **5**, 629–637 (1971)
20. Yntema, M.K.: Cap expressions for context-free languages. Inf. Control **8**, 311–318 (1971)

Distances, Norms and Error Propagation in Idempotent Semirings

Roland Glück(✉)

Deutsches Zentrum für Luft- und Raumfahrt, 86159 Augsburg, Germany
roland.glueck@dlr.de

Abstract. Error propagation and perturbation theory are well-investigated areas of mathematics dealing with the influence of errors and perturbations of input quantities on output quantities. However, these methods are restricted to quantities relying on real numbers under traditional addition and multiplication. We aim to present first steps of an analogous theory on idempotent semirings, so we define distances and norms on idempotent semirings and matrices over them. These concepts are used to derive inequalities characterizing the influence of changes in the input quantities on the output quantities of some often used semiring expressions.

1 Introduction

Idempotent semirings and related structures are widely used tools in mathematics and computer science. Their area of application ranges from graph theory [4,6,10,14,16,18], energy optimization [8] and language problems [3] over fuzzy logics [19] and software development and verification [1,11,23,24] to database theory [22]. In practice, one always has to be aware of defective input data. In the classical setting of finding a shortest path e.g. in a road network there will always be inaccurate measurements of distances. Similarly, faulty data about energy consumption will change the edge weights of automata used in [8] and hence lead to deviant results. Analogous observations hold for all other examples mentioned above. To deal with this problem, mathematics and physics have developed methods of estimating error propagation and perturbation theory [2,15]; however, they deal almost exclusively with numbers under traditional arithmetic operations. Although e.g. [9,12] introduce distances and norms on semirings there is apparently no work about consequences in the spirit of error propagation.

In this paper, we try to give the basics for future research in this direction. Throughout the paper, we illustrate our ideas with running examples concerning the maximum capacity and shortest path problems as well as finite languages. Some results may seem rather straightforward, and some estimations are rather rough but we expect deeper and tighter results in the future by sharper tools. However, as one reviewer pointed out, "Doing science is not quite an endless

J. Desharnais et al. (Eds.): RAMiCS 2018, LNCS 11194, pp. 53–69, 2018.
https://doi.org/10.1007/978-3-030-02149-8_4

sequence of fireworks, ecstasy and champaign; by nature, there must be more moments of toil and frustration than insight and epiphany".

The remainder of this paper is organized as follows: Sect. 2 gives a short overview over the used mathematical structures. The following Sect. 3 first introduces distances on idempotent semiring and subsequently expands this concept to matrices over idempotent semirings. Section 4 has a similar structure but deals with norms instead of distances. As usual, a conclusion and outlook to future work is given in Sect. 5. In Part A of the appendix we summarize some proofs and remarks of running examples. They are not of particularly deep argumentation but sometimes a little bit tedious and would disturb the flow of the presentation of the actual ideas.

2 Basic Structures

A basic structure we will use in the sequel is an idempotent semiring as in the following definition:

Definition 2.1. *An* idempotent semiring *is a structure* $S = (M, +, 0, \cdot, 1)$ *such that* $+$ *(the* addition *or* sum*) and* $\cdot$ *(the* multiplication *or* product*) are associative binary operations on a set* M *with neutral elements* 0 *(called* zero*) and* 1 *(called* one*), resp. Moreover, addition is commutative and idempotent, multiplication distributes from both sides over finite sums, and* 0 *is an annihilator of multiplication.*

For better readability it is standard to omit the multiplication sign in some places and to write xy instead of $x \cdot y$; however, we mostly keep the multiplication sign because we deal often with various semirings in parallel. On an idempotent semiring the relation $\sqsubseteq$, defined by $x \sqsubseteq y \Leftrightarrow_{def} x + y = y$, is a partial order (the *natural order*) with least element 0. Both addition and multiplication are isotone in both arguments with respect to the natural order. The fact that 0 is a multiplicative annihilator is already implied by distributivity of multiplication over finite sums (consider the empty sum); we included it into the definition for the sake of clarity. For further reading we recommend e.g. [5,17].

In many interpretations, addition corresponds to some kind of choice and multiplication to composition whereas the natural order models a kind of increase of information. There are numerous examples of idempotent semirings; we introduce three of them which will serve as running examples in this paper.

1. The *tropical semiring* $\mathbb{R}_{\geq 0}^{\min,+} =_{def} (\mathbb{R}_{\geq 0} \cup \{\infty\}, \min, \infty, +, 0)$.
2. The *max-min semiring* $\mathbb{R}_{\geq 0}^{\max,\min} =_{def} (\mathbb{R}_{\geq 0} \cup \{\infty\}, \max, 0, \min, \infty)$.
3. The *semiring of finite languages over* Σ, $LAN_{\Sigma}^{\mathbf{fin}} =_{def} (\Sigma^{\mathbf{fin}}, \cup, \emptyset, \cdot, \{\varepsilon\})$, where Σ is an alphabet, $\Sigma^{\mathbf{fin}}$ is the set of finite languages over Σ, and $\cdot$ denotes concatenation of languages.

These and other examples are investigated in detail in [13]. In the sequel, we will define distances and norms on idempotent semirings. The domain of distances and norms in traditional mathematics as e.g. in the classical textbook [7] are the real numbers, vectors and functions on them whereas their range are real numbers with conventional addition and multiplication. We intend to capture other structures, too, so we generalize the real numbers to a measure system:

Definition 2.2. *A* measure system *is a structure* $S_m = (M_m, +_m, 0_m, \cdot_m, \sqsubseteq_m)$ *such that* $+_m$ *(the* addition *or* sum*) and* $\cdot_m$ *(the* multiplication*) are binary associative operations on a set* M_m*. Addition is commutative and has* 0_m *as neutral element; multiplication distributes from both sides over addition. Moreover,* $\sqsubseteq_m$ *is an order on* M_m *with least element* 0_m *such that addition and multiplication are isotone in both arguments with respect to* $\sqsubseteq_m$*.*

Indeed, $(\mathbb{R}_{\geq 0}, +, 0, \cdot, \leq)$ is a measure system so our definition generalizes the traditional concept. Note in particular that we do not require a neutral element of $\cdot_m$ and that 0_m does not have to be an annihilator of $\cdot_m$. In the further course, we use the following measure systems:

1. $\mathbf{m}\mathbb{R}_{\geq 0}^{\max,+} =_{def} (\mathbb{R}_{\geq 0}, \max, 0, +, \leq)$
2. $\mathbf{m}\mathbb{R}_{\geq 0}^{\max,\max} =_{def} (\mathbb{R}_{\geq 0} \cup \{\infty\}, \max, 0, \max, \leq)$
3. $\mathbf{m}\mathbb{N}_0^{+,\cdot} =_{def} (\mathbb{N}_0, +, 0, \cdot, \leq)$
4. $\mathbf{m}LAN_{\Sigma}^{\mathbf{fin}} =_{def} (\Sigma^{\mathbf{fin}}, \cup, \emptyset, \cdot, \subseteq)$ with the same conventions as for $LAN_{\Sigma}^{\mathbf{fin}}$.

3 Distances

Based on the structures defined in the previous section we will now introduce our notion of a distance and extend it to matrices.

3.1 Basics

The intention of a distance is to measure in some way the difference between two objects. Given three objects x, y and z, a reasonable requirement similar to the triangle inequality could be that the distance between x and some kind of sum of y and z is not greater than the sum of the distance between x and y and the distance between x and z (recall that in an idempotent semiring there are no 'negative' elements). Extending this idea to both arguments of a distance function leads to an additive distance in Definition 3.1; the term multiplicative distance is motivated analogously. It will play an important role in Subsect. 3.2 where we consider distances on matrices. Another natural stipulation is that a distance respects the order of the measured idempotent semiring, i.e., if $x \sqsubseteq y \sqsubseteq z$ holds then y should have no greater distance from x than z has. This is captured by an order preserving distance. Finally, we expect that the distance between an object and itself is the least possible one, and that the only object with the least possible distance is the object itself. This is the reason for introducing strictness in the following definition.

Definition 3.1. *Let* $S_s = (M_s, +_s, 0_s, \cdot_s, 1_s)$ *be an idempotent semiring and let* $S_m = (M_m, +_m, 0_m, \cdot_m, \sqsubseteq_m)$ *be a measure system. A mapping* $\mathbf{d} : M_s \times M_s \to M_m$ *is called a distance. It is called an*

- additive distance *if it fulfills for all* x_1, x_2, y_1 *and* y_2 *the law of* additive subdistributivity $\boldsymbol{d}(x_1 +_s x_2, y_1 +_s y_2) \sqsubseteq_m \boldsymbol{d}(x_1, y_1) +_m \boldsymbol{d}(x_2, y_2)$,
- multiplicative distance *if it fulfills for all* x_1, x_2, y_1 *and* y_2 *the law of* multiplicative subdistributivity $\boldsymbol{d}(x_1 \cdot_s x_2, y_1 \cdot_s y_2) \sqsubseteq_m \boldsymbol{d}(x_1, y_1) \cdot_m \boldsymbol{d}(x_2, y_2)$,
- order preserving distance *if it is order preserving, i.e., if for all* x, y *and* z *the implication* $x \sqsubseteq_s y \wedge y \sqsubseteq_s z \Rightarrow \boldsymbol{d}(x, y) \sqsubseteq_m \boldsymbol{d}(x, z)$ *holds, and*
- strict distance *if the equivalence* $\boldsymbol{d}(x, y) = 0_m \Leftrightarrow x = y$ *holds for all* x *and* y.

A complete distance *is an additive, multiplicative, order preserving and strict distance.*

We also say that $\boldsymbol{d}$ is an (additive, multiplicative, order preserving, strict, complete) S_m-distance on S_s. Every kind of S_m-distance is called *idempotent* if addition of S_m is idempotent. Note that we do not require symmetry of $\boldsymbol{d}$. However, Lemma 3.7.1 gives a comparable property for a complete distance.

Assume that 0_m is an annihilator for $\cdot_m$ and consider a multiplicative and strict distance $\boldsymbol{d}$. Then we have $\boldsymbol{d}(x, y) = \boldsymbol{d}(1_s \cdot_s x, 1_s \cdot_s y) \sqsubseteq_m \boldsymbol{d}(1_s, 1_s) \cdot_m \boldsymbol{d}(x, y) = 0_m \cdot_m \boldsymbol{d}(x, y) = 0_m$ by neutrality of 1_s, multiplicative subdistributivity and strictness of $\boldsymbol{d}$, and annihilation by 0_m for all $x, y \in S_s$. This is the reason why we forwent annihilation of zero in Definition 2.2.

Additive subdistributivity extends easily to non-empty finite sums:

Lemma 3.2. *Let* $\boldsymbol{d}$ *be an additive* S_m*-distance on* S_s*. Then for all* $n \geq 1$ *the inequality* $\boldsymbol{d}(\sum_{i=1}^{n} x_i, \sum_{i=1}^{n} y_i) \sqsubseteq_m \sum_{i=1}^{n} \boldsymbol{d}(x_i, y_i)$ *holds.*

Proof: Doing induction on n we start with the generally valid inequality $\boldsymbol{d}(x_1, y_1) \sqsubseteq_m \boldsymbol{d}(x_1, y_1)$ as induction base for $n = 1$. As induction hypothesis we assume $\boldsymbol{d}(\sum_{i=1}^{n} x_i, \sum_{i=1}^{n} y_i) \sqsubseteq_m \sum_{i=1}^{n} \boldsymbol{d}(x_i, y_i)$ and finish with the induction step:

$$
\begin{aligned}
& \boldsymbol{d}(\sum_{i=1}^{n+1} x_i, \sum_{i=1}^{n+1} y_i) = \\
& \quad \{ \text{ splitting the sum } \} \\
& \boldsymbol{d}((\sum_{i=1}^{n} x_i) + x_{n+1}, (\sum_{i=1}^{n} y_i) + y_{n+1}) \sqsubseteq_m \\
& \quad \{ \text{ additive subdistributivity } \} \\
& \boldsymbol{d}(\sum_{i=1}^{n} x_i, \sum_{i=1}^{n} y_i) +_m \boldsymbol{d}(x_{n+1}, y_{n+1}) \sqsubseteq_m \\
& \quad \{ \text{ induction hypothesis, isotony of } +_m \} \\
& \sum_{i=1}^{n} \boldsymbol{d}(x_i, y_i) +_m \boldsymbol{d}(x_{n+1}, y_{n+1}) = \\
& \quad \{ \text{ rearranging the sum } \} \\
& \sum_{i=1}^{n+1} \boldsymbol{d}(x_i, y_i)
\end{aligned}
$$

■

In the sequel, we will refer to this property also as *additive subdistributivity*. Note that Lemma 3.2 holds for $n \geq 0$ if $\boldsymbol{d}$ is additionally strict.

Example 3.3. Consider the tropical semiring $\mathbb{R}_{\geq 0}^{\min,+}$ and the measure system $\mathbf{m}\,\mathbb{R}_{\geq 0}^{\max,+}$. Then the mapping $\boldsymbol{d}_s(x,y) =_{def} |x-y|$ is a complete $\mathbf{m}\,\mathbb{R}_{\geq 0}^{\max,+}$-distance on $\mathbb{R}_{\geq 0}^{\min,+}$. See also Appendix A for details. □

Example 3.4. Consider the max-min semiring $\mathbb{R}_{\geq 0}^{\max,\min}$ and the measure system $\mathbf{m}\,\mathbb{R}_{\geq 0}^{\max,\max}$. Then the mapping $\boldsymbol{d}_m(x,y) =_{def} |x-y|$ is a complete $\mathbf{m}\,\mathbb{R}_{\geq 0}^{\max,\max}$-distance on $\mathbb{R}_{\geq 0}^{\max,\min}$. Details are provided in Appendix A. □

Example 3.5. Consider for some alphabet Σ the semiring of finite languages $LAN_{\Sigma}^{\mathbf{fin}}$ and the measure system $\mathbf{m}LAN_{\Sigma}^{\mathbf{fin}}$. Then the mapping $\boldsymbol{d}_w(L_1,L_2) =_{def} |L_1 \triangle L_2|$ (where $L_1 \triangle L_2 =_{def} L_1 \backslash L_2 \cup L_2 \backslash L_1$ denotes the symmetric set difference of L_1 and L_2) is an additive, order preserving and strict $\mathbf{m}\,\mathbb{N}_0^{+,\cdot}$-distance on $LAN_{\Sigma}^{\mathbf{fin}}$. However, it is no multiplicative distance; see again Appendix A for details. □

Remark 3.6. In general, one can define more than one complete distance on an idempotent semiring. For example, for every complete $\mathbf{m}\,\mathbb{R}_{\geq 0}^{\max,+}$-distance $\boldsymbol{d}$ the mapping $c \cdot \boldsymbol{d}$ is again a complete $\mathbf{m}\,\mathbb{R}_{\geq 0}^{\max,+}$-distance for every $c > 0$. □

In traditional error analysis, one often has implications of the form $\boldsymbol{d}(x,x') \leq \varepsilon \Rightarrow \boldsymbol{d}(f(x),f(x')) \leq g(\varepsilon)$ where x denotes some actual value, x' some noisy value within a distance of ε of x, and g is a function which bounds the error between $f(x)$ and $f(x')$ for some function f. Here we will not reproduce this pattern fully but replace the ε by the distance of another value to x. This motivates Items 2 and 3 in the following lemma. The possible consequences of Part 4 are mentioned in the concluding Sect. 5.

Lemma 3.7. *Let* $\boldsymbol{d}$ *be a complete* S_m*-distance on* S_s*. Then the following properties hold for all* $x,y,z \in M_m$*,* $l,n \in \mathbb{N}_0$ *and mappings* $f,g,h : M_s \to M_s$*:*

1. $\boldsymbol{d}(x, y +_s z) \sqsubseteq_m \boldsymbol{d}(x,y) +_m \boldsymbol{d}(x,z)$
2. $\boldsymbol{d}(x,y) \sqsubseteq_m \boldsymbol{d}(x,z) \Rightarrow \boldsymbol{d}(x^n,y^n) \sqsubseteq_m \boldsymbol{d}(x,z)^n$
3. $\boldsymbol{d}(x,y) \sqsubseteq_m \boldsymbol{d}(x,z) \Rightarrow \boldsymbol{d}(\sum_{i=l}^{n} x^i, \sum_{i=l}^{n} y^i) \sqsubseteq_m \sum_{i=l}^{n} \boldsymbol{d}(x,z)^i$
4. *If* f *and* g *are isotone with respect to* $\sqsubseteq_s$ *and* $f(x) \sqsubseteq_s g(x) \sqsubseteq_s h(x)$ *holds for all* $x \in M_s$ *then we have* $\boldsymbol{d}(f^n(x), g^n(x)) \sqsubseteq_m \boldsymbol{d}(f^n(x), h^n(x))$.

Proof: 1: This follows simply from $\boldsymbol{d}(x, y +_s z) = \boldsymbol{d}(x +_s x, y +_s z) \sqsubseteq_m \boldsymbol{d}(x,y) +_m \boldsymbol{d}(x,z)$ by idempotence of $+_s$ and additive subdistributivity of $\boldsymbol{d}$.

2: By induction on n we have as induction base $\boldsymbol{d}(x^0,y^0) = \boldsymbol{d}(1_s,1_s) = 0_m \sqsubseteq_m \boldsymbol{d}(x,y)^0$ due to power properties, strictness of $\boldsymbol{d}$ and 0_m being the least element with respect to $\sqsubseteq_m$. As induction hypothesis we fix an arbitrary $n \in \mathbb{N}$ with $\boldsymbol{d}(x^n,y^n) \sqsubseteq_m \boldsymbol{d}(x,z)^n$ and finish with the following induction step:

$\boldsymbol{d}(x^{n+1}, y^{n+1}) =$
{ power properties }
$\boldsymbol{d}(x \cdot_s x^n, y \cdot_s y^n) \sqsubseteq_m$

{ multiplicative subdistributivity }
$\boldsymbol{d}(x,y) \cdot_m \boldsymbol{d}(x^n,y^n) \sqsubseteq_m$
{ $\boldsymbol{d}(x,y) \sqsubseteq_m \boldsymbol{d}(x,z)$, induction hypothesis, isotony of $\cdot_m$ }
$\boldsymbol{d}(x,z) \cdot_m \boldsymbol{d}(x,z)^n =$
{ power properties }
$\boldsymbol{d}(x,z)^{n+1}$

3: By Lemma 3.2 we have $\boldsymbol{d}(\sum_{i=l}^{n} x^i, \sum_{i=l}^{n} y^i) \sqsubseteq_m \sum_{i=1}^{n} \boldsymbol{d}(x^i,y^i)$. Part 2 ensures $\boldsymbol{d}(x^i,y^i) \sqsubseteq_m \boldsymbol{d}(x,z)^i$ for all i under consideration, so the claim follows from isotony of $\sum$.

4: Let us fix an arbitrary $x \in M_s$ and let f, g and h be mappings as described. For induction on n, the base case $n = 0$ becomes the trivially true inequality $\boldsymbol{d}(x,x) \sqsubseteq_m \boldsymbol{d}(x,x)$. Otherwise we have the induction hypothesis $f^n(x) \sqsubseteq_s g^n(x)$ and complete with the following induction step:
$f^{n+1}(x) =$
{ unfolding }
$f(f^n(x)) \sqsubseteq_s$
{ induction hypothesis, isotony of f }
$f(g^n(x)) \sqsubseteq_s$
{ $f \sqsubseteq_s g$, ununfolding }
$g^{n+1}(x)$
Analogously we can show $g^n(x) \sqsubseteq_s h^n(x)$ for all $n \in \mathbb{N}_0$, so the claim follows from order preservation of $\boldsymbol{d}$. ∎

3.2 Distances and Matrices

Matrices over semirings are widely used: [13] gives plenty of examples, the matrices from [25] as representation of relations can be considered as matrices over the Booleans, more recently [21] shows an application in the context of tropical optimization. Hence it makes sense to transfer our notion of distances to matrices and investigate the consequences.

Given two binary operations $+$ and $\cdot$ on a set M such that $+$ is associative and commutative, and $\cdot$ is associative and distributes over $+$ we define the multiplication of two matrices $A \in M^{l\times m}$ and $B \in M^{m\times n}$ as usual by $(A \cdot B)_{ij} =_{def} \sum_{k=1}^{m} A_{ik} \cdot B_{kj}$. Analogously, the sum $A + B$ of two matrices $A, B \in M^{m\times n}$ is defined entrywise by $(A + B)_{ij} =_{def} A_{ij} + B_{ij}$. From folklore it is known that matrix multiplication and addition are associative, that matrix addition is commutative, and that multiplication distributes from both sides over addition. If M is equipped additionally with an order $\sqsubseteq$ we lift this order entrywise to matrices of the same dimension by $A \sqsubseteq B \Leftrightarrow_{def} \forall i,j : A_{ij} \sqsubseteq B_{ij}$. In the case that $+$ and $\cdot$ are isotone in both arguments with respect to $\sqsubseteq$ this holds also for matrix addition and multiplication. For simplicity, we use the same signs for addition, multiplication and order on M and on matrices over M. If the operation $+$ is idempotent the so is the derived addition operation on matrices.

So matrices over semirings form a semiring, and matrices over a measure system form a measure system, provided they have suitable dimensions.

Given an idempotent semiring $S_s = (M_s, +_s, 0_s, \cdot_s, 1_s)$, a measure system $S_m = (M_m, +_m, 0_m, \cdot_m, \sqsubseteq_m)$, an S_m-distance $\boldsymbol{d}$ on S_s of any kind and two matrices $A, B \in M_s^{l\times n}$ we define the distance $\boldsymbol{d}^{l\times n}(A, B)$ entrywise as a matrix of the type $M_m^{l\times n}$, i.e., $(\boldsymbol{d}^{l\times n}(A, B))_{ij} =_{def} \boldsymbol{d}(A_{ij}, B_{ij})$. Clearly, $0_m^{l\times n}$, defined by $(0_m^{l\times n})_{ij} = 0_m$ for all i and j, is the least element of $S_m^{l\times n}$. Note that we do not yet know whether $\boldsymbol{d}^{l\times n}$ is also a distance. The following lemmata, culminating in Theorem 3.12, show the distance properties for $\boldsymbol{d}^{n\times n}$.

Lemma 3.8. *Let* $\boldsymbol{d}$ *be an additive and multiplicative* S_m*-distance on* S_s*, and consider two* $l\times m$*-matrices* A *and* A' *and two* $m\times n$*-matrices* B *and* B' *over* S_s*. Then we have the inequality* $\boldsymbol{d}^{l\times n}(A\cdot_s B, A'\cdot_s B') \sqsubseteq_m \boldsymbol{d}^{l\times m}(A, A')\cdot_m \boldsymbol{d}^{m\times n}(B, B')$.

Proof: By definition, we have to show the inequality $(\boldsymbol{d}^{l\times n}(A\cdot_s B, A'\cdot_s B'))_{ij} \sqsubseteq_m (\boldsymbol{d}^{l\times m}(A, A') \cdot_m \boldsymbol{d}^{m\times n}(B, B'))_{ij}$ for all relevant i, j, so we fix two arbitrary i, j and reason as follows:

$(\boldsymbol{d}^{l\times n}(A \cdot_s B, A' \cdot_s B'))_{ij} =$

{ definition of $\boldsymbol{d}^{l\times n}$, definition of matrix multiplication }

$\boldsymbol{d}(\sum_{k=1}^{m} A_{ik} \cdot_s B_{kj}, \sum_{k=1}^{m} A'_{ik} \cdot_s B'_{kj}) \sqsubseteq_m$

{ additive subdistributivity }

$\sum_{k=1}^{m} \boldsymbol{d}(A_{ik} \cdot_s B_{kj}, A'_{ik} \cdot_s B'_{kj}) \sqsubseteq_m$

{ multiplicative subdistributivity, isotony of addition }

$\sum_{k=1}^{m} \boldsymbol{d}(A_{ik}, A'_{ik}) \cdot_m \boldsymbol{d}(B_{kj}, B'_{kj}) =$

{ definition of $\boldsymbol{d}^{l\times m}$, $\boldsymbol{d}^{m\times n}$ and matrix multiplication }

$(\boldsymbol{d}^{l\times m}(A, A') \cdot_m \boldsymbol{d}^{m\times n}(B, B'))_{ij}$ ■

Lemma 3.9. *Consider an additive* S_m*-distance* $\boldsymbol{d}$ *on* S_s *and four* $l \times n$*-matrices* A, A', B *and* B' *over* S_s*. Then we have* $\boldsymbol{d}^{l\times n}(A +_s B, A' +_s B') \sqsubseteq_m \boldsymbol{d}^{l\times n}(A, A') +_m \boldsymbol{d}^{l\times n}(B, B')$.

Proof: This follows from the additivity of $\boldsymbol{d}$ and the definition of $\boldsymbol{d}^{l\times n}$. ■

Lemma 3.10. *Let* $\boldsymbol{d}$ *be an order preserving* S_m*-distance on* S_s*, and consider three* $l \times n$*-matrices* A, B *and* C *over* S_s *with* $A \sqsubseteq_m B \sqsubseteq_m C$*. Then we have* $\boldsymbol{d}^{l\times n}(A, B) \sqsubseteq_m \boldsymbol{d}^{l\times n}(A, C)$.

Proof: Similarly to the previous lemma, this is a simple consequence of the order preservation of $\boldsymbol{d}$ and the definition of $\boldsymbol{d}^{l\times n}$. ■

In an analogous manner we can prove the following lemma:

Lemma 3.11. *Let* $\boldsymbol{d}$ *be a strict* S_m*-distance on* S_s *and consider two arbitrary* $l \times n$*-matrix* A *and* B *over* S_s*. Then we have the equivalence* $\boldsymbol{d}^{l\times n}(A, B) = 0_m^{l\times n} \Leftrightarrow A = B$.

Putting the properties of matrix multiplication and Lemmata 3.8, 3.9, 3.10 and 3.11 together, we obtain the following theorem:

Theorem 3.12. *Let* $\mathbf{d}$ *be an* S_m*-distance on* S_s *and consider the idempotent semiring* $S_s^{n\times n}$ *of* $n \times n$*-matrices over* S_s *and let* $S_m^{n\times n}$ *be the set of* $n \times n$*-matrices over* S_m*. Then the following properties hold:*

1. *If* $\mathbf{d}$ *is additive then* $\mathbf{d}^{n\times n}$ *is an additive* $S_m^{n\times n}$*-distance on* $S_s^{n\times n}$*.*
2. *If* $\mathbf{d}$ *is order preserving then* $\mathbf{d}^{n\times n}$ *is an order preserving* $S_m^{n\times n}$*-distance on* $S_s^{n\times n}$*.*
3. *If* $\mathbf{d}$ *is strict then* $\mathbf{d}^{n\times n}$ *is a strict* $S_m^{n\times n}$*-distance on* $S_s^{n\times n}$*.*
4. *If* $\mathbf{d}$ *is both additive and multiplicative then* $\mathbf{d}^{n\times n}$ *is a multiplicative* $S_m^{n\times n}$*-distance on* $S_s^{n\times n}$*.*

So if $\mathbf{d}$ is a complete S_m-distance on S_s then $\mathbf{d}^{n\times n}$ is a complete $S_m^{n\times n}$-distance on $S_s^{n\times n}$. In particular, this means that we can apply the results from Lemma 3.7 what we will illustrate in the following examples.

Example 3.3 *(continued).* Consider the tropical semiring $\mathbb{R}_{\geq 0}^{\min,+}$ and let $\mathbf{d}_s$ be defined as in example 3.3. It is well-known that for an $n \times n$-matrix A over $\mathbb{R}_{\geq 0}^{\min,+}$ the entry $\sum_{k=0}^{n-1} A_{ij}^k$ equals the length of a shortest path from vertex i to vertex j in the directed graph corresponding to A. Assume now that $\hat{A}$ is an $n \times n$-matrix over $\mathbb{R}_{\geq 0}^{\min,+}$ which in each entry deviates from A by at most one, and let B be the $n \times n$-matrix defined by $B_{ij} =_{def} A_{ij} + 1$. Then we have $\mathbf{d}_s^{n\times n}(A, \hat{A}) \leq \mathbf{d}_s^{n\times n}(A, B) = 1^{n\times n}$ where $1^{n\times n}$ is the $n \times n$-matrix with only 1-entries on every position. Now application of Lemma 3.7.3 yields the inequality $\mathbf{d}_s^{n\times n}(\sum_{k=0}^{n-1} A^k, \sum_{k=0}^{n-1} \hat{A}^k) \leq \sum_{k=0}^{n-1} (1^{n\times n})^k$. Taking into account the definition of the used measure system $\mathbf{m}\,\mathbb{R}_{\geq 0}^{\max,+}$ this yields $\mathbf{d}_s^{n\times n}(\sum_{k=0}^{n-1} A^n, \sum_{k=0}^{n-1} \hat{A}^n) \leq (n-1)^{n\times n}$ where $(n-1)^{n\times n}$ is the $n\times n$-matrix with an entry of $n-1$ in every position. With other words, if we work with a perturbed matrix $\hat{A}$ representing edge length in a directed graph instead of the matrix A representing the actual values, but know additionally that $\hat{A}$ deviates from A on every entry by at most 1 then the error for the computed length of a shortest path is at most $n - 1$. □

Example 3.4 *(continued).* Consider the max-min semiring $\mathbb{R}_{\geq 0}^{\max,\min}$ and let $\mathbf{d}_m$ be defined as in Example 3.4. Let us chose A, $\hat{A}$, B and $1_{n\times n}$ analogously to the previous example. Then $\sum_{k=0}^{n-1} A_{ij}^k$ equals the capacity of a maximum capacity path from i to j. If we work with the noisy matrix $\hat{A}$ instead of A the error of the final result is again bounded by $\sum_{k=0}^{n-1} (1^{n\times n})^k$; however, due to the choice of $\mathbf{m}\,\mathbb{R}_{\geq 0}^{\max,\max}$ as measure system we obtain here $\sum_{k=0}^{n-1} (1^{n\times n})^k = 1^{n\times n}$. This means that for every pair (i, j) the error is at most 1. □

4 Norms

Analogously to Sect. 3 we first introduce norms on idempotent semirings and investigate subsequently the consequences of the expansion to matrices.

4.1 Basics

Similar considerations as before Definition 3.1 lead to the following definition of a norm which can be seen as a generalization of the absolute value on $\mathbb{R}$:

Definition 4.1. *Let $S_s = (M_s, +_s, 0_s, \cdot_s, 1_s)$ be an idempotent semiring and let $S_m = (M_m, +_m, 0_m, \cdot_m, \sqsubseteq_m)$ be a measure system. A mapping $\|\cdot\| : M_s \to M_m$ is called a* norm. *It is called an*

- additive norm *if it is additively subdistributive, i.e., if $\|x +_s y\| \sqsubseteq_m \|x\| +_m \|y\|$ holds for all x and y,*
- multiplicative norm *if it is multiplicatively subdistributive, i.e., if $\|x \cdot_s y\| \sqsubseteq_m \|x\| \cdot_m \|y\|$ holds for all x and y,*
- order preserving norm *if $x \sqsubseteq_s y \Rightarrow \|x\| \sqsubseteq_m \|y\|$ holds for all x and y, and*
- strict norm *if the equivalence $\|x\| = 0_m \Leftrightarrow x = 0_s$ holds.*

A complete norm *is an additive, multiplicative, order preserving and strict norm.*

We also say that $\|\cdot\|$ is an S_m-norm on S_s.

Example 4.2. For every idempotent semiring $S_s = (M_s, +_s, 0_s, \cdot_s, 1_s)$ the identity is a complete S_m-norm with $S_m = (M_s, +_s, 0_s, \cdot_s, \sqsubseteq_s)$. □

Example 4.3. The cardinality function on fuzzy relations as defined in [20] is an additive, order preserving and strict $\mathbf{m}\,\mathbb{R}_{\geq 0}^{\max,+}$-norm. However, it is not multiplicative.

Example 3.5 *(continued).* Consider the idempotent semiring of finite languages $LAN_{\Sigma}^{\mathbf{fin}}$ for some alphabet Σ, and the measure system $\mathbf{m}\,\mathbb{N}_0^{+,\cdot}$. Then the mapping $\|L\| =_{def} |L|$ is a complete $\mathbf{m}\,\mathbb{N}_0^{+,\cdot}$-norm on $LAN_{\Sigma}^{\mathbf{fin}}$; see Appendix A for details. □

Considering the traditional distance $\boldsymbol{d}(x, y) =_{def} |x - y|$ on the real numbers, we see that the mapping $\|\cdot\|_{\boldsymbol{d}}$, defined by $\|x\|_{\boldsymbol{d}} =_{def} \boldsymbol{d}(0, x)$ is a norm on the real numbers. Our definitions of distance and norm lead to an analogous property:

Theorem 4.4. *Let $S_s = (M_s, +_s, 0_s, \cdot_s, 1_s)$ be an idempotent semiring and let $\boldsymbol{d}$ by an (additive, multiplicative, order preserving, strict) S_m-distance on S_s for some measure system $S_m = (M_m, +_m, 0_m, \cdot_m, \sqsubseteq_m)$. Then the mapping $\|\cdot\|_{\boldsymbol{d}} : M_s \to M_m$, defined by $\|x\|_{\boldsymbol{d}} =_{def} \boldsymbol{d}(0_s, x)$, is an (additive, multiplicative, order preserving, strict) S_m-norm on S_s. It is called the* norm induced by $\boldsymbol{d}$.

Proof: Let us pick two arbitrary $x, y \in M_s$. Additive subdistributivity of $\boldsymbol{d}$ implies additive subdistributivity of $\|\cdot\|_{\boldsymbol{d}}$ as follows:

$\|x +_s y\|_{\boldsymbol{d}} =$
{ definition of $\|\cdot\|_{\boldsymbol{d}}$, additive neutrality of 0_s }
$\boldsymbol{d}(0_s +_s 0_s, x +_s y) \sqsubseteq_m$
{ additive subdistributivity of $\boldsymbol{d}$, definition of $\|\cdot\|_{\boldsymbol{d}}$ }
$\|x\|_{\boldsymbol{d}} +_m \|y\|_{\boldsymbol{d}}$

Multiplicative subdistributivity can be shown as follows:

$\|x \cdot_s y\|_{\boldsymbol{d}} =$
{ definition of $\|\cdot\|_{\boldsymbol{d}}$, $0_s = 0_s \cdot_s 0_s$ }
$\boldsymbol{d}(0_s \cdot_s 0_s, x \cdot_s y) \sqsubseteq_m$
{ multiplicative subdistributivity of $\boldsymbol{d}$, definition of $\|\cdot\|_{\boldsymbol{d}}$ }
$\|x\|_{\boldsymbol{d}} \cdot_m \|y\|_{\boldsymbol{d}}$

For order preservation, assume that $\boldsymbol{d}$ is order preserving and that $x \sqsubseteq_s y$ holds. Then we have $0_s \sqsubseteq_s x \sqsubseteq_s y$ by semiring properties, and order preservation of $\boldsymbol{d}$ implies $\boldsymbol{d}(0_s, x) \sqsubseteq_m \boldsymbol{d}(0_s, y)$. Definition of $\|\cdot\|_{\boldsymbol{d}}$ leads now to order preservation of $\|\cdot\|_{\boldsymbol{d}}$. Finally, assume that $\boldsymbol{d}$ is strict and pick an arbitrary $x \in M_s$. Then, by definition, $\|x\|_{\boldsymbol{d}} = 0_m$ is equivalent to $\boldsymbol{d}(0_s, x) = 0_m$ which in turn is equivalent to $x = 0_s$ due to strictness of $\boldsymbol{d}$. Hence, $\|\cdot\|_{\boldsymbol{d}}$ is strict, too. ■

Example 3.5 *(continued).* The identity norm on $LAN_{\Sigma}^{\mathbf{fin}}$ is the norm induced by the distance $\boldsymbol{d}_w$. □

A classical example in the traditional setting for an induced norm is the absolute value of a real number which is induced by the distance $\boldsymbol{d}(x, y) =_{def} |x - y|$. It has the additional property that it can be factored out of a distance, i.e., we have $\boldsymbol{d}(cx, cy) = |c| \cdot \boldsymbol{d}(x, y)$. This motivates the following definition.

Definition 4.5. *An S_m-distance $\boldsymbol{d}$ on S_s of any kind is called* norm-distributive *if its induced norm fulfills the property $\boldsymbol{d}(x \cdot_s y, x \cdot_s z) \sqsubseteq_m \|x\|_{\boldsymbol{d}} \cdot_m \boldsymbol{d}(y, z)$ for all x, y and z.*

Example 3.5 *(continued).* $\boldsymbol{d}_w$ is a norm-distributive distance on $LAN_{\Sigma}^{\mathbf{fin}}$. The same holds for the distance $\boldsymbol{d}_N$, defined by $\boldsymbol{d}_N(L_1, L_2) =_{def} |L_1 \triangle L_2|$. □

The mapping $\|\cdot\| : \mathbb{R}_{\geq 0} \to \mathbb{R}_{\geq 0}$, defined by $\|x\| =_{def} \frac{x}{2}$, is a norm on $\mathbb{R}_0$ in the traditional sense. Its application to the distance $\boldsymbol{d}(x, y) = |x - y|$ on $\mathbb{R}$ yields the function $\frac{|x-y|}{2}$ which in turn is a distance on $\mathbb{R}$ again. Here we have an analogous property, as stated in the following theorem (where we use the term 'norm' instead on semirings on measure systems, defined in a canonical way):

Theorem 4.6. *Consider an idempotent semiring $S_s = (M_s, +_s, 0_s, \cdot_s, 1_s)$ and two measure systems $S_m = (M_m, +_m, 0_m, \cdot_m, \sqsubseteq_m)$ and $S_n = (M_n, +_n, 0_n, \cdot_n, \sqsubseteq_n)$ Let $\boldsymbol{d}$ be an S_m-distance on S_s, let $\| \ \|$ be an S_n-norm on S_m, and consider the* *distance* $\mathbf{d}_{\|\ \|} =_{def} \|\ \| \circ \mathbf{d}$, *i.e.*, $\mathbf{d}_{\|\ \|}(x, y) =_{def} \|\mathbf{d}(x, y)\|$. *Then the following holds:*

1. *If* $\mathbf{d}$ *is additive and* $\|\ \|$ *is additive and order preserving then* $\mathbf{d}_{\|\ \|}$ *is additive.*
2. *If* $\mathbf{d}$ *is multiplicative and* $\|\ \|$ *is multiplicative and order preserving then* $\mathbf{d}_{\|\ \|}$ *is multiplicative.*
3. *If both* $\mathbf{d}$ *and* $\|\ \|$ *are order preserving then* $\mathbf{d}_{\|\ \|}$ *is order preserving.*
4. *If* $\mathbf{d}$ *and* $\|\ \|$ *are strict then* $\mathbf{d}_{\|\ \|}$ *is strict.*
5. *If* $\mathbf{d}$ *is norm-distributive and* $\|\ \|$ *is multiplicative and order preserving then* $\mathbf{d}_{\|\ \|}$ *is norm-distributive, too.*

Proof: We fix arbitrary x, y and z, and reason as follows:

1: Here we have:

$\boldsymbol{d}_{\|\ \|}(x_1 +_s x_2, y_1 +_s y_2) =$
 { definition of $\boldsymbol{d}_{\|\ \|}$ }
$\|\boldsymbol{d}(x_1 +_s x_2, y_1 +_s y_2)\| \sqsubseteq_n$
 { $\boldsymbol{d}$ is additive, $\|\ \|$ is order preserving }
$\|\boldsymbol{d}(x_1, y_1) +_m \boldsymbol{d}(x_2, y_2)\| \sqsubseteq_n$
 { additivity of $\|\ \|$, definition of $\boldsymbol{d}_{\|\ \|}$ }
$\boldsymbol{d}_{\|\ \|}(x_1, y_1) +_n \boldsymbol{d}_{\|\ \|}(x_2, y_2)$

2: This follows analogously to Part 1.

3: Assume that $x \sqsubseteq_s y \sqsubseteq_s z$ holds. By order preservation of $\boldsymbol{d}$ this implies $\boldsymbol{d}(x, y) \sqsubseteq_m \boldsymbol{d}(x, z)$, and because $\|\ \|$ is order preserving we have also $\|\boldsymbol{d}(x, y)\| \sqsubseteq_n \boldsymbol{d}(x, z)$. Definition of $\boldsymbol{d}_{\|\ \|}$ yields now $\boldsymbol{d}_{\|\ \|}(x, y) \sqsubseteq_n \boldsymbol{d}_{\|\ \|}(x, z)$, as desired.

4: Here we have $\boldsymbol{d}_{\|\ \|}(x, y) = 0_n \Leftrightarrow \|\boldsymbol{d}(x, y)\| = 0_n \Leftrightarrow \boldsymbol{d}(x, y) = 0_m \Leftrightarrow x = y$ by definition of $\boldsymbol{d}_{\|\ \|}$ and strictness of $\|\ \|$ and $\boldsymbol{d}$.

5: Here we show $\boldsymbol{d}_{\|\ \|}(x \cdot_s y, x \cdot_s z) \sqsubseteq_n \|x\|_{\boldsymbol{d}_{\|\ \|}} \cdot_n \boldsymbol{d}_{\|\ \|}(y, z)$ as follows:

$\boldsymbol{d}_{\|\ \|}(x \cdot_s y, x \cdot_s z) =$
 { definition of $\boldsymbol{d}_{\|\ \|}$ }
$\|\boldsymbol{d}(x \cdot_s y, x \cdot_s z)\| \sqsubseteq_n$
 { norm-distributivity of $\boldsymbol{d}$, order preservation of $\|\ \|$ }
$\|\ \|x\|_{\boldsymbol{d}} \cdot_m \boldsymbol{d}(y, z)\| \sqsubseteq_n$
 { definition of $\|\ \|_{\boldsymbol{d}}$, multiplicativity of $\|\ \|$ }
$\|\boldsymbol{d}(0_s, x)\| \cdot_n \|\boldsymbol{d}(y, z)\| =$
 { definition of $\boldsymbol{d}_{\|\ \|}$, definition of induced norm }
$\|x\|_{\boldsymbol{d}_{\|\ \|}} \cdot_n \boldsymbol{d}_{\|\ \|}(y, z)$ ■

Example 3.5 *(continued).* The cardinality function $|\ |$ is a complete $\mathbf{m}\mathbb{N}_0^{+,\cdot}$-norm on $\mathbf{m}LAN_\Sigma^{\mathbf{fin}}$. Hence, the mapping $\boldsymbol{d}_N$ is an additive, order preserving, strict and norm-distributive $\mathbf{m}\mathbb{N}_0^{+,\cdot}$-distance on $LAN_\Sigma^{\mathbf{fin}}$ (recall $\boldsymbol{d}_N(L_1, L_2) = |\boldsymbol{d}_w(L_1, L_2)|$, and that $\boldsymbol{d}_w$ is additive, order preserving, strict and norm-distributive). □

4.2 Norms and Matrices

Similarly to Subsect. 3.2 we investigate how norms can be transferred to matrices. Combination of Theorems 3.12 and 4.4 yields the following result:

Corollary 4.7. *Consider a measure system* $S_m = (M_m, +_m, 0_m, \cdot_m, \sqsubseteq_m)$ *and an idempotent semiring* $S_s = (M_s, +_s, 0_s, \cdot_s, 1_s)$. *Let* $\boldsymbol{d}$ *be an* S_m*-distance on* S_s *and consider the norm* $\| \; \|_{\boldsymbol{d}^{n\times n}}$, *i.e., the norm induced by* $\boldsymbol{d}^{n\times n}$ *for some* $n \in \mathbb{N}^+$. *Then the following holds:*

1. *If* $\boldsymbol{d}$ *is additive, then* $\| \; \|_{\boldsymbol{d}^{n\times n}}$ *is additive.*
2. *If* $\boldsymbol{d}$ *is order preserving, then* $\| \; \|_{\boldsymbol{d}^{n\times n}}$ *is order preserving.*
3. *If* $\boldsymbol{d}$ *is strict, then* $\| \; \|_{\boldsymbol{d}^{n\times n}}$ *is strict.*
4. *If* $\boldsymbol{d}$ *id additive and multiplicative, then* $\| \; \|_{\boldsymbol{d}^{n\times n}}$ *is multiplicative.*

Even norm-distributivity is inherited by matrix-valued distances:

Theorem 4.8. *Let* $\boldsymbol{d}$ *be a norm-distributive* S_m*-distance on* S_s. *Then* $\boldsymbol{d}^{n\times n}$ *is norm-distributive, too.*

Proof: Let us fix three arbitrary $n \times n$-matrices A, B and C over S_s and two indices i, j with $1 \leq i, j, \leq n$. By definition of $\sqsubseteq_m$ on matrices we have to show $(\boldsymbol{d}^{n\times n}(A \cdot_s B, A \cdot_s C))_{ij} \sqsubseteq_m (\|A\|_{\boldsymbol{d}^{n\times n}} \cdot_m \boldsymbol{d}^{n\times n}(B, C))_{ij}$. Therefore we reason as follows:

$$
\begin{aligned}
&(\boldsymbol{d}^{n\times n}(A \cdot_s B, A \cdot_s C))_{ij} = \\
&\quad \{ \text{ definition of } \boldsymbol{d}^{n\times n} \text{ and matrix multiplication } \} \\
&\boldsymbol{d}(\sum_{k=1}^{n} A_{ik} \cdot_s B_{kj}, \sum_{k=1}^{n} A_{ik} \cdot_s C_{kj}) \sqsubseteq_m \\
&\quad \{ \text{ additive subdistributivity } \} \\
&\sum_{k=1}^{n} \boldsymbol{d}(A_{ik} \cdot_s B_{kj}, A_{ik} \cdot_s C_{kj}) \sqsubseteq_m \\
&\quad \{ \text{ norm-distributivity of } \boldsymbol{d}, \text{ isotony of } +_m \} \\
&\sum_{k=1}^{n} \|A_{ik}\|_{\boldsymbol{d}} \cdot_m \boldsymbol{d}(B_{kj}, C_{kj}) = \\
&\quad \{ \text{ definition of } \boldsymbol{d}^{n\times n} \text{ and induced norm } \} \\
&\sum_{k=1}^{n} \boldsymbol{d}(0_s, A_{ik}) \cdot_m (\boldsymbol{d}^{n\times n}(B, C))_{kj} = \\
&\quad \{ \text{ definition of } \boldsymbol{d}^{n\times n} \text{ and } 0_s^{n\times n} \} \\
&\sum_{k=1}^{n} (\boldsymbol{d}^{n\times n}(0_s^{n\times n}, A))_{ik} \cdot_m (\boldsymbol{d}^{n\times n}(B, C))_{kj} = \\
&\quad \{ \text{ definition of induced norm and matrix multiplication } \} \\
&(\|A\|_{\boldsymbol{d}^{n\times n}} \cdot_m \boldsymbol{d}^{n\times n}(B, C))_{ij}
\end{aligned}
$$
■

Example 3.5 *(continued).* Matrices over languages are frequently used in the theory of finite automata. Let us consider three such $n \times n$-matrices L_1, L_2 and L_3, the distances $\boldsymbol{d}_w$ and $\boldsymbol{d}_N$ and the cardinality norm $|\,|$. Then we have $\boldsymbol{d}_w(L_1 \cdot L_2, L_1 \cdot L_3) \subseteq L_1 \cdot \boldsymbol{d}_w(L_2, L_3)$ (note that $\| \; \|_{\boldsymbol{d}_w} = \boldsymbol{id}$ holds) which gives us an (in the worst case sharp) estimation on which words $L_1 \cdot L_2$ and $L_1 \cdot L_3$ on the respective matrix entries disagree. Moreover, we have $\boldsymbol{d}_N(L_1 \cdot L_2, L_1 \cdot L_3) \leq |L_1| \cdot \boldsymbol{d}_N(L_2, L_3)$ which gives us an (also in the worst case sharp) upper bound on how much words $L_1 \cdot L_2$ and $L_1 \cdot L_3$ disagree in the respective entries. Note that both the matrices $|L_1|$ and $\boldsymbol{d}_N(L_2, L_3)$ are $n \times n$-matrices over $\mathbb{N}_0$. □

5 Conclusion and Outlook

The tools developed here demonstrated their ability to analyze error propagation and estimation on semirings and matrices over semirings. Although the methods may seem rather raw the results were tight in their respective worst cases. However, this work is thought to be only the first basic step of future research in this area.

An open issue is for example the investigation of the Kleene star. In the continuations of Examples 3.3 and 3.4 on Page 8 we considered actually the Kleene star of matrices; however, in this context it could be rewritten as a finite sum. To tackle the Kleene star in general one possibility would be to extend additive subdistributivity from Definitions 3.1 and 4.1 to arbitrary sums or suprema, possibly in combination with Lemma 3.7.4. Strongly related is the question how solutions of fixpoint equations of the form $ax + b = x$ behave with respect to changes of the parameters a or b.

These properties can be investigated both over basic idempotent semirings or over matrices over them. Considering matrices, it will be interesting to investigate the influence of changes of a matrix on the eigenvalues or bideterminants (as described in [13] and used in [21]) of this matrix. Also, the analysis of randomized changes should be of both theoretical and practical interest.

Acknowledgments. The author is grateful to the anonymous referees for helpful and enlightening remarks, especially to the fourth reviewer for his deep reflections about doing and selling science.

A Deferred Proofs and Remarks

Example 3.3 and Continuation. First we show additive subdistributivity of $\boldsymbol{d}_s$ which means $|\min\{x_1, x_2\} - \min\{y_1, y_2\}| \leq \max\{|x_1 - y_1|, |x_2 - y_2|\}$. W.l.o.g. we assume $x_1 \leq x_2$ and hence $|\min\{x_1, x_2\} - \min\{y_1, y_2\}| = |x_1 - \min\{y_1, y_2\}|$. Now we distinguish the following cases:

1. $y_1 \leq y_2$: Then we have $\min\{y_1, y_2\} = y_1$ and hence $|\min\{x_1, x_2\} - \min\{y_1, y_2\}| = |x_1 - y_1|$. Now the claim is obvious due to $|x_1 - y_1| \leq \max\{|x_1 - y_1|, |x_2 - y_2|\}$.
2. $y_2 < y_1$: Then we have $\min\{y_1, y_2\} = y_2$ and hence $|\min\{x_1, x_2\} - \min\{y_1, y_2\}| = |x_1 - y_2|$. We distinguish the following cases:
   (2a) $y_2 \leq x_1$: By assumption we have $x_1 \leq x_2$ and hence $|x_1 - y_2| \leq |x_2 - y_2|$. Together with $|x_2 - y_2| \leq \max\{|x_1 - y_1|, |x_2 - y_2|\}$ we obtain the claim.
   (2b) $y_2 > x_1$: By assumption we have $y_1 > y_2$ and hence $|x_1 - y_2| < |x_1 - y_1|$, and the claim follows analogously to above.

For multiplicative subdistributivity we have to show $|(x_1+x_2)-(y_1+y_2)| \leq |x_1-y_1|+|x_2-y_2|$. But this is an easy consequence of elementary calculus and the triangle inequality $|x+y| \leq |x|+|y|$.

To show order preservation we have to prove that $x \geq y$ and $y \geq z$ imply $|x-y| \leq |x-z|$ (note that the natural order of $\mathbb{R}_{\geq 0}^{\min,+}$ coincides with $\geq$). The proof of this obvious property and strictness is left to the reader. □

Example 3.4 and Continuation. Considering additive subdistributivity of $\boldsymbol{d}_m$ we have to show $|\max\{x_1,x_2\}-\max\{y_1,y_2\}| \leq \max\{|x_1-y_1|,|x_2-y_2|\}$. Here we assume w.l.o.g. $x_2 \leq x_1$ and hence $|\max\{x_1,x_2\}-\max\{y_1,y_2\}| = |x_1-\max\{y_1,y_2\}|$. Now we have the following cases:

1. $y_1 \geq y_2$: Then we have $\max\{y_1,y_2\} = y_1$ and hence $|\max\{x_1,x_2\}-\max\{y_1,y_2\}| = |x_1-y_1|$. This implies the claim due to $|x_1-y_1| \leq \max\{|x_1-y_1|,|x_2-y_2|\}$.
2. $y_1 < y_2$: Then we have $\max\{y_1,y_2\} = y_2$ and hence $|\max\{x_1,x_2\}-\max\{y_1,y_2\}| = |x_1-y_2|$. Again, we distinguish two cases:
   (2a) $y_2 \leq x_1$: Then we have $y_1 < y_2 \leq x_1$ and hence $|x_1-y_2| < |x_1-y_1|$, and $|x_1-y_1| \leq \max\{|x_1-y_1|,|x_2-y_2|\}$ implies the claim.
   (2b) $y_2 > x_1$: Recalling the assumption $x_2 \leq x_1$ we have $|x_1-y_2| \leq |x_2-y_2|$; here $|x_2-y_2| \leq \max\{|x_1-y_1|,|x_2-y_2|\}$ implies the claim.

This shows additive subdistributivity; multiplicative subdistributivity corresponds to additive subdistributivity from Example 3.3. Order preservation and strictness are also left to the reader. □

Example 3.5 and continuations First we show additive subdistributivity of $\boldsymbol{d}_w$ via the inclusion

$$(A \cup B)\triangle(C \cup D) \subseteq (A\triangle C) \cup (B\triangle D). \tag{1}$$

To this purpose, we expand the left side of Inequation 1 and obtain

$$((A \cup B) \cap \overline{(C \cup D)}) \cup ((C \cup D) \cap \overline{(A \cup B)}). \tag{2}$$

Analogously, its right side can be rewritten as

$$(A \cap \overline{C}) \cup (C \cap \overline{A}) \cup (B \cap \overline{D}) \cup (D \cap \overline{B}). \tag{3}$$

Due to symmetry of Expressions 2 and 3 in the involved variables, it suffices to show that $(A \cup B) \cap \overline{(C \cup D)}$ is contained in Expression 3. However, $(A \cup B) \cap \overline{(C \cup D)}$ equals $(A \cap \overline{(C \cup D)}) \cup (B \cap \overline{(C \cup D)})$, and $A \cap \overline{(C \cup D)}$ is contained in $A \cap \overline{C}$ (and hence in 3) due to antitony of the complement. An analogous argument shows that $(B \cap \overline{(C \cup D)})$ is contained in 3.

Strictness and order preservation are here a simple consequences of set theory so we give an example that this distance is not multiplicative. To this purpose, we consider the three languages $L_1 =_{def} \{a\}$, $L_2 =_{def} \{b\}$ and $L_3 =_{def} \{c\}$. Then we have $(L_1 \cdot L_2)\triangle(L_1 \cdot L_2) = \{ab, ac\} \not\subseteq \emptyset = (L_1\triangle L_1) \cdot (L_2\triangle L_3)$.

For norm-distributivity of $\boldsymbol{d}_w$ we first note that its induced norm corresponds to the identity (due to $\emptyset \triangle L = L$) so we have to show $(L_1L_2)\triangle(L_1L_3) \subseteq L_1(L_2\triangle L_3)$ (for better readability, we omit the $\cdot$ for concatenation). By expanding the symmetric difference and distributivity of concatenation over union this leads to the inclusion $(L_1L_2\backslash L_1L_3) \cup (L_1L_3\backslash L_1L_2) \subseteq L_1(L_2\backslash L_3) \cup L_1(L_3\backslash L_2)$. Obviously, it suffices to show $L_1L_2\backslash L_1L_3 \subseteq L_1(L_2\backslash L_3)$, so we reason as follows:

$w \in L_1L_2\backslash L_1L_3 \Leftrightarrow$
  { definition concatenation, set theory, logic }
$\exists w_1 \in L_1 \exists w_2 \in L_2 \forall w_3 \in L_1 \forall w_4 \in L_3 : w = w_1w_2 \wedge w \neq w_3w_4 \Rightarrow$
  { instantiation w_1 for w_3 }
$\exists w_1 \in L_1 \exists w_2 \in L_2 \forall w_4 \in L_3 : w = w_1w_2 \wedge w \neq w_1w_4 \Leftrightarrow$
  { cancellativity ($w_1w_2 = w_1w_4 \Leftrightarrow w_2 = w_4$) }
$\exists w_1 \in L_1 \exists w_2 \in L_2 \forall w_4 \in L_3 : w = w_1w_2 \wedge w_2 \neq w_4 \Rightarrow$
  { set theory, logic }
$\exists w_1 \in L_1 \exists w_2 \in L_2 : w = w_1w_2 \wedge w_2 \notin L_3 \Leftrightarrow$
  { definition concatenation, set theory, logic }
$w \in L_1(L_2\backslash L_3)$

Additivity, strictness and order preservation of the $\mathbf{m}\,\mathbb{N}_0^{+,\cdot}$-norm $|\cdot|$ on $\mathbf{m}LAN_\Sigma^{\mathbf{fin}}$ follow from basic facts about set cardinality. For multiplicative subdistributivity we consider two finite languages L_1 and L_2 and note that the mapping $\mathbf{concat} : L_1 \times L_2 \to L_1 \cdot L_2$, defined by $\mathbf{concat}((w_1, w_2)) =_{def} w_1 \cdot w_2$ is surjective. Now we have $|L_1 \cdot L_2| \leq |L_1 \times L_2| = |L_1| \cdot |L_2|$ by surjectivity of $\mathbf{concat}$ and elementary combinatorics. All properties of $\boldsymbol{d}_N$ follow now from Theorem 4.6. □

Example 4.3. Recall that a fuzzy relation according to [20] is a mapping $\alpha : X \times X \to [0,1]$ for some set X. Addition of two fuzzy relations α and β is defined by $(\alpha \oplus \beta)(x_1, x_2) =_{def} \max\{\alpha(x_1, x_2), \beta(x_1, x_2)\}$ and multiplication by $(\alpha \odot \beta)(x_1, x_2) =_{def} \max_{x_3 \in X} \min\{\alpha(x_1, x_3), \beta(x_3, x_2)\}$ whereas the cardinality $|\alpha|$ of a fuzzy relation α is given by $|\alpha| =_{def} \sum\limits_{x_1, x_2 \in X} \alpha(x_1, x_2)$.

Consider now the set $X = \{x_1, x_2\}$ and the two fuzzy relations α and β with $\alpha(x_1, x_2) = 1 = \beta(x_2, x_1)$ and $\alpha(x_i, x_j) = 0 = \beta(x_i, x_j)$ otherwise. Then $\alpha \oplus \beta$ is given by $(\alpha \oplus \beta)(x_1, x_2) = 1 = (\alpha \oplus \beta)(x_2, x_1)$ and $(\alpha \oplus \beta)(x_1, x_1) = 0 = (\alpha \oplus \beta)(x_2, x_2)$. Hence we have $|\alpha \oplus \beta| = 2$ but on the other hand $\max\{|\alpha|, |\beta|\} = \max\{1, 1\} = 1$, so the fuzzy cardinality is no additive $\mathbf{m}\,\mathbb{R}_{\geq 0}^{\max,+}$-norm.

Writing finite fuzzy relations in a canonical way as matrices we consider now the the fuzzy relations $\alpha = \begin{pmatrix} 0 & 0 & 0 \\ 0.9 & 0 & 0 \\ 0.9 & 0 & 0 \end{pmatrix}$ and $\beta = \begin{pmatrix} 0.9 & 0.9 & 0.9 \\ 0 & 0 & 0 \\ 0 & 0 & 0 \end{pmatrix}$. Then we have $|\alpha| + |\beta| = 4.5$ but also $\alpha \odot \beta = \begin{pmatrix} 0 & 0 & 0 \\ 0.9 & 0.9 & 0.9 \\ 0.9 & 0.9 & 0.9 \end{pmatrix}$ and hence $|\alpha \odot \beta| = 5.4$ so the fuzzy cardinality is no multiplicative $\mathbf{m}\,\mathbb{R}_{\geq 0}^{\max,+}$-norm either.

However, the fuzzy cardinality is a strict and order preserving norm which can be easily seen by isotony of addition and the equivalence $\sum_{x \in X} x = 0 \Leftrightarrow \forall x \in X : x = 0$, provided the fact $x \geq 0$ for all $x \in X$ (note that the range of a fuzzy relation is the interval $[0, 1]$). □

References

1. Armstrong, A., Struth, G., Weber, T.: Program analysis and verification based on Kleene Algebra in Isabelle/HOL. In: Blazy, S., Paulin-Mohring, C., Pichardie, D. (eds.) ITP 2013. LNCS, vol. 7998, pp. 197–212. Springer, Heidelberg (2013). https://doi.org/10.1007/978-3-642-39634-2_16
2. Avrachenkov, K., Filar, J.A., Howlett, P.G.: Analytic Perturbation Theory and Its Applications. SIAM (2013)
3. Backhouse, R.C.: Regular algebra applied to language problems. J. Log. Algebr. Program. **66**(2), 71–111 (2006)
4. Berghammer, R., Stucke, I., Winter, M.: Using relation-algebraic means and tool support for investigating and computing bipartitions. J. Log. Algebr. Meth. Program. **90**, 102–124 (2017)
5. Birkhoff, G.: Lattice Theory, 3rd edn. American Mathematical Society (1967)
6. Brunet, P., Pous, D., Stucke, I.: Cardinalities of finite relations in Coq. In: Blanchette, J.C., Merz, S. (eds.) ITP 2016. LNCS, vol. 9807, pp. 466–474. Springer, Cham (2016). https://doi.org/10.1007/978-3-319-43144-4_29
7. Courant, R., John, F.: Introduction to Calculus and Analysis I, 1st edn. Springer, Heidelberg (1999). https://doi.org/10.1007/978-3-642-58604-0
8. Ésik, Z., Fahrenberg, U., Legay, A., Quaas, K.: Kleene algebras and semimodules for energy problems. In: Van Hung, D., Ogawa, M. (eds.) ATVA 2013. LNCS, vol. 8172, pp. 102–117. Springer, Cham (2013). https://doi.org/10.1007/978-3-319-02444-8_9
9. Ghose, A., Harvey, P.: Metric SCSPs: partial constraint satisfaction via semiring CSPs augmented with metrics. In: McKay, B., Slaney, J. (eds.) AI 2002. LNCS (LNAI), vol. 2557, pp. 443–454. Springer, Heidelberg (2002). https://doi.org/10.1007/3-540-36187-1_39
10. Glück, R.: Algebraic investigation of connected components. In: Höfner, P., Pous, D., Struth, G. (eds.) RAMICS 2017. LNCS, vol. 10226, pp. 109–126. Springer, Cham (2017). https://doi.org/10.1007/978-3-319-57418-9_7
11. Glück, R., Krebs, F.B.: Towards interactive verification of programmable logic controllers using Modal Kleene Algebra and KIV. In: Kahl, W., Winter, M., Oliveira, J.N. (eds.) RAMICS 2015. LNCS, vol. 9348, pp. 241–256. Springer, Cham (2015). https://doi.org/10.1007/978-3-319-24704-5_15
12. Golan, J.S.: Semirings and their Applications, 1st edn. Springer, Dordrecht (1999). https://doi.org/10.1007/978-94-015-9333-5
13. Gondran, M., Minoux, M.: Graphs, Dioids and Semirings. Springer, New York (2008). https://doi.org/10.1007/978-0-387-75450-5
14. Guttmann, W.: Stone relation algebras. In: Höfner, P., Pous, D., Struth, G. (eds.) RAMICS 2017. LNCS, vol. 10226, pp. 127–143. Springer, Cham (2017). https://doi.org/10.1007/978-3-319-57418-9_8
15. Holmes, M.: Introduction to Perturbation Methods, 2nd edn. Springer, New York (2013). https://doi.org/10.1007/978-1-4614-5477-9

16. Jackson, M., McKenzie, R.: Interpreting graph colorability in finite semigroups. IJAC **16**(1), 119–140 (2006)
17. Jipsen, P., Rose, H.: Varieties of Lattices, 1st edn. Springer, Heidelberg (1992). https://doi.org/10.1007/BFb0090224
18. Kahl, W.: Graph transformation with symbolic attributes via monadic coalgebra homomorphisms. In: ECEASST, p. 71 (2014)
19. Kawahara, Y., Furusawa, H.: An algebraic formalization of fuzzy relations. Fuzzy Sets Syst. **101**(1), 125–135 (1999)
20. Kawahara, Y.: On the cardinality of relations. In: RelMiCS, pp. 251–265 (2006)
21. Krivulin, N.: Complete solution of an optimization problem in tropical semifield. In: Höfner, P., Pous, D., Struth, G. (eds.) RAMICS 2017. LNCS, vol. 10226, pp. 226–241. Springer, Cham (2017). https://doi.org/10.1007/978-3-319-57418-9_14
22. Litak, T., Mikulás, S., Hidders, J.: Relational lattices. In: Höfner, P., Jipsen, P., Kahl, W., Müller, M.E. (eds.) RAMICS 2014. LNCS, vol. 8428, pp. 327–343. Springer, Cham (2014). https://doi.org/10.1007/978-3-319-06251-8_20
23. Michels, G., Joosten, S., van der Woude, J., Joosten, S.: Ampersand - applying relation algebra in practice. In: de Swart, H. (ed.) RAMICS 2011. LNCS, vol. 6663, pp. 280–293. Springer, Heidelberg (2011). https://doi.org/10.1007/978-3-642-21070-9_21
24. Oliveira, J.N.: A relation-algebraic approach to the "hoare logic" of functional dependencies. J. Log. Algebr. Meth. Program. **83**(2), 249–262 (2014)
25. Schmidt, G., Ströhlein, T.: Relations and Graphs: Discrete Mathematics for Computer Scientists. Springer, Heidelberg (1993). https://doi.org/10.1007/978-3-642-77968-8

T-Norm Based Operations in Arrow Categories

Michael Winter(✉)

Department of Computer Science, Brock University,
St. Catharines, ON L2S 3A1, Canada
mwinter@brocku.ca

Abstract. Arrow categories have been shown to be a suitable categorical and algebraic framework for L-fuzzy relations. They axiomatize those relations and the usual relational operations such as meet, join, converse, and composition based on the operations given by the lattice of membership values L. However, an important tool when working with and applying fuzzy methods, are meet and composition operators that stem from a certain monoid operation on L (also called a t-norm on L). In this paper we investigate the relational properties of those operations in the abstract setting of arrow categories. In particular, we investigate properties similar to the well-known modular inclusion for regular composition.

1 Introduction

Allegories, and Dedekind categories in particular, are a fundamental tool to reason about relations. In addition to the standard model of binary relations these categories also cover L-fuzzy relations. Formally, an L-fuzzy relation R (or L-relation for short) between a set A and a set B is a function $R : A \times B \to L$ where L is a complete Heyting algebra. Two elements x and y are in relation R up to a certain degree $R(x, y)$ indicated by a membership value from L. If the degree $R(x, y)$ is equal to 0 (smallest element of L) or equal to 1 (greatest element of L), then R is called crisp. Crisp L-relations can be identified with regular, i.e., Boolean-valued, relations. Arrow categories [20,21] extend Dedekind categories by adding two additional operators mapping a relation R to the greatest crisp relation $R^{\downarrow}$ included in R and the smallest crisp relation $R^{\uparrow}$ containing R. These operations allow one to distinguish crisp and fuzzy relations, and, hence, allow one to apply fuzzy methods abstractly using these categories. In addition to crispness, so-called scalar relations as defined in Sect. 2 allow to identify the underlying lattice L in any abstract arrow category.

In applications of fuzzy methods, meet and composition operations based on some t-norm on L are very important. For example, t-norms are used to combine multiple components or results within a fuzzy or type-2 fuzzy controller [10,20,22,23]. This usually leads to a "smoother" behavior of the controller, and, hence, to a more desirable system. Another example is given by fuzzy databases and their query language FSQL [1,4–6]. This language allows combining conditions in queries based on a t-norm instead of the regular Boolean connectives. Finally, we would like to provide a more theoretical example. The definition of an L-fuzzy topological space [7] replaces

M. Winter—The author gratefully acknowledges support from the Natural Sciences and Engineering Research Council of Canada.

J. Desharnais et al. (Eds.): RAMiCS 2018, LNCS 11194, pp. 70–86, 2018.
https://doi.org/10.1007/978-3-030-02149-8_5

the requirement that open sets are closed under binary intersection by the fact that they are closet under a meet operation based on a t-norm. Any further study of L-fuzzy topological spaces requires using properties of the t-norm based meet operation.

In this paper we are interested in the basic properties of a meet $*$ and a composition operation $\overset{*}{;}$ based on a t-norm. Some investigations in the context of Goguen categories have already been made in [18,20]. Goguen categories are arrow categories with two additional axioms that make them representable in terms of cuts [17,19,20]. Both axioms depend heavily on the completeness of the lattice of relations. In addition, one of the axioms is a second-order axiom, i.e., an axiom not very suitable in the context of algebraic/equational reasoning. Our goal with the current paper is to provide a set of suitable axioms for $*$ and $\overset{*}{;}$. In particular, we are interested in versions of the well-known modular inclusion where some or all applications of the regular meet or composition are replaced by $*$ resp. $\overset{*}{;}$. We will show that our axioms imply that for every arrow category the lattice of scalar relations is equipped with a t-norm, i.e., that $*$ and $\overset{*}{;}$ stem from an operation on the membership values. Furthermore we show that if the so-called α-cut theorem is valid, then $*$ and $\overset{*}{;}$ are the operations as defined in [18,20]. This implies for concrete L-relations that both operations are indeed the componentwise defined operations as expected.

2 Mathematical Preliminaries

In this section we want to recall some basic notions from lattice, category, and allegory theory. In particular, we want to introduce arrow categories as the theoretical framework for working with L-relations and operations thereon. For further details we refer to [2,3].

A distributive lattice $\langle L, \sqcap, \sqcup \rangle$ with meet $\sqcap$, join $\sqcup$, and induced order $\sqsubseteq$ is called a complete Heyting algebra iff L is complete and $x \sqcap \bigsqcup M = \bigsqcup_{y \in M} (x \sqcap y)$ holds for all $x \in L$ and $M \subseteq L$. Since L is complete, L has a least element 0 and a greatest element 1. L also has relative pseudo-complements, i.e., for each pair $x, y \in L$ there is a greatest element $x \to y$ with $x \sqcap (x \to y) \sqsubseteq y$. The relative pseudo-complement can also be characterized as the residual of $\sqcap$ and, therefore, satisfies $x \sqcap y \sqsubseteq z$ iff $x \sqsubseteq y \to z$ for all $x, y, z \in L$.

If L is a complete lattice, then $\langle L, * \rangle$ with a binary operation $*$ on L is called a partially ordered (Abelian) monoid iff

1. $\langle L, *, 1 \rangle$ is a bounded Abelian monoid, i.e., $*$ is associative and commutative with the greatest element 1 of L as neutral element,
2. $*$ is monotonic in both parameters.

If $*$ distributes over arbitrary unions, then we call $*$ continuous. Please note that a continuous operation $*$ also provides a residual, i.e., there is a binary operation $\twoheadrightarrow\!\!\!\!*$ such that $x * y \sqsubseteq z$ iff $x \sqsubseteq y \twoheadrightarrow\!\!\!\!* z$ for all $x, y, z \in L$. A continuous partially ordered Abelian monoid is usually called a quantale.

In the following we will assume that $*$ binds tighter that meet or join, i.e., we write $x * y \sqcap z$ for $(x * y) \sqcap z$. The definition of partially ordered monoids above generalizes regular t-norms from the unit interval [0,1] to arbitrary complete lattices. Therefore, we will also refer to $*$ as a (generalized) t-norm in the remainder of the paper.

On every bounded lattice L we define the following operation:

$$x \circledast y := \begin{cases} x & \text{iff } y = 1, \\ y & \text{iff } x = 1, \\ 0 & \text{otherwise.} \end{cases}$$

It is easy to verify that $\circledast$ is a monoid operation as defined above. But please note that $\circledast$ is not necessarily continuous. As an example consider the lattice L_6 together with $\circledast$ of Fig. 1. The figure also defines a continuous monoid operation $*$ with its residual $\twoheadrightarrow$ on L_6 which we will use later. In this example we have $a \circledast (c \sqcup d) = a \circledast 1 = a$ and $a \circledast c \sqcup a \circledast d = 0 \sqcup 0 = 0$.

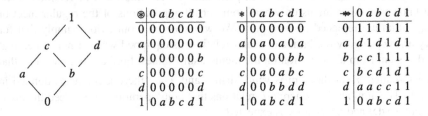

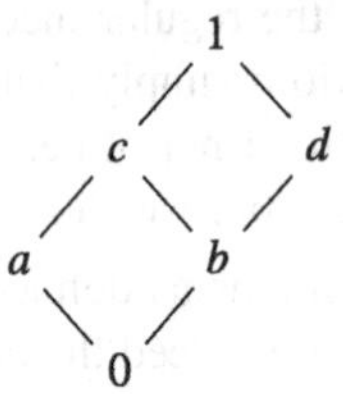

$\circledast$	0	a	b	c	d	1
0	0	0	0	0	0	0
a	0	0	0	0	0	a
b	0	0	0	0	0	b
c	0	0	0	0	0	c
d	0	0	0	0	0	d
1	0	a	b	c	d	1

$*$	0	a	b	c	d	1
0	0	0	0	0	0	0
a	0	a	0	a	0	a
b	0	0	0	0	b	b
c	0	a	0	a	b	c
d	0	0	b	b	d	d
1	0	a	b	c	d	1

$\twoheadrightarrow$	0	a	b	c	d	1
0	1	1	1	1	1	1
a	d	1	d	1	d	1
b	c	c	1	1	1	1
c	b	c	d	1	d	1
d	a	a	c	c	1	1
1	0	a	b	c	d	1

Fig. 1. Lattice L_6 with operation $\circledast$ and a continuous operation $*$ with its residual

We have the following lemma (see e.g. [20]).

Lemma 1. *Let $\langle L, \sqsubseteq, *, 1\rangle$ be a partially ordered monoid. Then we have for all $x, y \in L$*

1. $x * y \sqsubseteq x$ *and* $x * y \sqsubseteq y$,
2. $x * 0 = 0 * x = 0$,
3. $x * (y \sqcap z) \sqsubseteq x * y \sqcap x * z$ *and* $(x \sqcap y) * z \sqsubseteq x * z \sqcap y * z$,
4. $x \circledast y \sqsubseteq x * y \sqsubseteq x \sqcap y$,
5. $* = \sqcap$ *iff* $*$ *is idempotent, i.e.,* $x * x = x$ *for all* $x \in L$.

We will use the framework of Dedekind categories [11, 12] throughout this paper as a basic theory of relations. Categories of this type are called locally complete division allegories in [3].

We assume that the reader is familiar with the very basic notions from category theory. We will write $R : A \to B$ to indicate that a morphism R of a category $\mathcal{R}$ has source A and target B. Composition will be denoted by ; that has to be read from left to right, i.e., $R;S$ means R first, and then S. For the collection of all morphisms $R : A \to B$ (or hom-set) we write $\mathcal{R}[A, B]$, and the identity morphism on A is denoted by $\mathbb{I}_A$.

Definition 1. *A Dedekind category $\mathcal{R}$ is a category satisfying the following:*

1. *For all objects A and B the hom-set $\mathcal{R}[A, B]$ is a complete Heyting algebra. Meet, join, the induced ordering, the least and the greatest element are denoted by $\sqcap, \sqcup, \sqsubseteq, \perp\!\!\!\perp_{AB}, \top\!\!\!\top_{AB}$, respectively.*
2. *There is a monotone operation $\breve{}$ (called converse) mapping a relation $Q : A \to B$ to $Q\breve{} : B \to A$ such that for all relations $Q : A \to B$ and $R : B \to C$ the following holds: $(Q;R)\breve{} = R\breve{};Q\breve{}$ and $(Q\breve{})\breve{} = Q$.*

3. *For all relations* $Q : A \to B, R : B \to C$ *and* $S : A \to C$ *the modular law* $(Q;R) \sqcap S \sqsubseteq Q;(R \sqcap (Q^{\smile};S))$ *holds.*
4. *For all relations* $R : B \to C$ *and* $S : A \to C$ *there is a relation* $S/R : A \to B$ *(called the left residual of* S *and* R*) such that for all* $X : A \to B$ *the following holds:* $X;R \sqsubseteq S \iff X \sqsubseteq S/R$.

As already used in the previous definition, we will call a morphism of a Dedekind category a relation. If we refer to set-theoretic relations or L-relations, we will always explicitly mention this.

In the remainder we assume that ; binds tighter than the lattice operations so that the modular inclusion becomes $Q;R \sqcap S \sqsubseteq Q;(R \sqcap Q^{\smile};S)$. Based on the residual operation in the definition above it is possible to define a right residual by $R \backslash S = (S^{\smile}/R^{\smile})^{\smile}$. This operation is characterized by

$$R;X \sqsubseteq S \iff X \sqsubseteq R \backslash S.$$

In the remainder of this paper we will need the properties listed in the following lemma. A proof can be found in [14,15,20].

Lemma 2. *Let* $\mathcal{R}$ *be a Dedekind category,* $Q : A \to B$, $R : A \to D$, $S : B \to C$ *be relations and* $U : A \to A$ *be a partial identity, i.e.,* $U \sqsubseteq \mathbb{I}_A$. *Then we have*

1. $(Q \sqcap R;\mathbb{T}_{DB});S = Q;S \sqcap R;\mathbb{T}_{DC}$,
2. $U^{\smile} = U$,
3. $U;Q = U;\mathbb{T}_{AB} \sqcap Q$.

Please note that for every valid universally quantified equation or inclusion its dual (or opposite) by reversing the direction of composition is also valid. For example, the dual of Lemma 2(1), i.e., the equation $S;(\mathbb{T};R \sqcap Q) = \mathbb{T};R \sqcap S;Q$ for appropriate relations Q, R and S, is also valid.

As mentioned in the introduction, the collection of binary relations between sets as well as the collection of L-relations between sets form a Dedekind category. In the case of L-relations the lattice operations and constants are defined componentwise and composition of two L-relations $Q : A \to B$ and $R : B \to C$ is given by

$$(Q;R)(x,y) = \bigsqcup_{y \in B} Q(x,y) \sqcap R(y,z).$$

It is often desirable to represent and visualize L-relations between finite sets by L-valued matrices. For example, if $A = \{x,y,z\}$ and $B = \{1,2,3,4\}$, then the following matrices represent L_6-relations, where $R : A \to B$ and $\alpha : B \to B$:

$$R := \begin{pmatrix} 1 & a & c & 0 \\ 0 & d & a & 0 \\ 0 & 1 & b & d \end{pmatrix}, \quad \alpha := \begin{pmatrix} b & 0 & 0 & 0 \\ 0 & b & 0 & 0 \\ 0 & 0 & b & 0 \\ 0 & 0 & 0 & b \end{pmatrix}.$$

The matrix representation should be read as follows. The c in the first row of the matrix for R indicates that the first element of A (i.e. x) is in relation R with the third element of B (i.e. 3) to the degree c.

The relation α above is a so-called scalar. It is a partial identity, i.e., $\alpha \sqsubseteq \mathbb{I}_B$, so that all elements on the diagonal are identical. Such a matrix can be identified with the element on the diagonal so that the lattice of scalar relations on an object corresponds to the lattice L. In the example above α corresponds to the element b of L. In the following we will use α to denote the element of L as well as the corresponding scalar relation on any object, i.e., the diagonal matrix with α on its diagonal. In the abstract setting of Dedekind categories scalars can be defined as follows.

Definition 2. *A relation $\alpha : A \to A$ is called a scalar on A iff $\alpha \sqsubseteq \mathbb{I}_A$ and $\mathbb{T}_{AA} ; \alpha = \alpha ; \mathbb{T}_{AA}$.*

The notion of scalars was introduced by Furusawa and Kawahara [9] and is equivalent to the notion of ideals, i.e., relations $R : A \to B$ that satisfy $\mathbb{T}_{AA} ; R ; \mathbb{T}_{BB} = R$, which were introduced by Jónsson and Tarski [8].

The next definition introduces arrow categories, i.e., the basic theory for L-relations.

Definition 3. *An arrow category $\mathcal{A}$ is a Dedekind category with $\mathbb{T}_{AB} \neq \mathbb{L}_{AB}$ for all A, B and two operations $\uparrow$ and $\downarrow$ satisfying:*

1. *$R^{\uparrow}, R^{\downarrow} : A \to B$ for all $R : A \to B$*
2. *$({}^{\uparrow}, {}^{\downarrow})$ is a Galois correspondence, i.e., we have $Q^{\uparrow} \sqsubseteq R$ iff $Q \sqsubseteq R^{\downarrow}$ for all $Q, R : A \to B$.*
3. *$(R^{\smile} ; S^{\downarrow})^{\uparrow} = R^{\uparrow\smile} ; S^{\downarrow}$ for all $R : B \to A$ and $S : B \to C$*
4. *$(Q \sqcap R^{\downarrow})^{\uparrow} = Q^{\uparrow} \sqcap R^{\downarrow}$ for all $Q, R : A \to B$*
5. *If $\alpha_A \neq \mathbb{L}_{AA}$ is a non-zero scalar then $\alpha_A^{\uparrow} = \mathbb{I}_A$.*

For an L-relation $R : A \to B$ the arrow operations are defined by

$$R^{\uparrow}(x,y) = \begin{cases} 1 \text{ iff } R(x,y) \neq 0 \\ 0 \text{ iff } T(x,y) = 0 \end{cases} \qquad R^{\downarrow}(x,y) = \begin{cases} 1 \text{ iff } R(x,y) = 1 \\ 0 \text{ iff } T(x,y) \neq 1 \end{cases}$$

As one expects, L-relations together with the arrow operations above form an arrow category.

A relation that satisfies $R^{\uparrow} = R$, or equivalently $R^{\downarrow} = R$, is called crisp. Any arrow category is uniform, i.e., we have $\mathbb{T}_{AB} ; \mathbb{T}_{BC} = \mathbb{T}_{AC}$ for all objects A, B and C. As a consequence, if we denote the lattices of scalar relations on A by Sc[A], then Sc[A] and Sc[B] are in fact isomorphic for any pair of objects A and B via the mapping $\varphi_{AB}(\alpha) = \mathbb{T}_{BA} ; \alpha ; \mathbb{T}_{AB} \sqcap \mathbb{I}_B$. As mentioned in the concrete case, Sc[A] can, therefore, be taken as the membership values of the morphisms in $\mathcal{A}$. In the remainder of this paper we will use the following convention. With α we denote an abstract scalar, i.e., an element of the isomorphic lattices Sc[A] for each A. We write α_A for the corresponding scalar on A so that we have for α_A and α_B that $\varphi_{AB}(\alpha_A) = \alpha_B$ and $\varphi_{BA}(\alpha_B) = \alpha_A$. With this convention we have the following lemma. A proof can be found in [20].

Lemma 3. *Let $\mathcal{R}$ be an arrow category. Then $\alpha_A ; Q = Q ; \alpha_B$ for all relations $Q : A \to B$ and scalars α.*

If $\alpha_A : A \to A$ and $R : A \to B$ are L-relations and α_A a scalar, then the relation $(\alpha_A \backslash R)^{\downarrow}$ is characterized by

$$(\alpha_A \backslash R)^{\downarrow}(x,y) = \begin{cases} 1 & \text{iff } \alpha \sqsubseteq R(x,y), \\ 0 & \text{otherwise.} \end{cases}$$

i.e., the elements x and y are related in the crisp relation $(\alpha_A \backslash R)^{\downarrow}$ iff x and y are related in R by a degree greater than or equal to α. For this reason $(\alpha_A \backslash R)^{\downarrow}$ is called the α-cut of R.

It is easy to verify that $\alpha_A ; (\alpha_A \backslash R)^{\downarrow} \sqsubseteq R$ for all α, and, hence, that we have $\bigsqcup_{\alpha \text{ scalar}} \alpha_A ; (\alpha_A \backslash R)^{\downarrow} \sqsubseteq R$. The converse inclusion is not necessarily valid as an example in [20,21] shows. An arrow category is said to satisfy the α-cut theorem iff $\bigsqcup_{\alpha \text{ scalar}} \alpha_A ; (\alpha_A \backslash R)^{\downarrow} = R$. It is worth mentioning that concrete L-relations satisfy the α-cut theorem.

In fuzzy theory, t-norms based operations are essential in theoretical as well as practical applications. For example, the usual notion of L-fuzzy topological spaces replaces the requirement that open sets are closed under binary meets by a meet operation based on a t-norm.

In the case of L-relations and a given t-norm $*$ on L we may define two new operations on relations by

$$(Q * R)(x,y) = Q(x,y) * R(x,y),$$
$$(Q \mathbin{\overset{*}{;}} S)(x,z) = \bigsqcup_{y \in B} Q(x,y) * S(y,z).$$

In the general setting of arrow categories and a t-norm $*$ on the lattice of scalars, it is also possible to define such operations based on $*$.

Definition 4. *Let $Q, R : A \to B$ and $S : B \to C$ be relations, and $*$ a continuous monoid operation on the set of scalar relations. Then we define*

$$Q * R := \bigsqcup_{\alpha,\beta \text{ scalars}} (\alpha_A * \beta_A) ; ((\alpha_A \backslash Q)^{\downarrow} \sqcap (\beta_A \backslash R)^{\downarrow}),$$
$$Q \mathbin{\overset{*}{;}} R := \bigsqcup_{\alpha,\beta \text{ scalars}} (\alpha_A * \beta_A) ; (\alpha_A \backslash Q)^{\downarrow} ; (\beta_B \backslash S)^{\downarrow}.$$

Please note that we use $*$ to denote the operation on scalars as well as for the $*$ based meet operation on relations. It has been shown in [18,20] that for L-relations the abstract definition (Definition 4) and the componentwise definitions from above coincide. In addition, several properties of $*$ and $\overset{*}{;}$ from Definition 4 have been shown in the general setting of Goguen categories. Goguen categories are arrow categories with additional axioms. In such a category every relation can be uniquely represented by its α-cuts [19,20], i.e., Goguen categories satisfy the α-cut theorem plus an additional second order-axiom that makes the left-hand side of the α-cut theorem unique. Most of the proofs provided in [18,20] depend on those properties. For example, the proof of the associativity of both operations requires the α-cut theorem as well as the additional second-order axiom of Goguen categories. Furthermore, laws similar to the modular inclusion for $*$ and $\overset{*}{;}$ have not been investigated yet.

3 T-Norm Based Operations

In this section we will call $*$ resp. $\mathbin{;}$ concrete if they are defined for L-relations componentwise based on a continuous t-norm on L. We are interested in deriving a suitable set of axioms for arbitrary operations $*$ and $\mathbin{;}$ so that they can be seen as an abstract version of the concrete operations. Please note that we will assume, similar to regular composition, that $\mathbin{;}$ binds tighter than $*$. We start with $*$ and prove the following lemma if $*$ is concrete.

Lemma 4. *Suppose $Q, R : A \to B$ are L-relations. Then we have*

1. *(T1) $*$ is associative and commutative,*
2. *(T2) $*$ is continuous, and, hence, also monotonic,*
3. *(T3) $(Q * R)^{\smile} = Q^{\smile} * R^{\smile}$,*
4. *(T4) $Q * R^{\downarrow} = Q \sqcap R^{\downarrow}$.*

Proof. All properties follow immediately from the componentwise definition of $*$ where 4. uses $x * 1 = x = x \sqcap 1$ and $x * 0 = 0 = x \sqcap 0$ for all $x \in L$. □

For the rest of this section we assume that $*$ is an arbitrary operation that satisfies the Properties (T1)–(T4) of the previous Lemma.

Lemma 5. *Let $\mathcal{R}$ be an arrow category. Then we have for all $Q, R, S : A \to B$*

1. $(Q^{\downarrow} \sqcap R) \sqcup (Q \sqcap R^{\downarrow}) \sqsubseteq Q * R \sqsubseteq Q \sqcap R$,
2. $Q * (R \sqcap S) \sqsubseteq Q * R \sqcap Q * S$,
3. $Q * \perp\!\!\!\perp = \perp\!\!\!\perp$ *and* $Q * \top\!\!\!\top = Q$,
4. *if Q is crisp and $R \sqsubseteq Q$, then $Q * R = R$,*
5. $(Q * R)^{\downarrow} = Q^{\downarrow} * R^{\downarrow}$.

Proof. 1. and 2. follow analogously to Lemma 1(1) and (3).

3. $Q * \perp\!\!\!\perp = Q \sqcap \perp\!\!\!\perp = \perp\!\!\!\perp$ since $\perp\!\!\!\perp$ is crisp. The second property is shown analogously.
4. We have $Q * R = Q \sqcap R = R$ by (T4) since Q is crisp.
5. First of all, we have $(Q * R)^{\downarrow} \sqsubseteq (Q \sqcap R)^{\downarrow} = Q^{\downarrow} \sqcap R^{\downarrow} = Q^{\downarrow} * R^{\downarrow}$. Conversely, we have $(Q^{\downarrow} * R^{\downarrow})^{\uparrow} = (Q^{\downarrow} \sqcap R^{\downarrow})^{\uparrow} = Q^{\downarrow} \sqcap R^{\downarrow} = Q^{\downarrow} * R^{\downarrow} \sqsubseteq Q * R$ which implies $Q^{\downarrow} * R^{\downarrow} \sqsubseteq (Q * R)^{\downarrow}$. □

The operation $(Q^{\downarrow} \sqcap R) \sqcup (Q \sqcap R^{\downarrow})$ is the abstract counterpart of $\circledast$. As the example in Fig. 1 shows, this operation is usually not continuous.

Now, we consider $\mathbin{;}$ and provide the following lemma if $\mathbin{;}$ is concrete.

Lemma 6. *Suppose $P, Q : A \to B$ and $R, S : B \to C$ are L-relations. Then we have*

1. *(CT1) $\mathbin{;}$ is associative,*
2. *(CT2) $\mathbin{;}$ is continuous, and, hence, also monotonic,*
3. *(CT3) $(Q \mathbin{;} R)^{\smile} = Q^{\smile} \mathbin{;} R^{\smile}$,*

4. *(CT4)* $Q \mathbin{\overset{*}{;}} R^{\downarrow} = Q ; R^{\downarrow}$,
5. *(CT5)* $(P * Q) \mathbin{\overset{*}{;}} (R * S) \sqsubseteq P \mathbin{\overset{*}{;}} R * Q \mathbin{\overset{*}{;}} S$.

Proof. 1. to 4. follow immediately from the componentwise definition of $*$ where 4. again uses $x * 1 = x = x \sqcap 1$ and $x * 0 = 0 = x \sqcap 0$ for all $x \in L$.

5. The computation

$$\begin{aligned}
((P * Q) \mathbin{\overset{*}{;}} (R * S))(x,z) &= \bigsqcup_{y \in B} (P * Q)(x,y) * (R * S)(y,z) \\
&= \bigsqcup_{y \in B} P(x,y) * Q(x,y) * R(y,z) * S(y,z) \\
&= \bigsqcup_{y \in B} P(x,y) * R(y,z) * Q(x,y) * S(y,z) \\
&\sqsubseteq \left(\bigsqcup_{y \in B} P(x,y) * R(y,z) \right) * \left(\bigsqcup_{y \in B} Q(x,y) * S(y,z) \right) \\
&= (P \mathbin{\overset{*}{;}} R)(x,z) * (Q \mathbin{\overset{*}{;}} S)(x,z) \\
&= (P \mathbin{\overset{*}{;}} R * Q \mathbin{\overset{*}{;}} S)(x,z)
\end{aligned}$$

shows the assertion. □

Please note that the associativity of $\mathbin{\overset{*}{;}}$ requires $*$ to be continuous. For a counterexample consider the operation $\circledast$ from Fig. 1 and compute

$$\left(\begin{pmatrix} a & 0 \\ 0 & 0 \end{pmatrix} \mathbin{\overset{*}{;}} \begin{pmatrix} c & d \\ d & c \end{pmatrix} \right) \mathbin{\overset{*}{;}} \begin{pmatrix} 1 & 1 \\ 1 & 1 \end{pmatrix} = \begin{pmatrix} 0 & 0 \\ 0 & 0 \end{pmatrix} ; \begin{pmatrix} 1 & 1 \\ 1 & 1 \end{pmatrix} = \begin{pmatrix} 0 & 0 \\ 0 & 0 \end{pmatrix}$$

$$\begin{pmatrix} a & 0 \\ 0 & 0 \end{pmatrix} \mathbin{\overset{*}{;}} \left(\begin{pmatrix} c & d \\ d & c \end{pmatrix} \mathbin{\overset{*}{;}} \begin{pmatrix} 1 & 1 \\ 1 & 1 \end{pmatrix} \right) = \begin{pmatrix} a & 0 \\ 0 & 0 \end{pmatrix} ; \begin{pmatrix} 1 & 1 \\ 1 & 1 \end{pmatrix} = \begin{pmatrix} a & a \\ 0 & 0 \end{pmatrix}$$

For the rest of this section we assume that $\mathbin{\overset{*}{;}}$ is an arbitrary operation that satisfies the Properties (CT1)–(CT5) of the previous Lemma.

Lemma 7. *Let $\mathcal{R}$ be an arrow category. Then we have for all $Q : C \to A$, $R, S : A \to B$ and all scalars $\alpha, \beta : A \to A$*

1. $(Q^{\downarrow} ; R) \sqcup (Q ; R^{\downarrow}) \sqsubseteq Q \mathbin{\overset{*}{;}} R$,
2. $Q \mathbin{\overset{*}{;}} \mathbb{I}_A = Q$,
3. *if Q is crisp, then* $Q ; (R * S) \sqsubseteq Q ; R * Q ; S$,
4. $(R ; \mathbb{T}_{BC} * S ; \mathbb{T}_{BC}) ; \mathbb{T}_{CD} = R ; \mathbb{T}_{BD} * S ; \mathbb{T}_{BD}$,
5. $(\alpha * \beta) ; \mathbb{T}_{AB} = \alpha ; \mathbb{T}_{AB} * \beta ; \mathbb{T}_{AB}$,
6. *$\alpha * \beta$ is a scalar.*

Proof. 1. and 2. follow immediately from (CT4).

3. The first inclusion follows immediately from

$$\begin{aligned} Q;(R*S) &= Q\,\bar{;}\,(R*S) && \text{(CT4) since } Q \text{ is crisp}\\ &= (Q*Q)\,\bar{;}\,(R*S) && \text{(T4) since } Q \text{ is crisp}\\ &\sqsubseteq Q\,\bar{;}\,R * Q\,\bar{;}\,S && \text{(CT5)}\\ &= Q;R*Q;S. && \text{(CT4) since } Q \text{ is crisp} \end{aligned}$$

4. Using 2. twice we obtain

$$\begin{aligned} &(R;\mathbb{T}_{BC} * S;\mathbb{T}_{BC});\mathbb{T}_{CD}\\ &\sqsubseteq R;\mathbb{T}_{BC};\mathbb{T}_{CD} * S;\mathbb{T}_{BC};\mathbb{T}_{CD} && \text{dual of 3. since } \mathbb{T} \text{ is crisp}\\ &\sqsubseteq R;\mathbb{T}_{BD} * S;\mathbb{T}_{BD}\\ &\sqsubseteq (R;\mathbb{T}_{BD} * S;\mathbb{T}_{BD});\mathbb{T}_{DD} && \mathbb{I}_D \sqsubseteq \mathbb{T}_{DD}\\ &= (R;\mathbb{T}_{BD} * S;\mathbb{T}_{BD});\mathbb{T}_{DC};\mathbb{T}_{CD} && \mathcal{R} \text{ uniform}\\ &\sqsubseteq (R;\mathbb{T}_{BD};\mathbb{T}_{DC} * S;\mathbb{T}_{BD};\mathbb{T}_{DC});\mathbb{T}_{CD} && \text{dual of 3. since } \mathbb{T} \text{ is crisp}\\ &\sqsubseteq (R;\mathbb{T}_{BC} * S;\mathbb{T}_{BC});\mathbb{T}_{CD}. \end{aligned}$$

5. We immediately compute

$$\begin{aligned} (\alpha*\beta);\mathbb{T}_{AB} &= ((\alpha;\mathbb{T}_{AA} \sqcap \mathbb{I}_A) * (\beta;\mathbb{T}_{AA} \sqcap \mathbb{I}_A));\mathbb{T}_{AB} && \text{Lemma 2(3)}\\ &= (\alpha;\mathbb{T}_{AA} * \mathbb{I}_A * \beta;\mathbb{T}_{AA} * \mathbb{I}_A);\mathbb{T}_{AB} && \text{(T4) since } \mathbb{I}_A \text{ is crisp}\\ &= (\alpha;\mathbb{T}_{AA} * \beta;\mathbb{T}_{AA} \sqcap \mathbb{I}_A);\mathbb{T}_{AB} && \text{(T4) since } \mathbb{I}_A \text{ is crisp}\\ &= ((\alpha;\mathbb{T}_{AA} * \beta;\mathbb{T}_{AA});\mathbb{T}_{AA} \sqcap \mathbb{I}_A);\mathbb{T}_{AB} && \text{by 4.}\\ &= (\alpha;\mathbb{T}_{AA} * \beta;\mathbb{T}_{AA});\mathbb{T}_{AB} && \text{Lemma 2(1)}\\ &= \alpha;\mathbb{T}_{AB} * \beta;\mathbb{T}_{AB}. && \text{by 4.} \end{aligned}$$

6. First of all, we have $\alpha * \beta \sqsubseteq \alpha \sqsubseteq \mathbb{I}_A$. From

$$\begin{aligned} (\alpha*\beta);\mathbb{T}_{AA} &= \alpha;\mathbb{T}_{AA} * \beta;\mathbb{T}_{AA} && \text{by 5.}\\ &= \mathbb{T}_{AA};\alpha * \mathbb{T}_{AA};\beta && \alpha,\beta \text{ scalars}\\ &= \mathbb{T}_{AA};(\alpha*\beta) && \text{by dual of 5.} \end{aligned}$$

we conclude that $\alpha * \beta$ is indeed a scalar. □

1. of the previous lemma shows that the operation $(Q^{\downarrow};R) \sqcup (Q;R^{\downarrow})$ is the abstract counterpart of a composition based on ⊛. The reader might have expected that, similar to $*$ and $\sqcap$, regular composition is the greatest composition operation, i.e., that $Q\,\bar{;}\,R \sqsubseteq Q;R$. We have to defer this property to Lemma 9 since the proof requires a version of the modular inclusion for $\bar{;}$.

After these preparations we are ready to show that the lattice of scalar relations Sc[A] is indeed equipped with a t-norm operation. Furthermore, for two different objects A these structures are isomorphic.

Theorem 1. *Let $\mathcal{R}$ be an arrow category and* Sc[A] *the set of scalars on A. Then the partially ordered monoids* $\langle \mathrm{Sc}[A], * \rangle$ *are isomorphic.*

Proof. It was already shown in [20] that the complete Heyting algebras Sc[A] are isomorphic via the mappings $\varphi_{AB}(\alpha) = \mathbb{T}_{BA} ; \alpha ; \mathbb{T}_{AB} \sqcap \mathbb{I}_B$. Furthermore, Lemma 7(6) shows that scalars are closed under the operation $*$ so that it remains to show that φ_{AB} preserves $*$. The computation

$$\begin{aligned}
\varphi_{AB}(\alpha * \beta) &= \mathbb{T}_{BA} ; (\alpha * \beta) ; \mathbb{T}_{AB} \sqcap \mathbb{I}_B \\
&= \mathbb{T}_{BA} ; (\alpha ; \mathbb{T}_{AB} * \beta ; \mathbb{T}_{AB}) \sqcap \mathbb{I}_B && \text{Lemma 7(5)} \\
&= \mathbb{T}_{BA} ; (\alpha ; \mathbb{T}_{AA} ; \mathbb{T}_{AB} * \beta ; \mathbb{T}_{AA} ; \mathbb{T}_{AB}) \sqcap \mathbb{I}_B && \mathbb{T}_{AA} ; \mathbb{T}_{AB} = \mathbb{T}_{AB}
\end{aligned}$$

$$\begin{aligned}
&= \mathbb{T}_{BA} ; (\mathbb{T}_{AA} ; \alpha ; \mathbb{T}_{AB} * \mathbb{T}_{AA} ; \beta ; \mathbb{T}_{AB}) \sqcap \mathbb{I}_B && \alpha, \beta \text{ scalars} \\
&= (\mathbb{T}_{BA} ; \alpha ; \mathbb{T}_{AB} * \mathbb{T}_{BA} ; \beta ; \mathbb{T}_{AB}) \sqcap \mathbb{I}_B && \text{dual of Lemma 7(4)} \\
&= (\mathbb{T}_{BA} ; \alpha ; \mathbb{T}_{AB} * \mathbb{T}_{BA} ; \beta ; \mathbb{T}_{AB}) \sqcap \mathbb{I}_B \sqcap \mathbb{I}_B \\
&= \mathbb{T}_{BA} ; \alpha ; \mathbb{T}_{AB} * \mathbb{I}_B * \mathbb{T}_{BA} ; \beta ; \mathbb{T}_{AB} * \mathbb{I}_B && \text{(T4) since } \mathbb{I}_B \text{ is crisp} \\
&= (\mathbb{T}_{BA} ; \alpha ; \mathbb{T}_{AB} \sqcap \mathbb{I}_B) * (\mathbb{T}_{BA} ; \beta ; \mathbb{T}_{AB} \sqcap \mathbb{I}_B) && \text{(T4) since } \mathbb{I}_B \text{ is crisp} \\
&= \varphi_{AB}(\alpha) * \varphi_{AB}(\beta)
\end{aligned}$$

completes the proof. □

In the next theorem we want to show that the axioms so far determine $*$ uniquely if the α-cut theorem is valid. In addition, the theorem shows that $*$ is the operation defined in Definition 4 originating from a t-norm on the set of scalars.

Theorem 2. *Let $\mathcal{R}$ be an arrow category satisfying the α-cut theorem. Then we have*

$$Q * R = \bigsqcup_{\alpha,\beta \text{ scalars}} (\alpha * \beta) ; ((\alpha \backslash Q)^{\downarrow} \sqcap (\beta \backslash R)^{\downarrow}).$$

Proof.

$$\begin{aligned}
Q * R &= \left(\bigsqcup_{\alpha \text{ scalar}} \alpha_A ; (\alpha_A \backslash Q)^{\downarrow} \right) * \left(\bigsqcup_{\beta \text{ scalar}} \beta_A ; (\beta_A \backslash R)^{\downarrow} \right) && \alpha\text{-cut theorem} \\
&= \bigsqcup_{\alpha,\beta \text{ scalars}} (\alpha_A ; (\alpha_A \backslash Q)^{\downarrow}) * (\beta_A ; (\beta_A \backslash R)^{\downarrow}) && \text{(T2)} \\
&= \bigsqcup_{\alpha,\beta \text{ scalars}} (\alpha_A ; \mathbb{T}_{AB} \sqcap (\alpha_A \backslash Q)^{\downarrow}) * (\beta_A ; \mathbb{T}_{AB} \sqcap (\beta_A \backslash R)^{\downarrow}) && \text{Lemma 2(3)} \\
&= \bigsqcup_{\alpha,\beta \text{ scalars}} \alpha_A ; \mathbb{T}_{AB} * (\alpha_A \backslash Q)^{\downarrow} * \beta_A ; \mathbb{T}_{AB} * (\beta_A \backslash R)^{\downarrow} && \text{(T4)}
\end{aligned}$$

$$= \bigsqcup_{\alpha,\beta \text{ scalars}} \alpha_A ; \top\!\!\top_{AB} * \beta_A ; \top\!\!\top_{AB} * ((\alpha_A \backslash Q)^{\downarrow} \sqcap (\beta_A \backslash R)^{\downarrow}) \qquad \text{(T4)}$$

$$= \bigsqcup_{\alpha,\beta \text{ scalars}} \alpha_A ; \top\!\!\top_{AB} * \beta_A ; \top\!\!\top_{AB} \sqcap (\alpha_A \backslash Q)^{\downarrow} \sqcap (\beta_A \backslash R)^{\downarrow} \qquad \text{(T4)}$$

$$= \bigsqcup_{\alpha,\beta \text{ scalars}} (\alpha_A * \beta_A) ; \top\!\!\top_{AB} \sqcap (\alpha_A \backslash Q)^{\downarrow} \sqcap (\beta_A \backslash R)^{\downarrow} \qquad \text{Lemma 7(5)}$$

$$= \bigsqcup_{\alpha,\beta \text{ scalars}} (\alpha_A * \beta_A) ; ((\alpha_A \backslash Q)^{\downarrow} \sqcap (\beta_A \backslash R)^{\downarrow}) \qquad \text{Lemma 2(3)}$$

□

Please note that for concrete L-relations the previous theorem implies that $*$ is indeed the componentwise operation as defined in the previous section.

We want to investigate whether the modular inclusion $Q ; R \sqcap S \sqsubseteq Q ; (R \sqcap Q^{\smile} ; S)$ is valid whenever some or all of $\sqcap$ and $;$ are replaced by $*$ and $\mathbin{\overset{*}{;}}$, respectively. The result is summarized in Table 1 where + denotes that the corresponding inclusion is valid and – that it is not.

Table 1. Validity of the modular inclusions for.

$\sqsupseteq$	$Q;R \sqcap S$	$Q;R*S$	$Q \overset{*}{;} R \sqcap S$	$Q \overset{*}{;} R * S$
$Q;(R \sqcap Q^{\smile};S)$	+	+	+	+
$Q;(R \sqcap Q^{\smile} \overset{*}{;} S)$	–	+	–	+
$Q;(R * Q^{\smile};S)$	–	–	–	+
$Q;(R * Q^{\smile} \overset{*}{;} S)$	–	–	–	+
$Q \overset{*}{;} (R \sqcap Q^{\smile};S)$	–	–	–	–
$Q \overset{*}{;} (R \sqcap Q^{\smile} \overset{*}{;} S)$	–	–	–	–
$Q \overset{*}{;} (R * Q^{\smile};S)$	–	–	–	–
$Q \overset{*}{;} (R * Q^{\smile} \overset{*}{;} S)$	–	–	–	–

Obviously, if none of the operations on the right-hand side is replaced, then the corresponding inclusion follows immediately from the regular modular inclusion and the fact that the $*$-based operations are always smaller than or equal to the regular operations. This verifies the first row of Table 1. In order to investigate the remaining inclusions, we want to provide examples that show that certain replacements in the modular inclusion do not lead to valid inclusions. In fact, we will provide examples that verify or immediately imply the negative results listed in Table 1. All these examples are using the lattice L_6 from Fig. 1 with the continuous operation $*$. In addition, all examples will utilize 1×1-matrices, which are obviously equivalent to the lattice elements contained in the matrix. In addition, please note that for 1×1-matrices the operations $\overset{*}{;}$ and $*$ resp. $;$ and $\sqcap$ coincide and that converse is the identity operation. Consequently, the question becomes $(x \bullet y) \bullet z \sqsubseteq x \bullet (y \bullet (x \bullet z))$ where each $\bullet$ is chosen from $*$ and $\sqcap$.

We distinguish three cases based on the left-hand side of the inclusion. The three cases correspond to Column 3,2, and 4 of Table 1, respectively:

1. Case: $x{*}y{\sqcap}z$ as left-hand side. If the right-hand side is $x{*}(y{\sqcap}x{\sqcap}z)$ or $x{\sqcap}y{\sqcap}x{*}z$, then we have for $x = b, y = d$, and $z = b$ that $b{*}d{\sqcap}b = b{\sqcap}b = b$ but $b{*}(d{\sqcap}b{\sqcap}b) = b{*}b = 0$ and $b{\sqcap}d{\sqcap}b{*}b = b{\sqcap}d{\sqcap}0 = 0$, i.e., the two inclusions are not valid. If the right-hand side is $x{\sqcap}y{*}(x{\sqcap}z)$, then we have for $x = d, y = b$, and $z = b$ that $d * b{\sqcap}b = b{\sqcap}b = b$ but $d \sqcap b * (d \sqcap b) = d \sqcap b * b = d \sqcap 0 = 0$, i.e., the inclusion is again not valid. From the fact that all other potential right-hand-sides are smaller than or equal to the three we have just considered, we conclude that none of the modular inclusions with left-hand side $Q \mathbin{\hat{;}} R \sqcap S$ are valid.
2. Case: $(x{\sqcap}y){*}z$ as left-hand side. If the right-hand side is $x{*}(y{\sqcap}x{\sqcap}z)$ or $x{\sqcap}y{*}(x{\sqcap}z)$, then we have for $x = b, y = b$, and $z = d$ that $b{\sqcap}b{*}d = b{\sqcap}b = b$ but $b{*}(b{\sqcap}b{\sqcap}d) = b * b = 0$ and $b \sqcap b * (b \sqcap d) = b \sqcap b * b = b \sqcap 0 = 0$, i.e., the two inclusions are not valid.
3. Case: $x * y * z$ as left-hand side. If the right=hand side is $x * (y \sqcap x \sqcap z)$, then we have for $x = b, y = d$, and $z = d$ that $b * d * d = b * d = b$ but $b * (d \sqcap b \sqcap d) = b * b = 0$, i.e., the inclusion is not valid.

As mentioned above the previous examples verify or immediately imply all negative results listed in Table 1. The remaining two inclusions are valid and shown in the next lemma.

Lemma 8. *For L-relations $Q : A \to B, R : B \to C$ and $S : A \to C$ we have*

1. *(CT6) $Q;R * S \sqsubseteq Q;(R \sqcap Q^{\smile} \mathbin{\hat{;}} S)$,*
2. *(CT7) $Q \mathbin{\hat{;}} R * S \sqsubseteq Q;(R * Q^{\smile} \mathbin{\hat{;}} S)$,*

Proof. 1. We immediately compute

$$\begin{aligned}
(Q;R * S)(x,z) &= (Q;R)(x,z) * S(x,z) \\
&= \left(\bigsqcup_{y\in B} Q(x,y) \sqcap R(y,z)\right) * S(x,z) \\
&= \bigsqcup_{y\in B} (Q(x,y) \sqcap R(y,z)) * S(x,z) && * \text{ continuous} \\
&\sqsubseteq \bigsqcup_{y\in B} Q(x,y) * S(x,z) \sqcap R(y,z) * S(x,z) && \text{Lemma 1(3)} \\
&\sqsubseteq \bigsqcup_{y\in B} Q(x,y) \sqcap Q(x,y) * S(x,z) \sqcap R(y,z) && \text{Lemma 1(1)} \\
&\sqsubseteq \bigsqcup_{y\in B} Q(x,y) \sqcap \left(\bigsqcup_{x\in A} Q(x,y) * S(x,z)\right) \sqcap R(y,z) \\
&= \bigsqcup_{y\in B} Q(x,y) \sqcap (Q^{\smile} \mathbin{\hat{;}} S)(y,z) \sqcap R(y,z)
\end{aligned}$$

$$= \bigsqcup_{y \in B} Q(x,y) \sqcap (R \sqcap Q^{\smile} \mathbin{\hat{;}} S)(y,z)$$

$$= (Q ; (R \sqcap Q^{\smile} \mathbin{\hat{;}} S))(x,z)$$

2. Similar to 1. consider

$$\begin{aligned}
(Q \mathbin{\hat{;}} R * S)(x,z) &= (Q \mathbin{\hat{;}} R)(x,z) * S(x,z) \\
&= \left(\bigsqcup_{y \in B} Q(x,y) * R(y,z) \right) * S(x,z) \\
&= \bigsqcup_{y \in B} Q(x,y) * R(y,z) * S(x,z) && * \text{ continuous} \\
&= \bigsqcup_{y \in B} Q(x,y) \sqcap Q(x,y) * R(y,z) * S(x,z) && \text{Lemma 1(1)} \\
&\sqsubseteq \bigsqcup_{y \in B} Q(x,y) \sqcap R(y,z) * \left(\bigsqcup_{x \in A} Q(x,y) * S(x,z) \right) \\
&= \bigsqcup_{y \in B} Q(x,y) \sqcap R(y,z) * (Q^{\smile} \mathbin{\hat{;}} S)(y,z) \\
&= \bigsqcup_{y \in B} Q(x,y) \sqcap (R * Q^{\smile} \mathbin{\hat{;}} S)(y,z) \\
&= (Q ; (R * Q^{\smile} \mathbin{\hat{;}} S))(x,z)
\end{aligned}$$

showing the second assertion. □

As before, we will assume in the remainder of the paper that $*$ and $\mathbin{\hat{;}}$ satisfy (CT6) and (CT7).

As promised after Lemma 7 we will now show that regular composition is the greatest composition operator based on some t-norm.

Lemma 9. *Let $\mathcal{R}$ be an arrow category, $Q : A \to B$ and $R : B \to C$. Then we have $Q \mathbin{\hat{;}} R \sqsubseteq Q ; R$.*

Proof. The following computation

$$\begin{aligned}
Q \mathbin{\hat{;}} R &= Q \mathbin{\hat{;}} R \sqcap \top\!\!\top_{AC} \\
&= Q \mathbin{\hat{;}} R * \top\!\!\top_{AC} && \text{(T4) since } \top\!\!\top \text{ is crisp} \\
&\sqsubseteq Q ; (R * Q^{\smile} \mathbin{\hat{;}} \top\!\!\top_{AC}) && \text{(CT7)} \\
&\sqsubseteq Q ; (R * \top\!\!\top_{BC}) \\
&= Q ; (R \sqcap \top\!\!\top_{BC}) && \text{(T4) since } \top\!\!\top \text{ is crisp} \\
&= Q ; R
\end{aligned}$$

shows the assertion. □

It is well known that for partial identities the operations $\sqcap$ and $;$ coincide, i.e., that if $Q, R \sqsubseteq \mathbb{I}_A$, then $Q \sqcap R = Q;R$. We are now ready to show a corresponding property for $*$ and $\overset{*}{;}$.

Lemma 10. *Let $\mathcal{R}$ be an arrow category and $Q, R \sqsubseteq \mathbb{I}_A$. Then we have $Q * R = Q \overset{*}{;} R$.*

Proof. First, we show the inclusion $\sqsubseteq$ by

$$\begin{aligned} Q * R &= Q \overset{*}{;} \mathbb{I}_A * R && \text{Lemma 7(2)} \\ &\sqsubseteq Q;(\mathbb{I}_A * Q^{\smile} \overset{*}{;} R) && \text{(CT7)} \\ &\sqsubseteq \mathbb{I}_A;(\mathbb{I}_A * Q \overset{*}{;} R) && Q \text{ partial identity} \\ &= Q \overset{*}{;} R && \text{Lemma 5(4)} \end{aligned}$$

From the computation

$$\begin{aligned} Q \overset{*}{;} R &= Q \overset{*}{;} R * \mathbb{I}_A && \text{Lemma 5(4)} \\ &\sqsubseteq Q;(R * Q^{\smile} \overset{*}{;} \mathbb{I}_A) && \text{(CT7)} \\ &\sqsubseteq \mathbb{I}_A;(R * Q \overset{*}{;} \mathbb{I}_A) && Q \text{ partial identity} \\ &= Q * R && \text{Lemma 7(1)} \end{aligned}$$

we obtain the converse inclusion. □

Please note that the previous lemma also implies that $*$ and $\overset{*}{;}$ coincide for scalars, i.e., that they are both based on the same t-norm on Sc[A].

Now we are ready to verify that also $\overset{*}{;}$ is uniquely determined by the axioms in the case that the α-cut theorem is valid.

Theorem 3. *Let $\mathcal{R}$ be an arrow category satisfying the α-cut theorem. Then we have*

$$Q \overset{*}{;} R = \bigsqcup_{\alpha,\beta \text{ scalars}} (\alpha_A * \beta_A);(\alpha_A \backslash Q)^{\downarrow};(\beta_B \backslash R)^{\downarrow}.$$

Proof. The following computation

$$\begin{aligned} Q \overset{*}{;} R &= \left(\bigsqcup_{\alpha \text{ scalar}} \alpha_A;(\alpha_A \backslash Q)^{\downarrow} \right) \overset{*}{;} \left(\bigsqcup_{\beta \text{ scalar}} \beta_B;(\beta_B \backslash R)^{\downarrow} \right) && \alpha\text{-cut theorem} \\ &= \bigsqcup_{\alpha,\beta \text{ scalars}} (\alpha_A;(\alpha_A \backslash Q)^{\downarrow}) \overset{*}{;} (\beta_B;(\beta_B \backslash R)^{\downarrow}) && \text{(CT2)} \\ &= \bigsqcup_{\alpha,\beta \text{ scalars}} \alpha_A \overset{*}{;} (\alpha_A \backslash Q)^{\downarrow} \overset{*}{;} \beta_B \overset{*}{;} (\beta_B \backslash R)^{\downarrow} && \text{(CT4)} \\ &= \bigsqcup_{\alpha,\beta \text{ scalars}} \alpha_A \overset{*}{;} ((\alpha_A \backslash Q)^{\downarrow};\beta_B) \overset{*}{;} (\beta_B \backslash R)^{\downarrow} && \text{(CT4)} \\ &= \bigsqcup_{\alpha,\beta \text{ scalars}} \alpha_A \overset{*}{;} (\beta_A;(\alpha_A \backslash Q)^{\downarrow}) \overset{*}{;} (\beta_B \backslash R)^{\downarrow} && \text{scalar iso. \& Lemma 3} \end{aligned}$$

$$= \bigsqcup_{\alpha,\beta \text{ scalars}} \alpha_A \mathbin{\text{;}\!\ast} \beta_A \mathbin{\text{;}\!\ast} (\alpha_A\backslash Q)^{\downarrow} \mathbin{\text{;}\!\ast} (\beta_B\backslash R)^{\downarrow} \qquad \text{(CT4)}$$

$$= \bigsqcup_{\alpha,\beta \text{ scalars}} (\alpha_A \mathbin{\text{;}\!\ast} \beta_A) ; (\alpha_A\backslash Q)^{\downarrow} ; (\beta_B\backslash R)^{\downarrow} \qquad \text{(CT4)}$$

$$= \bigsqcup_{\alpha,\beta \text{ scalars}} (\alpha_A * \beta_A) ; (\alpha_A\backslash Q)^{\downarrow} ; (\beta_B\backslash R)^{\downarrow} \qquad \text{Lemma 10}$$

shows the assertion. □

As mentioned above the inclusion $Q \mathbin{\text{;}\!\ast} R * S \sqsubseteq Q \mathbin{\text{;}\!\ast} (R * Q^{\smile} \mathbin{\text{;}\!\ast} S)$ is not valid. This is mainly due to the fact that $*$ is not idempotent and Q appears once in the left-hand side but twice in the right-hand side of the inclusion. A simple modification of the inclusion that ensures that every relation occurs exactly once in both sides leads to a valid inclusion. Please note that a similar modification was used in [13] in order to show decidability of the equational theory of allegories based on labeled graphs and certain graph homomorphisms.

Lemma 11. *For L-relations $P, Q : A \to B, R : B \to C$ and $S : A \to C$ we have (CT8) $(P * Q) \mathbin{\text{;}\!\ast} R * S \sqsubseteq P \mathbin{\text{;}\!\ast} (R * Q^{\smile} \mathbin{\text{;}\!\ast} S)$.*

Proof. We immediately obtain

$$
\begin{aligned}
&((P * Q) \mathbin{\text{;}\!\ast} R * S)(x,z)\\
&= ((P * Q) \mathbin{\text{;}\!\ast} R)(x,z) * S(x,z)\\
&= \left(\bigsqcup_{y\in B}(P * Q)(x,y) * R(y,z)\right) * S(x,z)\\
&= \left(\bigsqcup_{y\in B} P(x,y) * Q(x,y) * R(y,z)\right) * S(x,z)\\
&= \bigsqcup_{y\in B} P(x,y) * Q(x,y) * R(y,z) * S(x,z) && * \text{ continuous}\\
&\sqsubseteq \bigsqcup_{y\in B} P(x,y) * R(y,z) * \left(\bigsqcup_{x\in A} Q(x,y) * S(x,z)\right)\\
&= \bigsqcup_{y\in B} P(x,y) * R(y,z) * (Q^{\smile} \mathbin{\text{;}\!\ast} S)(y,z)\\
&= \bigsqcup_{y\in B} P(x,y) * (R * Q^{\smile} \mathbin{\text{;}\!\ast} S)(y,z)\\
&= (P \mathbin{\text{;}\!\ast} (R * Q^{\smile} \mathbin{\text{;}\!\ast} S))(x,z)
\end{aligned}
$$

verifying (CT8). □

One may ask the question whether (CT6)–(CT8) are independent in the context of the other axioms. It is worth mentioning that we only used (CT7) in axiomatic proofs, i.e., in Lemmas 9 and 10 and Theorem 3. In particular, the uniqueness of $*$ and $\ast\!\!\!;$ in arrow categories with the α-cut theorem follows without (CT6) and (CT8). Therefore, these axioms follow from (CT7) for concrete L-relations. Consequently, finding an example satisfying (CT7) but neither (CT6) nor (CT8) must be based on non-representable arrow categories. The question remains open.

4 Conclusion and Future Work

In this paper we have provided a suitable set of axioms for t-norm based meet and composition operators. Future work will concentrate on at least two aspects. First of all, we would like to investigate under which circumstances the axioms (CT6)–(CT8) are independent. As mentioned above, finding suitable examples has to be based on non-representable arrow categories. Secondly, we would like to apply this theory to applications of arrow categories. In particular, we are interested in the relation-algebraic treatment of L-fuzzy topological spaces similar to the approach to regular topological spaces in [16]. This would lead to a fully algebraic treatment of L-fuzzy topology that clearly indicates differences to and common parts with regular topology.

References

1. Adjei, E., Chowdhury, W., Winter, M.: *L*-fuzzy databases in arrow categories. In: Kahl, W., Winter, M., Oliveira, J.N. (eds.) RAMICS 2015. LNCS, vol. 9348, pp. 295–311. Springer, Cham (2015). https://doi.org/10.1007/978-3-319-24704-5_18
2. Birkhoff, G.: Lattice Theory, vol. 25, 3rd edn. American Mathematical Society Colloquium Publications, New York (1940)
3. Freyd, P., Scedrov, A.: Categories. North-Holland, Allegories (1990)
4. Galindo, J., Medina, J.M., Pons, O., Cubero, J.C.: A server for fuzzy SQL queries. In: Andreasen, T., Christiansen, H., Larsen, H.L. (eds.) FQAS 1998. LNCS, vol. 1495, pp. 164–174. Springer, Heidelberg (1998). https://doi.org/10.1007/BFb0055999
5. Galindo, J.: New characteristics in FSQL, a fuzzy SQL for fuzzy databases. WSEAS Trans. Inf. Sci. Appl. **2**(2), 161–169 (2005)
6. Galindo, J., Urrutia, A., Piattini, M.: Fuzzy Databases: Modeling, Design and Implementation. Idea Group Publishing, Hershey (2006)
7. Höhle, U., Šostak, A.P.: Axiomatic foundation of fixed-basis fuzzy topology. In: Höhle, U., Rodabaugh, S.E.: Mathematics of Fuzzy Sets. Kluwer (1999)
8. Jónsson, B., Tarski, A.: Boolean algebras with operators, I, II, Am. J. Math. **73**, 891–939 (1951), **74**, 127–162 (1952)
9. Kawahara, Y., Furusawa, H.: Crispness and Representation Theorems in Dedekind Categories. DOI-TR 143, Kyushu University (1997)
10. Mamdani, E.H., Gaines, B.R.: Fuzzy Reasoning and its Application. Academic Press, London (1987)
11. Olivier, J.P., Serrato, D.: Catégories de Dedekind. Morphismes dans les Catégories de Schröder. C.R. Acad. Sci. Paris **290**, 939–941 (1980)
12. Olivier, J.P., Serrato, D.: Squares and rectangles in relational categories - three cases: semilattice, distributive lattice and Boolean non-unitary. Fuzzy Sets Syst. **72**, 167–178 (1995)
13. Pous, D., Vignudelli, V.: Allegories: decidability and graph homomorphisms. LICS (2018, accepted)

14. Schmidt, G., Ströhlein, T.: Relationen und Graphen. Springer, Heidelberg (1989); English version: Relations and Graphs. Discrete Mathematics for Computer Scientists, EATCS Monographs on Theoretical Computer Science. Springer, Heidelberg (1993)
15. Schmidt, G.: Relational Mathematics. Encyclopedia of Mathematics and Its Applications, vol. 132. Cambridge University Press, Cambridge (2011)
16. Schmidt, G., Winter, M.: Relational Topology. LNM, vol. 2208. Springer, Heidelberg (2018)
17. Winter, M.: A new algebraic approach to L-fuzzy relations convenient to study crispness. INS Inf. Sci. **139**, 233–252 (2001)
18. Winter, M.: Derived operations in Goguen categories. TAC Theory Appl. Categories **10**(11), 220–247 (2002)
19. Winter, M.: Representation theory of Goguen categories. Fuzzy Sets Syst. **138**, 85–126 (2003)
20. Winter, M.: Goguen Categories - A Categorical Approach to L-fuzzy Relations. Trends in Logic, vol. 25. Springer, Netherlands (2007)
21. Winter, M.: Arrow categories. Fuzzy Sets Syst. **160**, 2893–2909 (2009)
22. Winter, M.: Higher-order arrow categories. In: Höfner, P., Jipsen, P., Kahl, W., Müller, M.E. (eds.) RAMICS 2014. LNCS, vol. 8428, pp. 277–292. Springer, Cham (2014). https://doi.org/10.1007/978-3-319-06251-8_17
23. Winter, M.: Membership values in arrow categories. Fuzzy Sets Syst. **267**, 41–61 (2015)

Decidability of Equational Theories for Subsignatures of Relation Algebra

Robin Hirsch(✉)

Department of Computer Science, UCL, London, UK
r.hirsch@ucl.ac.uk

Abstract. Let S be a subset of the signature of relation algebra. Let $R(S)$ be the closure under isomorphism of the class of proper S-structures and let $F(S)$ be the closure under isomorphism of the class of proper S-structures over finite bases. Based on previous work, we prove that membership of $R(S)$ is undecidable when $S \supseteq \{\cdot, +, ;\}$, $S \supseteq \{\cdot, \smile, ;\}$ or $\smile \notin S \supseteq \{\leq, -, ;\}$, and for any of these signatures S if converse is excluded from S then membership of $F(S)$ is also undecidable, for finite S-structures.

We prove that the equational theories of $F(S)$ and $R(S)$ are undecidable when S includes composition and the signature of boolean algebra. If all operators in S are positive and it does not include negation, or if it can define neither domain, range nor composition, then the equational theory of either class is decidable. Open cases for decidability of the equational theory of $R(S)$ are when S can define negation but not meet (or join) and either domain, range or composition. Open cases for the decidability of the equational theory of $F(S)$ are (i) when S can define negation, converse and either domain, range or composition, and (ii) when S contains negation but not meet (or join) and either domain, range or composition.

1 Summary of Results

We collect, organise are marginally extend certain results from [HH01,HJ12, Neu16]. What is offered here is a summary and a bringing together of the main results in these and other papers, a significant simplification and shortening of some of the proofs and a slight extension to the range of signatures covered by the results. We consider subsignatures S of the signature of relation algebra $\{\leq, 0, 1, -, +, \cdot, 1', \smile, D, D', R, R', ;, \dagger\}$. Signatures covered here but not elsewhere (as far as we know) include various signatures involving dual operators (rather trivially), but also signatures containing $\{\cdot, +.;\}$, answering [Neu16, Problem 3.9].

A *proper S-structure* is an S-structure whose elements are binary relations over some non-empty base and where constants, operators and inequalities have natural set-theoretic definitions, see (1) below. $R(S)$ denotes the class of all S-structures isomorphic to proper S-structures, $F(S)$ denotes the class of all S-structures isomorphic to proper S-structures on finite bases.

J. Desharnais et al. (Eds.): RAMiCS 2018, LNCS 11194, pp. 87–96, 2018.
https://doi.org/10.1007/978-3-030-02149-8_6

We show, for any such signature S containing $\{\leq, -, ;\}$ but not including converse, or containing $\{\cdot, +, ;\}$ or $\{\cdot, \smile, ;\}$ that membership of $R(S)$ is undecidable, for finite S-structures. If S does not include converse and S contains either $\{\leq, -, ;\}$ or $\{\cdot, +, ;\}$, then membership of $F(S)$ is undecidable, for finite S-structures. It follows that the set of quantifier-free formulas valid over each of these representation classes is undecidable. Observe that atomic formulas $a = b$ or $a \leq b$ are valid over $R(S)$ if and only if $-a = -b$, respectively $-b \leq -a$, is valid over $R(S \cup \{-\})$, and by De Morgan's law plus the *duality* of swapping each pair in $\{(\cdot, +), (D, D'), (R, R'), (;, \dagger)\}$, this holds if and only if a dual formula in a signature S' obtained from S by this duality is valid over $R(S)$, hence each of these undecidablity results has a dual result for S'.

If S also contains the signature of boolean algebra then even the equational theory of the relevant class is undecidable (since the equational theory is contained in the quantifier-free theory, the undecidability of the equational theory entails the undecidability of the quantifier-free theory, but the converse entailment does not hold in general). On the other hand, for any signature S either not including negation or including neither domain, range, composition nor their duals, the equational theories of $R(S)$ and $F(S)$ are decidable.

2 Preliminaries

Let $S \subseteq \{\leq, 0, 1, -, +, \cdot, 1', \smile, D, D', R, R', ;, \dagger\}$ be a signature. [By an extension of a convention proposed by H. Andréka we list relations first, then booleans, then extra operators, within that we list constants then unary then binary operators, see (1) below for the intended meanings.] The *boolean part* of S is the intersection of S with $\{\leq, 0, 1, -, +, \cdot\}$. An *S-structure* $\mathcal{A}$ is a finite set A with a binary relation $\leq$ and elements named $0, 1, 1'$, unary functions $-, \smile, D, D', R, R'$ and binary functions $\cdot, +, ;, \dagger$ over A, if the symbol is included in S. A *term* of the language $L(S)$ is built from variables and constants in S using operators in S (note that we allow variables in terms). An *atomic formula* is either $s \leq t$ (if $\leq \in S$) or $s = t$, for terms s, t, the latter case is called an *equation*. Formulas of $L(S)$ are built from atomic formulas using first-order connectives. Given an $L(S)$-formula ϕ, an S-structure $\mathcal{A}$ and a variable assignment $v : vars \to \mathcal{A}$ we may evaluate $\mathcal{A}, v \models \phi$ using standard first-order semantics. For any class $\mathcal{K}$ of S-structures, the *quantifier-free theory* of $\mathcal{K}$ is the set of all quantifier-free $L(S)$-formulas ϕ such that for all $\mathcal{A} \in \mathcal{K}$ and all variable assignments $v : vars \to \mathcal{A} \in \mathcal{K}$ we have $\mathcal{A}, v \models \phi$, similarly the *equational theory* of $\mathcal{K}$ is the set of equations valid over $\mathcal{K}$. ϕ is satisfiable in an S-structure $\mathcal{A}$ if $\mathcal{A}, v \models \phi$, for some variable assignment v, and ϕ is satisfiable in some class $\mathcal{K}$ of S-structures if there is $\mathcal{A} \in \mathcal{K}$ and v such that $\mathcal{A}, v \models \phi$.

A *proper S-structure* $\mathcal{A}$ over the base X is an S-structure where each element of A is a binary relation over X, where $\leq$ is set inclusion $\subseteq$, 0 is the empty relation, $\cdot, +$ denote $\cap, \cup$, respectively, $1 = X \times X$ (these proper structures are often called *square*), negation is complement in $X \times X$ (called *universal complementation* in [Neu16]) and

$$\begin{aligned} 1' &= \{(x,x) : x \in X\} \\ a^{\smile} &= \{(y,x) : (x,y) \in a\} \\ D(a) &= \{(x,x) : \exists y (x,y) \in a\} \\ D'(a) &= -D(-a) \\ R(a) &= \{(y,y) : \exists x (x,y) \in a\} \\ R'(a) &= -R(-a) \\ a;b &= \{(x,y) : \exists z \in X\ (x,z) \in a \wedge (z,y) \in b\} \\ a\dagger b &= -((-a);(-b)) \end{aligned} \tag{1}$$

(in each case, only if the inequality, constant or operator is included in S). For any proper S-structure $\mathcal{A}$ and subset Y of the base of A, we write $\mathcal{A}{\restriction_Y}$ for the proper S-structure of binary relations generated by $\{a \cap (Y \times Y) : a \in A\}$ in the full S-algebra of binary relations over Y. There is a *boolean duality* that swaps each of the pairs $\{(\cdot, +), (D, D'), (R, R'), (;, \dagger)\}$.

Other connectives are definable, e.g. $a \backslash b = -(a^{\smile}; -b)$, $A(a) = 1' \cdot D'(-a) = 1' \cdot -D(a)$, but not considered separately here. All operators considered here (other than negation) are *positive*. Our signatures include redundant symbols, e.g. $a \leq b \iff a+b = b$ is valid over proper $\{\leq, +\}$-structures, $D(a) = 1' {\cdot} a; 1$ is valid over proper $\{1, \cdot, 1', D, ;\}$-structures, $a\dagger b = -(-a; -b)$ is valid over proper $\{-, \dagger, ;\}$-structures, etc. The ordering $\leq$ is only needed as a primitive relation symbol for signatures with neither $\cdot$ nor $+$. Henceforth, we work modulo inter-definability so, e.g., the signatures $\{-, ;\}, \{-, \dagger\}$ are considered equal, and we may write $\{-, +\}$ for the signature of boolean algebra as $\leq, 0, 1, \cdot$ are definable.

3 Decidablity

Lemma 1. *Let S be a signature without negation. Let $s = t$ be an equation, let k be the number of occurences of operators from $\{D, R, ;\}$ in s or t. If $s = t$ is falsifiable in some proper S-structure $\mathcal{A}$ then it is falsifiable in some proper S structure with at most $2 + k$ points.*

Proof. See [KNSS18, Theorem 1]. Suppose $s = t$ is falsified in some proper S-structure $\mathcal{A}$. So, so s, t denote distinct binary relations over the base X of $\mathcal{A}$ and there are $x, y \in X$ such that (x, y) is in the symmetric difference of s and t, without loss $(x, y) \in s \setminus t$. Since all operators are positive, we have $(x, y) \notin t^{\mathcal{A}\restriction_Y}$ for any subset Y of X. By structured induction on the term s, since $(x, y) \in s$, there is a finite set $F \subseteq X$ with at most k points, and for any Y with $\{x, y\} \cup F \subseteq Y \subseteq X$ we have $(x, y) \in s^{\mathcal{A}\restriction_Y}$. Then $|F| \leq k$ and $s = t$ is falsifiable in $\mathcal{A}{\restriction_{\{x,y\} \cup F}}$.

See [AB95, Corollary 4, Theorem 5] for the decidability of the equational theory of $R(S)$ for $S = \{\cdot, +, 0, 1, 1', ^{\smile}, ;\}$ and $S = \{+, 1', ^{\smile}, ;\}$.

The proof of the next lemma is trivial and omitted.

Lemma 2. *Let S be a signature with neither domain, range, composition nor their duals. If a quantifier-free formula ϕ is satisfiable in some S-structure $\mathcal{A}$ over a base X then there are $x, y \in X$ such that ϕ is satisfiable in $\mathcal{A}\restriction_{\{x,y\}}$.*

4 Undecidability by Tiling

For the undecidabiliy of the equational theory of $R(S)$ where S contains the lattice operators and composition we use a tiling algebra. An instance τ of the version of the tiling problem considered here is a finite set of tiles and adjacencies. We may write $Rt(S) = Lt(T)$, $Top(S) = Bot(T)$ if the adjacencies allow S to be placed immediately on the left (respectively, below) T. It is a yes-instance if for each $T \in \tau$ there is a tiling function $f : \mathbb{Z}\times\mathbb{Z} \to \tau$ such that $f(0,0) = T$ and for all $i, j \in \mathbb{Z}$, $Lt(f(i+1, j)) = Rt(f(i, j))$ and $Bot f(i, j+1) = Top(f(i, j))$, it is a no-instance otherwise. Though hardly used here we assume there is one tile $T^* \in \tau$ which may be adjacent to itself vertically or horizontally but may not be adjacent to any other tile (this tile is needed for the proof of Lemma 3 below, see [HH01, Sect. 6.4], the addition of such a tile will clearly not change a yes to a no instance or vice versa).

A partial tiling is a partial function $f : \mathbb{Z} \times \mathbb{Z} \to \tau$ such that whenever $(i, j), (i + 1, j) \in \mathsf{dom}(f)$ we have $Rt(f(i, j)) = Lt(f(i + 1, j))$ and similarly for vertical adjacencies within the domain of f. An instance (τ, T^{00}) of the deterministic tiling function consists of a finite set of tiles and adjacencies and a specified tile from the set, where there is a bijection $I : \mathbb{N} \to \mathbb{Z}\times\mathbb{Z}$ and a function $f : \mathbb{Z}\times\mathbb{Z} \to \tau$ (not necessarily a tiling function) such that $I(0) = (0, 0)$, $f(0, 0) = T^{00}$ and the set of tiles T such that the partial map $\{(I(i), f(I(i))) : i \leq n\} \cup \{(I(n+1), T)\}$ is a partial tiling is either the singleton $\{f(I(n+1))\}$ or empty. It is a yes-instance of the deterministic tiling problem if and only if there is a tiling f of the plane where $f(0, 0) = T^{00}$. Both the tiling problem and the deterministic tiling problem are known to be undecidable [Ber66, HJ12].

From an instance τ of the tiling problem, a relation algebra $RA(\tau)$ is constructed in [HH01, Sect. 4] (also, see the appendix, here). The following was first presented at the 1997 Relmics conference in Tunisia, proved in [HH01, Theorems 3 and 4].

Lemma 3. *τ is a yes-instance of the tiling problem if and only if $RA(\tau)$ is a representable relation algebra.*

A modified construction in [HJ12, Theorem 7.6] produced a relation algebra $RA(\tau, T^{00})$ from an instance (τ, T^{00}) of the deterministic tiling problem, and the following was proved.

Lemma 4. *Let (τ, T^{00}) be an instance of the deterministic tiling problem. It is a yes-instance if and only if $RA(\tau, T^{00})$ is a representable relation algebra.*

Based on results in [HJ12, Neu16] these results may be strengthened further.

Lemma 5.

1. *Let τ be an instance of the tiling problem. If the $\{\cdot, +, ;\}$-reduct of $RA(\tau)$ is in $R(\{\cdot, +, ;\})$ then τ is a yes-instance of the tiling problem.*
2. *Let (τ, T^{00}) be an instance of the deterministic tiling problem. If the $\{\cdot, \smile, ;\}$-reduct of $RA(\tau, T^{00})$ is in $R(\{\cdot, \smile, ;\})$ then (τ, T^{00}) is a yes-instance of the deterministic tiling problem.*

Converse implications to both parts follow from Lemmas 3 and 4.

Proof. Part (2) is proved in [HJ12, Theorem 8.6]. For part 1, the case $S \supseteq \{\cdot, +, 1', ;\}$ is covered in [HJ12, Theorem 6.1], the case $S \supseteq \{\cdot, +, ;\}$ answers [Neu16, Problem 3.9] and is proved in Lemma 12, in the appendix.

Hence

Lemma 6. *Membership of $R(S)$ is undecidable, for finite S-structures, for signatures $S \supseteq \{\cdot, +, ;\}$ or $S \supseteq \{\cdot, \smile, ;\}$.*

5 Undecidability by Partial Groups

The known representations of tiling algebras $RA(\tau)$ are all infinite — whether they can be adapted to provide finite representations remains unknown. For results on finite representability we may use a different construction based on partial groups, provided converse is not included in the signature. A *finite partial group* $\mathcal{A} = (A, e, \sqrt{A}, *)$ consists of a finite set A including some element e, a subset $\sqrt{A} \ni e$ of A and a total binary surjective function $* : \sqrt{A} \times \sqrt{A} \to A$ such that $e * a = a * e = a$, for all $a \in \sqrt{A}$. It is a yes-instance of the partial group embedding problem if there is a group $(G, e, \circ)$ where $G \supseteq A$ and for $a, b \in \sqrt{A}$ we have $a \circ b = a * b$, it is a no-instance otherwise. The partial group finite embedding problem has the same set of instances $\mathcal{A}$, it is a yes-instance if there is a finite group $(G, e, \circ)$ where $G \supseteq A$ such that $a, b \in \sqrt{A} \Rightarrow a \circ b = a * b$, otherwise it is a no instance. Both problems are known to be undecidable, see [Eva53, JV09].

From a finite partial group $\mathcal{A}$, a finite $\{-, +, 1', ;\}$-structure $M(\mathcal{A})$ is defined in [HJ12, Sect. 3], whose signature contains the whole signature of relation algebra except converse is excluded. The map that sends $\mathcal{A}$ to $M(\mathcal{A})$ is shown to be a reduction of the partial group embedding problem to $R(\{-, +, 1', ;\})$ and of the partial group finite embedding problem to $F(\{-, +, 1', ;\})$ [HJ12, Proposition 5.1]. Without $1'$ in the signature, if $M(A) \in R(S)$ we cannot be sure that any reflexive edges (x, x) belong to 1 in a proper S-structure $\mathcal{B}$ isomorphic to $M(\mathcal{A})$, and such points are needed in the proof of undecidability. However, $M(\mathcal{A})$ includes an atom e_0 where $e_0; e_0 = e_0$, so if $\mathcal{B}$ has a finite base X then there must be points $x \in X$ where $(x, x) \in e_0^{\mathcal{B}}$, and when negation is in the signature, since it is complementation relative to $X \times X$, every reflexive edge (x, x) belongs to some element of $\mathcal{B}$. These reflexive points are needed for the proof of the following to work [Neu16, Theorem 2.5].

Lemma 7. *Let S be a signature without converse. Membership of $F(S)$ is undecidable when $\{\cdot, +, ;\} \subseteq S$ and membership of either $R(S)$ and $F(S)$ are undecidable when $\{\leq, -, ;\} \subseteq S$, for finite S-structures.*

Problem 1. Is membership of $F(S)$ decidable for finite S-structures, when $\{\cdot, \smile, ;\} \subseteq S$?

6 Quantifier-Free Theory and Equational Theory

Lemma 8. *If S does not include negation then the set of equations valid over $R(S)$ is* **co-NP-complete***, the same holds for $F(S)$. If S does not include domain, range, composition nor their boolean duals then the validity of quantifier-free formulas is* **co-NP-complete***.*

Proof. For the first part, by Lemma 1 a falsifiable equation is falsifiable in an S-structure of size at most two plus the number of occurrences of $\{D, R, ;\}$ in the equation. To prove that the complementary problem (satisfiability of inequations problem) belongs to **NP** consider a non-deterministic algorithm that, given a inequality $s \neq t$, guesses a proper S-structure $\mathcal{A}$ whose base size is at most two plus the number of occurrences of $\{D, R, ;\}$ in s or t, guesses two points x, y in the base, and verifies that (x, y) is in the symmetric difference of s and t. Since **PSAT** trivially reduces to the complement of either problem in the lemma (without using any symbol from S), they are both **co-NP-hard**.

The proof of the second part is similar, this time based on Lemma 2.

For any signatures S and finite S-structures $\mathcal{A}$, let the diagram of $\mathcal{A}$, $\Delta(\mathcal{A})$ be the quantifier-free formula

$$\bigwedge_{a \leq b} x_a \leq x_b \wedge \bigwedge_{a \not\leq b} \neg(x_a \leq x_b) \wedge \bigwedge_{a \neq b \in \mathcal{A}} \neg(x_a = x_b) \wedge \bigwedge_{o \in S} o(\bar{x_a}) = x_{o(\bar{a})} \tag{2}$$

where the first two conjunctions are only included when $S \cap \{\cdot, +\} = \emptyset$, $\leq \in S$, $\bar{a}$ ranges over tuples over $\mathcal{A}$ of the same arity as $o \in S$ and $\bar{x_a}$ is a tuple of corresponding variables. Then $\Delta(\mathcal{A})$ is satisfiable in $\mathcal{B}$ if and only if $\mathcal{A}$ embeds into $\mathcal{B}$, and since $R(S)$ and $F(S)$ are closed under substructures, $\Delta(\mathcal{A})$ is satisfiable in $R(S)$ if and only if $\mathcal{A} \in R(S)$ and $\Delta(\mathcal{A})$ is satisfiable in $F(S)$ if and only if $\mathcal{A} \in F(S)$. Hence,

Lemma 9. *Let $\mathcal{K} = R(S)$ or $F(S)$. The map $\mathcal{A} \mapsto \Delta(\mathcal{A})$ is a reduction from memberhips of $\mathcal{K}$ for finite S-structures to the satisfiability of quantifier-free formulas over $\mathcal{K}$.*

Lemma 10. *The set $QF(R(S))$ of quantifier-free formulas valid over $R(S)$ is undecidable, for signatures S containing $\{\cdot, +, ;\}$, $\{\cdot, \smile, ;\}$, $\{\cdot, +, \dagger\}$, $\{+, \smile, \dagger\}$ or $\{\leq, -, ;\}$. The set of quantifier-free formulas satisfiable over $F(S)$ is undecidable, for signatures S without converse, but containing $\{\cdot, +, ;\}$, $\{\leq, -, ;\}$, or $\{\cdot, +, \dagger\}$.*

Proof. By Lemmas 6, 7 and 9.

Lemma 11. *Let S contain the signature of boolean algebra and include composition, let $\mathcal{K}$ be the class of all S-structures such that the term $1;x;1$ is a unary discriminator (i.e. $1;0;1 = 0$ and $1;x;1 = 1$ for $x \neq 0$ is valid in $\mathcal{K}$). For each quantifier-free formula ϕ there is an equation $s = 0$, effectively computable from ϕ, and equivalent to ϕ over $\mathcal{K}$.*

Proof. Make the following replacements of atomic formulas in ϕ: $s \leq t \mapsto s+t = t$, $s = t \mapsto s\cdot(-t)+(-s)\cdot t = 0$, so that all atomic formulas have the form $s = 0$. Since S contains the signature of boolean algebra, this equation is equivalent to ϕ over $\mathcal{K}$. We may assume that the propositional connectives in ϕ are $\wedge, \neg$. Replace $s = 0 \wedge t = 0$ by $s + t = 0$ and replace $\neg(s = 0)$ by $-(1;s;1) = 0$.

We now restrict to algebraic signatures (signatures without $\leq$) and summarise the results on the equational theories of these representation classes.

Theorem 1. *The equational theory of $R(S)$ is undecidable if $\{-,+,;\} \subseteq S$, the equational theory of $R(S)$ is decidable if S does not include negation, and even the quantifier-free theory is decidable if S includes neither domain, range, composition nor their duals. [The open cases for decidability of equational theory of $R(S)$ are $\{-,x\} \subseteq S \subseteq \{-,1',\smile,D,R,;\}$ where $x \in \{D,R,;\}$.]*

The equational theory of $F(S)$ is undecidable if S does not include converse and $\{-,+,;\} \subseteq S$, it is decidable if either (i) negation is not included or (ii) neither domain, range, composition nor their duals are included in S. [The open cases are $\{-,x\} \subseteq S \subseteq \{-,1',D,R,;\}$ and $\{-,\smile,x\} \subseteq S$ where $x \in \{D,R,;\}$.]

Proof. Let $S \supseteq \{-,+,;\}$. Let τ be an instance of the tiling problem. The term $1;x;1$ is a discriminator for the S-reduct of the simple relation algebra $RA(\tau)$. By Lemma 11 there is an equation $s(\tau) = 0$ equivalent over $\{RA(\tau') : \tau' \text{ is an instance of the tiling problem}\}$ to $\Delta(RA(\tau))$. By Lemma 9, the map $\tau \mapsto (s(\tau) = 0)$ is a reduction from the tiling problem to the equational theory of $R(S)$, so this is undecidable. For cases where S excludes converse, note that $1;x;1$ is a discriminator for the S-structure $M(\mathcal{A})$, where $\mathcal{A}$ is a partial group, so a similar argument shows that the equational theory of $F(S)$ is undecidable. If S does not include negation or is disjoint from $\{D,R,;\}$ then the equational theories of $R(S)$ and $F(S)$ are decidable, by Lemma 8.

7 Conclusion

The table below summarises results on the decidability of membership of $R(S)$, $F(S)$ for finite S-structures, quantifier-free theories of $R(S)$, $F(S)$, and equational theories of these classes, $u, d, ?$ denote undecidable, decidable and unknown/variable, respectively.

Signature	$m(R(S))$	$m(F(S))$	$QF(R(S))$	$QF(F(S))$
$S \supseteq \{\cdot, +, ;\}$ or $S \supseteq \{\cdot, +, \dagger\}$	u	u if $\smile \notin S$	u	u if $\smile \notin S$
$S \supseteq \{\cdot, \smile, ;\}$ or $S \supseteq \{\cdot, \smile, \dagger\}$	u	?	u	?
$\smile \notin S$, $S \supseteq \{\leq, -, ;\}$	u	u	u	u
$\{D, D', R, R', ;, \dagger\} \cap S = \emptyset$	d	d	d	d

Signature	$EqTh(R(S))$	$EqTh(F(S))$
$\{-, +, ;\} \subseteq S$, $\smile \notin S$	u	u
$\{-, +, \smile, ;\} \subseteq S$	u	?
$- \notin S$	d	d
$\{D, D', R, R', ;, \dagger\} \cap S = \emptyset$	d	d

For the membership problems, the cases not covered by the table are cases involving domain, range or duals but not composition or relative sum, cases where converse and composition are in S but not intersection, or where composition is in S but not converse or intersection and S omits either $\leq$ or $-$, also the single case $S = \{\cdot, ;\}$, plus duals of these cases. For one signature under the second case $S = \{\leq, \smile, D, R, ;\}$ we know that $m(R(S))$ and $m(F(S))$ are both decidable [HM13], other cases remain open.

Appendix: The tiling algebra $RA(\tau)$.

Let τ be an instance of the tiling problem, recall that $T^* \in \tau$ is a special tile that may be adjacent to itself but to no other tile. The following construction of a relation algebra $RA(\tau)$ is from [HH01, Sect. 4]. The boolean part of $RA(\tau)$ is finite and has the following atoms:

$$At = \{e_i, w_{ij}, c_{0k}, c_{k0}, +1_k, -1_k, T_{12}, T_{21} : i, j < 3,\ c \in \{g, u, v\},\ 1 \leq k \leq 2,\ T \in \tau\}.$$

The identity is $e_0 + e_1 + e_2$, the converse of an atom x_{ij} is x_{ji}, the converse of $+1_k$ is -1_k. To define composition we list the forbidden triples of atoms F and then let $\alpha;\beta = \sum\{c \in At : \exists a \leq \alpha,\ b \leq \beta,\ (a, b, c) \notin F\}$, for $\alpha, \beta \in RA(\tau)$.

Any triple where the indices do not match is forbidden, e.g. $(x_i, y_{i'j'}, z_{i^*, j^*})$ is forbidden if $i \neq i'$, $j' \neq j^*$ or $i \neq i^*$. Any triple of atoms (e_i, a, b) where $a \neq b$ is forbidden. The following are also forbidden.

$$(g_{10}, g_{02}, w_{12}) \tag{3}$$

$$(S_{12}, T_{21}, +1_1) \text{ any } S, T \in \tau, \text{ unless } Rt(S) = Lt(T) \tag{4}$$

$$(u_{10}, g_{02}, T_{12}) \text{ any } T \in \tau \setminus \{T^*\} \tag{5}$$

$$(v_{10}, g_{01}, +1_1),\ (v_{10}, g_{01}, -1_1) \tag{6}$$

There are three dual rules for forbidden triples, obtained from 4, 5 and 6 by swapping the subscripts 1 and 2 throughout and replacing Lt, Rt by Bot, Top, respectively. The Peircean transforms of a triple of atoms (a, b, c) may be obtained by a sequence of up to three maps $(a, b, c) \mapsto (c^\smile, b^\smile, a^\smile)$ or $(a, b, c) \mapsto (b, c^\smile, a^\smile)$. F is the set of Peircean transforms of the triples of atoms forbidden above.

Lemma 12. *Let τ be an instance of the tiling problem and $S \supseteq \{\cdot, +, ;\}$. If $RA(\tau) \in R(S)$ then τ is a yes-instance.*

Proof. Suppose $RA(\tau)$ is isomorphic to a proper S-structure over a base X. Without loss, the isomorphism is the identity map, all elements of $RA(\tau)$ are binary relations over X, all operators in S are defined set-theoretically, but note that if $0, 1' \notin S$ we cannot assume they are represented as the empty set and identity, and for each $a \in RA(\tau)$ the element $\bar{a} \in RA(\tau)$ might not be the true complement of a. But $+ \in S$ so $a \leq b \iff a \subseteq b$. For each tile $T^{00} \in \tau$, since $T^{00}_{12} \not\leq 0$ there are $x_0, y_0 \in X$ such that $(x_0, y_0) \in T^{00}_{12}$ and $(x_0, y_0) \notin 0$, see the Fig. 1.

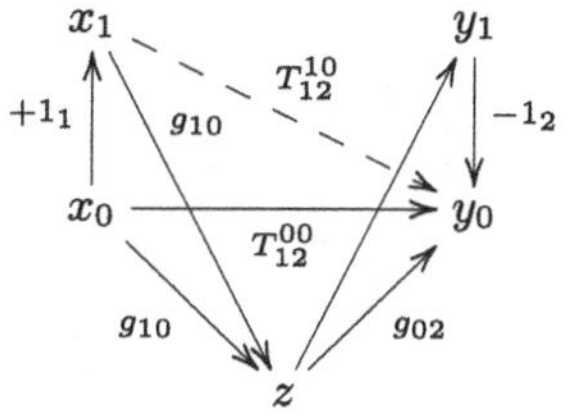

Fig. 1. If $RA(\tau) \in R\{\cdot, +, ;\}$ then points $z, \ldots, x_{-1}, x_0, x_1, \ldots, y_0, y_1, \ldots$ must exist in any representation, as shown.

Since $(g_{10}, g_{02}, T^{00}_{12})$ is not forbidden and ; is composition of binary relations, there must be $z \in X$ with $(x_0, z) \in g_{10}$, $(z, y_0) \in g_{02}$. Since $(+1_1, g_{10}, g_{10})$ is not forbidden there must be a point x_1 such that $(x_0, x_1) \in +1_1$ and $(x_1, z) \in g_{10}$. Continuing in this way there are $x_2, x_3, \ldots \in X$ where $(x_i, x_{i+1}) \in +1_1$, $(x_{i+1}, z) \in g_{10}$ for $i \in \mathbb{N}$. Since $(-1_1, g_{10}, g_{10})$ is not forbidden there are also $x_{-1}, x_{-2}, \ldots \in X$ where $(x_i, x_{i-1}) \in -1_1$ and $(x_{i-1}, z) \in g_{10}$, for $0 \geq i \in \mathbb{Z}$. Dually, there are points y_j for $j \in \mathbb{Z}$ where $(z, y_j) \in g_{02}$ and $(y_j, y_{j+1}) \in +1_2$ $(j \geq 0)$, $(y_j, y_{j-1}) \in -1_2$ $(j \leq 0)$.

From (x_i, z, y_j) we have $(x_i, y_j) \in g_{10}; g_{02} = \sum_{T \in \tau} T_{12}$ and from $\alpha; 0; \beta = 0$ for $\alpha, \beta \in \mathbf{RA}(\tau)$ we have $(x_i, y_j) \notin 0$, for $i, j \in \mathbb{Z}$. Since $+ \in S$ there is a tile $T^{ij} \in \tau$ such that $(x_i, y_j) \in T^{ij}_{12}$ and since $\cdot \in S$ this tile is unique. From (x_i, x_{i+1}, y_j) we have $(x_i, y_j) \in T^{ij}_{12} \cdot (+1_1; T^{i+1,j}_{12})$ so $Rt(T^{ij}) = Lt(T^{i+1,j})$ for $i \geq 0$, similarly, using -1_1 the same holds for $i < 0$ and dually $Top(T^{ij}) = Bot(T^{i,j+1})$ for $i, j \in \mathbb{Z}$, so the map $(i, j) \mapsto T^{ij}$ is a tiling of the plane mapping $(0, 0)$ to T^{00}. Since $T^{00} \in \tau$ was arbitrary, τ is a yes-instance of the tiling problem.

References

[AB95] Andréka, H., Bredikhin, D.: The equational theory of union-free algebras of relations. Algebra Universalis **33**, 516–532 (1995)

[Ber66] Berger, R.: The undecidability of the domino problem, volume 66 of Memoirs. Amer. Math. Soc., Providence, Rhode Island (1966)

[Eva53] Evans, T.: Embedability and the word problem. J. London Math. Soc. **28**, 76–80 (1953)

[HH01] Hirsch, R., Hodkinson, I.: Representability is not decidable for finite relation algebras. Trans. Amer. Math. Soc. **353**, 1403–1425 (2001)

[HJ12] Hirsch, R., Jackson, M.: Some undecidable problems on representability as binary relations. J. Symbolic Logic **77**(4), 1211–1244 (2012)

[HM13] Hirsch, R., Mikulas, S.: Ordered domain algebras. J. Appl. Logic (2013)

[JV09] Jackson, M., Volkov, M.: Undecidable word problems for completely 0-simple semigroups. J. Pure Appl. Algebra **213**, 1961–1978 (2009)

[KNSS18] Kurucz, A., Nemeti, I., Sain, I., Simon, A.: Decidable and undecidable modal logics with a binary modality. J. Logic Lang. Inf. **4**, 191–206 (2018)

[Neu16] Neuzerling, M.: Undecidability of representability for lattice-ordered semigroups and ordered complemented semigroups. Algebra Universalis **76**, 431–443 (2016)

Composition of Different-Type Relations via the Kleisli Category for the Continuation Monad

Koki Nishizawa[1(✉)] and Norihiro Tsumagari[2]

[1] Department of Information Systems Creation, Faculty of Engineering, Kanagawa University, Yokohama, Japan
nishizawa@kanagawa-u.ac.jp
[2] Center for Education and Innovation, Sojo University, Kumamoto, Japan
tsumagari@ed.sojo-u.ac.jp

Abstract. We give the way of composing different types of relational notions under certain condition, for example, ordinary binary relations, up-closed multirelations, ordinary (possibly non-up-closed) multirelations, quantale-valued relations, and probabilistic relations. Our key idea is to represent a relational notion as a generalized predicate transformer based on some truth value in some category and to represent it as a Kleisli arrow for some continuation monad. The way of composing those relational notions is given via identity-on-object faithful functors between different Kleisli categories. We give a necessary and sufficient condition to have such identity-on-object faithful functor.

1 Introduction

There are various relations to describe systems with certain behaviours, for example, ordinary binary relations for non-determinism, multirelations for different two-types of non-determinism, probabilistic relations for probabilistic behaviours. These are studied in general setting, for example, in Kleisli categories a monad represents certain behaviour of transition systems. However not much is well known to compose systems carrying each different type of behaviours. In this paper, we give the way of composing a type of relational notions with another type of relational notions under certain condition.

First of all, we describe a relational notion as a generalized predicate transformer based on some truth value in a certain category and represent it as a Kleisli arrow for appropriate continuation monad. Concretely, we define the notion of V-relation in L where L is a category with powers [8] and V is an object of L. Table 1 shows examples of V-relation in L. **CjLat** is the category of all complete join semilattices and homomorphisms among them, and **mSLat** is the category of all bounded meet semilattices and homomorphisms among them, and K-**LMod** is the category of left K-modules for a quantale K, and **EMod** is the category of effect modules. Effect module [5,7] have been introduced for the

J. Desharnais et al. (Eds.): RAMiCS 2018, LNCS 11194, pp. 97–112, 2018.
https://doi.org/10.1007/978-3-030-02149-8_7

foundation of quantum mechanics. Effect relation is defined as a probabilistic relation notion in this paper.

We give the way of composing those different relational arrows via identity-on-object faithful functors between different Kleisli categories. We also give a necessary and sufficient condition to have such identity-on-object faithful functor. Our composition includes various constructions of relational notions, for example, a construction of an up-closed multirelation from two ordinary binary relations, a construction of a K^{12}-valued relation from a K^4-valued relation and a K^6-valued relation for a quantale K, and a construction of a new relational arrow from an ordinary multirelation and a $\mathbf{2}^n$-valued relation for a natural number n.

Table 1. Examples of V-relation in L (A, B are sets)

V-relation in L	$A \to L(V^B, V)$ in **Set**	L : category with powers, V : object of L
Multirelation	$R \subseteq A \times \wp(B)$	$L = \mathbf{Set}$, $V = 2 = \{0, 1\}$
Up-closed multirelation	$R \subseteq A \times \wp(B)$ s.t. $(a, X) \in R$ and $X \subseteq Y$ imply $(a, Y) \in R$	$L = \mathbf{Poset}$, $V = \mathbf{2} = (2, \leq)$
Finite meet closed multirelation	$R \subseteq A \times \wp(B)$ s.t. $(a, X), (a, Y) \in R$ imply $(a, X \cap Y) \in R$ and $(a, B) \in R$	$L = \mathbf{mSLat}$, $V = \mathbf{2}$
Relation	$R \subseteq A \times B$	$L = \mathbf{CjLat}$ $V = \mathbf{2}$
K-valued relation (K:quantale)	$R\colon A \times B \to K$	$L = K\text{-}\mathbf{LMod}$ $V = K$
Effect relation	$R\colon A \times [0, 1]^B \to [0, 1]$ s.t. $R(a, p \oplus q) = R(a, p) + R(a, q)$ and $R(a, \mathbf{1}) = 1$	$L = \mathbf{EMod}$ $V = [0, 1]$

This paper is organized as follows. Section 2 gives the definition of relational arrows as Kleisli arrows for some continuation monad. Section 3 gives the way of composing of different types of such relational arrows under certain condition. In Sect. 4, we show examples of our relational arrow. Section 5 shows examples of the composition of different types of relational arrows. Section 6 summarizes this work and future work.

2 Representation of Relational Notions as Generalized Predicate Transformers

In this section, we define the notion of a V-relation in L with two parameters V, L. For that, we recall notions of powers and Kleisli category [9].

For a category C and objects c,c' of C, we write $C(c, c')$ for the hom-set, that is to say, the set of all arrows from c to c'. A category L *has powers (or cotensors)* [8], if the contravariant hom-set functor $L(-, V)\colon L^{op} \to \mathbf{Set}$ has a left adjoint for any object V. In other words, a category L has powers, if and only if for any object V of L, a functor $V^{(-)}\colon \mathbf{Set} \to L^{op}$ is given and for a set X and an object V' of L, the following isomorphisms $\varphi_{X,V'}$ are given, which are natural in X and V'.

$$\varphi_{X,V'}\colon \mathbf{Set}(X, L(V', V)) \cong L^{op}(V^X, V')$$

V^X is equal to the product of X copies of V.

The object part of the monad induced from this adjunction on **Set** maps a set X to the hom-set $L(V^X, V)$. This monad is called a *continuation monad* in the context of functional programming [12]. We write $L(V^{(-)}, V)$ for the continuation monad. Using the monad, we define a category of generalized relational notions as follows.

Definition 1 (The category of V-relations in L). *If a category L has powers and V is an object of L, then V-$\mathbf{Rel}(L)$ is defined to be the Kleisli category of the continuation monad $L(V^{(-)}, V)$.*

By the definition of Kleisli category of a monad, an object of V-$\mathbf{Rel}(L)$ is a set. An arrow from A to B in V-$\mathbf{Rel}(L)$ is a map from A to the set $L(V^B, V)$. We call such arrow *V-relation in L from A to B*.

$$V\text{-}\mathbf{Rel}(L)(A, B) \stackrel{\text{def}}{=} \mathbf{Set}(A, L(V^B, V))$$

Therefore, we can regard a V-relation in L as an arrow of L^{op} by the correspondence φ_{A,V^B} which depends on powers of L.

$$\varphi_{A,V^B}\colon V\text{-}\mathbf{Rel}(L)(A, B) \stackrel{\text{def}}{=} \mathbf{Set}(A, L(V^B, V)) \cong L^{op}(V^A, V^B) = L(V^B, V^A)$$

The composition of V-relations in L is defined by construction of the Kleisli category, while it is also explained by using the above correspondence φ and the composition $\circ_{L^{op}}$ in L^{op}. The composition of $f\colon A \to B$ and $g\colon B \to C$ in V-$\mathbf{Rel}(L)$ is given by $\varphi^{-1}_{A,V^C}(\varphi_{B,V^C}(g) \circ_{L^{op}} \varphi_{A,V^B}(f))\colon A \to C$. Similarly, the identity arrow on A in V-$\mathbf{Rel}(L)$ is given by $\varphi^{-1}_{A,V^A}(\mathrm{Id}_{V^A})\colon A \to A$, where Id_{V^A} is the identity arrow on V^A in L^{op}.

By the general theory of Kleisli categories, we have the following functor, which is called the *comparison functor.*

Proposition 1. *The following data $K_{V,L}$ gives a fully faithful functor from V-$\mathbf{Rel}(L)$ to L^{op}.*

- *For an object A of V-$\mathbf{Rel}(L)$, $K_{V,L}(A) \stackrel{\text{def}}{=} V^A$*
- *For an arrow $f\colon A \to B$ in V-$\mathbf{Rel}(L)$, $K_{V,L}(f) \stackrel{\text{def}}{=} \varphi_{A,V^B}(f)\colon V^A \to V^B$ in L^{op}*

A functor $F\colon C \to D$ is called *identity-on-object*, if the object part of F is the identity map, that is to say, the object class of C is equal to the object class of D and $Fc = c$ for each object c of C. By the general theory of Kleisli category, we also have the following identity-on-object functor.

Proposition 2. *The following data $J_{V,L}$ gives an identity-on-object functor from* **Set** *to V-*$\mathbf{Rel}(L)$ *satisfying $K_{V,L} \circ J_{V,L} = V^{(-)}$.*

- *For a set A, $J_{V,L}(A) \stackrel{\mathrm{def}}{=} A$*
- *For a map $f \colon A \to B$, $J_{V,L}(f) \stackrel{\mathrm{def}}{=} \varphi^{-1}_{B,V^B}(\mathrm{Id}_{V^B}) \circ_{\mathbf{Set}} f$, where $\circ_{\mathbf{Set}}$ is the composition in* **Set**.

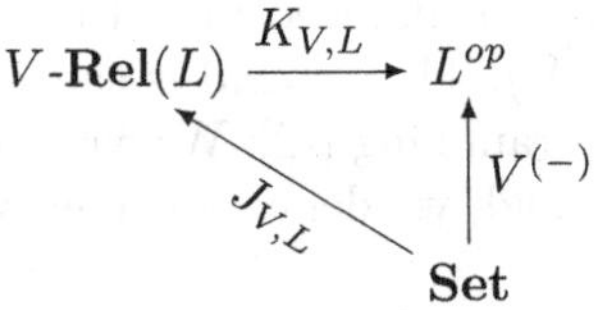

Proof. The above diagram commutes by naturality of $\varphi_{X,V'}$ in X. □

3 Composition of Different Types of Relational Arrows

In this section, we give the way of composing different types of relational notions under certain condition. We compose V_1-relation $R_1 \colon A \to B$ in L_1 and V_2-relation $R_2 \colon B \to C$ in L_2, by regarding both of R_1 and R_2 as V_3-relations in L_3 such that V_1-$\mathbf{Rel}(L_1)$ and V_2-$\mathbf{Rel}(L_2)$ are identity-on-object subcategories of V_3-$\mathbf{Rel}(L_3)$. In fact, Propositions 3 and 4 are provable for an arbitrary monad. Since this paper needs results for only the continuation monad, we prove these propositions for the monad.

Proposition 3. *If a faithful functor $F \colon L_1^{op} \to L_2^{op}$ and a natural isomorphism $\alpha \colon F \circ V_1^{(-)} \cong V_2^{(-)}$ are given, then there exists a unique pair $(\tilde{F}, \tilde{\alpha})$ of an identity-on-object faithful functor $\tilde{F}$ from V_1-*$\mathbf{Rel}(L_1)$ *to V_2-*$\mathbf{Rel}(L_2)$ *and a natural isomorphism $\tilde{\alpha} \colon F \circ K_{V_1,L_1} \cong K_{V_2,L_2} \circ \tilde{F}$ satisfying $\tilde{F} \circ J_{V_1,L_1} = J_{V_2,L_2}$ and $\tilde{\alpha}_{J_{V_1,L_1}X} = \alpha_X$ for any $X \in$* **Set**.

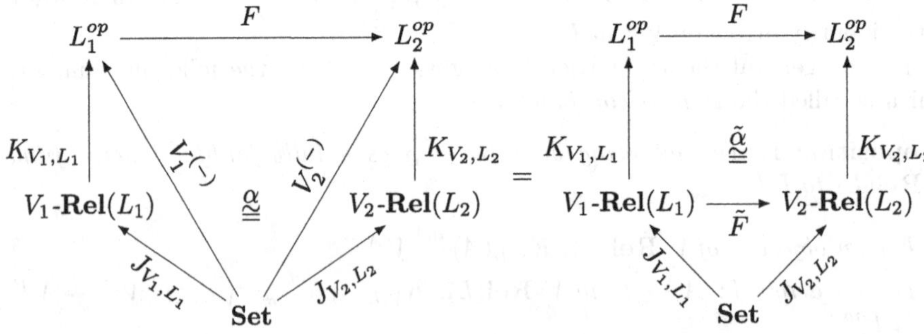

Proof. We abbreviate $\circ_{L_2^{op}}$ to $\circ$ in this proof. We give $\tilde{F}$ by $\tilde{F}(R) \stackrel{\mathrm{def}}{=} \varphi^{-1}_{A,V_2^B}(\alpha_B \circ F(K_{V_1,L_1}(R)) \circ \alpha_A^{-1})$ for $R \colon A \to B$ in V_1-$\mathbf{Rel}(L_1)$ and give $\tilde{\alpha}$ by $\tilde{\alpha}_X \stackrel{\mathrm{def}}{=} \alpha_X \colon FV_1^X \to V_2^X$ in L_2^{op} for X in V_1-$\mathbf{Rel}(L_1)$. $\tilde{F}$ is faithful, that is to say, its arrow part is injective, since the arrow part of F is so. $\tilde{\alpha}_X$ is an isomorphism, since α_X is so. The naturality condition of $\tilde{\alpha}$ is the following commutative diagram in L_2^{op} for any $R\colon A \to B$ in $V_1\text{-}\mathbf{Rel}(L_1)$.

$$
\begin{array}{ccc}
FV_1^B & \xrightarrow{\;\tilde{\alpha}_B\;} & V_2^B \\
{\scriptstyle F(K_{V_1,L_1}(R))}\Big\uparrow & & \Big\uparrow{\scriptstyle K_{V_2,L_2}(\tilde{F}(R))} \\
FV_1^A & \xrightarrow[\;\tilde{\alpha}_A\;]{} & V_2^A
\end{array}
$$

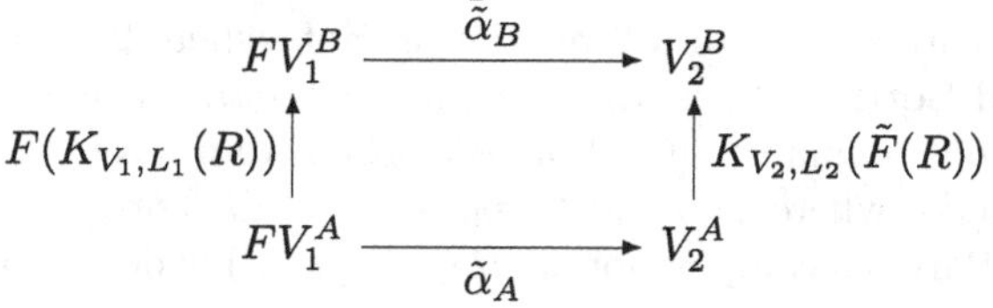

The following equation implies the above commutativity and uniqueness of $\tilde{F}$.

$$
\begin{aligned}
& K_{V_2,L_2}(\tilde{F}(R)) \circ \tilde{\alpha}_A = \tilde{\alpha}_B \circ F(K_{V_1,L_1}(R)) \\
\Leftrightarrow\; & K_{V_2,L_2}(\tilde{F}(R)) \circ \alpha_A = \alpha_B \circ F(K_{V_1,L_1}(R)) \\
\Leftrightarrow\; & K_{V_2,L_2}(\tilde{F}(R)) = \alpha_B \circ F(K_{V_1,L_1}(R)) \circ \alpha_A^{-1} \\
\Leftrightarrow\; & \varphi_{A,V_2^B}(\tilde{F}(R)) = \alpha_B \circ F(K_{V_1,L_1}(R)) \circ \alpha_A^{-1} \\
\Leftrightarrow\; & \tilde{F}(R) = \varphi_{A,V_2^B}^{-1}(\alpha_B \circ F(K_{V_1,L_1}(R)) \circ \alpha_A^{-1})
\end{aligned}
$$

$\tilde{\alpha}$ is unique by the condition $\tilde{\alpha}_{J_{V_1,L_1}X} = \alpha_X$, since J_{V_1,L_1} is identity-on-object. The equality $\tilde{F}(J_{V_1,L_1}(f)) = J_{V_2,L_2}(f)$ is shown by the following equation for any $f\colon A \to B$ in $\mathbf{Set}$.

$$
\begin{aligned}
\tilde{F}(J_{V_1,L_1}(f)) &= K_{V_2,L_2}^{-1}(\alpha_B \circ F(K_{V_1,L_1}(J_{V_1,L_1}(f))) \circ \alpha_A^{-1}) \\
&= K_{V_2,L_2}^{-1}(\alpha_B \circ F(V_1^{(-)}(f)) \circ \alpha_A^{-1}) = K_{V_2,L_2}^{-1}(V_2^{(-)}(f) \circ \alpha_A \circ \alpha_A^{-1}) \\
&= K_{V_2,L_2}^{-1}(V_2^{(-)}(f)) = J_{V_2,L_2}(f)
\end{aligned}
$$
$\hfill\square$

By Proposition 3, we obtain the sufficient condition to compose different types of relational notions.

Proposition 4. *When $R_1\colon A \to B$ in $V_1\text{-}\mathbf{Rel}(L_1)$ and $R_2\colon B \to C$ in $V_2\text{-}\mathbf{Rel}(L_2)$ are given, they are composable as $\tilde{F}_2(R_2) \circ \tilde{F}_1(R_1)\colon A \to C$ in $V_3\text{-}\mathbf{Rel}(L_3)$, if faithful functors $F_1\colon L_1^{op} \to L_3^{op}$, $F_2\colon L_2^{op} \to L_3^{op}$ and natural isomorphisms $\alpha_1\colon F_1 \circ V_1^{(-)} \cong V_3^{(-)}$, $\alpha_2\colon F_2 \circ V_2^{(-)} \cong V_3^{(-)}$ are given.*

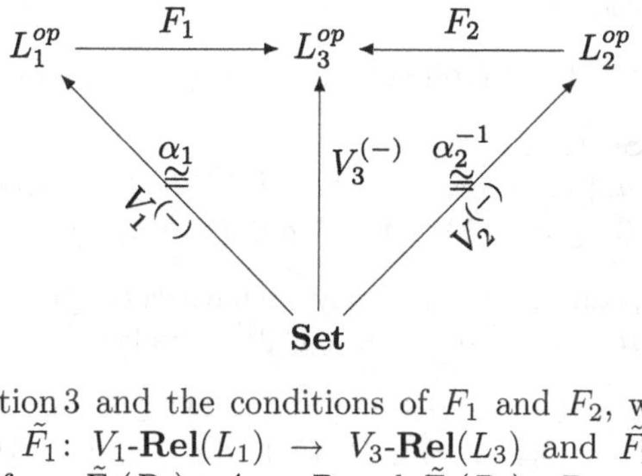

Proof. By Proposition 3 and the conditions of F_1 and F_2, we obtain identity-on-object functors $\tilde{F}_1\colon V_1\text{-}\mathbf{Rel}(L_1) \to V_3\text{-}\mathbf{Rel}(L_3)$ and $\tilde{F}_2\colon V_2\text{-}\mathbf{Rel}(L_2) \to V_3\text{-}\mathbf{Rel}(L_3)$. Therefore, $\tilde{F}_1(R_1)\colon A \to B$ and $\tilde{F}_2(R_2)\colon B \to C$ are composable in $V_3\text{-}\mathbf{Rel}(L_3)$. $\hfill\square$

4 Examples of V-Relation in L

4.1 Multirelations

A multirelation corresponds to a 2-relation in **Set**, where $2 = \{0, 1\}$. The continuation monad $\mathbf{Set}(2^{(-)}, 2)$ is known as a *neighbourhood monad* in the coalgebra/modal logic community [6]. A *multirelation* from a set A to a set B is a subset of $A \times \wp(B)$, where $\wp(B)$ is the powerset of B. For $Q \subseteq A \times \wp(B)$ and $R \subseteq B \times \wp(C)$, Parikh's composition $Q; R \subseteq A \times \wp(C)$ is defined as follows [14].

$$(a, Z) \in Q; R \Leftrightarrow \exists Y \subseteq B.(a, Y) \in Q \,\text{and}\, \forall b \in Y.(b, Z) \in R.$$

Parikh's composition is associative for up-closed multirelations, while it is known that the composition is not associative for ordinary (non-up-closed) multirelations. So, we use another composition of multirelations [2]. The associativity of the composition and the unit law hold.

Definition 2. *We define the category* **MRel** *of multirelations by the following data.*

- *Its objects are sets.*
- *Its arrow* $R\colon A \to B$ *is a multirelation from* A *to* B*, that is to say, a subset of* $A \times \wp(B)$*.*
- *For* $Q\colon A \to B$ *and* $R\colon B \to C$*, the composition* $R \circ Q\colon A \to C$ *is defined as follows.*
$$R \circ Q = \{(a, Z) \mid (a, \{b \mid (b, Z) \in R\}) \in Q\}$$
- *An identity* $\mathrm{Id}\colon A \to A$ *is* $\{(a, X) \mid X \subseteq A, a \in X\}$*.*

Proposition 5. *A multirelation corresponds to a 2-relation in* **Set***, that is to say, we have that* $\mathbf{MRel} \cong 2\text{-}\mathbf{Rel}(\mathbf{Set})$*.*

Proof. **Set** has the power 2^X in **Set** which is the set of all maps from a set X to 2 with the following natural isomorphism $\varphi_{A,2^B}$, which is called *flipping* [4].

$$\begin{array}{l}\varphi_{A,2^B} : \mathbf{Set}(A, \mathbf{Set}(2^B, 2)) \cong \mathbf{Set}^{op}(2^A, 2^B) = \mathbf{Set}(2^B, 2^A)\\ \varphi_{A,2^B}(f)(g)(a) = f(a)(g) \quad (f\colon A \to \mathbf{Set}(2^B, 2), g \in 2^B, a \in A)\\ \varphi^{-1}_{A,2^B}(h)(a)(g) = h(g)(a) \quad (h\colon 2^B \to 2^A, a \in A, g \in 2^B)\end{array}$$

The isomorphism $\mathbf{MRel} \cong 2\text{-}\mathbf{Rel}(\mathbf{Set})$ is given by φ^{-1} and the following ψ.

$$\begin{array}{l}\psi\colon \wp(A \times \wp(B)) \cong \mathbf{Set}(2^B, 2^A)\\ \psi(R)(g)(a) = 1 \Leftrightarrow (a, \{b \in B \mid g(b) = 1\}) \in R \quad (R \subseteq A \times \wp(B), g \in 2^B, a \in A)\\ \psi^{-1}(h) = \{(a, \{b \in B \mid g(b) = 1\}) \mid a \in A, g \in 2^B, h(g)(a) = 1\} \quad (h\colon 2^B \to 2^A)\end{array}$$

ψ preserves the composition, that is to say, we have $\psi(R \circ Q)(g) = \psi(Q)(\psi(R)(g))$ for $Q \subseteq A \times \wp(B)$, $R \subseteq B \times \wp(C)$, and $g \in 2^C$ as follows.

$$\begin{aligned}\psi(R \circ Q)(g)(a) = 1 &\Leftrightarrow (a, \{c \in C \mid g(c) = 1\}) \in R \circ Q\\ &\Leftrightarrow (a, \{b \in B \mid (b, \{c \in C \mid g(c) = 1\}) \in R\}) \in Q\\ &\Leftrightarrow (a, \{b \in B \mid \psi(R)(g)(b) = 1\}) \in Q\\ &\Leftrightarrow \psi(Q)(\psi(R)(g))(a) = 1\end{aligned}$$

ψ preserves the identities on A, that is to say, $\psi(\mathrm{Id})(g) = g$ for $g \in 2^A$ as follows.

$$\psi(\mathrm{Id})(g)(a) = 1 \Leftrightarrow (a, \{a' \in A \mid g(a') = 1\}) \in \mathrm{Id} \Leftrightarrow g(a) = 1 \qquad \square$$

4.2 Up-Closed Multirelations

An up-closed multirelation corresponds to a **2**-relation in **Poset**, where **Poset** is the category of all partially ordered sets and monotone maps among them. **2** is the two element partially ordered set $(\{0,1\}, \leq)$ where $0 \leq 1$. A Kleisli arrow for the continuation monad $\mathbf{Poset}(\mathbf{2}^{(-)}, \mathbf{2})$ is known as an *up-closed multifunction*, which corresponds to a *monotone predicate transformer* [10].

A multirelation $R \subseteq A \times \wp(B)$ is called *up-closed*, if $(a, X) \in R$ and $X \subseteq Y$ imply $(a, Y) \in R$ for each $a \in A$ and $X, Y \subseteq B$. We define the category of up-closed multirelations by using Parikh's composition [14].

Definition 3. *We define the category* **UMRel** *of up-closed multirelations by the following data.*

- *Its objects are sets.*
- *Its arrow* $R\colon A \to B$ *is an up-closed multirelation from* A *to* B.
- *For* $Q\colon A \to B$ *and* $R\colon B \to C$, *the composition* $R \circ Q\colon A \to C$ *is defined as follows.*

$$(a, Z) \in R \circ Q \Leftrightarrow \exists Y \subseteq B.(a, Y) \in Q \,\text{and}\, \forall b \in Y.(b, Z) \in R.$$

- *An identity* $\mathrm{Id}\colon A \to A$ *is* $\{(a, X) \mid X \subseteq A, a \in X\}$.

The associativity of the above composition and the unit law are known [14]. Note that the composition of **MRel** is not always equal to the composition of **UMRel** for arrows of **MRel**. However, **UMRel** is a subcategory of **MRel**, that is to say, for arrows of **UMRel**, the composition of **MRel** gives the same composition as **UMRel**.

It is well-known that a monotone predicate transformer corresponds to an up-closed multirelation [15].

Proposition 6. *An up-closed multirelation corresponds to a* **2***-relation in* **Poset**, *that is to say, we have that* **UMRel** $\cong$ **2-Rel(Poset)**.

Proof. **Poset** has the power $\mathbf{2}^X$ which is the pointwise ordered set of all maps from a set X to **2**. The following $\varphi_{A,\mathbf{2}^B}$ and ψ are defined in the same way as the case of multirelation. The isomorphism **UMRel** $\cong$ **2-Rel(Poset)** is given by φ^{-1} and ψ.

$$\varphi_{A,\mathbf{2}^B}\colon \mathbf{Set}(A, \mathbf{Poset}(\mathbf{2}^B, \mathbf{2})) \cong \mathbf{Poset}^{op}(\mathbf{2}^A, \mathbf{2}^B) = \mathbf{Poset}(\mathbf{2}^B, \mathbf{2}^A)$$
$$\psi\colon \mathbf{UMRel}(A, B) \cong \mathbf{Poset}(\mathbf{2}^B, \mathbf{2}^A)$$

$\square$

4.3 Finite Meet Closed Multirelation

A finite meet closed multirelation corresponds to a **2**-relation in **mSLat**, where **mSLat** is the category of all posets with all finite meets (called *bounded meet semilattices*) and finite meets preserving maps among them. The continuation monad $\mathbf{mSLat}(\mathbf{2}^{(-)}, \mathbf{2})$ is known as *filter monad* [6].

A up-closed multirelation $R \subseteq A \times \wp(B)$ is called *finite meet closed*, if $(a, X) \in R$ and $(a, Y) \in R$ imply $(a, X \cap Y) \in R$ for each $a \in A$ and $\forall a \in A.(a, B) \in R$. We define the category of finite meet closed multirelations by using the same composition as up-closed one.

Definition 4. *We define the category* **FMRel** *of finite meet closed multirelations by the subcategory of* **UMRel**, *whose arrow* $R\colon A \to B$ *is a finite meet closed multirelation and whose composition is the same as up-closed one.*

Proposition 7. *A finite meet closed multirelation corresponds to a* **2**-*relation in* **mSLat**, *that is to say, we have that* $\mathbf{FMRel} \cong \mathbf{2}\text{-}\mathbf{Rel}(\mathbf{mSLat})$.

4.4 Binary Relations

A binary relation corresponds to a **2**-relation in **CjLat**, where **CjLat** is the category of all posets with all joins (called *complete join semilattices*) and join-preserving (so, monotone) maps among them.

Definition 5. *We define the category* **Rel** *of relations by the following data. The associativity of the composition and the unit law are known.*

- *Its objects are sets.*
- *Its arrow* $R\colon A \to B$ *is a subset of* $A \times B$
- *For* $Q\colon A \to B$ *and* $R\colon B \to C$, *the composition* $R \circ Q\colon A \to C$ *is defined as follows.*
$$(a, c) \in R \circ Q \Leftrightarrow \exists b \in B.(a, b) \in Q \text{ and } (b, c) \in R.$$
- *An identity* $\mathrm{Id}\colon A \to A$ *is* $\{(a, a) \mid a \in A\}$.

It is well-known that a binary relation corresponds to the existential operator (or possibility operator) $\diamond$ of modal logics. The next proposition means that.

Proposition 8. *A binary relation corresponds to a* **2**-*relation in* **CjLat**, *that is to say, we have that* $\mathbf{Rel} \cong \mathbf{2}\text{-}\mathbf{Rel}(\mathbf{CjLat})$.

Proof. **CjLat** has the power $\mathbf{2}^X$ which is the pointwise ordered set of all maps from a set X to **2**. The following natural isomorphism $\varphi_{A,\mathbf{2}^B}$ is defined in the same way as the Proposition 5. The isomorphism $\mathbf{Rel} \cong \mathbf{2}\text{-}\mathbf{Rel}(\mathbf{CjLat})$ is given by its φ^{-1} and the following ψ. Here, $\delta_b \in \mathbf{2}^B$ is defined to be $\delta_b(b) = 1$ and $\delta_b(x) = 0$ if $x \neq b$.

$$\varphi_{A,\mathbf{2}^B} : \mathbf{Set}(A, \mathbf{CjLat}(\mathbf{2}^B, \mathbf{2})) \cong \mathbf{CjLat}^{op}(\mathbf{2}^A, \mathbf{2}^B) = \mathbf{CjLat}(\mathbf{2}^B, \mathbf{2}^A)$$
$$\psi\colon \mathbf{Rel}(A, B) \cong \mathbf{CjLat}(\mathbf{2}^B, \mathbf{2}^A)$$
$$\psi(R)(g)(a) = 1 \Leftrightarrow \exists b \in B.g(b) = 1 \text{ and } (a, b) \in R \quad (R \subseteq A \times B, g \in \mathbf{2}^B, a \in A)$$
$$\psi^{-1}(h) = \{(a, b) \mid b \in B, h(\delta_b)(a) = 1\} \quad (h \in \mathbf{CjLat}(\mathbf{2}^B, \mathbf{2}^A))$$
□

On the other hand, a binary relation also corresponds to a **2**-relation in **CmLat**, where **CmLat** is the category of all posets with all meets (called *complete meet semilattices*) and meet-preserving (so, monotone) maps among them. It is well-known that a binary relation corresponds to the universal operator (or necessity operator) $\Box$ of modal logics. The next proposition means that.

Proposition 9. *A binary relation corresponds to a* **2***-relation in* **CmLat***, that is to say, we have that* **Rel** $\cong$ **2-Rel(CmLat)**.

Proof. **CmLat** has the power $\mathbf{2}^X$ which is the pointwise ordered set of all maps from a set X to **2**. The following natural isomorphism $\varphi_{A,\mathbf{2}^B}$ is defined in the same way as the Proposition 5. The isomorphism **Rel** $\cong$ **2-Rel(CmLat)** is given by its φ^{-1} and the following ψ.

$$\varphi_{A,\mathbf{2}^B}: \mathbf{Set}(A, \mathbf{CmLat}(\mathbf{2}^B, \mathbf{2})) \cong \mathbf{CmLat}^{op}(\mathbf{2}^A, \mathbf{2}^B) = \mathbf{CmLat}(\mathbf{2}^B, \mathbf{2}^A)$$
$$\psi\colon \mathbf{Rel}(A, B) \cong \mathbf{CmLat}(\mathbf{2}^B, \mathbf{2}^A)$$
$$\psi(R)(g)(a) = 1 \Leftrightarrow \forall b \in B.(a, b) \in R \,\text{imply}\, g(b) = 1 \quad (R \subseteq A \times B, g \in \mathbf{2}^B, a \in A)$$
$$\psi^{-1}(h) = \{(a, b) \mid \forall g \in \mathbf{2}^B.h(g)(a) = 1 \Rightarrow g(b) = 1\} \qquad (h \in \mathbf{CmLat}(\mathbf{2}^B, \mathbf{2}^A))$$

$\Box$

4.5 Quantale-Valued Relations

A quantale-valued relation (called multi-valued relation) is also an example of V-relation in L.

Definition 6 (quantale). *A* quantale *[13, 16, 17] is a tuple* $(K, \leq, \bigvee, \cdot, 1)$ *with the following properties:*

(a) $(K, \cdot, 1)$ *is a monoid.*
(b) $(K, \leq, \bigvee)$ *is a complete join semilattice.*
(c) $(\bigvee S) \cdot a = \bigvee\{b \cdot a | b \in S\}$ *for each element* a *and each subset* S *of* K.
(d) $a \cdot (\bigvee S) = \bigvee\{a \cdot b | b \in S\}$ *for each element* a *and each subset* S *of* K.

Example 1. The complete join semilattice $\mathbf{2}^A$ of all maps from A to **2** forms a quantale where $a \cdot b$ is given by the pointwise minimum.

Definition 7. *For a quantale* $(K, \leq, \bigvee, \cdot, 1)$*, we define the category* K-**vRel** *of* K*-valued relations by the following data. The associativity of the composition and the unit law are known.*

- *Its objects are sets.*
- *Its arrow* $R\colon A \to B$ *is a function from* $A \times B$ *to* K.
- *For* $Q\colon A \to B$ *and* $R\colon B \to C$*, the composition* $R \circ Q\colon A \to C$ *is defined as follows.*
$$(R \circ Q)(a, c) = \bigvee\{R(b, c) \cdot Q(a, b) \mid b \in B\}$$
- *An identity* $\mathrm{Id}\colon A \to A$ *is defined to be* $\mathrm{Id}(a, a) = 1$ *and* $\mathrm{Id}(a, a') = \bot$ *for* $a \neq a'$. $\bot$ *is the least element* $\bigvee \emptyset$.

For a quantale K, we define the category K-**LMod** as follows.

Definition 8. *For a quantale $(K, \leq, \bigvee, \cdot, 1)$, we define the category K-**LMod** by the following data.*

- *Its object is a left K-module, that is, a tuple $(G, \bigvee_G, \star)$ of $(G, \bigvee_G) \in$ **CjLat** and a function $\star\colon K \times G \to G$ satisfying*
  (a) $1 \star g = g$ for $g \in G$,
  (b) $(k \cdot k') \star g = k \star (k' \star g)$ for $k, k' \in K, g \in G$, and
  (c) $k \star (\bigvee X) = \bigvee\{k \star g \mid g \in X\}$ for $k \in K, X \subseteq G$.
- *Its arrow $h\colon (G, \bigvee_G, \star_G) \to (G', \bigvee_{G'}, \star_{G'})$ is $h\colon (G, \bigvee_G) \to (G', \bigvee_{G'}) \in$ **CjLat** satisfying $h(k \star_G g) = k \star_{G'} h(g)$ for $k \in K, g \in G$.*
- *The composition $\circ$ is the same as the composition of **CjLat**.*
- *Its identities* Id *are the same as the identities of **CjLat**.*

Proposition 10. *For a quantale $(K, \leq, \bigvee, \cdot, 1)$, K-**LMod** has $K = (K, \bigvee, \cdot)$ as an object and powers $K^B = (K^B, \bigvee, \star)$ which consists of the complete join semilattice of maps from B to K and the function $\star\colon K \times K^B \to K^B$, which is defined to be $(k \star g)(b) = k \cdot g(b)$ for $k \in K$, $g \in K^B$, and $b \in B$.*

Proof. K satisfies the condition (a), (b) for objects in Definition 8, since $(K, \cdot, 1)$ is a monoid, and K satisfies the condition (c) by the condition (d) of Definition 6.

Similarly, K^B satisfies the condition (a), (b), (c) for objects in Definition 8. The following natural isomorphism φ_{A,K^B} is defined in the same way as the Proposition 5.

$$\varphi_{A,K^B} : \mathbf{Set}(A, K\text{-}\mathbf{LMod}(K^B, K)) \cong K\text{-}\mathbf{LMod}(K^B, K^A)$$

□

Since K-**LMod** has the object K and powers, K-**Rel**(K-**LMod**) is the category whose arrow is a K-relation in K-**LMod**.

Proposition 11. *For a quantale $(K, \leq, \bigvee, \cdot, 1)$, a K-valued relation corresponds to a K-relation in K-**LMod**, that is to say, we have the following isomorphism.*

$$K\text{-}\mathbf{vRel} \cong K\text{-}\mathbf{Rel}(K\text{-}\mathbf{LMod})$$

Proof. The isomorphism K-**vRel** $\cong$ K-**Rel**(K-**LMod**) is given by φ^{-1} of Proposition 10 and the following ψ. Here, $\delta_b \in K^B$ is defined to be $\delta_b(b) = 1$ and $\delta_b(x) = 0$ if $x \neq b$.

$$\psi\colon K\text{-}\mathbf{vRel}(A, B) \cong K\text{-}\mathbf{LMod}(K^B, K^A)$$
$$\psi(R)(g)(a) = \bigvee\{g(b) \cdot R(a, b) \mid b \in B\} \qquad (R \in K\text{-}\mathbf{vRel}(A, B), g \in K^B, a \in A)$$
$$\psi^{-1}(h)(a, b) = h(\delta_b)(a) \qquad (h \in K\text{-}\mathbf{LMod}(K^B, K^A), a \in A, b \in B)$$

$R \in K$-**vRel**(A, B) implies $\psi(R) \in K$-**LMod**(K^B, K^A) by the following proof for $k \in K, g \in K^B, a \in A$.

$$\begin{aligned}\psi(R)(k \star g)(a) &= \bigvee\{(k \star g)(b) \cdot R(a, b) \mid b \in B\}\\ &= \bigvee\{k \cdot g(b) \cdot R(a, b) \mid b \in B\}\\ &= k \cdot \bigvee\{g(b) \cdot R(a, b) \mid b \in B\}\\ &= k \cdot \psi(R)(g)(a)\\ &= (k \star \psi(R)(g))(a)\end{aligned}$$

We prove $\psi(\psi^{-1}(h)) = h$ and $\psi^{-1}(\psi(R)) = R$ as follows. We use the equality $(\bigvee\{g(b) \star \delta_b \mid b \in B\})(b') = \bigvee\{g(b) \cdot \delta_b(b') \mid b \in B\} = g(b')$ for $b' \in B$.

$$\begin{aligned}\psi(\psi^{-1}(h))(g)(a) &= \bigvee\{g(b) \cdot \psi^{-1}(h)(a, b) \mid b \in B\} \\ &= \bigvee\{g(b) \cdot h(\delta_b)(a) \mid b \in B\} \\ &= \bigvee\{h(g(b) \star \delta_b)(a) \mid b \in B\} \\ &= h(\bigvee\{g(b) \star \delta_b \mid b \in B\})(a) \\ &= h(g)(a)\end{aligned}$$

$$\psi^{-1}(\psi(R))(a, b) = \psi(R)(\delta_b)(a) = \bigvee\{\delta_b(b') \cdot R(a, b') \mid b' \in B\} = R(a, b)$$

ψ preserves the composition, that is to say, we have $\psi(R \circ Q) = \psi(Q) \circ \psi(R)$ for $Q\colon A \to B$ and $R\colon B \to C$ in K-**vRel** as follows.

$$\begin{aligned}\psi(R \circ Q)(g)(a) &= \bigvee\{g(c) \cdot (R \circ Q)(a, c) \mid c \in C\} \\ &= \bigvee\{g(c) \cdot \bigvee\{R(b, c) \cdot Q(a, b) \mid b \in B\} \mid c \in C\} \\ &= \bigvee\{g(c) \cdot R(b, c) \cdot Q(a, b) \mid b \in B, c \in C\} \\ &= \bigvee\{\bigvee\{g(c) \cdot R(b, c) \mid c \in C\} \cdot Q(a, b) \mid b \in B\} \\ &= \bigvee\{\psi(R)(g)(b) \cdot Q(a, b) \mid b \in B\} \\ &= \psi(Q)(\psi(R)(g))(a)\end{aligned}$$

ψ preserves the identities on A, that is to say, we have $\psi(\mathrm{Id})(g) = g$ as follows.

$$\psi(\mathrm{Id})(g)(a) = \bigvee\{g(a') \cdot \mathrm{Id}(a, a') \mid a' \in A\} = g(a)$$

Therefore, we obtain the isomorphism ψ: K-**vRel** $\cong$ K-**Rel**(K-**LMod**). □

4.6 Effect Relations as Probabilistic Relations

Effect algebra [3] and effect module [5,7] have been introduced for the foundation of quantum mechanics and moreover studied in investigations of fuzzy probability theory and quantum logic. We start by recalling the definitions of effect algebra and effect module.

Definition 9. *An effect module is a tuple $(E, \oplus, {}^\perp, 0)$ where $\oplus$ is a partial binary operator, ${}^\perp$ is a unary operator, $0 \in E$ is special element and the following holds:*

1. *$(E, \oplus, {}^\perp, 0)$ is an effect algebra, i.e.*
   - *(a) If $x \oplus y$ is defined, then so is $y \oplus x$ and $x \oplus y = y \oplus x$ holds.*
   - *(b) If $x \oplus y$ and $(x \oplus y) \oplus z$ are defined, then so are $y \oplus z$ and $x \oplus (y \oplus z)$, and $(x \oplus y) \oplus z = x \oplus (y \oplus z)$ holds.*
   - *(c) $0 \oplus x$ is always defined, and equal to x.*
   - *(d) ${}^\perp$ is the orthocomplement, i.e. for each $x \in E$, $x \oplus x^\perp = 1$ where $1 = 0^\perp$.*
   - *(e) $x \oplus 1$ is defined if and only if $x = 0$.*
2. *E carries scalar multiplication $\cdot$, that is to say function $[0, 1] \times E \to E$ satisfying the following conditions:*
   - *(a) $(rs) \cdot x = r \cdot (s \cdot x)$ holds.*

(b) If $r+s \leq 1$ *then* $(r \cdot x) \oplus (s \cdot x)$ *is defined and* $(r+s) \cdot x = (r \cdot x) \oplus (s \cdot x)$ *holds.*
(c) If $x \oplus y$ *is defined, then so is* $(r \cdot x) \oplus (r \cdot y)$ *and* $r \cdot (x \oplus y) = (r \cdot x) \oplus (r \cdot y)$ *holds.*
(d) $1 \cdot x = x$ *holds.*

Each effect algebra is a poset with $\leq$ given by $x \leq y$ iff there exists z such that $x \oplus z = y$. A *map of effect algebras* $f : E \to D$ is a function that preserves 1 and $\oplus$ if defined, i.e. (a) $f(1) = 1$, (b) If $x \oplus y$ is defined then so is $f(x) \oplus f(y)$ and $f(x \oplus y) = f(x) \oplus f(y)$ holds. A *map of effect modules* $f : E \to D$ is a map of effect algebras that preserves scalar multiplication.

A typical example of effect algebras is the unit interval $[0,1]$: if $x+y \leq 1$ then $x \oplus y$ is defined and equal to $x+y$, and $x^{\perp} = 1 - x$. The set of functions $[0,1]^X$ for a set X is also an effect algebra: if $p(x) + q(x) \leq 1$ for all $x \in X$ then $p \oplus q$ is defined by $(p \oplus q)(x) = p(x) + q(x)$, the zero element $\mathbf{0}$ and the orthocomplement $p^{\perp}$ of p are defined by $\mathbf{0}(x) = 0$, $p^{\perp}(x) = 1 - p(x)$ respectively. Effect algebras $[0,1]$ and $[0,1]^X$ have scalar multiplication, so these are effect modules.

Definition 10. *We define the category* **ERel** *of effect relations by the following data*

- *Its objects are sets.*
- *Its arrow* $R : A \to B$ *is a fuzzy relation between a set* A *and an effect module* $[0,1]^B$*, that is to say function* $A \times [0,1]^B \to [0,1]$*, satisfying the following conditions:*
  - *for* $p, q \in [0,1]^B$*, if* $p \oplus q$ *is defined then* $R(a, p \oplus q) = R(a,p) + R(a,q)$*.*
  - $R(a, \mathbf{1}) = 1$ *where* $\mathbf{1} = \mathbf{0}^{\perp}$*.*
- *For* $Q : A \to B$ *and* $R : B \to C$*, the composition* $R \circ Q : A \to C$ *is defined by* $(R \circ Q)(a,q) = Q(a,p)$ *where* $p(b) = R(b,q)$*.*
- *An identity* $\mathrm{Id} : A \to A$ *is defined to be* $\mathrm{Id}(a,p) = p(a)$*.*

An effect relation corresponds to $[0,1]$-relation in **EMod**, where **EMod** is the category of all effect modules and maps among them.

Proposition 12. *An effect relation corresponds to a* $[0,1]$*-relation in* **EMod***, that is to say, we have that* **ERel** $\cong [0,1]$-**Rel**(**EMod**)

Proof. **EMod** has the power $[0,1]^X$. The following natural isomorphism $\varphi_{A,[0,1]^B}$ is defined in the same way as the Proposition 5.

$$\begin{aligned}\varphi_{A,[0,1]^B} : \mathbf{Set}(A, \mathbf{EMod}([0,1]^B, [0,1])) &\cong \mathbf{EMod}^{op}([0,1]^A, [0,1]^B) \\ &= \mathbf{EMod}([0,1]^B, [0,1]^A)\end{aligned}$$

The isomorphism **ERel** $\cong [0,1]$-**Rel**(**EMod**) is given by the above φ^{-1} and the following ψ.

$$\begin{aligned}&\psi : \mathbf{ERel}(A,B) \cong \mathbf{EMod}([0,1]^B, [0,1]^A) \\ &\psi(R)(p)(a) = R(a,p) \qquad (R \in \mathbf{ERel}(A,B), p \in [0,1]^B, a \in A) \\ &\psi^{-1}(h)(a,p) = h(p)(a) \qquad (h \in \mathbf{EMod}([0,1]^B, [0,1]^A), a \in A, p \in [0,1]^B)\end{aligned}$$

□

Actually $\mathcal{E}(X) = \mathbf{EMod}([0,1]^X, [0,1])$ is known as *the expectation monad.* If a set X is finite, $\mathcal{E}(X) = \mathcal{D}(X)$ for the subdistribution monad $\mathcal{D}$. For details, see [5,7].

5 Examples of Composition of Different Relational Notions

5.1 Up-Closed Multirelation as Composition of Two Binary Relations

An interesting example of an up-closed multirelation is a 2-player game. Here, a 2-player game has two players A, B and it consists of two state sets S_A, S_B and two binary relations $R_A \subseteq S_A \times S_B$ and $R_B \subseteq S_B \times S_A$. Assume that S_A and S_B are disjoint. The whole state set of the game is $S_A \cup S_B$. When a current state s of the game belongs to S_A, the player A can choose the next state in S_B which is related from s by R_A. Conversely, when a current state belongs to S_B, the player B can choose the next reachable state by R_B. A 2-player game is often used to model a kind of system. Then, the player A represents the system itself and the player B represents the external environment. Therefore, R_A is called an *internal choice* and R_B is called an *external choice.* Modelling a system by a 2-player game, one can model the expected property of the system by the uniform statement whether the system can eventually reach to the desirable states by choosing states according to the internal choice relation, regardless of states chosen by the external choice relation. For example, we define a 2-player game by $S_A = \{a_1, a_2, a_3\}$, $S_B = \{b_1, b_2\}$, $R_A = \{(a_1, b_1), (a_1, b_2)\}$, and $R_B = \{(b_1, a_1), (b_1, a_2), (b_2, a_2), (b_2, a_3)\}$. The above 2-player game can be represented by the following up-closed multirelation $R_{A,B} \subseteq S_A \times \wp(S_A)$.

$$R_{A,B} = \{(a_1, \{a_1, a_2\}), (a_1, \{a_2, a_3\}), (a_1, \{a_1, a_2, a_3\})\}$$

Here, let a_1 be the initial state and let a_2, a_3 be desirable states for A. This system can eventually reach to the desirable states by choosing states according to the internal choice relation, regardless of states chosen by the external choice relation, since the pair of the initial state and the set of desirable states $(a_1, \{a_2, a_3\})$ is an element of $R_{A,B}$.

The above construction of $R_{A,B}$ from R_A and R_B can be modeled by Proposition 4, that is to say, our way of composing different types of relational notions. There are forgetful functors U_1, U_2 satisfying the following commutative diagrams. Two necessary natural isomorphisms are identities.

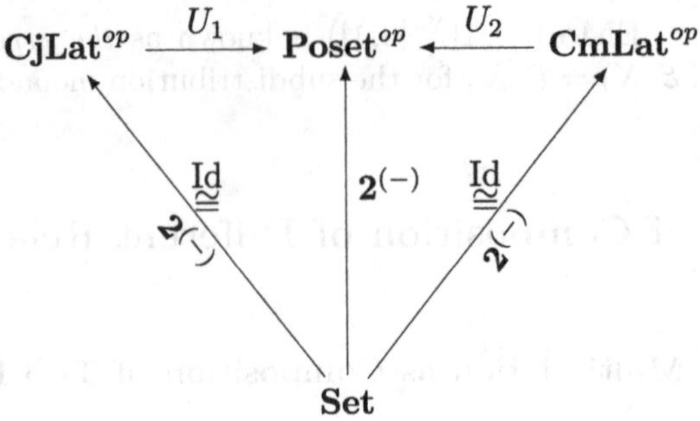

We write χ_1 for the isomorphism **Rel** $\cong$ **2-Rel**(**CjLat**) of Proposition 8, χ_2 for the isomorphism **Rel** $\cong$ **2-Rel**(**CmLat**) of Proposition 9, and χ_3 for the isomorphism **UMRel** $\cong$ **2-Rel**(**Poset**) of Proposition 6. The functor $\chi_3^{-1} \circ \tilde{U_1} \circ \chi_1$: **Rel** $\to$ **UMRel** is known as the *angelic lifting* and the functor $\chi_3^{-1} \circ \tilde{U_2} \circ \chi_2$: **Rel** $\to$ **UMRel** is known as the *demonic lifting* [1,11]. By Proposition 4, we obtain $\tilde{U_2}(\chi_2(R_B)) \circ \tilde{U_1}(\chi_1(R_A)) \colon S_A \to S_A$ in **2-Rel**(**Poset**), which is equal to $\chi_3(R_{A,B})$ for the above up-closed multirelation $R_{A,B}$.

5.2 Composition of Quantale-Valued Relations for Different Quantales

Composition of different types of quantale-valued relations can be modeled by our composition, too. We write K^n for the the product of n copies of a quantale K. We can get a K^{12}-valued relation as the composition of a K^4-valued relation $R_1 \colon A \times B \to K^4$ and a K^6-valued relation $R_2 \colon B \times C \to K^6$, by embedding them with 3 times pairing $K^4 \to K^{12}$ and 2 times pairing $K^6 \to K^{12}$, respectively.

This composition is an example of Proposition 4. By Sect. 4.5, we have K-**vRel** $\cong$ K-**Rel**(K-**LMod**). There are two faithful functors $(-)^3, (-)^2$ and natural isomorphisms $\alpha_X \colon ((K^4)^X)^3 \cong (K^{12})^X$, $\beta_X \colon ((K^6)^X)^2 \cong (K^{12})^X$ as follows.

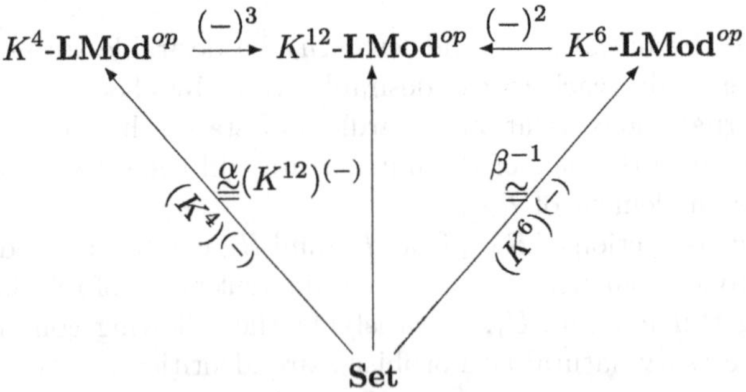

The functor $(-)^n$ maps an object $(G, \bigvee, \star) \in K^m$-**LMod** to $(G^n, \bigvee^n, \star^n)$ where $\star^n$ is regarded as $\star^n \colon K^{m \times n} \times G^n \cong (K^m \times G)^n \to G^n$. On the arrow part, $(-)^n$ maps h to h^n. Therefore, it is faithful. By Proposition 4, we obtain $(\tilde{R_2})^2 \circ (\tilde{R_1})^3 \colon A \to C$ in K^{12}-**Rel**(K^{12}-**LMod**), which corresponds to the above composition of R_1 and R_2 as K^{12}-valued relations.

Note that the composition depends on two natural isomorphisms α, β. For example, let $A = B = C = \{*\}$, $R_1(*, *) = (k_1, \cdots, k_4)$, $R_2(*, *) = (k'_1, \cdots, k'_6)$, and $\alpha((x_1, \cdots, x_4), (x_5, \cdots, x_8), (x_9, \cdots, x_{12})) = (x_1, \cdots, x_{12})$.

If $\beta((x'_1, \cdots, x'_6), (x'_7, \cdots, x'_{12})) = (x'_1, \cdots, x'_{12})$, then $((\tilde{R_2})^2 \circ (\tilde{R_1})^3)(*, *) = (k'_1 \cdot k_1, k'_2 \cdot k_2, k'_3 \cdot k_3, k'_4 \cdot k_4, k'_5 \cdot k_1, k'_6 \cdot k_2, k'_1 \cdot k_3, k'_2 \cdot k_4, k'_3 \cdot k_1, k'_4 \cdot k_2, k'_5 \cdot k_3, k'_6 \cdot k_4)$. On the other hand, if $\beta((x'_1, \cdots, x'_6), (x'_7, \cdots, x'_{12})) = (x'_{12}, \cdots, x'_1)$, then $((\tilde{R_2})^2 \circ (\tilde{R_1})^3)(*, *) = (k'_6 \cdot k_1, k'_5 \cdot k_2, k'_4 \cdot k_3, k'_3 \cdot k_4, k'_2 \cdot k_1, k'_1 \cdot k_2, k'_6 \cdot k_3, k'_5 \cdot k_4, k'_4 \cdot k_1, k'_3 \cdot k_2, k'_2 \cdot k_3, k'_1 \cdot k_4)$.

5.3 Composition of Multirelation and Quantale-Valued Relation

Our theory can model composition of an ordinary multirelation and $\mathbf{2}^n$-valued relation. There are the forgetful functor U, the n-product functor $(-)^n$, a natural isomorphism $\alpha_X \colon (2^X)^n \cong (2^n)^X$, and the identity $\mathrm{Id}_X \colon U((\mathbf{2}^n)^X) \cong (2^n)^X$.

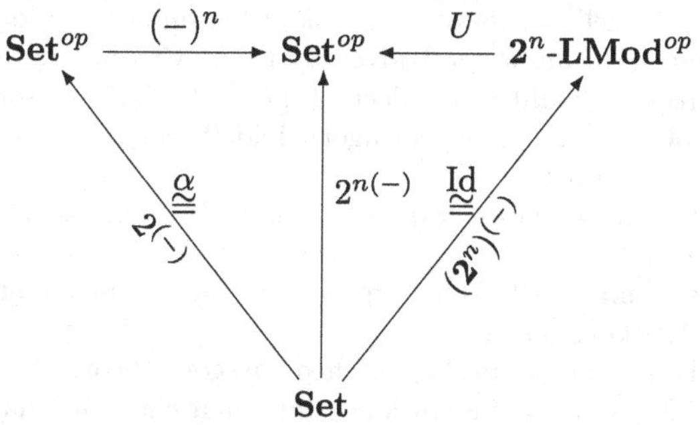

By Proposition 4, we obtain 2^n-relation in **Set** from an ordinary multirelation and $\mathbf{2}^n$-valued relation. For example, from $R_1 \subseteq A \times \wp(B)$ and $R_2 \colon B \times C \to \mathbf{2}^3$, we obtain a 2^3-relation in **Set**, that is to say, a function $A \to \mathbf{Set}((2^3)^C, 2^3)$ which depends on R_1, R_2, α.

6 Conclusion and Future Work

The contribution of this paper is to give the way of composing different types of relational notions, for example, ordinary binary relations, up-closed multirelations, ordinary multirelations, quantale-valued relations, and effect relations. Our key idea is to represent a relational notion as a Kleisli arrow for some continuation monad and to compose those different relational arrows via identity-on-object faithful functors between different Kleisli categories.

In fact, it is already known that each monad on **Set** can be regarded as a continuation monad, since its Eilenberg-Moore category has a right adjoint to **Set** and any right adjoint to **Set** is representable. However, a monad has multiple representations as a continuation monad. It is future work to clear what is the best representation as a continuation monad for each relational notion.

Acknowledgements. The authors thank Tatsuya Abe, Hitoshi Furusawa, Walter Guttmann, Masahito Hasegawa, Shinya Katsumata, and Georg Struth for valuable comments. The second author would like to thank Ichiro Hasuo and members of his group at the University of Tokyo for their generous support. This work was supported by JSPS KAKENHI Grant Numbers JP24700017, JP16K21557.

References

1. Back, R.J., von Wright, J.: Refinement Calculus: A Systematic Introduction. Graduate Texts in Computer Science. Springer, New York (1998). https://doi.org/10.1007/978-1-4612-1674-2
2. Berghammer, R., Guttmann, W.: A relation-algebraic approach to multirelations and predicate transformers. In: Hinze, R., Voigtländer, J. (eds.) MPC 2015. LNCS, vol. 9129, pp. 50–70. Springer, Cham (2015). https://doi.org/10.1007/978-3-319-19797-5_3
3. Foulis, D., Bennett, M.K.: Effect algebras and unsharp quantum logics. Found. Phys. **24**(10), 1331–1352 (1994)
4. Hudak, P.: The Haskell School of Expression: Learning Functional Programming Through Multimedia. Cambridge University Press, Cambridge (2000)
5. Jacobs, B.: Convexity, duality and effects. In: Calude, C.S., Sassone, V. (eds.) TCS 2010. IAICT, vol. 323, pp. 1–19. Springer, Heidelberg (2010). https://doi.org/10.1007/978-3-642-15240-5_1
6. Jacobs, B.: A recipe for state-and-effect triangles. Log. Methods Comput. Sci. **13**(2), 1–26 (2017)
7. Jacobs, B., Mandemaker, J.: The expectation monad in quantum foundations. CoRR abs/1112.3805 (2011)
8. Kelly, G.M.: The basic concepts of enriched category theory, January 2005
9. Mac Lane, S.: Categories for the Working Mathematician. Graduate Texts in Mathematics. Springer, New York (1971). https://doi.org/10.1007/978-1-4612-9839-7
10. Martin, C.E., Curtis, S.A.: The algebra of multirelations. Math. Struct. Comput. Sci. **23**, 635–674 (2013)
11. Martin, C., Curtis, S., Rewitzky, I.: Modelling angelic and demonic nondeterminism with multirelations. Sci. Comput. Program. **65**(2), 140–158 (2007)
12. Moggi, E.: Notions of computation and monads. Inf. Comput. **93**(1), 55–92 (1991)
13. Mulvey, C.J.: & . In: Second Topology Conference. Rendiconti del Circolo Matematico di Palermo, ser. 2, supplement no. 12, pp. 99–104 (1986)
14. Parikh, R.: The logic of games. Ann. Descrete Math. **24**, 111–140 (1985)
15. Rewitzky, I., Brink, C.: Monotone predicate transformers as up-closed multirelations. In: Schmidt, R.A. (ed.) RelMiCS 2006. LNCS, vol. 4136, pp. 311–327. Springer, Heidelberg (2006). https://doi.org/10.1007/11828563_21
16. Rosenthal, K.I.: Quantales and Their Applications. Pitman Research Notes in Mathematical Series. Longman Scientific & Technical, New York, Harlow (1990)
17. Stubbe, I.: Q-modules are Q-suplattices. Theory Appl. Categ. **19**(4), 50–60 (2007)

Axiomatizing Discrete Spatial Relations

Giulia Sindoni[1(✉)], Katsuhiko Sano[2], and John G. Stell[1]

[1] School of Computing, University of Leeds, Leeds LS2 9JT, UK
{scgsi,j.g.stell}@leeds.ac.uk
[2] Department of Philosophy, Graduate School of Letters,
Hokkaido University, Sapporo, Japan
v-sano@let.hokudai.ac.jp

Abstract. Qualitative spatial relations are used in artificial intelligence to model commonsense notions such as regions of space overlapping, touching only at their boundaries, or being separate. In this paper we extend earlier work on qualitative relations in discrete space by presenting a bi-intuitionistic modal logic with universal modalities, called UBiSKt. This logic has a semantics in which formulae are interpreted as subgraphs. We show how a variety of qualitative spatial relations can be defined in UBiSKt. We make essential use of a sound and complete axiomatisation of the logic and an implementation of a tableau based theorem prover to establish novel properties of these spatial relations. We also explore the role of UBiSKt in expressing spatial relations at more than one level of detail. The features of the logic allow it to represent how a subgraph at a detailed level is approximated at a coarser level.

Keywords: Spatial relations · Discrete space
Intuitionistic-modal logic · Qualitative representation and reasoning

1 Introduction

1.1 The RCC and the Problem of Discrete Space

Qualitative spatial relations are used in artificial intelligence to model commonsense notions such as regions of space overlapping, touching only at their boundaries, or being separate. Qualitative approaches, as opposed to quantitative approaches, abstract from numerical information, which is often unnecessary and sometimes unavailable at the human level. These approaches have become popular in areas like AI and robot navigation, GIS (Geographical Information Systems) and Image Understanding. For a survey on the qualitative representation of spatial knowledge and examples of the problems that can be addressed

K. Sano—Partially supported by JSPS KAKENHI Grant-in-Aid for Young Scientists (B) Grant Number 15K21025 and Grant-in-Aid for Scientific Research (B) Grant Number 17H02258, and JSPS Core-to-Core Program (A. Advanced Research Networks).

J. Desharnais et al. (Eds.): RAMiCS 2018, LNCS 11194, pp. 113–130, 2018.
https://doi.org/10.1007/978-3-030-02149-8_8

using these approaches we refer the reader to [4]. Various spatial calculi have been developed including the Region-Connection-Calculus (RCC) [13] and the 9-intersection model [7]. The RCC is a first-order logical theory where a primitive predicate of Connection, C, between regions of the space. From Connection a notion of Parthood is defined by $P(x, y)$ iff $\forall z(C(x, z) \Rightarrow C(y, z))$. Using Connection and Parthood, a set of eight Jointly Exhaustive and Pairwise Disjoint Spatial Relations between regions is obtained. This is known as the RCC-8.

The RCC-8 can distinguish Non-Tangential Proper Part ($NTPP$) from Tangential Proper Part (TPP). RCC-8 can express the relation of sharing only a part, or Partial Overlapping (PO), as well as connection on the boundaries, or External Connection (EC) as well as disjointness, or Disconnection (DC). Equality (EQ), and the inverses of TPP and $NTPP$ are also included. Cohn and Varzi in [5] show that the RCC can be interpreted in a topological space where regions of the space are certain non-empty subsets of the space, and an operator of Kuratowski closure c is used to define Connection. Three notions of connection are proposed there: $C_1(x, y) \Leftrightarrow x \cap y \neq \varnothing$, $C_2(x, y) \Leftrightarrow c(x) \cap y \neq \varnothing$ or $x \cap c(y) \neq \varnothing$ and $C_3(x, y) \Leftrightarrow c(x) \cap c(y) \neq \varnothing$. Although the RCC aims to be neutral about whether space is dense, that is whether space can be repeatedly sub-divided *ad infinitum*, it is well known that, once atomic regions are allowed, giving non-empty regions of the space without any proper parts, the RCC theory becomes contradictory [13]. Moreover, as noticed in [18], the use of Kuratowski topological closure prevents the expression of a natural form of connection in a discrete space even in some simple examples.

The ability to reason about discrete space is important in many fields and applications. Any kind of transport network (road networks, railway networks, airlines network) is naturally represented by discrete structures such as graphs. Images in image processing lie in discrete spaces in the form of pixel arrays. Geographical data represented digitally are essentially discrete both in vector and raster formats as ultimately there is a limit to resolution.

1.2 Related Work

Galton [8] studied a notion of connection between subsets of a particular kind of discrete space, known as Adjacency Space. This is a set N together with a relation of adjacency $\alpha \subseteq N \times N$. N can be thought of as a set of pixels, following the approach of Rosenfeld's Digital Topology [15]. A single pixel is the atomic region of the space. The relation α is symmetric and reflexive, but not necessarily transitive. Connection, C_α, is defined for subsets $X, Y \subseteq N$ by $C_\alpha(X, Y)$ if there are $a \in X$ and $b \in Y$ such that $(a, b) \in \alpha$. From this eight spatial relations between regions are obtained in a first-order logical theory, a discrete version of the RCC-8, and employed in the context of correction of segmentation errors in histological images [14].

Galton's discrete space (N, α) can be regarded as a graph where N is the set of nodes and the relation $\alpha \subset N \times N$ gives the edges. There is a notable difference between theory of adjacency space and graph theory [9]. A substructure of an adjacency space can be specified just in terms of nodes, two nodes being connected by only one edge, or relation of adjacency. This is not true in the general setting of a multigraph, where multiple edges may occur between two nodes, and, therefore, different subgraphs sharing the same set of nodes may be considered. Cousty et al. [6] argue that edges need to play a more central role, and make the key observation that sets of nodes which differ only in their edges need to be regarded as distinct. This generality appears also important in examples such as needing to model two distinct roads between the same endpoints, or distinct rail connections between the same two stations. The logic used in the present paper has a semantics in which formulae are interpreted as down-closed sets arising from a set U together with a pre-order H. A special case of a set with a pre-order is a graph, and formulae are interpreted as its subgraphs. Discrete regions are seen as subgraphs and they are more general than Galton's construction. We allow graphs to have multiple edges between the same pair of nodes, thus using a structure sometime called a multigraph.

This paper extends the work in [18] where spatial relations between discrete regions are expressed in a logic, which here we denote **UBiSKt**. Here we provide a sound, complete and decidable axiomatization of the logic **UBiSKt** and a tableau calculus, extending that in [19], thus providing a computational tool for performing discrete spatial reasoning, which we have proved to be equivalent to the axiomatic proof-system. We begin here to use these new tools by exploring the role of **UBiSKt** in expressing spatial relations at more than one level of detail. The features of the logic and its connection with mathematical morphology are essential in this. This opens many directions for future work, including studying the different notions of approximation expressible in the logic, and being able to reason with spatial relations between regions at different levels of detail.

The paper is structured as follows. Section 2 introduces **UBiSKt**, a bi-intuitionistic modal logic with universal modalities, and a sound and complete axiomatization is given. Section 3 presents **UBiSKt** as a logic for graphs and shows a set of Spatial Relations between subgraphs expressed as formulae in the logic. Then we prove spatial entailments between properties of subgraphs using the axiomatization. In Sect. 4, we show that, given a subgraph at a detailed level, the logic allows to approximate it at a coarser level. Connection and other spatial relations between these approximated regions are also expressed. Finally we provide conclusions and further works.

2 The Logic UBiSKt

2.1 Kripke Semantics for UBiSKt

Let Prop be a countable set of propositional variables. Our syntax $\mathcal{L}$ for bi-intuitionistic stable tense logic with universal modalities consists of all logical connectives of bi-intuitionistic logic, i.e., two constant symbols $\bot$ and $\top$, disjunction $\vee$, conjunction $\wedge$, implication $\rightarrow$, coimplication $\prec$, and a finite set $\{\blacklozenge, \square, \mathsf{A}, \mathsf{E}\}$ of modal operators. The set $\mathsf{Form}_{\mathcal{L}}$ of all formulae in $\mathcal{L}$ is defined inductively as follows:

$$\varphi ::= \top \mid \bot \mid p \mid \varphi \wedge \varphi \mid \varphi \vee \varphi \mid \varphi \rightarrow \varphi \mid \varphi \prec \varphi \mid \blacklozenge \varphi \mid \Box \varphi \mid \mathsf{E}\varphi \mid \mathsf{A}\varphi \quad (p \in \mathsf{Prop}).$$

We define the following abbreviations:

$$\neg\varphi := \varphi \rightarrow \bot, \quad \lrcorner\varphi := \top \prec \varphi, \quad \varphi \leftrightarrow \psi := (\varphi \rightarrow \psi) \wedge (\psi \rightarrow \varphi),$$
$$\Diamond\varphi := \lrcorner\Box\neg\varphi, \quad \blacksquare\varphi := \neg\blacklozenge\lrcorner\varphi.$$

Definition 1 ([19]). *Let H be a preorder on a set U. We say that $X \subseteq U$ is an H-*set *if X is closed under H-successors, i.e., uHv and $u \in X$ jointly imply $v \in X$ for all elements u, $v \in U$. Given a preorder (U, H), a binary relation $R \subseteq U \times U$ is* stable *if it satisfies $H; R; H \subseteq R$.*

It is easy to see that a relation R on U is stable, if and only if, $R; H \subseteq R$ and $H; R \subseteq R$. Given any binary relation R on U, $\breve{R}$ is defined as the converse of R in the usual sense. Even if R is a stable relation on U, its converse $\breve{R}$ may be not stable.

Definition 2 ([19]). *The* left converse $\circlearrowleft R$ *of a stable relation R is $H; \breve{R}; H$.*

Definition 3. *We say that $F = (U, H, R)$ is an H-*frame *if U is a nonempty set, H is a preorder on U, and R is a* stable *binary relation on U. A* valuation *on an H-frame $F = (U, H, R)$ is a mapping V from* Prop *to the set of all H-sets on U. $M = (F, V)$ is an H-*model *if $F = (U, H, R)$ is an H-frame and V is a valuation. Given an H-model $M = (U, H, R, V)$, a state $u \in U$ and a formula φ, the satisfaction relation $M, u \models \varphi$ is defined inductively as follows:*

$$\begin{array}{lll}
M,u \models p & \iff & u \in V(p), \\
M,u \models \top, & & \\
M,u \not\models \bot, & & \\
M,u \models \varphi \vee \psi & \iff & M,u \models \varphi \text{ or } M,u \models \psi, \\
M,u \models \varphi \wedge \psi & \iff & M,u \models \varphi \text{ and } M,u \models \psi, \\
M,u \models \varphi \rightarrow \psi & \iff & \text{For all } v \in U\ ((uHv \text{ and } M,v \models \varphi) \text{ imply } M,v \models \psi), \\
M,u \models \varphi \prec \psi & \iff & \text{For some } v \in U\ (vHu \text{ and } M,v \models \varphi \text{ and } M,v \not\models \psi), \\
M,u \models \blacklozenge\varphi & \iff & \text{For some } v \in U\ (vRu \text{ and } M,v \models \varphi), \\
M,u \models \Box\varphi & \iff & \text{For all } v \in U\ (uRv \text{ implies } M,v \models \varphi), \\
M,u \models \mathsf{E}\varphi & \iff & \text{For some } v \in U\ (M,v \models \varphi), \\
M,u \models \mathsf{A}\varphi & \iff & \text{For all } v \in U\ (M,v \models \varphi).
\end{array}$$

The truth set *$[\![\varphi]\!]_M$ of a formula φ in an H-model M is defined by $[\![\varphi]\!]_M := \{u \in U \mid M,u \models \varphi\}$. If the underlying model M in $[\![\varphi]\!]_M$ is clear from the context, we drop the subscript and simply write $[\![\varphi]\!]$. We write $M \models \varphi$ (read: 'φ is valid in M') to mean that $[\![\varphi]\!]_M = U$ or $M,u \models \varphi$ for all states $u \in U$. For a set Γ of formulae, $M \models \Gamma$ means that $M \models \gamma$ for all $\gamma \in \Gamma$. Given any H-frame $F = (U, H, R)$, we say that a formula φ is* valid *in F (written: $F \models \varphi$) if $(F, V) \models \varphi$ for any valuation V and any state $u \in U$, i.e., $[\![\varphi]\!]_{(F,V)} = U$.*

As for the abbreviated symbols, we may derive the following satisfaction conditions:

$$
\begin{array}{lcl}
M,u \models \neg\varphi & \iff & \text{For all } v \in U \; (uHv \text{ implies } M,v \not\models \varphi),\\
M,u \models \lrcorner\varphi & \iff & \text{For some } v \in U \; (vHu \text{ and } M,v \not\models \varphi),\\
M,u \models \Diamond\varphi & \iff & \text{For some } v \in U \; ((v,u) \in \curvearrowright R \text{ and } M,v \models \varphi),\\
M,u \models \blacksquare\varphi & \iff & \text{For all } v \in U \; ((u,v) \in \curvearrowright R \text{ implies } M,v \models \varphi).
\end{array}
$$

Proposition 1. *Given any H-model M, the truth set* $[\![\varphi]\!]_M$ *is an H-set.*

Proof. By induction on φ. When φ is of the form $\mathsf{E}\,\psi$, $\mathsf{A}\,\psi$, we remark that $[\![\varphi]\!]_M = U$ or $\varnothing$, which are both trivially H-sets. □

Definition 4. *Given a set* $\Gamma \cup \{\varphi\}$ *of formulae,* φ *is a* semantic consequence *of* Γ *(notation:* $\Gamma \models \varphi$*) if, whenever* $M,u \models \gamma$ *for all* $\gamma \in \Gamma$, $M,u \models \varphi$ *holds, for all H-models* $M = (U,H,R,V)$ *and all states* $u \in U$*. When* Γ *is a singleton* $\{\psi\}$ *of formulae, we simply write* $\psi \models \varphi$ *instead of* $\{\psi\} \models \varphi$*. When both* $\varphi \models \psi$ *and* $\psi \models \varphi$ *hold, we use* $\varphi \mathrel{=\!|\!\models} \psi$ *to mean that they are equivalent with each other. When* Γ *is empty, we also simply write* $\models \varphi$ *instead of* $\varnothing \models \varphi$.

Table 1. Hilbert system **HUBiSKt**

Axioms and Rules for Intuitionistic Logic			
(A0)	$p \to (q \to p)$		
(A1)	$(p \to (q \to r)) \to ((p \to q) \to (p \to r))$		
(A2)	$p \to (p \vee q)$	(A3)	$q \to (p \vee q)$
(A4)	$(p \to r) \to ((q \to r) \to (p \vee q \to r))$	(A5)	$(p \wedge q) \to p$
(A6)	$(p \wedge q) \to q$	(A7)	$(p \to (q \to p \wedge q))$
(A8)	$\bot \to p$	(A9)	$p \to \top$
(MP)	From φ and $\varphi \to \psi$, infer ψ		
(US)	From φ, infer a substitution instance φ' of φ		
Additional Axioms and Rules for Bi-intuitionistic Logic			
(A10)	$p \to (q \vee (p \prec q))$	(A11)	$((q \vee r) \prec q) \to r$
(Mon-$\prec$)	From $\delta_1 \to \delta_2$, infer $(\delta_1 \prec \psi) \to (\delta_2 \prec \psi)$		
Additional Axioms and Rules for Tense Operators			
(A12)	$p \to \Box\blacklozenge p$	(A13)	$\blacklozenge\Box p \to p$
(Mon$\Box$)	From $\varphi \to \psi$, infer $\Box\varphi \to \Box\psi$	(Mon$\blacklozenge$)	From $\varphi \to \psi$, infer $\blacklozenge\varphi \to \blacklozenge\psi$
Additional Axioms and Rules for Universal Modalities			
(A14)	$p \to \mathsf{A}\mathsf{E}p$	(A15)	$\mathsf{E}\mathsf{A}p \to p$
(A16)	$\mathsf{A}p \to p$	(A17)	$\mathsf{A}p \to \mathsf{A}\mathsf{A}p$
(A18)	$\mathsf{A}\neg p \leftrightarrow \neg\,\mathsf{E}p$	(A19)	$(\mathsf{A}p \wedge \mathsf{E}q) \to \mathsf{E}(p \wedge q)$
(A20)	$\mathsf{A}p \to \Box p$	(A21)	$(\mathsf{A}p \wedge \blacklozenge q) \to \blacklozenge(p \wedge q)$
(A22)	$(\mathsf{A}p \wedge (q \prec r)) \to ((p \wedge q) \prec r)$		
(Mon A)	From $\varphi \to \psi$, infer $\mathsf{A}\varphi \to \mathsf{A}\psi$	(Mon E)	From $\varphi \to \psi$, infer $\mathsf{E}\varphi \to \mathsf{E}\psi$

2.2 Hilbert System of Bi-intuitionistic Stable Tense Logic with Universal Modalities

Table 1 provides the Hilbert system **HUBiSKt**. Roughly speaking, it is a bi-intuitionistic tense analogue of a Hilbert system for the ordinary modal logic with the universal modalities [10] (see also [1, p. 417]). In what follows in this paper, we assume that the reader is familiar with theorems and derived inference rules in intuitionistic logic. We define the notion of *theoremhood* in **HUBiSKt** as usual and write $\vdash_{\mathbf{HUBiSKt}} \varphi$ to mean that φ is a theorem of **HUBiSKt**. We say that φ is *provable* from Γ (notation: $\Gamma \vdash_{\mathbf{HUBiSKt}} \varphi$) if there is a finite set $\Gamma' \subseteq \Gamma$ such that $\vdash_{\mathbf{HUBiSKt}} \bigwedge \Gamma' \to \varphi$, where $\bigwedge \Gamma'$ is the conjunction of all elements of Γ' and $\bigwedge \Gamma' := \top$ when Γ' is an emptyset. When no confusion arises, we often simply write $\vdash \varphi$ and $\Gamma \vdash \varphi$ instead of $\vdash_{\mathbf{HUBiSKt}} \varphi$ and $\Gamma \vdash_{\mathbf{HUBiSKt}} \varphi$, respectively.

Theorem 1 (Soundness). *Given any formula φ, $\vdash_{\mathbf{HUBiSKt}} \varphi$ implies $\models \varphi$.*

Proof. Since **HUBiSKt** *without* universal modalities are already shown to be sound in [16], we focus on some of the new axioms and rules. Let $M = (U, H, R, V)$ be an H-model. Validity of axioms (A19), (A20) and (A22) are shown by the fact that $M, x \models p$ implies $[\![\mathsf{A} p]\!]_M = U$ for every $x \in U$. Let us check the validity of (A18) in detail. To show $\models \mathsf{A} \neg p \leftrightarrow \neg \mathsf{E} p$, it suffices to show $\mathsf{A} \neg p \,=\!\!|\!\models \neg \mathsf{E} p$. Fix any $x \in U$. Assume that $M, x \models \mathsf{A} \neg p$, which implies $[\![\neg p]\!]_M = U$. To show $M, x \models \neg \mathsf{E} p$, fix any $y \in U$ such that xHy. Our goal is to show $M, y \not\models \mathsf{E} p$, i.e., $V(p) = \varnothing$. But this is an easy consequence from $[\![\neg p]\!]_M = U$. Conversely, assume that $M, x \models \neg \mathsf{E} p$. Then $M, x \not\models \mathsf{E} p$ by xHx. This implies $V(p) = \varnothing$. To show $M, x \models \mathsf{A} \neg p$, fix any $y \in U$. Our goal is to establish $M, y \models \neg p$. But this is easy from $V(p) = \varnothing$. □

Our proofs of the following theorems and proposition can be found at [17].

Proposition 2. *All the following hold for* **HUBiSKt***.*

1. $\vdash (\psi \prec \gamma) \to \rho$ *iff* $\vdash \psi \to (\gamma \vee \rho)$.
2. *If* $\vdash \varphi \leftrightarrow \psi$ *then* $\vdash (\gamma \prec \varphi) \leftrightarrow (\gamma \prec \psi)$.
3. $\vdash \neg(\varphi \prec \varphi)$.
4. $\vdash \varphi \vee \neg\!\!\!\!\!\!\lrcorner\, \varphi$.
5. $\vdash \lrcorner \lrcorner \varphi \to \varphi$.
6. $\vdash \neg \varphi \to \lrcorner \varphi$.
7. $\vdash \varphi \to \neg \psi$ *iff* $\vdash \psi \to \neg \varphi$.
8. $\vdash \lrcorner \varphi \to \psi$ *iff* $\vdash \lrcorner \psi \to \varphi$.
9. $\vdash \lrcorner \neg \varphi \to \psi$ *iff* $\vdash \varphi \to \neg \lrcorner \psi$.
10. $\vdash \varphi \to \lrcorner \neg \varphi$.
11. $\vdash \neg \lrcorner \varphi \to \varphi$.
12. *If* $\vdash \varphi \to \psi$ *then* $\vdash \lrcorner \psi \to \lrcorner \varphi$.
13. $\vdash \lrcorner(\varphi \wedge \lrcorner \varphi)$.
14. $\vdash \mathsf{E} \varphi \to \psi$ *iff* $\vdash \varphi \to \mathsf{A} \psi$.
15. $\vdash \varphi \to \mathsf{E} \varphi$.
16. $\vdash \mathsf{E} \mathsf{E} \varphi \to \mathsf{E} \varphi$.
17. $\vdash \mathsf{A} \mathsf{E} \varphi \leftrightarrow \mathsf{E} \varphi$.
18. $\vdash \neg \mathsf{A} \varphi \leftrightarrow \lrcorner \mathsf{A} \varphi$.
19. $\vdash \mathsf{A} \varphi \vee \neg \mathsf{A} \varphi$.
20. $\vdash \neg \mathsf{E} \varphi \leftrightarrow \lrcorner \mathsf{E} \varphi$.
21. $\vdash \mathsf{E} \varphi \vee \neg \mathsf{E} \varphi$.
22. $\vdash \mathsf{E} \varphi \leftrightarrow \neg \mathsf{A} \neg \varphi$.
23. $\vdash \mathsf{A}(\lrcorner \varphi \to \psi) \leftrightarrow \mathsf{A}(\lrcorner \psi \to \varphi)$.
24. $\vdash \mathsf{E}(\lrcorner \neg \varphi \wedge \psi) \leftrightarrow \mathsf{E}(\varphi \wedge \lrcorner \neg \psi)$.

Theorem 2 (Strong Completeness of HUBiSKt). *If $\Gamma \models \varphi$ then $\Gamma \vdash_{\mathbf{HUBiSKt}} \varphi$, for every set $\Gamma \cup \{\varphi\}$ of formulae.*

Theorem 3 (Decidability of HUBiSKt). *For every non-theorem φ of* HUBiSKt*, there is a finite frame F such that $F \not\models \varphi$. Therefore,* HUBiSKt *is decidable.*

2.3 Tableau-System for UBiSKt

TabUBiSKt is a tableau-system for **UBiSKt**. It has been also implemented using the theorem-prover generator *MetTel* [21]. Our implementation of TabUBiSKt is available at [17]. We are going to show that TabUBiSKt is equipollent with HUBiSKt, and so the tableau with its implementation can be seen as a computational tool for reasoning with **UBiSKt**. Expressions in the calculus have one of these forms:

$$s : S\varphi \qquad \bot \qquad sHt \qquad sRt \qquad s \approx t \qquad s \not\approx t$$

where S denotes a sign, either T for true or F for false, and s, t are names or labels from a fixed set Label in the tableau language whose intended meaning are elements of U.

Let TabUBiSKt be the extension of TabBiSKt, as described in [19] plus the following rules (for the full tableau calculus, see the manuscript at [17]):

$$\frac{s : T(\mathsf{A}\,\varphi), \quad t : S\psi}{t : T\varphi}\ (T\,\mathsf{A}) \qquad \frac{s : F(\mathsf{A}\,\varphi)}{m : F\varphi}\ (F\,\mathsf{A})\ m \text{ is fresh in the branch}$$

$$\frac{s : T(\mathsf{E}\,\varphi)}{m : T\varphi}\ (T\,\mathsf{E})\ m \text{ is fresh in the branch} \qquad \frac{s : F(\mathsf{E}\,\varphi), \quad t : S\psi}{t : F\varphi}\ (F\,\mathsf{E})$$

As in ordinary tableau calculi, rules in TabUBiSKt are used to decompose formulae analyzing their main connective. Since some rules are branching or splitting, the tableau derivation process constructs a tree. If a branch in the tableau derivation ends with $\bot$, then the branch is said to be *closed.* If a branch is not closed, then it is *open.* If a branch is open and no more rules can be applied to it then the branch is *fully-expanded.* A tableau is closed when all its branches are closed, it is open otherwise. The derivation process stops when all the branches in the tableau derivation are either closed or fully expanded. An open fully expanded branch will give the information for building model for a set of tableau expressions given as derivation input. A formula φ is a *theorem* in TabUBiSKt if a tableau derivation for the input set $\{a : F\varphi\}$, where 'a' is a constant label which is intended to represent the initial world, will give a closed tableau. A formula φ is provable from a finite set Γ of formulae if a tableau derivation for the input set $\{a : T\Gamma\} \cup \{a : F\varphi\}$ will give a closed tableau, where $a : T\Gamma$ means $(a : T\gamma)$, for all $\gamma \in \Gamma$.

For the proofs of the following two theorems see manuscript at [17].

Theorem 4 (Soundness of TabUBiSKt). *Given a finite set $\Gamma \cup \{\varphi\}$ of formulae, if φ is provable from Γ in* **TabUBiSKt** *then $\Gamma \models \varphi$.*

Theorem 5. *Given a formula $\varphi \in \mathsf{Form}_{\mathcal{L}}$ the following are equivalent: (1) φ is a theorem in* **HUBiSKt**, *(2) φ is a theorem in* **TabUBiSKt**, *(3) φ is valid in all H-models.*

Theorem 5 shows that the proof systems **HUBiSKt** and **TabUBiSKt** capture the same set of theorems. Since **HUBiSKt** is decidable (Theorem 3), the tableau-system **TabUBiSKt** can be seen as the specification of a concrete algorithm for deciding whether a formula $\varphi \in \mathsf{Form}_{\mathcal{L}}$ is a theorem in **HUBiSKt**.

3 Reasoning with Spatial Relations in UBiSKt

3.1 UBiSKt as a Logic for Graphs

The logic **UBiSKt** is an expansion of the logic **BiSKt**, introduced in [19] and also studied in [16]. As is already noted in [19], a special case of an H-model is where the set U is the set of all edges and nodes of a multigraph, and H is the incidence relation as follows.

Definition 5. *A multigraph G consists of two disjoints sets E and N called the edges and the nodes, together a function $i : E \to \mathcal{P}(N)$ such that for all $e \in E$ the cardinality of $i(e)$ is either 1 or 2, and where $\mathcal{P}(N)$ is the powerset of the set of nodes. Note that these are undirected multigraphs and that edges may be loops incident only with a single node. A subgraph K of G is a subset of G such that given $u \in K$, if $v \in i(u)$ then $v \in K$.*

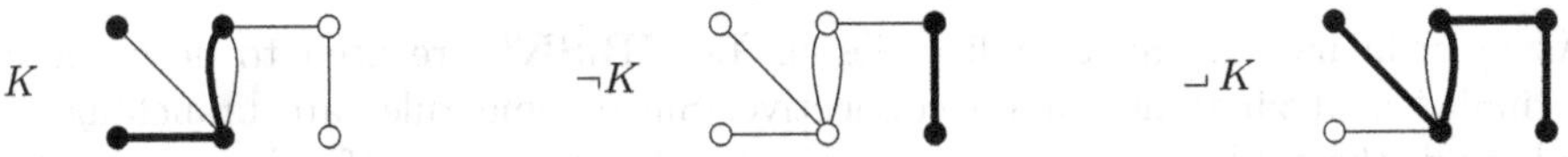

Fig. 1. The two kinds of complement of a subgraph K.

In [11] multigraphs are also called pseudographs. Any multigraph gives rise to a pre-order from which the structure of edges and nodes can be re-captured. Let $G = (E, N, i)$ be a multigraph. Define $U = E \cup N$ and define a relation $H \subseteq U \times U$ by $(u, v) \in H$ if and only if either (1) u is an edge and $v \in i(u)$, or (2) $u = v$. It is clear that H is reflexive and transitive. A structure (U, H) obtained from a multigraph in this way, uniquely determines the original multigraph, as the nodes are those elements $u \in U$ such that for all $v \in U$, $(u, v) \in H$ implies $u = v$.

Figure 2 shows a multigraph and the associated pre-order. From now on we will refer to multigraphs simply as graphs. It is easy to see that the subgraphs of a graph $G = (U, H)$ are exactly the subsets $K \subseteq U$ that are closed under H-successor. Therefore the notion of H-set as in Definition 1 corresponds to the

Fig. 2. The multi-graph on the left has four nodes, a, b, c, d, and four edges w, x, y, z. The corresponding pre-order for this multi-graph is the reflexive closure of the relation on the set $\{a, b, c, d, w, x, y, z\}$ shown on the right hand side.

notion of subgraph. Since any formula φ in the logic is interpreted as the H-set $[\![\varphi]\!]_M$, formulae in the logic can be regarded as names for subgraphs of an underlying graph $G = (U, H)$. Similarly, operations in the logic provide operations on subgraphs following the semantics defined in Sect. 2.1. Figure 1 shows a subgraph, K, of a graph, G, and the effect of the two complement operations $\neg$ and $\lrcorner$ on this subgraph. We note, in Fig. 1, that $\neg K$ is the largest subgraph disjoint from K and $\lrcorner K$ is the smallest subgraph whose union with K gives all the underlying graph G.

In the next section we are going to use two negations and universal modalities in **UBiSKt** to encode spatial relations between subgraphs, where subgraphs are naturally thought of as discrete regions of the space, i.e., sets of single nodes and edges between them.

3.2 Topological Notions in UBiSKt

Definition 6. *Let X be a Heyting Algebra with bottom element 0 and top element 1, and let $c : X \to X$ be a function. We say that (X, c) is a* Čech closure algebra *if for all x, $y \in X$:*

$$c(0) = 0, \quad x \leq c(x), \quad c(x \vee y) = c(x) \vee c(y).$$

Given a function $i : X \to X$, We say that (X, i) is a Čech interior algebra *if for all x, $y \in X$:*

$$i(1) = 1, \qquad i(x) \leq x, \quad i(x \wedge y) = i(x) \wedge i(y).$$

Let M be an H-model. Since **UBiSKt** is an expansion of intuitionistic logic, it is easy to see that $\{[\![\varphi]\!]_M \mid \varphi \in \mathsf{Form}_{\mathcal{L}}\}$ forms a Heyting algebra by interpreting $\bot$ as the bottom element 0 and $\top$ as the top element 1. Then, as we already noted in [18], $\lrcorner\neg$ enables us to define a Čech closure algebra. This can be also verified by our axiomatization. Since the adjunction "$\lrcorner\neg \dashv \neg\lrcorner$", the combination $\lrcorner\neg$ preserves finite disjunctions and the combination $\neg\lrcorner$ preserves finite conjunctions (due to item 9 of Proposition 2), we can easily obtain the first and the third conditions for a Čech closure algebra by soundness of **HUBiSKt**. Moreover the second condition follows from item 10 of Proposition 2 and soundness of **HUBiSKt**. Dually, we can also similarly verify that the combination $\neg\lrcorner$ gives rise to a Čech interior algebra on $\{[\![\varphi]\!]_M \mid \varphi \in \mathsf{Form}_{\mathcal{L}}\}$.

We may regard $\lrcorner\neg$ and $\neg\lrcorner$ as $\Diamond$ and $\blacksquare$ arising from the left converse $\circlearrowleft H$ of H, respectively. This is explained as follows. When we restrict our attention to the class of H-models $M = (U, H, R, V)$ satisfying $R = H$, we note that the modal operators $\Diamond$ and $\blacksquare$ arising from the left converse $\circlearrowleft R$ of R are equivalent with $\lrcorner\neg$ and $\neg\lrcorner$, respectively, while the modal operators $\blacklozenge$ and $\Box$ become trivial in the sense that $\blacklozenge\varphi \leftrightarrow \varphi$ and $\Box\varphi \leftrightarrow \varphi$ are valid in the model.

Definition 7. *Given an H-model $M = (U, H, R, V)$, an H-set $K \subseteq U$ is* representable *in the syntax $\mathcal{L}$ of* **UBiSKt** *if there is a formula $\varphi \in \mathsf{Form}_{\mathcal{L}}$ such that $K = [\![\varphi]\!]_M$.*

With the help of two kinds of negations $\neg$ and $\lrcorner$, we can talk about the notions of boundary and exterior of a representable subgraph.

$\partial^N(\varphi) := \varphi \wedge \lrcorner\varphi$ represents the nodes-boundary of a subgraph $[\![\varphi]\!]_M$.

$\partial(\varphi) := \neg\neg(\varphi \wedge \lrcorner\varphi) \wedge \varphi$ represents the general boundary of a subgraph $[\![\varphi]\!]_M$. This is the node-boundary plus the edges in the subgraph between these nodes.

$\neg\varphi$ represents the exterior of the subgraph $[\![\varphi]\!]_M$.

We also remark the following: the formula $\mathsf{A}\varphi$ represents $[\![\varphi]\!]_M = U$ and $\mathsf{E}\varphi$ represents $[\![\varphi]\!]_M \neq \varnothing$ and $\mathsf{A}\neg\varphi$ or $\neg\mathsf{E}\varphi$ represent $[\![\varphi]\!]_M = \varnothing$.

Using the closure operator the spatial relation of connection between subgraphs $[\![\varphi]\!]$ and $[\![\psi]\!]$ can be expressed by an appropriate formula in **UBiSKt**:

$$C(\varphi, \psi) := \mathsf{E}(\lrcorner\neg\varphi \wedge \psi).$$

The formula states that $[\![\lrcorner\neg\varphi]\!]_M \cap [\![\psi]\!]_M \neq \varnothing$. This means that the two subgraphs are connected if they are an edge apart, in the limit case. This notion of connection is the equivalent of the notion of adjacency found in Galton [8], and is one of the notions of connection expressed by closure in [5], with the difference that the operation $\lrcorner\neg$ is not a Kuratowski closure, but a Čech closure.

Beside connection the following Spatial Relations can be defined inside **UBiSKt**: Part, non-Part, Proper Part, Non-tangential Proper Part, Tangential Proper Part, External Connection, Disconnection, Partial overlapping, Equality, and the Inverse of Non-tangential Proper Part and Tangential Proper Part respectively. We list each relation with its correspondent formula in Table 2.

3.3 Reasoning on Spatial Entailments in UBiSKt

In this section we are going to show some interesting entailments between spatial properties of subgraphs, that can be derived syntactically in **UBiSKt**. Indeed all the following has been proved using **HUBiSKt**. For these axiomatic proofs the reader is referred to the manuscript at [17], where a wider list of properties of spatial relations between subgraphs has been included. We remark that Propositions 3–5 are also mechanically verified in our implementation of Tab**UBiSKt** in terms of *MetTel* [21]. The implemented prover with instructions on how to use it and how to prove any of the following propositions can be found at [17].

Table 2. Spatial relations and the corresponding formulae

Spatial relation	Formula	Spatial relation	Formula
$P(\varphi,\psi)$	$\mathsf{A}(\varphi\rightarrow\psi)$	$DC(\varphi,\psi)$	$\mathsf{A}\neg(\lrcorner\neg\varphi\wedge\psi)$
$non\text{-}P(\varphi,\psi)$	$\mathsf{E}(\varphi\prec\psi)$	$PO(\varphi,\psi)$	$\mathsf{E}(\varphi\wedge\psi)\wedge$ $non\text{-}P(\varphi,\psi)$ $\wedge\, non\text{-}P(\psi,\varphi)$
$PP(\varphi,\psi)$	$P(\varphi,\psi)\wedge non\text{-}P(\psi,\varphi)$	$EQ(\varphi,\psi)$	$\mathsf{A}(\varphi\leftrightarrow\psi)$
$NTPP(\varphi,\psi)$	$PP(\varphi,\psi)\wedge P(\lrcorner\neg\varphi,\psi)$	$NTPP^i(\varphi,\psi)$	$NTPP(\psi,\varphi)$
$TPP(\varphi,\psi)$	$PP(\varphi,\psi)\wedge non\text{-}P(\lrcorner\neg\varphi,\psi)$	$TPP^i(\varphi,\psi)$	$TPP(\psi,\varphi)$
$EC(\varphi,\psi)$	$C(\varphi,\psi)\wedge\mathsf{A}(\neg(\varphi\wedge\psi))$		

Proposition 3. $\vdash_{\mathbf{HUBiSKt}} \mathsf{E}(\lrcorner\neg\varphi\wedge\psi)\leftrightarrow\mathsf{E}(\varphi\wedge\lrcorner\neg\psi)$. *If the closure of a region intersects another region, then the closure of this latter region will intersect the former.*

This holds due to item 24 of Proposition 2. From Proposition 3 we can infer that the spatial relation of Connection, $C(\varphi,\psi)$ can also be expressed by the formula $\mathsf{E}(\varphi\wedge\lrcorner\neg\psi)$, showing that our formulation is equivalent to the notion of connection C_2 found in [5].

Proposition 4. (i) $\vdash_{\mathbf{HUBiSKt}} P(\lrcorner\neg\varphi,\psi)\leftrightarrow P(\varphi,\neg\lrcorner\psi)$. *If the closure of a region is part of another region then the former region will be part of the interior of the latter, and vice versa.* (ii) $\vdash_{\mathbf{HUBiSKt}} NTP(\neg\lrcorner\varphi,\varphi)$ *The interior of a region is always Non-tangential part of the region.*

Proposition 5. $\vdash_{\mathbf{HUBiSKt}} \partial^N(\varphi)\leftrightarrow\partial^N(\varphi)\wedge\lrcorner\partial^N(\varphi)$. *The Nodes-boundary of a region is always boundary of itself.*

4 Granular Spatial Relations

The idea of zooming out, or viewing a situation in a less detailed way, is commonplace. Intuitively, zooming out on an image (a set of pixels) we expect narrow cracks to fuse and narrow spikes to become invisible. This intuitive expectation is bourne out in the formalisation due to mathematical morphology. The idea here is that instead of being able to see individual pixels, only groups of pixels can be seen. This is illustrated in Fig. 3 using the operations of opening and closing by a structuring element. For details of mathematical morphology see [12], but here it is sufficient to know that the opening consists of the image formed by (overlapping) copies of the structuring element within the original, and that closing consists of the complement of the (overlapping) copies of the structuring element but rotated by half a turn, that can be placed wholly outside the original.

As explained in [12] the operations of mathematical morphology are not restricted to approximating subsets of a grid of pixels by a structuring element,

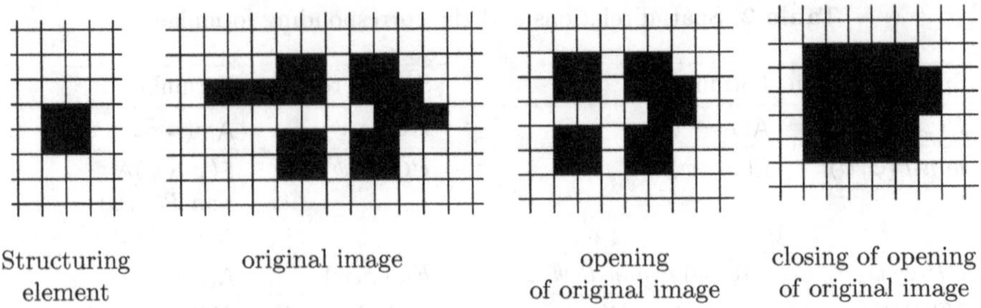

Fig. 3. Approximation of a subset of $\mathbb{Z}^2$ by a 2×2 structuring element.

but apply in the context of any subset of set with an arbitrary binary relation on the set instead of a structuring element. As [19] shows, we can extend this to a pre-order (U, H) and approximate H-sets in this structure by means of a stable relation R. Given $X \subseteq U$, we use $X \oplus R$ (dilation of X) to denote $\{u \in U \mid \exists v(v\ R\ u \wedge v \in X)\}$, and use $R \ominus X$ (erosion of X) to denote $\{u \in U \mid \forall v(u\ R\ v \Rightarrow v \in X)\}$. It is well known that for R fixed the operations $\_ \oplus R$ and $R \ominus \_$ form an adjunction from the lattice $\mathcal{P}(U)$ to itself, with $\_ \oplus R$ left adjoint to $R \ominus \_$. From adjunction, some properties of dilation and erosion follow, for example, given two sets A and B and a relation R, $A \oplus R \subseteq B$ is equivalent to $A \subseteq R \ominus B$. The opening of X by R is denoted $X \circ R$ and defined as $(R \ominus X) \oplus R$ and the closing is $X \bullet R = R \ominus (X \oplus R)$. The connection between mathematical morphology and modal logic has been studied in [2] in the set based case, and extended to the graph based case in [19]. Here, the modalities $\Diamond$, $\blacklozenge$, $\Box$ and $\blacksquare$ function as semantic operators taking H-sets to H-sets, with $\Diamond$ associated to $X \mapsto X \oplus \smallsmile R$, $\blacklozenge$ associated to $X \mapsto X \oplus R$, $\Box$ associated to $R \ominus X$ and $\blacksquare$ associated to $\smallsmile R \ominus X$. So, given a propositional variable p representing an H-set, opening and closing of the H-set are expressible in the logic by the formulae $\blacklozenge \Box p$ and $\Box \blacklozenge p$ respectively. This extends to **UBiSKt**, as it is an extension of the logic studied in [19]. In this setting, the idea of opening as fitting copies of a structuring element inside an image remains meaningful. Copies of the structuring element correspond to R-dilates in the following sense.

Definition 8. *A subset $X \subseteq U$ is an R-dilate if $X = \{u\} \oplus R$ for some $u \in U$.*

Stability implies that R-dilates are always H-sets. It is straightforward to check that opening and closing can be expressed in terms of dilates:

$$X \circ R = \bigcup \{\{u\} \oplus R \mid \{u\} \oplus R \subseteq X\},$$

$$X \bullet R = \{u \in U \mid \{u\} \oplus R \subseteq \bigcup \{\{x\} \oplus R \mid x \in X\}\}.$$

To give concrete examples, let (U, H) be the graph with $\mathbb{Z}^2$ for nodes and two nodes are connected by an edge if exactly one of their coordinates differs by 1. We refer to this as the graph $\mathbb{Z}^2$, visualized as in Fig. 4. The dilates by H

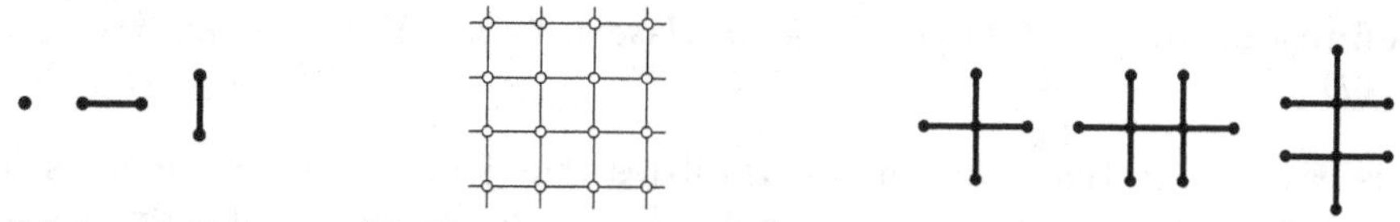

Dilates of H (U, H) is infinite grid, $\mathbb{Z}^2$ Dilates of $\looparrowright H$

Fig. 4. Shapes of the dilates of H and of $\looparrowright H$ when (U, H) is the graph shown.

and by $\looparrowright H$ of a node, a horizontal edge, and a vertical edge are shown in the figure.

We can think of $(X \circ R) \bullet R$ as a granular version of X in which we cannot 'see' arbitrary H-sets, but only ones that can be described in terms of the R-dilates. As we have seen, opening and closing correspond to specific sequences of modalities in the logic. So, given a representable H-set, we can capture its granular version by a formula in the logic.

Definition 9. *Given a propositional variable p representing an H-set, 'coarsely p' is defined by* $\mathbb{G}p := \Box\blacklozenge\blacklozenge\Box p$.

We notice that the closing of the opening of a region is known in mathematical morphology as an alternating filter. This gives a way of zooming-out for a region, but how should we define connection between coarse regions? The issue is that the space underlying the regions should become coarser – regions disconnected may become connected for example. In the same way that coarse regions are described in terms of dilates, a coarse version of connection can be formulated using dilates. To motivate this consider Fig. 5 which shows the idea that coarse regions are coarsely connected if there is a dilate intersecting both, or visually and informally that the gap between can be bridged by a dilate. Requiring an R-dilate joining two regions seems a suitable notion of coarse connection, as it extends the intuition of connection at the detailed level. Indeed two H-sets X and Y are connected at the detailed level (see Table 2) if the gap between them can be bridged by an H-dilate, so if they are an edge apart, in the limit case. Going to the granular level, single H-dilates are no longer "visible", and the space has coarser atomic parts: copies of the structuring element, i.e. R-dilates.

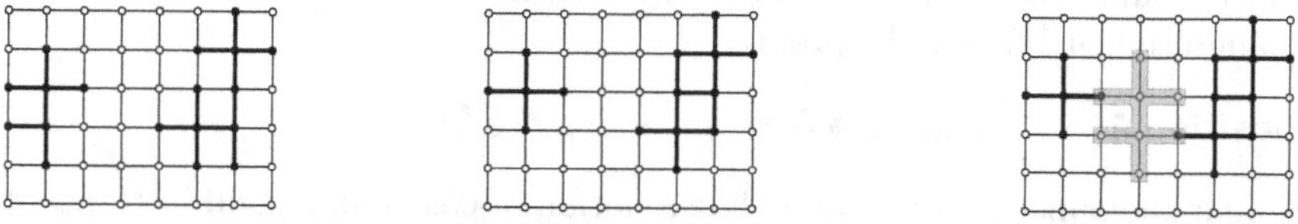

Two H-sets Their approximations by opening then closing by $\looparrowright H$ The approximations can be joined by a dilate

Fig. 5. Granular view by relation $\looparrowright H$

Definition 10. *An R-dilate, D, joins H-sets X and Y if* $X \cap D \neq \varnothing$ *and* $Y \cap D \neq \varnothing$.

It is easy to see that requiring an R-dilates that joins X and Y amounts to require that, given the union of the R-dilates intersecting X, at least one of them intersects Y.

Lemma 1 ([19]). *If R and S are relations on a set U and* $X \subseteq U$ *then* $X \oplus (R; S) = (X \oplus R) \oplus S$.

Lemma 2. *Let X be an H-set and R a stable relation. The union of the R-dilates intersecting X is* $X \oplus (\circlearrowleft R; R)$.

Proof. First we show that the union of the R-dilate intersecting X is $X \oplus \breve{R} \oplus R$. If $\{u\} \oplus R$ intersects X, for some $u \in U$, then there is a $x \in X$ such that $\{u\} \subseteq \{x\} \oplus \breve{R}$. Hence $\{u\} \oplus R \subseteq \{x\} \oplus \breve{R} \oplus R \subseteq X \oplus \breve{R} \oplus R$. In the other direction, if $y \in X \oplus \breve{R} \oplus R$, then there is some $u \in U$ and $x \in X$ such that uRy and uRx, so that $y \in \{u\} \oplus R$ with $\{u\} \oplus R$ intersecting X. Now, since $\breve{R} \subseteq \circlearrowleft R$ (see Definition 2), $X \oplus \breve{R} \oplus R \subseteq X \oplus \circlearrowleft R \oplus R = X \oplus \circlearrowleft R; R$. Also $X \oplus \circlearrowleft R; R = X \oplus H; \breve{R}; H; R = X \oplus \breve{R}; H; R \subseteq X \oplus \breve{R}; R = X \oplus \breve{R} \oplus R$ because X is an H-set and R is stable. So $X \oplus \breve{R} \oplus R = X \oplus \circlearrowleft R; R$. □

Proposition 6. *There is an R-dilate joining H-sets X and Y iff* $(X \oplus (\circlearrowleft R; R)) \cap Y \neq \varnothing$.

Proof. The union of R-dilates intersecting X is $X \oplus (\circlearrowleft R; R)$ from Lemma 2. This intersects Y iff $(X \oplus (\circlearrowleft R; R)) \cap Y \neq \varnothing$.

The above discussion provides a semantic justification for the following definition.

Definition 11 (Coarse connection). $C_G(p, q) := \mathsf{E}(\blacklozenge\Diamond\mathbb{G}p \wedge \mathbb{G}q)$.

Note that when $R = H$, then $\mathbb{G}p$ is equivalent to p and $C_G(p, q)$ is equivalent to $C(p, q)$. Indeed, as noticed in Sect. 3.2, $\lrcorner\neg$ can be regarded as $\Diamond$ arising from the left converse of H, $\circlearrowleft H$, and $\blacklozenge\varphi \leftrightarrow \varphi$ and $\Box\varphi \leftrightarrow \varphi$ are valid in a model where $R = H$. Another special case is when H is the identity relation on a set, and R is an equivalence relation. In this case $[\![\mathbb{G}p]\!]$ will correspond to the lower approximation, in the sense of rough-set theory, of $[\![p]\!]$.

As we would expect, our notion of coarse connection is symmetric as follows. See manuscript at [17] for the proof.

Proposition 7. $\vdash_{\mathbf{HUBiSKt}} \mathsf{E}(\blacklozenge\Diamond\varphi \wedge \psi) \leftrightarrow \mathsf{E}(\varphi \wedge \blacklozenge\Diamond\psi)$.

Similar to connection, we can define a notion of coarse parthood in terms of R-dilates. The standard notion of parthood at the detailed level (Table 2) says that, given H-sets X and Y, X is part of Y if and only if all the atomic H-dilates in X lie in Y. A suitable notion of coarse parthood will require that X is coarse part of Y if and only if all the R-dilates in X lie also in Y.

Proposition 8. *Let X and Y be H-sets, and R a stable relation. The following are equivalent: (1) all the R-dilates in X lie in Y and (2) $R \ominus (X) \subseteq R \ominus (Y)$.*

Proof. The union of all the R-dilates in X is the opening of X: $X \circ R = (R \ominus X) \oplus R$. Hence, requiring the all the R-dilates in X lie in Y amounts to require that $(R \ominus X) \oplus R \subseteq Y$. By properties of adjunction this is equivalent to $R \ominus X \subseteq R \ominus Y$.

Lemma 3 ([19]). *Let M be and H-model and let φ, $\psi \in \mathsf{Form}_{\mathcal{L}}$ with $[\![\varphi]\!]_M$ and $[\![\psi]\!]_M$ associated H-sets. Then $[\![\varphi]\!]_M \subseteq [\![\psi]\!]_M$ iff $M \models \mathsf{A}(\varphi \rightarrow \psi)$.*

The above reasoning together with Lemma 3 provide a semantic justification for the following definition of coarse parthood between coarse regions.

Definition 12 (Coarse parthood). $P_G(p, q) := \mathsf{A}(\Box \mathbb{G} p \rightarrow \Box \mathbb{G} q)$.

The negation of the notion of coarse parthood will give a notion of coarse non-parthood: this requires that there is at least an R-dilate in X such that it is not in Y. From Proposition 8, we know that this is equivalent to $R \ominus X \not\subseteq R \ominus Y$.

Lemma 4. *Let M be and H-model and let φ, $\psi \in \mathsf{Form}_{\mathcal{L}}$ with $[\![\varphi]\!]_M$ and $[\![\psi]\!]_M$ associated H-sets. Then $[\![\varphi]\!]_M \not\subseteq [\![\psi]\!]_M$ iff $M \models \mathsf{E}(\varphi \prec \psi)$.*

For the proof of Lemma 4, see manuscript at [17]. Because of Lemma 4 we propose the following definition.

Definition 13 (Coarse non-parthood). *non-*$P_G(p, q) := \mathsf{E}(\Box \mathbb{G} p \prec \Box \mathbb{G} q)$.

We now analyze how to extend the spatial relation of overlapping to the granular level. Two H-sets X and Y overlaps at the detailed level if and only if there is at least a non-empty H-dilate that lies both in X and Y. Following this idea, a suitable notion of coarse overlapping requires a non-empty R-dilate that lies both in X and Y.

Proposition 9. *Let X and Y be H-sets and R a stable relation. The following are equivalent:* (1) *there is a non-empty R-dilate that lies both in X and in Y and* (2) $(X \cap Y) \circ R \neq \varnothing$.

Proof. $(X \cap Y) \circ R$ is the opening of $X \cap Y$, so union of all R-dilates both in X and in Y. Hence requiring that there is a non empty R-dilate that lies both in X and in Y amounts to require that the opening of $X \cap Y$ is non-empty: $(X \cap Y) \circ R \neq \varnothing$. □

Thus we define coarse overlapping between coarse regions as follows.

Definition 14 (Coarse overlapping). $O_G(p, q) := \mathsf{E}(\blacklozenge \Box (\mathbb{G} p \wedge \mathbb{G} q))$.

As an example, in Fig. 6 on the left we show two H-sets (red and black) that intersect, but an R-dilate will not fit inside the region of intersection ($R = \circlearrowleft H$). Therefore the spatial relation O_G does not hold. If the region of intersection is at least as big as an R-dilate, as happens on the right of the figure, then the relation O_G does hold.

Given H-sets X and Y, X is non-tangential part of Y at the detailed level if X is part of Y and the closure of X, $\lrcorner\neg X$, is still part of Y. This means that all the H-dilates that intersect X lie in Y. Hence, a suitable notion of coarse non-tangential part between H-sets X and Y is obtained by requiring that X is coarse part of Y and all the R-dilates intersecting X lie in Y.

Proposition 10. *Let X and Y be H-sets and R a stable relation. The following are equivalent: (1) all the R-dilates overlapping X lie in Y, and (2) $X \oplus \circlearrowleft R \subseteq R \ominus Y$.*

Proof. The union of the R-dilates overlapping X lie in Y is $(X \oplus \circlearrowleft R \oplus R) \subseteq Y$ by Lemma 2. This is equivalent to $X \oplus \circlearrowleft R \subseteq R \ominus Y$ by properties of adjunctions. □

The above reasoning provides a semantic justification for the following definition.

Definition 15 (Coarse non-tangential part). $NTP_G(p,q) := \mathsf{A}(\Box\mathbb{G}p \rightarrow \Box\mathbb{G}q) \wedge \mathsf{A}(\Diamond\mathbb{G}p \rightarrow \Box\mathbb{G}q)$.

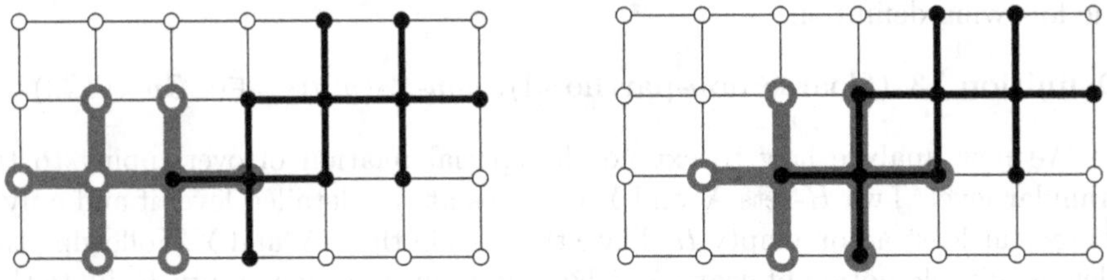

Two H-sets not sharing a whole R-dilate   Two H-sets sharing a whole R-dilate

Fig. 6. Cases of coarse non-overlapping and of coarse overlapping, where R is $\circlearrowleft H$. (Color figure online)

Finally, we analyze the notion of coarse tangential part. At the detailed level, an H-set X is tangential part of Y if it is its part and there is at least an H-dilate intersecting X that does not lie in Y. This is obtained by requiring that the closure of X is not part of Y. Hence, at the granular level we will require that the union of all R-dilates intersecting X does not lie in Y. This means that we have to negate the requirement for NTP_G: by Proposition 10 this is $X \oplus \circlearrowleft R \not\subseteq R \ominus Y$. Because of this and Lemma 4 we propose the following.

Definition 16 (Coarse tangential part). $TP_G(p,q) := \mathsf{A}(\Box\mathbb{G}p \rightarrow \Box\mathbb{G}q) \wedge \mathsf{E}(\Diamond\mathbb{G}p \prec \Box\mathbb{G}q)$.

5 Conclusions and Further Work

We have provided a sound and complete axiomatisation for the logic **UBiSKT** and used this to prove a number of results in Sect. 3.3 demonstrating that the definitions of discrete spatial relations have properties appropriate to the spatial concepts involved. We have also provided a tableau calculus for the logic, and proved that this is equivalent to the Hilbert-style axiomatisation. While spatial relations in discrete space have been studied before, the novelty in our work here is the use of reasoning in a formal logic together with an implementation of a theorem-proving procedure for the logic.

There are several directions for further work. Our use of **UBiSKT** to formulate a notion of coarsening fits in with existing work observing that both rough set theory and mathematical morphology are closely connected with modal logic [3]. Our definitions of coarse spatial relations in this setting are, however, a novelty. We have been able to indicate the semantic basis and some basic properties of these relations. In future work we will investigate the use of the axiomatisation in establishing more general forms of connection. For example, by measuring the connection between two regions at two levels of detail, that is the value of $(C(p,q), C_G(p,q))$, we anticipate based on the evidence in [20] (which considered granularity but not the connection relation) that spatial relations able to make finer distinctions can be obtained.

References

1. Blackburn, P., de Rijke, M., Venema, Y.: Modal Logic. Cambridge Tracts in Theoretical Computer Science. Cambridge University Press, Cambridge (2001)
2. Bloch, I.: Modal logics based on mathematical morphology for qualitative spatial reasoning. J. Appl. Non-Cl. Log. **12**(3–4), 399–423 (2002). https://doi.org/10.3166/jancl.12.399-423
3. Bloch, I.: Spatial reasoning under imprecision using fuzzy set theory, formal logics and mathematical morphology. Int. J. Approx. Reason. **41**(2), 77–95 (2006)
4. Chen, J., Cohn, A.G., Liu, D., Wang, S., Ouyang, J., Yu, Q.: A survey of qualitative spatial representations. Knowl. Eng. Rev. **30**(1), 106–136 (2015)
5. Cohn, A.G., Varzi, A.C.: Mereotopological connection. J. Philos. Log. **32**, 357–390 (2003)
6. Cousty, J., Najman, L., Dias, F., Serra, J.: Morphological filtering on graphs. Comput. Vis. Image Underst. **117**, 370–385 (2013)
7. Egenhofer, M.J., Herring, J.: Categorizing binary topological relations between regions, lines, and points in geographic databases. Department of Surveying Engineering, University of Maine, Orono, ME, vol. 9(94-1), p. 76 (1991)
8. Galton, A.: The mereotopology of discrete space. In: Freksa, C., Mark, D.M. (eds.) COSIT 1999. LNCS, vol. 1661, pp. 251–266. Springer, Heidelberg (1999). https://doi.org/10.1007/3-540-48384-5_17
9. Galton, A.: Discrete mereotopology. In: Calosi, C., Graziani, P. (eds.) Mereology and the Sciences. SL, vol. 371, pp. 293–321. Springer, Cham (2014). https://doi.org/10.1007/978-3-319-05356-1_11
10. Goranko, V., Passy, S.: Using the universal modality: gains and questions. J. Log. Comput. **2**(1), 5–30 (1992). https://doi.org/10.1093/logcom/2.1.5

11. Harary, F.: Graph Theory. Addison-Wesley, Reading (1969)
12. Najman, L., Talbot, H.: Mathematical Morphology: From Theory to Applications. Wiley, Hoboken (2010)
13. Randell, D.A., Cui, Z., Cohn, A.G.: A spatial logic based on regions and connection. In: Nebel, B., Rich, C., Swartout, W. (eds.) Principles of Knowledge Representation and Reasoning. Proceedings of the Third International Conference (KR 1992), pp. 165–176. Morgan Kaufmann (1992)
14. Randell, D.A., Galton, A., Fouad, S., Mehanna, H., Landini, G.: Mereotopological correction of segmentation errors in histological imaging. J. Imaging **3**(4), 63 (2017)
15. Rosenfeld, A.: Digital topology. Am. Math. Mon. **86**, 621–630 (1979)
16. Sano, K., Stell, J.G.: Strong completeness and the finite model property for bi-intuitionistic stable tense logics. In: Electronic Proceedings in Theoretical Computer Science. Open Publishing Association (2016)
17. Sindoni, G., Sano, K., Stell, J.G.: Axiomatizing Discrete Spatial Relations (Extended Version with Omitted Proofs) (2018). https://doi.org/10.5518/427
18. Sindoni, G., Stell, J.G.: The logic of discrete qualitative relations. In: COSIT 2017 Proceedings, vol. 86. Schloss Dagstuhl-Leibniz-Zentrum fuer Informatik (2017)
19. Stell, J., Schmidt, R., Rydeheard, D.: A bi-intuitionistic modal logic: foundations and automation. J. Log. Algebr. Methods Program. **85**(4), 500–519 (2016)
20. Stell, J.G.: Granular description of qualitative change. In: Rossi, F. (ed.) Proceedings of the 23rd International Joint Conference on Artificial Intelligence, pp. 1111–1117 (2013)
21. Tishkovsky, D., Schmidt, R.A., Khodadadi, M.: MetTeL2: towards a tableau prover generation platform. In: PAAR@ IJCAR, pp. 149–162 (2012)

A Modal and Relevance Logic for Qualitative Spatial Reasoning

Pranab Kumar Ghosh and Michael Winter(✉)

Department of Computer Science, Brock University, St. Catharines, ON, Canada
{pg15kq,mwinter}@brocku.ca

Abstract. Boolean contact algebras constitute a suitable algebraic theory for qualitative spatial reasoning. They are Boolean algebras with an additional contact relation grasping the topological aspect of spatial entities. In this paper we present a logic that combines propositional logic, relevance logic, and modal logic to reason about Boolean contact algebras. This is done in two steps. First, we use the relevance logic operators to obtain a logic suitable for Boolean algebras. Then we add modal operators that are based on the contact relation. In both cases we present axioms that are equivalent to requirement that every frame for the logic is indeed a Boolean algebra resp. Boolean contact algebra. We also provide a natural deduction system for this logic by defining introduction and eliminations rules for each logical operator. The system is shown to be sound. Furthermore, we sketch an implementation of the natural deduction system in the functional programming language and interactive theorem prover Coq.

1 Introduction

In mathematics and computer science, qualitative spatial reasoning (QSR) deals with qualitative features of spatial entities. It is an alternative technique that represents so-called regions and their spatial relations without using numbers or other quantitative measures. Usually the relationship between regions has two aspects. The first aspect is based on the so-called "part-of" relationship, i.e., one region being included in another region. This aspect is mathematically covered by mereology. The second aspect focuses on topological nature, i.e., whether they are in"contact" without necessarily having a common part. Mereotopology is a mathematical theory that covers these two aspects. We will use the theory of Boolean contact algebras (BCAs) as the concrete mathematical theory of mereotopology. A BCA is a Boolean algebra with a suitable binary contact relation C. The order of the Boolean algebra provides the part-of relationship between regions and the contact relation C the topological relationship between them.

M. Winter—Gratefully acknowledges support from the Natural Sciences and Engineering Research Council of Canada.

J. Desharnais et al. (Eds.): RAMiCS 2018, LNCS 11194, pp. 131–147, 2018.
https://doi.org/10.1007/978-3-030-02149-8_9

In this paper, our aim is to introduce a modal and relevance logic to reason about spatial entities. This logic allows reasoning about regions in any application that establishes a BCA. For example, let us assume that the different spots of interest in an apartment, here called locations, are regions of a BCA. Certain locations are rooms potentially containing several smaller locations that are not necessarily rooms themselves. Furthermore, assume that there are some locations that contain a water outlet. If we use the propositional variables r indicating that a location is a room and w that a location has a water outlet, then the scenario requires that if w is true for a location l, then w must be also true for any locations that is greater than or equal to l. Similarly, we might be interested to express that "every room is connected to a room with an water outlet". Our logic is capable of expressing both properties using relevance and modal operators. The natural deduction system can be used to derive properties of the problem at hand. For example, if every location with an water outlet has also a drain, then we will be able to derive that every room is connected to a room with a drain. First, we will present the logic and its basic features and then we will sketch an implementation of the logic in the functional programming language and interactive theorem prover Coq.

Our logic is a combination of modal and relevance logic. Modal logic is a class of logics extending propositional logic. It adds new operators that provide access to a restricted version of quantification. In our application these operators will be used to formalize the topological relationship between regions. Relevance logic is non-classical logic that was developed to represent aspects of implication which are ignored in classical propositional logic, i.e., it requires that the antecedent is relevant to the consequent of an implication. A more general approach to relevance logic defines the implication based on a ternary relation on possible worlds similar to the corresponding binary relation in modal logics. Modern relevance logics usually also define a relevance negation that is based on regular negation and a unary function. In this paper, we will use a relevance implication that is based on the sum (or union) operation of regions and one negation based on the complement of a region. Consequently, the relevance logic part will cover the mereological aspect of a BCA. First, we will concentrate on this portion of the logic and present a set of axioms that is equivalent to fact that the frame provides the structure of a Boolean algebra. Then we will introduce modal operators based on the contact relation together with a set of appropriate axioms forcing each frame to be a BCA. In addition, we will provide a natural deduction system for our logic that we will prove to be sound. Please note that our logic differs from the logic presented in [6]. Models of our logic are Boolean contact algebras, i.e., the elements of the universe is the set of regular closed sets of a topological space. Therefore, formulas represent properties of regions and is true if the property holds for all regions. On the other hand, the models of the logic presented in [6] are general topological spaces, i.e., the elements of the universe are points, and formulas are interpreted as open/closed sets (not necessarily regular open/closed). The logical operations correspond to set-theoretic and/or topological operations, and a formula is true iff the set described by the formula is equal to the universe. Therefore, a formula expresses a relationship between the open/closed sets represented by the propositional variables or constants appearing in the formula.

Finally, we will use Coq to provide an implementation of the natural deduction system.

2 Mathematical Preliminaries

In this section, we will focus on regular closed sets as a topological model for regions, Boolean algebras (BA) and Boolean contact algebras (BCA) as an abstract theory for regions as well as propositional, relevance and modal logic. We begin by recalling some basic definitions and properties of the first three structures mentioned above. For more details, we refer to [4,5,7,11,12,15,17].

Definition 2.1 (Boolean algebra). *A Boolean algebra* $\mathcal{B} = \langle B, +, \cdot, ^*, 0, 1\rangle$ *is a structure with a set* B*, two binary operators* $+$ *and* $\cdot$*, a unary operator* * *and two elements* $0, 1 \in B$ *satisfying the following axioms for all* $x, y, z \in B$*:*

Commutativity	$x + y = y + x$	$x \cdot y = y \cdot x$
Identity	$x + 0 = x$	$x \cdot 1 = x$
Distributivity	$x + y \cdot z = (x + y) \cdot (x + z)$	$x \cdot (y + z) = x \cdot y + x \cdot z$
Complement	$x + x^* = 1$	$x \cdot x^* = 0$

Please note that the definition above is sufficient. Properties such as associativity of both operations and the absorption laws follow (see e.g. [7]).

If $\langle X, \tau\rangle$ is a topological space, then a set $x \subseteq X$ is called regular closed iff $x = \mathrm{cl}(\mathrm{int}(x))$ where cl(.) and int(.) denote the closure and interior operation of the topological space. Intuitively, a regular closed set is a closed set that does not have any isolated points.

Lemma 2.1 ([11]). *Let* $RegCL(X)$ *be the set of regular closed sets of* $\langle X, \tau\rangle$*. Then* $RegCL(X)$ *together with* $0 := \emptyset$*,* $1 := X$*, and the operations*

$$x + y := x \cup y, \qquad x \cdot y := \mathrm{cl}(\mathrm{int}(x \cap y)), \qquad x^* = \mathrm{cl}(X \setminus x)$$

is a Boolean algebra.

In our context regular closed sets are the standard example for regions. The Boolean algebra as defined above provides the mereological structure of those regions. A Boolean contact algebra adds the topological aspect to the theory by requiring a suitable binary relation on B [11,12]. If R is a binary relation, we will use the notation xRy to denote that $(x, y) \in R$, i.e., that x and y are in relation R.

Definition 2.2 (Boolean contact algebra). *A binary relation C on a BA $\mathcal{B}$ is called contact relation if it satisfies the following axioms for all $x, y, z \in B$:*

(C_0) *Nulldisconnectedness*	$xCy \Rightarrow x, y \neq 0$
(C_1) *Reflexivity*	$x \neq 0 \Rightarrow xCx$
(C_2) *Symmetry*	$xCy \Leftrightarrow yCx$
(C_3) *Compatibility*	xCy *and* $y \leq z \Rightarrow xCz$
(C_3) *Summation*	$xC(y+z) \Rightarrow xCy$ *or* xCz

A Boolean algebra together with a contact relation defined on B, i.e., the structure $\langle B, C, +, \cdot, ^*, 0, 1\rangle$ is called a Boolean contact algebra (BCA).

The contact relation on regular closed sets is defined by xCy iff $x \cap y \neq \emptyset$. Please note that $x \cdot y$ differs from $x \cap y$ since $x \cdot y \leq x \cap y$ but not necessarily equal. Consequently, the contact relation cannot be defined in terms of the operations of the Boolean algebra. It is easy to verify that the Boolean algebra of regular closed sets with the contact relation above defines a Boolean contact algebra. The theory of Boolean contact algebras is also sufficient to reason about regular closed sets since every BCA can be represented by a BCA of regular closed sets [12]. This implies that our logic can also be used to reason about regular closed sets of an arbitrary topological space.

2.1 Propositional Logic (PL)

In this section, we will define the syntax and semantics of propositional logic in slightly different way than usual. Our definition is equivalent to the usual definition but allows an immediate application of propositional logic in the context of modal and relevance logic, i.e., we will define a Kripke-style semantics for propositional logic.

Definition 2.3. *Let P be a set of propositional variables, then the set Prop of propositional formulas is recursively defined by the following rules:*

PropL.1: Each variable $p \in P$ is a propositional formula, i.e., $P \subseteq Prop$,
PropL.2: $\bot$ is a propositional formula, i.e., $\bot \in Prop$,
PropL.3: If $\varphi, \psi \in Prop$, then $\varphi \to \psi \in Prop$.

We will be using the same set of the operators and formulas defined by the rules PropL.1–PropL.3 in the extensions later by just referring to the rules above. Obviously, the set of logical operations defined above are sufficient to represent all other operators such as $\neg, \wedge, \vee$ and $\top$ known in propositional logic. In the remainder of the paper we will use those operator as an abbreviation of the corresponding formula in Prop.

Definition 2.4. *A propositional logic frame (PL-frame) $\mathcal{F}$ is just a non-empty set W. A (propositional logic) model $\mathcal{M} = \langle \mathcal{F}, v\rangle$ is a PL-frame together with a valuation function $v : P \to \mathcal{P}(W)$.*

As usual we call the elements of W states or worlds.

Definition 2.5 (Propositional logic semantics). *Let $\mathcal{M}$ be a model, $x \in W$, and $\varphi \in Prop$. Then the satisfaction relation $\mathcal{M}, x \models \varphi$ (φ is true at x in $\mathcal{M}$) is recursively defined by*

SemPropL.1: $\mathcal{M}, x \models p$ *iff* $x \in v(p)$,
SemPropL.2: $\mathcal{M}, x \not\models \bot$,
SemPropL.3: $\mathcal{M}, x \models \varphi \rightarrow \psi$ *iff* $\mathcal{M}, x \models \varphi$ *implies* $\mathcal{M}, x \models \psi$.

We will apply the following definitions for all logics defined in the paper without explicitly repeating them. We define $\mathcal{M} \models \varphi$ (φ is true in the model $\mathcal{M}$) iff $\mathcal{M}, x \models \varphi$ for all $x \in W$. Furthermore, $\mathcal{F} \models \varphi$ (φ is true in the frame $\mathcal{F}$) iff $\mathcal{M} \models \varphi$ for all models based on $\mathcal{F}$, i.e., for all models $\mathcal{M}$ with $\mathcal{M} = \langle \mathcal{F}, v \rangle$ for some valuation function v. Finally, we define $\Gamma \models \varphi$ for a set Γ of formulas (Γ implies/entails φ) iff $\mathcal{M} \models \psi$ for all $\psi \in \Gamma$ implies $\mathcal{M} \models \varphi$ for all models $\mathcal{M}$. We obtain a special case of the previous definition if $\Gamma = \emptyset$. In this case $\models \varphi$ we say that φ is true (or valid).

Please note that our definition of $\models \varphi$ for propositional logic is equivalent to the usual definition since each state x and valuation v of our approach leads to a truth assignment a by defining $a(p) =$ true iff $x \in v(p)$ and the satisfaction relation does not change the state x.

Last but not least, we want to mention that we will use axiom schemas. An axiom schema is a formula that uses meta variables $\varphi, \psi, \ldots$ for arbitrary formulas of the logic. We say that a schema is true at x in $\mathcal{M}$ if any instantiation of the schema with concrete formulas of the logic is true at x in $\mathcal{M}$. The derived notations as defined above generalize similarly to formula schemas.

2.2 Modal Logic (ML)

In this section, we will recall syntax and semantics of modal logic (cf. [3,14]).

Definition 2.6. *The set Mod of modal logic formulas is recursively defined by the rules PropL.1–PropL.3 and*

ModL.1: If $\varphi \in Mod$, then $[R]\varphi \in Mod$.

As usual we will introduce the diamond operator as abbreviation based on the box operator, i.e., we define $\langle R \rangle \varphi := \neg [R] \neg \varphi$.

Definition 2.7. *A modal logic frame (ML-frame) $\mathcal{F} = \langle W, R \rangle$ is a non-empty set W together with a binary relation R on W, i.e., $R \subseteq W \times W$. A (modal logic) model is a ML-frame together with a valuation function.*

The following definition of the semantics of modal logic is standard.

Definition 2.8. *Let $\mathcal{M}$ be a model, $x \in W$, $\varphi \in Mod$. Then the satisfaction relation $\mathcal{M}, x \models \varphi$ is recursively defined by SemPropL.1–SemPropL.3 and*

SemModL.1: $\mathcal{M}, x \models [R]\varphi$ *iff* xRy *implies* $\mathcal{M}, y \models \varphi$ *for all* $y \in W$.

2.3 Relevance Logic (RL)

In this section, we will recall syntax and semantics of relevance logic. For more details, we refer the reader to [1,10,16]. In order to avoid notational conflicts with the "classical" implication ($\rightarrow$), we will use $\twoheadrightarrow$ to denote"relevance" implication. Please note that the relevance negation, denoted by $\sim$ in this paper, cannot be defined in terms of the relevance implication so that we include this operator explicitly in the syntax.

Definition 2.9. *The set RL of relevance logic formulas is recursively defined by the rules PropL.1 and*

RelL.1: *If* $\varphi, \psi \in RL$*, then* $\varphi \twoheadrightarrow \psi \in RL$*,*
RelL.2: *If* $\varphi \in RL$*, then* $\sim\varphi \in RL$*.*

Urquhart's original definition of the semantics of the relevance implication is based on a so-called fusion function $\circ$ that"clues" two pieces of information together [10]. The actual definition is $\mathcal{M}, x \models \varphi \twoheadrightarrow \psi$ iff $y \models \varphi$ implies $\mathcal{M}, x \circ y \models \psi$ for all $y \in W$, i.e., the information x is needed for the implication iff whenever information y is needed for φ, then $x \circ y$ is needed for ψ. A more general approach to the semantics of this kind of implication is based on a ternary relation S on W. In this generalization the definition becomes $\mathcal{M}, x \models \varphi \twoheadrightarrow \psi$ iff $y \models \varphi$ implies $\mathcal{M}, z \models \psi$ for all $y, z \in W$ with $(x, y, z) \in S$. Obviously, the original definition is a spacial case by defining S as $(x, y, z) \in S$ iff $z = x \circ y$. In this paper we will also use a binary function to instantiate S but differently from Urquhart.

Definition 2.10. *A relevance logic frame (RL-frame)* $\mathcal{F} = \langle W, f, g\rangle$ *is a non-empty set* W *together with a binary function* f *and a unary function* g *on* W*. A (relevance logic) model is a RL-frame together with a valuation function.*

Now we are ready to define the version of relevance logic that we will use throughout this paper. Please note that the definition of the relevance negation is standard.

Definition 2.11. *Let* $\mathcal{M}$ *be a model,* $x \in W$*, and* $\varphi \in RL$*. Then the satisfaction relation* $\mathcal{M}, x \models \varphi$ *is recursively defined by SemPropL.1 and*

SemRelL.1: $\mathcal{M}, x \models \varphi \twoheadrightarrow \psi$ *iff* $\mathcal{M}, y \models \varphi$ *implies* $\mathcal{M}, z \models \psi$ *for all* $y, z \in W$ *with* $x = f(y, z)$*,*
SemRelL.2: $\mathcal{M}, x \models \sim\varphi$ *iff* $\mathcal{M}, g(x) \not\models \varphi$*.*

In our definition the function f is applied to y and z rather than to x and y. This will allow us to split an element x of Boolean algebra into two parts y and z, i.e., $x = y + z$, so that we can use the implication to state properties of $+$.

2.4 Propositional Relevance Logic with E (PRLE)

Our first goal is to provide a logic suitable for Boolean algebras. More precisely, we want to establish a logic so that every model/frame of the logic is a Boolean algebra. For this purpose we use the combination of propositional and relevance logic as introduced so far extended by an additional constant.

Definition 2.12. *The set PRLE of propositional relevance logic formulas with E is recursively defined by the rules PropL.1–PropL.3, RelL.1–RelL.2 and*

PRLE.1: $E \in PRLE$.

Informally, the formula E will only be true at the least element 0 of a Boolean algebra. For a general interpretation of E in an arbitrary frame we need an additional element.

Definition 2.13. *A propositional relevance logic with E frame (PRLE-frame)* $\mathcal{F} = \langle W, e, f, g\rangle$ *so that* $\langle W, f, g\rangle$ *is a RL-frame and* $e \in W$. *A PRLE-model is a PRLE-frame together with a valuation function.*

This leads to the following definition.

Definition 2.14. *Let* $\mathcal{M}$ *be a model,* $x \in W$ *be a state, and* $\varphi \in PRLE$. *Then the satisfaction relation* $\mathcal{M}, x \models \varphi$ *is recursively defined by SemPropL.1–SemPropL.3, SemRelL.1–SemRelL.2 and*

SemPRLE.1: $\mathcal{M}, x \models E$ *iff* $x = e$.

In order to establish axioms that force any frame of the logic to be a Boolean algebra we start with the property $g(g(x)) = x$ indicating that complement in a Boolean algebra is involutive.

Lemma 2.2. *The formula schema* $\varphi \to {\sim\sim}\varphi$ *is true in a PRL-frame* $\mathcal{F}$, *iff* $g(g(x)) = x$ *for all* $x \in W$.

Proof. First, consider the following computation for an arbitrary formula φ:

$$\begin{aligned} \mathcal{M}, x \models {\sim\sim}\varphi &\Leftrightarrow \mathcal{M}, g(x) \not\models {\sim}\varphi && \text{by Def. SemRlL.2} \\ &\Leftrightarrow \mathcal{M}, g(g(x)) \models \varphi && \text{by Def. SemRlL.2} \end{aligned}$$

$\Rightarrow$: Assume the schema is true in $\mathcal{F}$. Then we have $\mathcal{M}, x \models p \to {\sim\sim}p$ where $\mathcal{M}$ is the model based on $\mathcal{F}$ with $v(p) = \{x\}$. Since $\mathcal{M}, x \models p$ we obtain $\mathcal{M}, x \models {\sim\sim}p$. By the computation above the latter is equivalent to $\mathcal{M}, g(g(x)) \models p$, which implies $g(g(x)) \in v(p)$, i.e., $g(g(x)) = x$.

$\Leftarrow$: Assume $g(g(x)) = x$ for all $x \in W$ and that $\mathcal{M}, x \models \varphi$ for some formula φ, $x \in W$, and a model $\mathcal{M}$ based on $\mathcal{F}$. Then we have $\mathcal{M}, g(g(x)) \models \varphi$, and by the computation above $\mathcal{M}, x \models {\sim\sim}\varphi$. Therefore, $\mathcal{M}, x \models \varphi \to {\sim\sim}\varphi$ and we conclude $\mathcal{F} \models \varphi \to {\sim\sim}\varphi$. □

Before we proceed it will be convenient to introduce some abbreviations. We start with the dual function (with respect to g) of f.

Definition 2.15. *Let $\mathcal{F}$ be a PRLE-frame and $x, y \in W$. The dual operator f^d of f is defined by $f^d(x, y) := g(f(g(x), g(y)))$.*

If g is involutive, then f^d is indeed dual to f as the next lemma shows. The proof is straight-forward and, therefore, omitted.

Lemma 2.3. *If $g(g(x)) = x$ for all $x \in W$, then $f(x, y) = g(f^d(g(x), g(y)))$ for all $x, y \in W$.*

In addition to f^d we will use the following abbreviations:

$\varphi \mathbin{\wedge\!\!\!\wedge} \psi$	$:=$	$\neg(\varphi \twoheadrightarrow \neg\psi)$	PRLEAbbr1
$\varphi \mathbin{\vee\!\!\!\vee} \psi$	$:=$	$\neg\varphi \twoheadrightarrow \psi$	PRLEAbbr2
$N\varphi$	$:=$	$\sim\neg\varphi$	PRLEAbbr3
$\varphi \multimap \psi$	$:=$	$N(N\varphi \twoheadrightarrow N\psi)$	PRLEAbbr4
$\varphi \parr \psi$	$:=$	$\neg(\varphi \multimap \neg\psi)$	PRLEAbbr5
$\varphi \bindnasrepma \psi$	$:=$	$\neg\varphi \multimap \psi$	PRLEAbbr6
U	$:=$	NE	PRLEAbbr7

The following lemma is an immediate consequence of the definition above.

Lemma 2.4. *Let $\mathcal{M}$ be a model and $x \in W$. Then we have:*

1. $\mathcal{M}, x \models \varphi \mathbin{\wedge\!\!\!\wedge} \psi$ *iff there are y, z with $x = f(y, z)$ and $\mathcal{M}, y \models \varphi$ and $\mathcal{M}, z \models \psi$.*
2. $\mathcal{M}, x \models \varphi \mathbin{\vee\!\!\!\vee} \psi$ *iff $x = f(y, z)$ implies $\mathcal{M}, y \models \varphi$ or $\mathcal{M}, z \models \psi$ for all y, z.*
3. $\mathcal{M}, x \models N\varphi$ *iff $\mathcal{M}, g(x) \models \varphi$.*
4. $\mathcal{M}, x \models \varphi \multimap \psi$ *iff $x = f^d(y, z)$ and $\mathcal{M}, y \models \varphi$ implies $\mathcal{M}, z \models \psi$ for all y, z.*
5. $\mathcal{M}, x \models \varphi \parr \psi$ *iff there are y, z with $x = f^d(y, z)$, $\mathcal{M}, y \models \varphi$ and $\mathcal{M}, z \models \psi$.*
6. $\mathcal{M}, x \models \varphi \bindnasrepma \psi$ *iff $x = f^d(y, z)$ implies $\mathcal{M}, y \models \varphi$ or $\mathcal{M}, z \models \psi$ for all y, z.*
7. $\mathcal{M}, x \models U$ *iff $x = g(e)$.*

Now we are ready to present the axioms forcing f and f^d to be commutative functions with identity e resp. $g(e)$ that distribute over each other. Please note that all formulas in the next lemma are actually formula schemas.

Lemma 2.5. *Let $\mathcal{F}$ be a PRLE-frame. Then we have:*

1. $\mathcal{F} \models \varphi \mathbin{\wedge\!\!\!\wedge} \psi \rightarrow \psi \mathbin{\wedge\!\!\!\wedge} \varphi$ *iff f is commutative.*
2. $\mathcal{F} \models \varphi \parr \psi \rightarrow \psi \parr \varphi$ *iff f^d is commutative.*
3. $\mathcal{F} \models \varphi \rightarrow \varphi \mathbin{\wedge\!\!\!\wedge} E$ *iff $x = f(x, e)$ for all $x \in W$.*
4. $\mathcal{F} \models \varphi \rightarrow (\varphi \parr U)$ *iff $x = f^d(x, g(e))$ for all $x \in W$.*
5. $\mathcal{F} \models \varphi \mathbin{\wedge\!\!\!\wedge} (\psi \parr \chi) \rightarrow (\varphi \mathbin{\wedge\!\!\!\wedge} \psi) \parr (\varphi \mathbin{\wedge\!\!\!\wedge} \chi)$ *iff f distributes over f^d.*
6. $\mathcal{F} \models \varphi \parr (\psi \mathbin{\wedge\!\!\!\wedge} \chi) \rightarrow (\varphi \parr \psi) \mathbin{\wedge\!\!\!\wedge} (\varphi \parr \chi)$ *iff f^d distributes over f.*

The proof of the previous lemma is similar to the proof of Lemma 2.2 and, therefore, omitted.

By assuming two additional properties of the frame we obtain axioms for the remaining properties of f resp. f^d.

Lemma 2.6. *Let $\mathcal{F}$ be a PRLE-frame with $g(g(x)) = x$ and $x = f(x, e)$ for all $x \in W$. Then we have:*

1. $\mathcal{F} \models \varphi \to \top \parr (U \wedge (\varphi \Wedge N\varphi))$ *iff* $f(x, g(x)) = g(e)$ *for all* $x \in W$.
2. $\mathcal{F} \models \varphi \to \top \Wedge (E \wedge (\varphi \parr N\varphi))$ *iff* $f^d(x, g(x)) = e$ *for all* $x \in W$.

Proof. We only prove the first property.

$\Rightarrow$: Assume the schema is true in $\mathcal{F}$ and let $x \in W$. Then we have $\mathcal{M}, x \models p \to \top \parr (U \wedge (p \Wedge Np))$ where $\mathcal{M}$ is the model based on $\mathcal{F}$ with $v(p) = \{x\}$. Since $\mathcal{M}, x \models p$ we obtain $\mathcal{M}, x \models \top \parr (U \wedge (p \Wedge Np))$. This implies that there are y, z with $x = f^d(y, z)$ and $\mathcal{M}, y \models \top$ and $\mathcal{M}, z \models U \wedge (p \Wedge Np)$. From the letter we obtain that $z = g(e)$ by Lemma 6 and, hence, $\mathcal{M}, g(e) \models p \Wedge Np$. Again, the latter implies that there are u, v with $g(e) = f(u, v)$ and $\mathcal{M}, u \models p$ and $\mathcal{M}, v \models Np$. From the definition of $v(p)$ we get $u = x$, and, hence, $g(e) = f(x, v)$. From $\mathcal{M}, v \models Np$, which is equivalent to $\mathcal{M}, g(v) \models p$, and the definition of $v(p)$ we obtain $g(v) = x$. This implies $v = g(g(v)) = g(x)$ by the assumption and, hence, $g(e) = f(x, g(x))$.

$\Leftarrow$: Assume $f(x, g(x)) = g(e)$ for all $x \in W$ and that $\mathcal{M}, x \models \varphi$. Then $\mathcal{M}, g(x) \models N\varphi$ so that the assumption implies $\mathcal{M}, g(e) \models \varphi \Wedge N\varphi$. Using Lemma 6 we get $\mathcal{M}, g(e) \models U \wedge (\varphi \Wedge N\varphi)$. Since $\mathcal{M}, x \models \top$ and $f^d(x, g(e)) = g(f(g(x), g(g(e)))) = g(f(g(x), e)) = g(g(x)) = x$ by using both assumptions we obtain $\mathcal{M}, x \models \top \parr (U \wedge (\varphi \Wedge N\varphi))$. □

Please note that under the assumption $g(g(x)) = x$ the commutativity axioms, the identity axioms, the distributivity axioms, and the complement axioms for f resp. f^d are equivalent. We summarize our results so far in the following theorem.

Theorem 2.1. *A PRLE-frame $\mathcal{F}$ together with the definitions*

$$x + y := f(x, y), \quad x \cdot y := f^d(x, y), \quad x^* := g(x), \quad 0 := e, \quad 1 := g(e)$$

is a Boolean algebra iff the axiom schemas

1. $\varphi \to \sim\sim\varphi$,
2. $\varphi \Wedge \psi \to \psi \Wedge \varphi$,
3. $\varphi \to \varphi \Wedge E$,
4. $\varphi \Wedge (\psi \parr \chi) \to (\varphi \Wedge \psi) \parr (\varphi \Wedge \chi)$,
5. $\varphi \to \top \parr (U \wedge (\varphi \Wedge N\varphi))$

are true in $\mathcal{F}$.

Due to the previous lemma, the frames of the logic PRLE with the axiom schemas above are Boolean algebras. The relevance implication resp. the relevance negation are based on the join resp. complement operations of the Boolean algebra, and the basic formula E is true at the bottom element. Therefore, this logic is capable of reasoning about arbitrary Boolean algebras, i.e., the mereological part of a BCA.

2.5 Modal Relevance Logic (MRL)

In this section we will add modal operators to PRLE in order to cover Boolean contact algebras.

Definition 2.16. *The set MRL of modal relevance logic formulas is recursively defined by the rules PropL.1–PropL.3, RelL.1–RelL.2, PRLE.1 and ModL.1.*

Frames and models are defined as expected.

Definition 2.17. *A modal relevance logic frame (MRL-frame) is a structure* $\mathcal{F} = \langle W, e, R, f, g\rangle$ *so that* $\langle W, e, f, g\rangle$ *is a PRLE-frame and* R *is a binary relation on* W. $\mathcal{F}$ *is called a Boolean modal relevance logic frame (BMRL-frame) if* W *with the definitions of Theorem 2.1 is a Boolean algebra. A MRL/BMRL-model is a MRL/BMRL-frame together with a valuation function* v.

Based on the previous definitions we define the semantics as follows.

Definition 2.18. *Let* $\mathcal{M}$ *be a model,* $x \in W$, $\varphi \in MRL$. *The satisfaction relation* $\mathcal{M}, x \models \varphi$, *is recursively defined by SemPropL.1–SemPropL.3, SemRelL.1–SemRelL.2, SemPRLE.1 and SemModL.1.*

The next lemma presents axiom schemas that are equivalent to null disconnectedness, reflexivity, symmetry, compatibility, and summation property of a Boolean contact algebra.

Lemma 2.7. *Let* $\mathcal{F}$ *be a BMRL-frame. Then we have:*

1. $\mathcal{F} \models [R]\neg E$ *iff* xRy *implies* $y \neq 0$ *for all* $x, y \in W$.
2. $\mathcal{F} \models \neg E \to ([R]\varphi \to \varphi)$ *iff* R *satisfies* C_1.
3. $\mathcal{F} \models \varphi \to [R]\langle R\rangle\varphi$ *iff* R *satisfies* C_2.
4. $\mathcal{F} \models [R]\varphi \to [R](\top \multimap \varphi)$ *iff* R *satisfies* C_3.
5. *If* R *satisfies* C_2, *then* $\mathcal{F} \models \varphi \to [R](\neg\langle R\rangle\varphi \twoheadrightarrow \langle R\rangle\varphi)$ *iff* R *satisfies* C_4.

The proof of the previous lemma is omitted due to lack of space. For details we refer to [13]. Obviously, a BRML-frame with the relation R is a BCA iff the axiom schemas of the previous lemma are true in $\mathcal{F}$. We will call such a frame a BCMRL-frame.

Similar to the previous section, the logic BMRL with the axioms above is capable of reasoning about arbitrary BCAs.

3 Natural Deduction

In this section we want to sketch a natural deduction system for the logic MRL with the axioms from Theorem 2.1 and Lemma 2.7. This calculus will actually utilize annotated MRL-formulas. In order to define these formulas we first need to define terms.

Definition 3.1. *Let X be a set of variables. Then the set Term is recursively defined by:*

1. *Every variable is a term, i.e., $X \subseteq Term$.*
2. *If $\alpha \in Term$, then $\alpha^* \in Term$.*
3. *If $\alpha, \beta \in Term$, then $\alpha + \beta \in Term$ and $\alpha \cdot \beta \in Term$.*

As already indicated in the previous definition we will use $x, y, z, \ldots$ to denote variables and $\alpha, \beta, \gamma, \ldots$ to denote terms.

If $\mathcal{F}$ is a BMRL-frame, any function $n : X \to W$ can naturally be extended to $Term$ by mapping the symbols $+, \cdot, ^*$ to the corresponding operation in W. We will call n an environment and use the notation $n[\alpha]$ to denote the application of the extension of n to the term α.

An annotated formula is of the form φ_α where φ is a MRL-formula and α is a term. Given a BCMRL-model $\mathcal{M}$, we define the satisfaction relation for annotated formulas by $\mathcal{M}, n \models \varphi_\alpha$ iff $\mathcal{M}, n[\alpha] \models \varphi$. Similar to regular formulas we define $\mathcal{M} \models \varphi_\alpha$ iff $\mathcal{M}, n \models \varphi_\alpha$ for all n. Obviously, we have for any variable $x \in X$ that $\mathcal{M} \models \varphi$ iff $\mathcal{M} \models \varphi_x$.

Please note that we will skip the annotation for the formula $\bot$ since $\bot$ is never true.

The calculus actually uses three different kind of formulas, annotated formulas, equation of the form $\alpha = \beta$, and statements of the form $\alpha C \beta$ for terms α, β. We will use $\chi, \ldots$ to denote either one of these kinds. Besides the introduction and elimination rules for each logical operator and certain "proof by contradiction"-style rules, the calculus has some specific rules and axioms. Commutativity and distributivity of $+$ and $\cdot$ are used as axioms, i.e., assumptions in natural deduction that do not need to be discarded. Please note that the identity and complement axioms are not used since the are covered by the introduction and elimination rule of E (resp. U) and N (resp. $\sim$). In addition, we use the following rules that are an immediate consequence of the axioms listed in Lemma 2.7:

$$\frac{\alpha C \beta}{(\neg E)_\alpha}\ (BCA_{0l}) \qquad \frac{\alpha C \beta}{(\neg E)_\alpha}\ (BCA_{0r}) \qquad \frac{(\neg E)_\alpha}{\alpha C \alpha}\ (BCA_1) \qquad \frac{\beta C \alpha}{\alpha C \beta}\ (BCA_2)$$

$$\frac{\alpha C \beta \quad \beta = \beta \cdot \gamma}{\alpha C \gamma}\ (BCA_3) \qquad \frac{\alpha C(\beta + \gamma) \quad \begin{array}{c}[\alpha C \beta] \\ \vdots \\ \chi\end{array} \quad \begin{array}{c}[\alpha C \gamma] \\ \vdots \\ \chi\end{array}}{\chi}\ (BCA_4)$$

In the calculus we treat all logical operations as basic and not as an abbreviation as we did in the previous sections. We have 17 logical operators (listed below).

$$=, \neg, \to, \wedge, \vee, \sim, \twoheadrightarrow, \mathbin{\wedge\!\!\!\wedge}, \mathbin{\vee\!\!\!\vee}, N, \multimap, ⅋, \between, E, U, [C], \langle C \rangle$$

Three of the operators are negation-style operators ($\neg, \sim, N$) that require additionally a "proof by contradiction"-style rule. Therefore, we have in total 37 rules. For lack of space we cannot list all of them in this paper. As examples we present the introduction and elimination rules for $\mathbin{\wedge\!\!\!\wedge}$ and $[C]$.

$$\dfrac{\varphi_\beta \quad \psi_\gamma \quad \alpha = \beta + \gamma}{(\varphi \mathbin{\wedge\!\!\!\wedge} \psi)_\alpha}\ (\mathbin{\wedge\!\!\!\wedge} I) \qquad \dfrac{(\varphi \mathbin{\wedge\!\!\!\wedge} \psi)_\alpha \qquad \begin{matrix}[\varphi_x] \quad [\psi_y] \quad [\alpha = x + y] \\ \vdots \\ \chi\end{matrix}}{\chi}\ (\mathbin{\wedge\!\!\!\wedge} E)^*$$

$$\dfrac{\begin{matrix}[\alpha C y] \\ \vdots \\ \varphi_y\end{matrix}}{([C]\varphi)_\alpha}\ ([C]I)^{**} \qquad \dfrac{([C]\varphi)_\alpha \quad \alpha C \beta}{\varphi_\beta}\ ([C]E)$$

Following typical notation we have denoted by $[\varphi]$ the fact that the rule discards the assumption φ. In particular, the $(\mathbin{\wedge\!\!\!\wedge} E)$ and $([C]I)$ discard the assumptions φ_x, ψ_y, and $\alpha = x + y$ resp. $\alpha C y$. As indicated by * resp. ** the rules $(\mathbin{\wedge\!\!\!\wedge} E)$ and $([C]I)$ have a side-condition similar to the $\forall$-introduction and the $\exists$-elimination rules in the calculus for first-order logic.

*: The variables x, y do not occur in χ and in any open assumption of the right subtree except φ_x, ψ_y, and $\alpha = x + y$.

**: The variables y does not occur in any open assumption of the subtree except $\alpha C y$.

In the following we will write $\Gamma \vdash \chi$ if there is a derivation in the calculus with open assumptions among Γ and conclusion χ. We present two examples of derivations in our calculus. The first example, shows one of the implications related to the definition of $\mathbin{\wedge\!\!\!\wedge}$ by $\neg$ and $\twoheadrightarrow$:

$$\dfrac{\dfrac{\dfrac{[(\varphi \mathbin{\wedge\!\!\!\wedge} \psi)_z]^1 \qquad \dfrac{\dfrac{[(\varphi \twoheadrightarrow \neg\psi)_z]^2 \quad [\varphi_x]^3 \quad [z = x + y]^3}{(\neg\psi)_y}\ (\twoheadrightarrow E^5) \qquad [\psi_y]^3}{\bot}\ (\neg E^4)}{\bot}\ (\mathbin{\wedge\!\!\!\wedge} E^3)}{\neg(\varphi \twoheadrightarrow \neg\psi)_z}\ (\neg I^2)}{((\varphi \mathbin{\wedge\!\!\!\wedge} \psi) \rightarrow \neg(\varphi \twoheadrightarrow \neg\psi))_z}\ (\rightarrow I^1)$$

Please note that side-condition in the application of the $(\mathbin{\wedge\!\!\!\wedge} E)$ rule is satisfied. Furthermore, the proof has no open assumption so that its conclusion is valid by the correctness theorem. The second example verifies the commutativity of $\mathbin{\wedge\!\!\!\wedge}$:

$$\dfrac{\dfrac{[(\varphi \mathbin{\wedge\!\!\!\wedge} \psi)_z]^1 \qquad \dfrac{[\varphi_x]^2 \quad [\psi_y]^2 \qquad \dfrac{\dfrac{}{x + y = y + x} \quad [z = x + y]^2}{z = y + x}\ (= E)}{(\psi \mathbin{\wedge\!\!\!\wedge} \varphi)_z}\ (\mathbin{\wedge\!\!\!\wedge} I^3)}{(\psi \mathbin{\wedge\!\!\!\wedge} \varphi)_z}\ (\mathbin{\wedge\!\!\!\wedge} E^2)}{(\varphi \mathbin{\wedge\!\!\!\wedge} \psi \rightarrow \psi \mathbin{\wedge\!\!\!\wedge} \varphi)_z}\ (\rightarrow I^1)$$

Again, the side-condition of $(\mathbin{\wedge\!\!\!\wedge} E)$ is satisfied and the proof has no open assumptions. Please note that this proof uses the commutativity of $+$ as an axiom, i.e., an assumption that does not have to be discarded. We have indicated this by using the notation of a rule without a subtree.

Theorem 3.1. *The calculus is sound, i.e.,* $\Gamma \vdash \varphi_z$ *implies* $\Gamma \models \varphi$.

The proof of the previous theorem is done by induction on the derivation tree. This is straight-forward and, therefore, omitted. For details, please refer to [13].

4 Implementation in Coq

In this section, we will discuss the implementation of our calculus in Coq. For more information about Coq we refer to [8,9,18]. First we define the theory of Boolean algebras as a class (cf. [8,18]). In order to be able to use appropriate syntactical notation we define the signature of a Boolean algebra first:

```
Class BASig (A : Type) := {
join    :   A -> A -> A;
meet    :   A -> A -> A;
zero    :   A;
one     :   A;
comp    :   A -> A }.
```

Here `A` is the underlying type or set of the algebra, `join` and `meet` are two binary operators, `comp` is a unary operator, and `zero` and `one` are two elements of `A`. As mentioned above we now define appropriate syntax for this structure:

```
Infix "+"          := (join).
Infix "*"          := (meet).
Notation "0"       := zero.
Notation "1"       := one.
Notation "x  ^* " := (comp x) (at level 30).
```

The next class definition adds the axioms of a Boolean algebra as listed in Definition 2.1. Obviously, we will have to add the signature to the structure.

```
Class BA (A : Type) := {
sig           : > BASig A;
join_comm     : forall x y, x + y = y + x;
zero_ident    : forall x, x + 0 = x;
join_distr    : forall x y z, x + y * z = (x + y) * (x + z);
join_comp     : forall x, x + x^* = 1;
meet_comm     : forall x y, x * y = y * x;
one_ident     : forall x, x * 1 = x;
meet_distr    : forall x y z, x * (y + z) = x * y + x * z;
meet_comp     : forall x, x * x^* = 0 }.
```

Our actual implementation provides proofs of several theorems related to Boolean algebras. For example, we show that associativity of both operations follows from the definition. Due to lack of space, we omit these theorems and proofs in this paper.

In addition to the operations of a Boolean algebra we define the order relation and appropriate syntax for them as follows:

```
Definition leBA {A : Type} {ba : BA A} : A -> A -> Prop := fun x
    y => x * y = x.
Infix "<=" := leBA.
Definition geBA {A : Type} {ba : BA A} : A -> A -> Prop := fun x
    y => x + y = x.
Infix ">=" := geBA.
```

A Boolean contact algebra is now defined by extending the class of a Boolean algebra. Obviously, we add the binary contact relation with the appropriate axioms (cf. Definition 2.2).

```
Class BCA (A : Type) := {
  ba :> BA A;
  C  : relation A;
  c0 : forall x y,     C x y                -> x<>0 /\ y<>0;
  c1 : forall x,       x<>0                 -> C x x;
  c2 : forall x y,     C x y                <-> C y x;
  c3 : forall x y z, (C x y /\ y <= z)      -> C x z;
  c4 : forall x y z,  C x (y+z)             -> (C x y \/ C x z)
}.
```

In order to implement the calculus we follow the same idea as in [2] for modal logics. A proposition (or formula) becomes a proposition on the underlying type of the Boolean algebra. The function `V` below applies an MRL-formula to a specific element resulting in a proposition. We say that the formula is valid iff it is true for all elements and all Boolean algebras. This is defined as the function `MRLValid` below. In addition, we introduce proper syntax for annotated formulas. Finally, the tactic `start` will convert a MRL-formula into an annotated formula and start the proof within our calculus.

```
Definition MRLProp := forall (A : Type) (bca : BCA A), A -> Prop.
Definition V (p : MRLProp) (A : Type) (bca : BCA A) (x : A) :
    Prop := p A bca x.
Definition MRLValid (p : MRLProp) : Prop := forall (A : Type)
    (bca : BCA A) (x : A), V p A bca x.
Notation "[ p ] x" := (V p _ _ x) (at level 70).
Ltac start := unfold MRLValid; let A:= fresh "A" in intro A;let
    bca:= fresh "bca" in intro bca;let z:= fresh "z" in intro z.
```

Every logical operator of MRL is implemented as function on `MRLProp`. As an example we provide the definition of the $\mathbin{⩓}$ operator. Please note that the definition follows the satisfaction relation for this operator as verified in Lemma 1. As before we also define proper syntax for each operator.

```
Definition MRL_RJ_And  (p q : MRLProp) : MRLProp := fun (A :
    Type) (bca : BCA A) x => exists y z, x = y + z /\ V p A bca y
    /\ V q A bca z.
Infix      "//\\"      := (MRL_RJ_And)      (at level 60).
```

Every rule of the calculus is implemented as tactic in Coq. As an example we provide the implementation of the ($\owedge I$) and ($\owedge E$) rule without explaining the details of their implementation. We just want to mention that Coq will automatically take care of the side-condition of ($\owedge E$) by using new variables (application of `fresh` in the tactic below).

```
Ltac MRL_RJ_And_Intro x y :=
  match goal with
    | |- V (?p //\\ ?q) ?A ?bca ?z
          => replace (V (p //\\ q) A bca z)
               with ( MRL_RJ_And p q A bca z)
               by (unfold V; unfold MRL_RJ_And; trivial);
             exists x; exists y; repeat split; repeat assumption
    | _ => fail 1 "Goal is not an MRL_RJ_And formula"
  end.

Ltac MRL_RJ_And_Elim H0 :=
  match type of H0 with
    | V (?p //\\ ?q) ?A ?bca ?z
      => replace (V (p //\\ q) A bca z)
             with ( MRL_RJ_And p q A bca z)
             by (unfold V; unfold MRL_RJ_And in H0; trivial);
           let x := fresh "x" in destruct H0 as [x H0];
           let y := fresh "y" in destruct H0 as [y H0];
           let H1 := fresh "H" in destruct H0 as [H1 H0];
           let H2 := fresh "H" in destruct H0 as [H2 H0];
             repeat assumption
    | _ => fail 1 "Hypothesis is not an MRL_RJ_And formula"
  end.
```

Last but not least we list the two proof trees from the previous section as Coq lemmas. A close investigation of the sequence of rules applied in the proof will show that the Coq proofs are a one-to-one translation of the proof trees from the previous section.

```
Lemma Abbreviations_Axiom_MRL_Logic_1 (φ ψ : MRLProp) : MRLValid
    (φ //\\ ψ -> ¬ (φ ->> ¬ ψ)).
Proof.
    start.
    MRL_P_Impl_Intro. MRL_P_Not_Intro.
    MRL_RJ_And_Elim H. assume ([¬ ψ] y).
    MRL_P_Not_Elim H5 H. MRL_RJ_Impl_Elim H0 H4 H3.
Qed.
```

Please note that the tactic `assume` is used like similar to a cut in the sequent calculus. It adds the formula as an assumption and simultaneously generates a proof obligation with the formula as a goal. This is convenient when stitching together separate proof trees.

```
Lemma Boolean_Algebra_Axiom_MRL_Logic_1 (φ ψ : MRLProp) :
    MRLValid ((φ //\\ ψ) -> (ψ //\\ φ)).
```

```
Proof.
  start.
  MRL_P_Impl_Intro. MRL_RJ_And_Elim H.
  MRL_RJ_And_Intro y x. rewrite join_comm in H0; assumption.
Qed.
```

The tactic `rewrite` implements the rule $(= E)$, i.e., `rewrite join_comm in H0` corresponds to the $(= E)$ in the right subtree of the previous section.

5 Conclusion and Future Work

In this paper we have introduced a modal relevance logic for mereotopology, i.e., we presented a logic for QSR. The logic utilizes a modal operator to cover the topological aspects and relevance operators to cover the mereological aspects. We have provide a set of axioms that a necessary and sufficient for this application. Furthermore, we presented a natural deduction system for this logic, which has been shown to be sound. An implementation of the calculus in Coq has been presented and will be accessible via [13].

Future work will mainly focus on three aspects. First, we would like to show that the calculus is also complete. We will apply standard techniques for this purpose and do not foresee any problems in doing so. Secondly, we will investigate the decidability of this logic. We would like to show a filtration theorem similar to those in modal logic showing that any satisfiable formula will be satisfiable within a model of a certain size determined by the size of the formula. Last but not least, we would like to extend the logic by adding additional axioms known from the theory of Boolean contact algebras. In particular, we are interested in the extensionality of C (C_5), the interpolation axiom (C_6), and the connection axiom (C_7). To our knowledge it is currently unknown whether these extended theories are decidable. Any attempt to show decidability will definitely not involve a filtration argument since these extended theories do not have finite models.

References

1. Anderson, A.R., Belnap, N., Dunn, J.: Entailment: The Logic of Relevance and Necessity, vol. 1. Princeton University Press (1976)
2. Benzmüller, C., Woltzenlogel Paleo, B.: Interacting with modal logics in the Coq proof assistant. In: Beklemishev, L.D., Musatov, D.V. (eds.) CSR 2015. LNCS, vol. 9139, pp. 398–411. Springer, Cham (2015). https://doi.org/10.1007/978-3-319-20297-6_25
3. Blackburn, P., de Rijke, M., Venema, Y.: Modal Logic. Cambridge Tracts in TCS 8 (2002)
4. Barmak, J.A.: Algebraic Topology of Finite Topological Spaces and Applications. LNM, vol. 2032. Springer, Heidelberg (2011). https://doi.org/10.1007/978-3-642-22003-6
5. Baum, J.D.: Elements of Point Set Topology. Dover (1991)
6. Bennett, B.: Modal logics for qualitative spatial reasoning. Logic J. IGPL **4**(1), 23–45 (1996)

7. Boolean Algebra (structure). https://en.wikipedia.org/wiki/Boolean_algebra_(structure), Accessed 15 Apr 2017
8. Chlipala, A.: An introduction to programming and proving with dependent types in Coq. J. Formalized Reasoning **3**(2), 1–93 (2010)
9. The Coq Proof Assistant. https://coq.inria.fr/. Accessed 15 Apr 2017
10. Dunn, J.M., Restall, G.: Relevance logic. In: Gabbay, G., Guenthner, F. (eds.) Handbook of Philosophical Logic, pp. 1–128 (2002)
11. Düntsch, I., Winter, M.: Algebraization and representation of mereotopological structures. JoRMiCS **1**, 161–180 (2004)
12. Düntsch, I., Winter, M.: A representation theorem for boolean contact algebras. TCS **347**(3), 498–512 (2005)
13. Ghosh, P.K.: A Modal Relevance Logic for Qualitative Spatial Reasoning. MSc. Thesis, Brock University (tbf) (2018)
14. Huth, M., Ryan, M.: Logic in Computer Science: Modelling and Reasoning About Systems. Cambridge University Press (2004)
15. Munkres, J.R.: Topology: A First Course. Prentice-Hall (1974)
16. Read, S.: Relevant Logic: A Philosophical Examination of Inference. Basil Blackwell (1989)
17. Renz, J. (ed.): Qualitative Spatial Reasoning with Topological Information. LNCS (LNAI), vol. 2293. Springer, Heidelberg (2002). https://doi.org/10.1007/3-540-70736-0
18. Sozeau, M., Oury, N.: First-class type classes. In: Mohamed, O.A., Muñoz, C., Tahar, S. (eds.) TPHOLs 2008. LNCS, vol. 5170, pp. 278–293. Springer, Heidelberg (2008). https://doi.org/10.1007/978-3-540-71067-7_23

On the Structure of Generalized Effect Algebras and Separation Algebras

Sarah Alexander, Peter Jipsen(✉), and Nadiya Upegui

Chapman University, Orange, CA, USA
jipsen@chapman.edu

Abstract. *Separation algebras* are models of separation logic and *effect algebras* are models of unsharp quantum logics. We investigate these closely related classes of partial algebras as well as their noncommutative versions and the subclasses of (generalized) (pseudo-)orthoalgebras. We present an orderly algorithm for constructing all nonisomorphic generalized pseudoeffect algebras with n elements and use it to compute these algebras with up to 10 elements.

1 Introduction

Separation algebras were introduced by Calcagno, O'Hearn and Yang [3] as semantics for separation logic, and *effect algebras* were defined by Foulis and Bennett [6] as an abstraction of unsharp measurements in quantum mechanics. Detailed definitions are recalled in the next section, but we note here that they are both cancellative commutative partial monoids and that every effect algebra is a separation algebra. Hence results about separation algebras automatically apply to effect algebras, and the in-depth study of effect algebras over the past two decades provides insight into this particular subclass of separation algebras. Lattice effect algebras, MV-effect algebras, orthoalgebras, orthomodular posets, orthomodular lattices and Boolean effect algebras are all well known subclasses of effect algebras, so positioning effect algebras as a subclass of separation algebras provides many interesting algebraic models for separation logic.

In this paper, we are primarily interested in finite partial algebras since they can be computed for small cardinalities, and browsing models with up to a dozen elements is useful for investigating the structure of these finite algebras. To this end, we develop an algorithm for computing finite effect algebras and some of their noncommutative generalizations.

One of the aims of this paper is to increase awareness of the model theory of partial algebras since it has been developed quite extensively, but is not necessarily widely known. In the next section, we recall the basic notions of weak/full/closed homomorphisms, subalgebras and congruences for partial algebras. Examples from the classes of separation algebras and effect algebras are discussed in Sect. 3. Many of the results about these algebras do not depend on the commutativity of the partial binary operation $+$, hence we mostly consider noncommutative versions and often write the operation as $x \cdot y$ or xy rather than

J. Desharnais et al. (Eds.): RAMiCS 2018, LNCS 11194, pp. 148–165, 2018.
https://doi.org/10.1007/978-3-030-02149-8_10

$x+y$. Effect algebras without the assumption of commutativity were introduced by Dvurecenski and Vetterlein [5] under the name *pseudoeffect algebras* and have been studied extensively since then.

Every partial algebra $\mathbf{A}$ can be easily *lifted* to a total algebra $\hat{A} = A \cup \{\mathbf{u}\}$, where the element $\mathbf{u}$ denotes undefined, and an operation on A produces $\mathbf{u}$ as output whenever the operation is undefined. In particular, if any of the inputs to the lifted operation are $\mathbf{u}$, then the output is $\mathbf{u}$. This map from partial to total algebras is a functor between the respective categories, but it does not preserve direct products of algebras, which means that the universal algebraic theory of partial algebras is not subsumed by total algebras. Given a partial monoid $\mathbf{A} = (A, \cdot, e)$, its lifted version is a well known total algebra called a *monoid with zero* $\hat{\mathbf{A}} = (A, \cdot, e, 0)$. These algebras occur, for example, as reducts of rings when $+$ and $-$ are removed from the signature. A monoid with zero is *cancellative* if all nonzero elements can be cancelled on the left and right of the multiplication operation. For partial algebras, a binary operation is *left-cancellative* if whenever $xy = xz$ are defined, then $y = z$. *Right-cancellativity* is defined analogously. Hence a partial monoid is cancellative if and only if the corresponding lifted total monoid with zero is cancellative.

Note that in the finite case these cancellative partial monoids are quite close to groups. For example, given any element x in a finite cancellative **total** monoid without a zero, the sequence $x, x^2, x^3, \dots, x^n, \dots$ must contain a duplicate when n exceeds the cardinality of the monoid. Hence, $x^i = x^j$ for some $i > j$ and by cancellativity $xx^{i-j-1} = e$, so x has an inverse. This well known argument shows that the class of finite cancellative monoids coincides with the class of finite groups. The significance of group theory in mathematics and its numerous fundamental applications in the sciences are well established, and allowing partiality of the binary operation leads to the class of finite cancellative partial monoids that properly contains all finite groups, thus making it an important class to study.

Complex algebras of separation algebras provide models of Boolean bunched implication logic and, in the noncommutative case, models of Boolean residuated lattices, also called residuated monoids in [8–10]. This indicates that separation algebras are functional Kripke structures, and in the past decade the field of modal logics and their Kripke semantics has been recognized as a branch of coalgebra. This meshes well with recent approaches to separation algebras [4] and effect algebroids [12]. We also highlight a method of [13] (Prop. 20) that converts a generalized pseudoeffect algebra to a total residuated partially ordered monoid by adding two elements $\bot, \top$, and we note that this totalization method preserves the property of being involutive.

In Sect. 2 we give basic definitions of partial algebras, homomorphisms, subalgebras, congruences and related concepts. Section 3 contains definitions and results about generalized separation algebras and (generalized pseudo-)effect algebras, and we map out some of the subclasses and implications between various axioms. Section 4 covers the results leading to the orderly algorithm for constructing all finite generalized pseudoeffect algebras up to isomorphism.

In the subsequent section we prove new structural results about certain effect algebras that were suggested by the output of our enumeration program, and Sect. 6 concludes with some remarks and open problems.

2 Background on Partial Algebras

To facilitate our discussion of separation algebras and effect algebras, we begin with a brief summary of partial algebras. More details can be found in [1,2]. A *partial operation* g of arity n on a set A is a function from a subset $D(g)$ of A^n to A. The set $D(g)$ is the *domain* of g, and $(a_1, \dots, a_n) \in D(g)$ is also written as $g(a_1, \dots, a_n) \neq \mathbf{u}$ (but this is just convenient notation; $\mathbf{u}$ is not an element of A). Two partial operations g, h on A are equal if $D(g) = D(h)$ and $g(a_1, \dots, a_n) = h(a_1, \dots, a_n)$ for all $(a_1, \dots, a_n) \in D(g)$. The notation $g : A^n \multimap A$ is used to indicate that g is an n-ary partial function on A. If $D(g) = A^n$, then g is a *total operation*, or simply an *operation*. If $n = 0$ then g is a *constant operation*, which we always assume to be total. A *signature* is a function $\sigma : \mathcal{F} \to \mathbb{N}$ where $\mathcal{F}$ is a set. The members of $\mathcal{F}$ are called *(partial) operation symbols*.

A *partial algebra of type* τ is a pair $\mathbf{A} = (A, \mathcal{F}^{\mathbf{A}})$ where A is a set and $\mathcal{F}^{\mathbf{A}} = \{f^{\mathbf{A}} : A^{\sigma(f)} \multimap A \mid f \in \mathcal{F}\}$ is a set of partial operations on A. If every partial operation in $\mathcal{F}^{\mathbf{A}}$ is in fact total, then $\mathbf{A}$ is a *total algebra*. Examples of partial algebras abound since any subset B of a total algebra $\mathbf{A}$ is the universe of an induced partial algebra $\mathbf{B}$, with *partial* operations $f^{\mathbf{B}}$ given by $f^{\mathbf{A}}$ restricted to B, so for $b_1, \dots, b_n \in B$, $f^{\mathbf{B}}(b_1, \dots, b_n)$ is undefined if and only if $f^{\mathbf{A}}(b_1, \dots, b_n) \notin B$. $\mathbf{B}$ is called a *relative subalgebra* of $\mathbf{A}$, and $\mathbf{A}$ is a *total extension* of $\mathbf{B}$. More natural examples are given by any field, such as the rational, real or complex numbers, with a signature that includes $^{-1}$ or division $/$, since $0^{-1} = \mathbf{u} = x/0$.

Terms, equations (= atomic formulas) and first-order formulas over a set of variables $X = \{x_1, x_2, \dots\}$ are defined inductively as for total algebras, but for a term t we also write $t = \mathbf{u}$ or $t \neq \mathbf{u}$ depending on whether t is undefined or defined. For a partial algebra $\mathbf{A}$ and an assignment $\mathbf{a} : X \to A$, the semantic interpretation of a term t as a *term function* $t^{\mathbf{A}} : A^n \to A$ is defined inductively by $t^{\mathbf{A}}(\mathbf{a}) = \mathbf{a}(t)$ if $t \in X$, and for $t = f(t_1, \dots, t_n)$,

$$t^{\mathbf{A}}(\mathbf{a}) = \begin{cases} f^{\mathbf{A}}(t_1^{\mathbf{A}}(\mathbf{a}), \dots, t_n^{\mathbf{A}}(\mathbf{a})) \text{ if } t_i^{\mathbf{A}}(\mathbf{a}) \neq \mathbf{u} \text{ for all } i = 1, \dots, n \\ \mathbf{u} \text{ otherwise.} \end{cases}$$

Hence if any subterm is undefined under the assignment, then the whole term is undefined. An *identity* (i.e. universally quantified equation with no free variables) $s = t$ is satisfied by an algebra $\mathbf{A}$, written $\mathbf{A} \models s = t$ if $s^{\mathbf{A}} = t^{\mathbf{A}}$, i.e., if the term functions are equal. Note that this means both sides have to be defined or both sides have to be undefined for any given input tuple. This interpretation of an equation in partial algebras is called a *strong identity* or *Kleene identity*. An even stronger form of satisfaction is given by *existence identities*: $\mathbf{A} \models s \stackrel{e}{=} t$ if $s^{\mathbf{A}} = t^{\mathbf{A}}$ and $D(s^{\mathbf{A}}) = A$. Note that for an identity of the form $x = t$ the concept of strong identity and existence identity coincide since $x^{\mathbf{A}}$ is always defined for a variable x.

A quasi-identity is a formula $s_1 = t_1 \,\&\ldots\&\, s_m = t_m \implies s = t$, and is satisfied in $\mathbf{A}$ if any assignment to the variables that satisfies $s_1 = t_1, \ldots, s_m = t_m$ (both sides defined) also satisfies $s = t$ (both sides defined). Under this interpretation a quasi-identity with no premises ($m = 0$) is equivalent to an existence identity. A *(strong/existence/quasi)equational class* $\mathcal{K}$ of partial algebras is a class of algebras of the same signature such that for some set $\mathcal{I}$ of (strong/quasi/existence) identities we have $\mathcal{K} = \{\mathbf{A} : \mathbf{A} \models \varepsilon \text{ for all } \varepsilon \in \mathcal{I}\}$.

Direct products $\prod_{i\in I} \mathbf{A}_i$ are defined for partial algebras in exactly the same way as for total algebras, with pointwise fundamental operations $f(\mathbf{x}_1, \ldots, \mathbf{x}_n) = (\ldots, f^{\mathbf{A}_i}(x_{1i}, \ldots, x_{ni}), \ldots)$ that are defined iff $f^{\mathbf{A}_i}(x_{1i}, \ldots, x_{ni})$ is defined for all $i \in I$. For a class $\mathcal{K}$ of partial algebras the class of products of members of $\mathcal{K}$ is denoted by $\mathbf{P}\mathcal{K}$.

There are three notions of homomorphism, with the weakest one being standard relational homomorphism. A function $h : \mathbf{A} \to \mathbf{B}$ is

- a *(weak) homomorphism* if for all $f \in \mathcal{F}$,

$$(a_1, \ldots, a_n) \in D(f^{\mathbf{A}}) \text{ implies } h(f^{\mathbf{A}}(a_1, \ldots, a_n)) = f^{\mathbf{B}}(h(a_1), \ldots, h(a_n)),$$

- *full* if for all $f \in \mathcal{F}$, $f^{\mathbf{B}}(h(a_1), \ldots, h(a_n)) = h(a_0)$ implies there exists $(a'_1, \ldots, a'_n) \in D(f^{\mathbf{A}})$ such that $h(a_i) = h(a'_i)$ for $i = 0, \ldots n$,
- *closed* if for all $f \in \mathcal{F}$,

$$(h(a_1), \ldots, h(a_n)) \in D(f^{\mathbf{B}}) \text{ implies } (a_1, \ldots, a_n) \in D(f^{\mathbf{A}}).$$

Note that if h is a closed homomorphism, then it is a full homomorphism. The *category of partial algebras* with signature σ has morphisms given by the first (weak) notion of homomorphism. For a class $\mathcal{K}$ of partial algebras the class of homomorphic images is

$$\mathbf{H}\mathcal{K} = \{\mathbf{B} \mid h : \mathbf{A} \to \mathbf{B} \text{ is a surjective homomorphism for some } \mathbf{A} \in \mathcal{K}\}.$$

The class of full or closed homomorphic images of $\mathcal{K}$ are denoted by $\mathbf{H}_f\mathcal{K}$ and $\mathbf{H}_c\mathcal{K}$, respectively.

There are also three notions of a subalgebra $\mathbf{A}$ of $\mathbf{B}$. Assuming $A \subseteq B$, a partial algebra $\mathbf{A}$ is

- a *weak subalgebra* if for all $f \in \mathcal{F}$,

$$(a_1, \ldots, a_n) \in D(f^{\mathbf{A}}) \text{ implies } f^{\mathbf{A}}(a_1, \ldots, a_n) = f^{\mathbf{B}}(a_1, \ldots, a_n),$$

- a *relative subalgebra* if for all $f \in \mathcal{F}$, $f^{\mathbf{A}} = f^{\mathbf{B}}{\restriction_{A^n}}$, and
- a *(closed) subalgebra* if for all $f \in \mathcal{F}$,

$$(a_1, \ldots, a_n) \in A^n \cap D(f^{\mathbf{B}}) \text{ implies } f^{\mathbf{A}}(a_1, \ldots, a_n) = f^{\mathbf{B}}(a_1, \ldots, a_n).$$

These notions correspond to the injection map $i : A \to B$ being a weak/full/closed homomorphism. The class of weak, relative or closed subalgebras of a class $\mathcal{K}$ of partial algebras are denoted $\mathbf{S}_w\mathcal{K}$, $\mathbf{S}_r\mathcal{K}$ and $\mathbf{S}\mathcal{K}$ respectively. As the notation indicates, closed subalgebras are the standard concept for partial algebras.

A *congruence* θ on a partial algebra $\mathbf{A}$ is an equivalence relation such that for all $f \in \mathcal{F}$, $a_1\theta b_1, \ldots, a_n\theta b_n$ and $(a_1, \ldots, a_n), (b_1, \ldots, b_n) \in D(f^{\mathbf{A}})$ imply $f^{\mathbf{A}}(a_1, \ldots, a_n)\theta f^{\mathbf{A}}(b_1, \ldots, b_n)$. A congruence is *closed* if $(a_1, \ldots, a_n) \in D(f^{\mathbf{A}})$ implies $(b_1, \ldots, b_n) \in D(f^{\mathbf{A}})$. The *quotient algebra* $\mathbf{A}/\theta$ is defined on the set $A/\theta = \{[a]_\theta \mid a \in A\}$ of equivalence classes by

$$f^{\mathbf{A}/\theta}([a_1]_\theta, \ldots, [a_n]_\theta) = [f^{\mathbf{A}}(a'_1, \ldots, a'_n)]_\theta$$

if $(a'_1, \ldots, a'_n) \in D(f^{\mathbf{A}})$ and $a_1\theta a'_1, \ldots, a_n\theta a'_n$ for some $a'_1, \ldots, a'_n \in A$, and $f^{\mathbf{A}/\theta}$ is undefined otherwise. The *canonical map* $\gamma : \mathbf{A} \to \mathbf{A}/\theta$ given by $\gamma(a) = [a]_\theta$ is a full homomorphism and, if θ is closed, then γ is a closed homomorphism. We often write $[a]$ rather than $[a]_\theta$ when the confusion is unlikely.

For a set I, a *filter* F on I is a collection of subsets of I that is closed under finite intersection and if $X \in F$ and $X \subseteq Y$ then $Y \in F$. On a product $\mathbf{A} = \prod_{i \in I} \mathbf{A}_i$, a filter F on I induces a congruence θ_F by $a\theta_F b$ if and only if $\{i \in I \mid a_i = b_i\} \in F$. The resulting quotient algebra $\mathbf{A}/\theta_F$ is called a *reduced product*. For a class $\mathcal{K}$ of partial algebras the class of all reduced products is denoted by $\mathbf{P}_r\mathcal{K}$.

For total algebras Birkhoff's variety theorem says that $\mathcal{V}$ is an equational class if and only if $\mathcal{V}$ is a *variety*, i.e., $\mathcal{V} = \mathbf{HSP}\mathcal{K}$ for some class $\mathcal{K}$. Generalizations of this result to partial algebras are summarized below.

Theorem 1 ([2]). *Let $\mathcal{V}$ be a class of partial algebras.*

1. *$\mathcal{V}$ is an existence equational class if and only if $\mathcal{V} = \mathbf{HSP}\mathcal{K}$ for some $\mathcal{K}$.*
2. *If $\mathcal{V}$ is a strong equational class then $\mathcal{V} = \mathbf{H}_c\mathbf{SP}_r\mathcal{K}$ for some class $\mathcal{K}$.*
3. *$\mathcal{V}$ is a quasiequational class if and only if $\mathcal{V} = \mathbf{SP}_r\mathcal{K}$ for some class $\mathcal{K}$.*

A characterization for strong equational classes can be found in [15]. Since an equational class is a variety, the three classes above are also referred to as existence varieties, strong varieties and quasivarieties of partial algebras.

3 Generalized Separation Algebras and Some Subclasses

A *partial semigroup* $(S, \cdot)$ is a partial algebra with a binary operation that is *associative*, i.e., the identity $(xy)z = x(yz)$ holds. A *partial monoid* $(M, \cdot, e)$ is a partial semigroup with an identity element e such that $xe = x = ex$ holds. In fact, every variety of total algebras gives rise to a strong variety of partial algebras, simply by reinterpreting the same defining identities. However, some equational axioms only have total algebras as models. For example, the class of groups can be axiomatized as monoids that satisfy $xx^{-1} = e$, where $^{-1}$ is a unary operation symbol. Then $x = xe = x(yy^{-1}) = (xy)y^{-1}$ is defined for all values of x, y, hence the subterm xy is always defined. Consequently, the class of all partial algebras that satisfy these group axioms is simply the class of all (total) groups.

A *generalized separation algebra*, or GS-algebra, is a partial monoid that is *cancellative* and *conjugative*, i.e., satisfies the axioms

left cancellativity $xy = xz \implies y = z$
right cancellativity $xz = yz \implies x = y$
conjugation $\exists v(vx = y) \iff \exists w(xw = y)$

A *separation algebra* [3] is a commutative GS-algebra, i.e., the identity $xy = yx$ holds, making the conjugation axiom redundant. The category of generalized separation algebras with partial algebra homomorphisms has the category of (total) groups as a full subcategory, and the same is true for the category of total cancellative conjugative monoids. This includes all free commutative monoids such as the natural numbers with addition, but does not include any noncommutative free monoid.

Theorem 2. *The conjugation axiom is preserved by reduced products, but not by subalgebras, even in the presence of cancellative monoid axioms. Therefore the class of GS-algebras is not a quasivariety.*

Proof. Let $\mathbf{A} = \prod_{i \in I} \mathbf{A}_i$ be a product of GS-algebras, F a filter on I, and assume $[c][a] = [b]$ for some $a, b, c \in A$, where $[x] = [x]_{\theta_F}$. Therefore $c_i a_i = b_i$is true for all i in some set $X \in F$. Since A_i is conjugative, $a_i d_i = b_i$ for some $b_i \in A_i$ and all $i \in X$. Let $d_j = e$ for $j \in I \setminus X$ and define b' by $b'_i = b_i$ if $i \in X$, and $b'_i = a_i$ otherwise. Then $[b'] = [b]$ and $ad = b'$, hence $[a][d] = [b'] = [b]$. The reverse implication is similar, so the reduced product $\mathbf{A}/\theta_F$ satisfies the conjugation axiom.

Let $A = \{e, a, b, c, d\}$ and define $\cdot$ on A by $ex = x = xe$ for $x \in A$, $ab = bc = ca = d$ and in all other cases xy is undefined. It is easy to check that $\mathbf{A} = (A, \cdot, e)$ is the smallest noncommutative generalized separation algebra. It has a closed subalgebra given by $B = \{e, a, b, d\}$ in which $ab = d$, but there is not element $x \in B$ such that $xa = d$, hence conjugation fails.

By the characterization theorem of quasivarieties for partial algebras, stated in Theorem 1, the class of GS-algebras is not a quasivariety. □

Note that the class of separation algebras **is** a quasivariety of partial algebras.

The following example demonstrates that conjugation is not preserved by weak homomorphisms. Consider the following GE-algebra **G** and partial algebra **A**,

G

$\cdot_G$	e	a	b
e	e	a	b
a	a	–	–
b	b	–	–

A

$\cdot_A$	e	a	b
e	e	a	b
a	a	–	–
b	b	e	–

The mapping that sends e, a and b in $\mathbf{G}$ to e, a and b in $\mathbf{A}$, respectively, is a weak homomorphism from $\mathbf{G}$ to $\mathbf{A}$ that does not preserve conjugation.

A binary relation $\leq$ is defined by $x \leq y \iff \exists v(vx = y)$, and the conjugation axiom ensures that this binary relation could have also been defined by $\exists w(xw = y)$. An equivalent form of this axiom is $xy = z \implies \exists v, w(vx = yw = z)$. Reflexivity of $\leq$ follows from $ex = x$, and $x \leq y$, $y \leq z$ imply $vx = y$, $wy = z$ for some v, w and therefore $wvx = z$, which proves transitivity. Hence $\leq$ is a preorder, and its *symmetrization* is defined by $x \equiv y \iff x \leq y$ and $y \leq x$. As usual, the equivalence classes $[x]$ of $\equiv$ are partially ordered by $[x] \leq [y] \iff x \leq y$. An element v is *invertible* if there exists w such that $vw = e = wv$, and the set of invertible elements of a GS-algebra $\mathbf{A}$ is denoted by A^*. The inverse of v, if it exists, is unique and is denoted by v^{-1}.

Lemma 3. *Let* $\mathbf{A}$ *be a generalized separation algebra. Then*

1. A^* *is the bottom equivalence class* $[e]$ *of the poset* $A/{\equiv} = (\{[x] : x \in A\}, \leq)$,
2. $\mathbf{A}^* = (A^*, \cdot, e, ^{-1})$ *is a (total) group and is a closed subalgebra of* $\mathbf{A}$,
3. $x \equiv y$ *holds if and only if* $x \in yA^*$, *and*
4. $\equiv$ *is the identity relation if and only if* e *is the only invertible element.*

Proof. 1. If $x \equiv e$ then $vx = e$ for some $v \in A$, so $(vx)v = ev = ve$. By associativity, $v(xv) = ve$, and from cancellativity we conclude that x is invertible. Conversely, we always have $e \leq x$, and if x is invertible then $x^{-1}x = e$, hence $x \leq e$, which proves that $[e] = A^*$ is the bottom element of $A/{\equiv}$. 2. It suffices to show that $\cdot$ restricted to A^* is a total operation. Given $v, w \in A^*$ there exists $u \in A$ such that $uv = e$. Thus $(uv)w = w$, and by associativity we get $u(vw) = w$, which implies that vw is defined. 3. From $x \equiv y$ we have $xv = y$ and $yw = x$ for some v, w. Therefore $ywv = y$ and $xvw = x$. By cancellativity it follows that $wv = e = vw$, so $w \in A^*$. Now $yw = x$ implies $x \in \{yz : z \in A^*\} = yA^*$. Conversely, assume $x \in yA^*$, whence $x = yv$ for some invertible element v. Then $xv^{-1} = y$, so $x \equiv y$. 4. Note that $\equiv$ is not the identity relation if and only if $x \equiv y$ for some $x \neq y$. By 2. this is equivalent to $|A^*| > 1$. □

A *generalized pseudoeffect algebra*, or GPE-algebra, is a GS-algebra that is *positive*, i.e., $xy = e \implies x = e$, in which case $y = e$ follows from $ey = y$. Equivalently, a GPE-algebra is a GS-algebra in which $\leq$ is antisymmetric and hence a partial order. A commutative GPE-algebra is called a *generalized effect algebra*, or GE-algebra. As mentioned before, the conjugation axiom always holds in commutative partial algebras, hence separation algebras and GE-algebras are quasivarieties.

The following theorem shows that there is a close relationship between generalized separation algebras and generalized pseudoeffect algebras. In particular, the result shows that every separation algebra can be collapsed in a unique way to a largest generalized effect algebra. Hence a substantial part of the structure theory of separation algebras is covered by results about generalized effect algebras.

Theorem 4. *For a GS-algebra* $\mathbf{A}$,

1. *the relation* $\equiv$ *is a closed congruence,*
2. $\mathbf{A}/\equiv$ *is a GPE-algebra,*
3. *the congruence classes of* $\equiv$ *all have the same cardinality, and*
4. *if* $h : \mathbf{A} \to \mathbf{B}$ *is a homomorphism and* $\mathbf{B}$ *is a GPE-algebra then there exists a unique homomorphism* $g : \mathbf{A}/\equiv \to \mathbf{B}$ *such that* $g \circ \gamma = h$ *(where* $\gamma : \mathbf{A} \to \mathbf{A}/\equiv$ *is the canonical homomorphism* $\gamma(x) = [x]$*).*

Proof. Let $x \equiv y$, $z \equiv w$ and assume yw is defined. We want to show that xz is defined and $xz \equiv yw$. Using the assumptions and conjugation, we obtain $ux = y$ and $zv = w$ for some u, v, and, since yw is defined, $(ux)(zv) = yw$. By associativity it follows that $(u(xz))v = yw$, hence xz is defined. Using conjugation again, there exists r such that $r(u(xz)) = yw$, hence $xz \leq yw$. Given that xz is now known to be defined, a similar argument shows $yw \leq xz$, so $xz \equiv yw$.

The quotient algebra is positive since if $[x][y] = [e]$, then $xy \equiv e$. This gives $xyv = e$ for some v and therefore $x \leq e$, from which $[x] = [e]$ follows. Therefore $A/\equiv$ is a GPE-algebra. For $x \in A$, $x = x + 0$ and the map $x \mapsto x + v$ for $v \in \mathbf{A}^*$ is a bijection between $[x]$ and $[0]$. Hence the congruence classes have the same cardinality.

Now assume $h : \mathbf{A} \to \mathbf{B}$ is a homomorphism and $\mathbf{B}$ is a GPE-algebra. Define $g : \mathbf{A}/\equiv \to \mathbf{B}$ by $g([a]) = h(a)$. To prove that g is well defined, assume $[a'] = [a]$, or equivalently $a \equiv a'$. This means $va = a'$, so $h(v)h(a) = h(va) = h(a')$, whence $h(a) \leq h(a')$. Similarly $h(a') \leq h(a)$, and since $\leq$ is a partial order in any GPE-algebra, $h(a) = h(a')$ follows.

Now suppose g' is a homomorphism that also satisfies $g' \circ \gamma = h$. This means $g'([a]) = h(a)$ for all $a \in A$, so $g' = g$. □

Theorem 5. *Let* $\mathbf{G}$ *be a group and* $\mathbf{B}$ *a GPE-algebra. Then* $\mathbf{A} = \mathbf{G} \times \mathbf{B}$ *is a GS-algebra with* $\mathbf{A}^* = \mathbf{G} \times \{e\}$.

Proof. The product of GS-algebras is again a GS-algebra since by, Theorem 2, this class of algebras is closed under reduced products. The element $(g, e) \in A$ has inverse (g^{-1}, e), and there are no other inverses by Lemma 3.4. □

Several of the prominent subclasses of GPE-algebras extend the signature of these algebras with a constant 1 and unary operations $^{\sim}, ^{-}$ or $'$. In the case of commutativity it is also traditional to replace $\cdot, e$ with $+, 0$ (or $\oplus, 0$).

Starting from GPE-algebras using the $+, 0$ signature, seven subclasses are obtained by adding combinations of the following three independent axioms:

(comm) $x + y = y + x$ (*commutativity*)
(orth) $x + y = 1 \iff y = x^{\sim} \iff x = y^{-}$ (*orthocomplementation*)
(cons) $x + x \neq \mathbf{u} \implies x = 0$ (*consistency*)

In particular, adding these different combinations of the above axioms to a GPE-algebra produces the following subclasses:

In a pseudoeffect algebra, $x^{\sim}$ and x^{-} are called the *right* and *left complement* of x, and 1 is the top element. In fact any GPE-algebra with a top element,

Axioms	Name	Abbrev.
(comm)	*generalized effect algebra*	GE
(cons)	*generalized pseudo-orthoalgebra*	GPO
(orth)	*pseudoeffect algebra*	PE
(comm), (cons)	*generalized orthoalgebra*	GO
(comm), (orth)	*effect algebra*	E
(cons), (orth)	*pseudo-orthoaglebra*	PO
(comm), (cons), (orth)	*orthoaglebra*	O

denoted by 1, can be extended with these two unary operations such that (orth) holds, and it is easy to check that they are total operations. For commutative subclasses such as effect algebras, we always have $x^{\sim} = x^{-}$ and in this case we write x' for the complement of x. From (orth) it follows that $x^{\sim -} = x = x^{-\sim}$, so for effect algebras and orthoalgebras this is written as $x'' = x$.

Below is a diagram that depicts the containment between these subclasses of GPE-algebras. The initial addition of the three independent axioms is shown as well as the larger classes of (generalized) separation algebras.

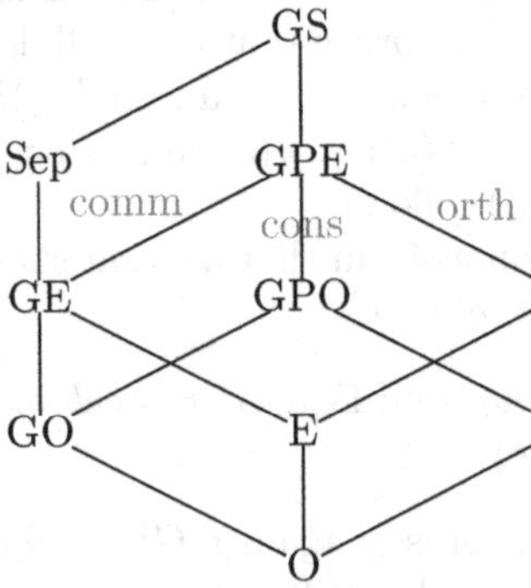

The two most studied subclasses of GPE-algebras are *effect algebras* (EA) which satisfy (comm) and (orth), and *orthoalgebras* (OA) which satisfy (comm), (orth) and (cons). The signature for effect algebras and orthoalgebras is $+, ', 0, 1$. Some examples of effect algebras are given below:

1. *One-element effect algebra* $(\{0\}, +, ', 0, 0)$
2. *Two-element effect algebra* $(\{0,1\}, +, ', 0, 1)$ where $0 + x = x = x + 0$, $1 + 1 =$ undefined, $0' = 1$, $1' = 0$.
3. The *standard MV-effect algebra* $[0,1]_E = ([0,1], +, ', 0, 1)$ where $x + y$ is addition, but *undefined* if the result is bigger than 1, $x' = 1 - x$.
4. For any MV-algebra $(A, \oplus, \neg, 0)$ define $x + y = \begin{cases} x \oplus y & \text{if } x \leq \neg y \\ \text{undefined} & \text{otherwise} \end{cases}$ Then $(A, +, \neg, 0, \neg 0)$ is called an *MV-effect algebra.*

5. Let $(G, \cdot, ^{-1}, e, \leq)$ be a *partially ordered group* and $u \in G$ such that $u \geq e$. Then $([0,u], \cdot, ^{\sim}, ^{-}, e, u)$ is an *interval pseudoeffect algebra* where $\cdot$ is undefined if the result is outside of $[0,u]$, $x^{\sim} = x^{-1}u$ and $x^{-} = ux^{-1}$. If G is abelian this construction produces an *interval effect algebra.*
6. Let $(L, \vee, \wedge, ', 0, 1)$ be an *orthomodular lattice*, i.e., a lattice $(L, \vee, \wedge)$ that satisfies $x \wedge x' = 0$, $x \vee x' = 1$, $x'' = x$, $(x \wedge y)' = x' \vee y'$ and $x \leq y \implies x \vee (x' \wedge y) = y$. Define $x + y = \begin{cases} x \vee y & \text{if } x \leq y' \\ \text{undefined} & \text{otherwise} \end{cases}$. Then $(L, +, ', 0, 1)$ is an orthoalgebra since it is *consistent*: if $x + x$ is defined then $x = 0$.

Examples of generalized separation algebras (that are not GPE-algebras) can be constructed using Theorem 5.

All classes defined here are closed under *products*, but some of them are also closed under certain amalgamated disjoint unions. The *horizontal sum* $\mathbf{A} + \mathbf{B}$ of PE-algebras $\mathbf{A}$ and $\mathbf{B}$ is the disjoint union of $A - \{0,1\}$ and $B - \{0,1\}$ with new bottom and top added. The new operations $\cdot, '$ agree with $\cdot, '$ on $\mathbf{A}$ and $\mathbf{B}$, and, for $a \in A - \{0,1\}, b \in B - \{0,1\}$, the value of ab is undefined. The result is again a PE-algebra, and, if $\mathbf{A}, \mathbf{B}$ are effect algebras or orthoalgebras, the same is true for $\mathbf{A} + \mathbf{B}$. Clearly horizontal sums can also be defined for arbitrary families of PE-algebras. For the class of GPE-algebras or its subclasses one can define a *bottom sum* that takes the disjoint union of the factors and identifies all the bottom elements. The result is again a GPE-algebra, or a GE-algebra if all summands are commutative. In Fig. 1 we give some diagrams of finite effect algebras and GPE-algebras to indicate the range of possible examples. A black dot is used for elements that are equal to their complements, and other elements are represented by open circles.

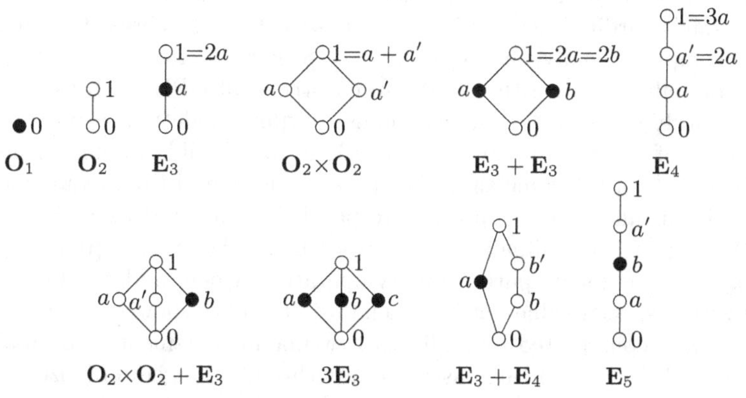

Fig. 1. Effect algebras with up to 5 elements

For a (pseudo)effect algebra $\mathbf{A}$, let $\bar{\mathbf{A}}$ denote the $'$-free reduct. Applying this to the 2-element orthoalgebra we obtain $\bar{\mathbf{O}}_2 = (\{0,1\}, +, 0)$, the 2-element GE-algebra.

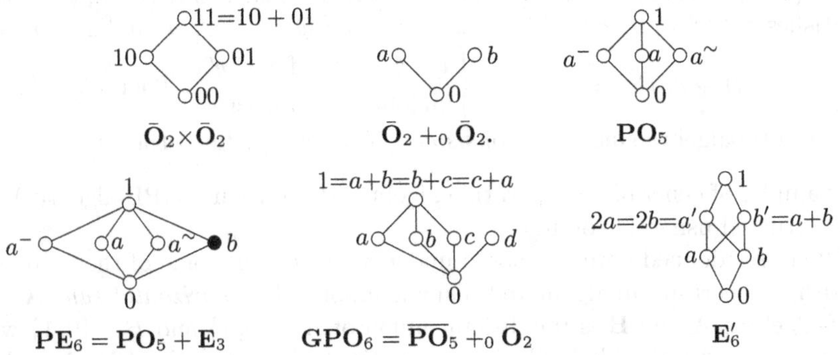

Fig. 2. Examples of GO-, GPO-, E-, GE- and GPE-algebras

4 An Orderly Algorithm for Constructing Generalized Pseudoeffect Algebras

Examples of GPE-algebras can provide insight into their structure that may not be apparent simply from studying their axioms. General purpose model generators such as Mace4 [11] can be used to find all models of cardinality n of a finitely axiomatized first-order theory. However, if a class has many models even for small cardinalities, as is the case with GPE-algebras, this approach becomes computationally unfeasible for $n > 8$. For this reason, it is helpful to have a more efficient algorithm that can construct all GPE-algebras of a given size. An orderly algorithm constructs nonisomorphic models of cardinality $n+1$ from models of cardinality n without checking for possible isomorphisms with all other models of the same size. This reduces space and time requirements of such algorithms and makes it possible to parallelize the model search.

A GPE-algebra of cardinality n is represented on the set $A = \{0, 1, \ldots, n-1\}$ by the $n \times n$ table for its partial binary operation, where undefined entries are marked with a special value not in A. The results below show how a GPE-table of size $n \times n$ is constructed by adding a new maximal element n to an already existing $(n-1) \times (n-1)$ table associated with a GPE-algebra of size $n-1$. A subset B of a poset is a *downset* if $x \leq y \in B$ implies $x \in B$.

Lemma 6. *Let* $\mathbf{A} = (A, +, 0)$ *be a GS-algebra and* B *a nonempty downset with respect to the preorder* $\leq$*. Then* $\mathbf{B} = (B, +\restriction_B, 0)$ *is a GS-algebra. If* $\mathbf{A}$ *is a GPE-algebra, the same holds for* $\mathbf{B}$.

Proof. Let B be a downset of A. For simplicity of notation, we will denote $+\restriction_B$, the restriction of $+$ onto B, as $+_B$. We need to check that $\mathbf{B}$ satisfies the axioms for GS-algebras.

Identity element: Since 0 is the bottom element of $\mathbf{A}$ it follows that 0 is in the downset B, hence it is also the identity element of $\mathbf{B}$ under the restriction of $+$ to B.

Cancellativity: Assume $x +_B y = x +_B z$ is defined for some $x, y, z \in B$. Then $x + y = x + z$ is defined in $\mathbf{A}$ and $x = y$ by cancellativity in $\mathbf{A}$. Similarly, if $x +_B y = z +_B y$, then $x + y = z + y$ in $\mathbf{A}$ and $x = z$ by cancellativity in $\mathbf{A}$.

Associativity: Assume $x +_B y$ and $(x +_B y) +_B z$ are defined in $\mathbf{B}$. Then $x + y$ and $(x + y) + z$ are defined in $\mathbf{A}$. By associativity in $\mathbf{A}$, it follows that $y + z$ and $x + (y + z)$ are defined in $\mathbf{A}$, and $(x + y) + z = x + (y + z)$. By substitution, $x + (y + z) = (x +_B y) +_B z$. It remains to show that $y +_B z$ is defined in $\mathbf{B}$. By conjugation in $\mathbf{A}$, there exists a $w \in A$ such that $x + (y + z) = (y + z) + w = (x +_B y) +_B z$. Therefore, $y + z \leq (x +_B y) +_B z \in B$, so $y +_B z \in B$ since $\mathbf{B}$ is a downset, and we can conclude that $(x +_B y) +_B z = x +_B (y +_B z)$.

Conjugation: Assume $x +_B y$ is defined in $\mathbf{B}$. Then $x + y$ is defined in $\mathbf{A}$ and by conjugation in $\mathbf{A}$, there exist $u, v \in A$ such that $x + y = u + x = y + v$. Hence $u \leq x + y = x +_B y \in B$, so $u \in B$ and $u +_B x = x +_B y$ in $\mathbf{B}$. Similarly, there exists $w \in A$ such that $y + v = v + w = x + y$ by conjugation in $\mathbf{A}$. Thus, $v \leq x + y = x +_B y \in B$, so $v \in B$ and $y +_B v = x +_B y$ in $\mathbf{B}$. □

The next result shows what needs to be checked to extend a GPE-algebra with a new maximal element n. The forward direction of the proof follows from the assumption that $\mathbf{A}$ is a relative subalgebra of $\mathbf{A}'$, and for the reverse direction it suffices to check that the GPE-axioms hold in $\mathbf{A}'$. The final statement of the theorem follows from Lemma 6

Theorem 7. *Let $\mathbf{A} = (A, \oplus, 0)$ be a GPE-algebra and let $A' = A \cup \{n\}$ for $n \notin A$. Then $\mathbf{A}' = (A', +, 0)$ is a GPE-algebra with $\mathbf{A}$ as a relative subalgebra and n as maximal element if and only if the following conditions hold for all $x, y, z \in A$*

1. *$x + y \in A$ if and only if $x \oplus y$ is defined, in which case $x + y = x \oplus y$,*
2. *$n + 0 = n = 0 + n$,*
3. *$x \neq 0 \implies n + x$ and $x + n$ are undefined, and $n + n$ is undefined,*
4. *$x + y = n = x + z \implies y = z$ and $x + y = n = z + y \implies x = z$,*
5. *$x + y = n \implies u + x = n = y + v$ for some $u, v \in A$, and*
6. *$(x + y) + z = n \iff x + (y + z) = n$.*

Furthermore, every GPE-algebra of cardinality $n + 1$ has a relative subalgebra of cardinality n.

Our algorithm uses the preceding result to construct all GPE-algebras of cardinality n starting with the one-element GPE-algebra. This is done by a backtracking search, ensuring that all possible one-point extensions of each algebra are computed. To remove isomorphic copies efficiently, the binary operation is coded as a directed graph and a canonical labeling algorithm is used to map to a unique fixed representative of the isomorphism class of the directed graph. Some optimizations are used for the cancellativity, conjugation and associativity checks. The algorithm was implemented in Python and uses a canonical labeling algorithm from the Sage computer algebra systems [14]. The number of algebras computed up to isomorphism in each subclass of GPE-algebras are summarized in Table 1 and the partial algebras can be downloaded from https://github.com/jipsen/Effect-algebras. For the class Sep of separation algebras and for the class of GS-algebras the Mace4 model finder [11] was used.

Table 1. Number of partial algebras in each class

n	O	PO	GO	GPO	E	PE	GE	Sep	GPE	GS
2	1	1	1	1	1	1	1	2	1	2
3	0	0	1	1	1	1	2	3	2	3
4	1	1	2	2	3	3	5	8	5	8
5	0	1	2	3	4	5	12	13	13	14
6	1	2	4	7	10	12	35	39	42	48
7	0	2	8	19	14	19	119	120	171	172
8	2	5	18	68	40	52	496	507	1020	
9	0	4	42	466	60	84	2699		11742	
10	2	10	156	8740	172	240	21888		322918	
11	0	9	834		282		292496			

As indicated by Theorem 4, there are only a small number of GS-algebras that are not GPE-algebras since the structure of a generalized separation algebra is highly restricted by its group of invertible elements and the GPE-quotient determined by this group.

5 Further Results About GPE-algebras

The *height* of an element a in a finite GPE-algebra is the length of a maximal path from 0 to a in the Hasse diagram of the partial order. A set of elements of the same height make up a *level*. The *atoms* of a GPE-algebra are the elements in level 1, i.e., they only have the bottom element 0 below them.

Lemma 8. *Associativity holds automatically for naturally ordered partial algebras that have two levels or less.*

Proof. Level 1: If $(x + y) + z$ is defined in a partial algebra with 1 level, then $x + y = 0$ or $z = 0$. If $x + y = 0$, then $x, y = 0$ by positivity and so $(0 + 0) + z = z = 0 + (0 + z)$. If $z = 0$, then $(x + y) + 0 = x + y = x + (y + 0)$. In either case, associativity holds.

Level 2: Now assume $(x + y) + z$ is defined in a partial algebra with level 2. If $(x + y) + z$ has height 1, then it satisfies associativity by the same reasoning as the first part of this proof. If $(x + y) + z$ has height 2, then there are three possibilities:

(i) $x+y$ has height 2 and z has height 0, in which case $(x+y)+0 = x+y = x + (y + 0)$. (ii) $x + y$ has height 0 and z has height 2, in which case $x, y = 0$ and so $(0 + 0) + z = z = 0 + (0 + z)$. (iii) $x + y$ and z both have height 1. $x + y$ of height 1 implies that either $x = 0$ or $y = 0$. This means we either have $(0 + y) + z = y + z = 0 + (y + z)$ or $(x + 0) + z = x + z = x + (0 + z)$. □

Lemma 9. *A GPE-algebra is a GE-algebra if and only if it has a generating set in which all elements commute.*

Proof. Since GE-algebras are by definition commutative, the elements of any generating set will commute trivially. Thus we only need to prove the reverse implication, which we do by induction on the level n.

Let $\mathbf{A}$ be a GPE-algebra with a set of generators X such that for all $x, y \in X$ either $x + y, y + x$ are both undefined or $x + y = y + x$. $P(n)$: All levels up to and including n are commutative.

$P(2)$: Let $x, y \in A$ and w.l.o.g., let x have height 2 and y have any height. Then there exist atoms $a, b \in X$ such that $a+b = x$ and elements $c, d \in X \cup \{0\}$ such that $c+d = y$. If $x+y$ is defined, then by commutativity and associativity of $X \cup \{0\}$, we get that $x + y = (a + b) + (c + d) = (c + d) + (a + b) = y + x$.

Now assume $P(k)$ holds for all $2 \leq n \leq k$.

$P(k + 1)$: Let $x, y \in A$ and w.l.o.g., let x have height $k + 1$ and y have any height. Then there exist $a, b, c, d \in A$ with heights less than $k + 1$ such that $a+b = x$ and $c+d = y$. If $x+y$ is defined, then by the inductive hypothesis and associativity we have $(a + b) + (c + d) = (c + d) + (a + b)$ and thus $x + y = y + x$ for any x on level $k + 1$ with $x + y$ defined. □

It is an elementary result in group theory that every 1-generated group is commutative. For GPE-algebras a similar result holds for 1- and 2-generated algebras.

Theorem 10. *Every 2-generated GPE-algebra is commutative.*

Proof. Let $\mathbf{A}$ be a 2-generated GPE-algebra with atoms $a \neq b$. By symmetry, it suffices to show that if $a + b$ is defined, then $b + a$ is defined and the two values are equal. So assume $a+b$ is defined. Then, by conjugation, there exists a $w \in A$ such that $a + b = w + a$. It follows from the last equation that $w \leq a + b$, which means w is either $a+b, a, b$ or 0, since a, b are atoms. By cancellativity, w cannot be 0, a or $a + b$, hence $w = b$ and we have that $a + b = b + a$. Since the atoms commute, the previous lemma implies that $\mathbf{A}$ is commutative. □

Let $L(n_1, n_2, ..., n_k)$ denote the number of GPE-algebras (up to isomorphism) with level structure $(n_1, n_2, ..., n_k)$ and $n = 1+\sum_{i=1}^{k} n_i$ number of elements. The number of *integer partitions* for a positive integer n, also called the *partition function*, is the number of ways positive integers can sum to n, ignoring order, and is denoted by $p(n)$.

We now show that the number of GPE-algebras of height ≤ 2 with cardinality n is given by the sum of the partition function from 1 to n. We first observe that a partial operation $+$ can be viewed as a coalgebra $\alpha : A \to \mathcal{P}(A^2)$ where $\alpha(x) = \{(y,z) \in A^2 \mid x = y+z\}$.

Lemma 11. *For a GPE-algebra* $\mathbf{A}$ *and* $x \in A$, *the binary relation* $\alpha(x)$ *is a permutation of its domain, hence in the finite case the domain is partitioned into disjoint finite cycles.*

Proof. The relation $\alpha(x)$ is a function on its domain since $\cdot$ is left-cancellative, and injective since $\cdot$ is right-cancellative. By conjugation $\alpha(x)$ is surjective, hence it is a permutation. Letting α act on the domain gives the partition into cycles. □

Lemma 12. *For any GPE-algebra of size* $n \geq 3$, $L(n-2,1) = L(n-3,1) + p(n-2)$.

Proof. Consider an algebra $\mathbf{A} = (A, +, 0)$ of size $n-1$ with the level structure given by $(n-3,1)$. Define a new algebra $\mathbf{A}' = (A', \oplus, 0)$ of size n by $A' = A +_0 \bar{O}_2$ (see Fig. 2 for examples of the bottom sum $+_0$). Then A' is a GPE-algebra of size n with a level structure given by $(n-2,1)$. This means that at the very least there are $L(n-3,1)$ GPE-algebras of size n with level structure $(n-2,1)$.

Now let $\mathbf{A}$ be a pseudoeffect algebras with level structure $(n-2,1)$, and let x be the top element. For every element y in the first level, there exists y' such that $y + y' = x$, hence the domain of α is $\mathbf{A}$. By Lemma 11, A is partitioned into disjoint cycles, with one of the cycles being $\{0, x\}$. Therefore the remaining $n-2$ elements in level 1 are partitioned into cycles, and there are $p(n-2)$ different possible partitions up to isomorphism. □

Theorem 13. *The number of GPE-algebra of cardinality* n *with level structure* $(n-2,1)$ *is* $\sum_{k=1}^{n} p(k)$.

Recall that the partial order on a GPE-algebra is given by $a \leq b \iff \exists z\,(a + z = b) \iff \exists w\,(w + a = b)$. By cancellativity, z, w are unique, so we denote $z = a\backslash b$ and $w = b/a$. Rump and Yang [13] define a *two-point extension* for a GPE-algebra $\mathbf{A}$ that produces a total algebra $\mathbf{A}_\bot^\top = (A \cup \{\bot, \top\}, \leq, \cdot, e, \backslash, /, \bot, \top)$ such that $\leq$ is the natural order, extended with $\bot \leq x \leq \top$, let $e = 0$,

$$a \cdot b = \begin{cases} a + b & \text{if } a + b \text{ is defined} \\ \bot & \text{if } a = \bot \text{ or } b = \bot \\ \top & \text{otherwise} \end{cases}$$

$\bot\backslash x = x/\bot = x\backslash\top = \top/x = \top$ and if $a \not\leq b$ then define $a\backslash b = \bot = b/a$.

A *residuated partially ordered monoid* $(A, \leq, \cdot, e, \backslash, /)$ is a poset $(A, \leq)$ and a monoid $(A, \cdot, e)$, and for all $x, y, z \in A$, $xy \leq z \iff y \leq x\backslash z \iff y \leq z/x$.

Theorem 14. ([13]). *Let* $\mathbf{A}$ *be a GPE-algebra. Then* $\mathbf{A}_{\perp}^{\top}$ *is a residuated po-monoid, i.e.,* $(A, \leq)$ *is a poset,* $(A, \cdot, e)$ *is a monoid and* $xy \leq z \iff y \leq x\backslash z \iff y \leq z/x$.

The preceeding result shows that every GPE-algebra is an interval in a total residuated po-monoid that has a unit e as its unique atom.

A residuated po-monoid is *involutive* [7] if there exists an element d such that the terms ${\sim}x = x\backslash d$ and $-x = d/x$ satisfy $-{\sim}x = x = {\sim}{-x}$. Then $-d = e = {\sim}d$ and $x\backslash y = {\sim}((-y)x)$, $x/y = -(y({\sim}x))$. Equivalently, $(A, \leq, \cdot, e, d, \sim, -)$ is an involutive residuated po-monoid if $-{\sim}x = x = {\sim}{-x}$ and $xy \leq d \iff x \leq -y$.

Theorem 15. *The two-point totalization of PE/PO-algebras, effect algebras and orthoalgebras produces involutive residuated po-monoids.*

Recall that a groupoid is a (small) category in which every morphism is an isomorphism. While groups capture the symmetries of individual mathematical objects, groupoids model symmetries of systems of related objects. For example, the fundamental groupoid of a topological space captures more information about the space than the fundamental group determined by a choice of base point.

We end this section with a recent generalization of effect algebras that is similar to modifications of separation algebras in [4] that allow several local identity elements.

A pseudoeffect algebra is *symmetric* if $x^{\sim} = x^{-}$. Roumen [12] has taken the important step of generalizing symmetric pseudoeffect algebras to effect algebroids. Here the concept is reformulated for a unisorted partial algebra.

An *effect algebroid* is a partial algebra $(A, +, ')$ such that

(asso) $(x + y) + z = x + (y + z)$
(idenL) $(x + x')' + x = x$
(orthL) $x + y$ defined and $x + y = x + x'$ implies $y = x'$
(orthR) $x + y$ defined and $x + y = y' + y$ implies $x = y'$
(dbl) $x'' = x$
(0-1) if $x + (x' + x)$ is defined then $x = (x' + x)'$.

For comparison with effect algebras and pseudoeffect algebras, we computed the number of effect algebroids of cardinality n.

$n =$	1	2	3	4	5	6	7	8
Effect algebroids	1	2	3	7	12	27	49	114

An effect algebroid is a symmetric pseudoeffect algebra if and only if it satisfies $x + x' = y + y'$. It is an effect algebra if, in addition, it satisfies $x + y = y + x$.

6 Conclusion

Partial algebras are considerably more general than total algebras. The class of generalized separation algebras and its subclass of generalized pseudoeffect algebras are closely related, but so far have been studied separately since they arose in the unrelated areas of separation logic and quantum logic. We proved that there is a canonical map from separation algebras to GPE-algebras and computed finite GPE-algebras up to 10 elements (up to 11 elements for GE-algebras). Insight from these finite models was used to prove that all 2-generated GPE-algebras are commutative and to describe all GPE-algebras with a single element on the second level.

References

1. Burmeister, P.: A Model Theoretic Oriented Approach to Partial Algebras, vol. 32. Akademie-Verlag, Berlin (1986). http://www2.mathematik.tu-darmstadt.de/Math-Net/Lehrveranstaltungen/Lehrmaterial/SS2002/AllgemeineAlgebra/download/pa86.pdf
2. Burmeister, P.: Lecture notes on universal algebra, many-sorted partial algebras (2002). http://www2.mathematik.tu-darmstadt.de/Math-Net/Lehrveranstaltungen/Lehrmaterial/SS2002/AllgemeineAlgebra/download/LNPartAlg.pdf
3. Calcagno, C., O'Hearn, P.W., Yang, H.: Local action and abstract separation logic. In: Proceedings of the 22nd Annual IEEE Symposium on Logic in Computer Science, LICS 2007, pp. 366–378, Washington, DC (2007)
4. Dockins, R., Hobor, A., Appel, A.W.: A fresh look at separation algebras and share accounting. In: Hu, Z. (ed.) APLAS 2009. LNCS, vol. 5904, pp. 161–177. Springer, Heidelberg (2009). https://doi.org/10.1007/978-3-642-10672-9_13
5. Dvurecenskij, A., Vetterlein, T.: Pseudoeffect algebras. I. Basic properties. Int. J. Theor. Phys. **40**(3), 685–701 (2001)
6. Foulis, D.J., Bennett, M.K.: Effect algebras and unsharp quantum logics. Found. Phys. **24**, 1325–1346 (1994)
7. Galatos, N., Jipsen, P., Kowalski, T., Ono, H.: Residuated lattices: an algebraic glimpse at substructural logics. In: Studies in Logic and the Foundations of Mathematics. Elsevier B. V., Amsterdam (2007)
8. Jipsen, P.: Computer-aided investigations of relation algebras. ProQuest LLC, Ann Arbor, MI, Ph.D. thesis, Vanderbilt University (1992)
9. Jipsen, P.: Discriminator varieties of Boolean algebras with residuated operators. In: Algebraic Methods in Logic and in Computer Science, vol. 28, pp. 239–252. Banach Center Publications, Warsaw (1993). Polish Acad. Sci. Inst. Math, Warsaw
10. Jónsson, B., Tsinakis, C.: Relation algebras as residuated Boolean algebras. Algebra Universalis **30**(4), 469–478 (1993)
11. McCune, W.: Prover9 and Mace4 (2005–2010). https://www.cs.unm.edu/~mccune/mace4/
12. Roumen, F.A.: Effect algebroids. Ph.D. thesis. Radboud University (2017)

13. Rump, W., Yang, Y.C.: Non-commutative logical algebras and algebraic quantales. Ann. Pure Appl. Logic **165**, 759–785 (2014)
14. The Sage Developers. Sagemath, The Sage Mathematics Software System (Version 8.1) (2017). http://www.sagemath.org
15. Staruch, B.: HSP-type characterization of strong equational classes of partial algebras. Studia Logica **93**(1), 41–65 (2009)

Counting Finite Linearly Ordered Involutive Bisemilattices

Stefano Bonzio[1], Michele Pra Baldi[2], and Diego Valota[3](✉)

[1] Dipartimento di Scienze Biomediche e Sanità Pubblica,
Università Politecnica delle Marche,
Via Tronto 10/a, 60200 Torrette di Ancona, Italy
stefano.bonzio@gmail.com

[2] Dipartimento F.I.S.P.P.A., Università di Padova, Padova, Italy
m.prabaldi@gmail.com

[3] Dipartimento di Informatica, Università degli Studi di Milano,
Via Comelico 39, 20135 Milano, Italy
valota@di.unimi.it

Abstract. The class of involutive bisemilattices plays the role of the algebraic counterpart of paraconsistent weak Kleene logic. Involutive bisemilattices can be represented as *Płonka sums* of Boolean algebras, that is semilattice direct systems of Boolean algebras. In this paper we exploit the Płonka sum representation with the aim of counting, up to isomorphism, finite involutive bisemilattices whose direct system is given by totally ordered semilattices.

Keywords: Finite involutive bisemilattices · Weak Kleene logic
Płonka sums

1 Introduction

The class of involutive bisemilattices plays the role of the algebraic counterpart among one of the three-valued logics introduced by Kleene in [22], namely paraconsistent weak Kleene logic – PWK for short. PWK, essentially introduced by Halldén [16], can be defined as the logic induced by a matrix given by the weak Kleene tables with $\{1, n\}$ as truth set:

$\wedge$	0	n	1
0	0	n	0
n	n	n	n
1	0	n	1

$\vee$	0	n	1
0	0	n	1
n	n	n	n
1	1	n	1

$\neg$	
1	0
n	n
0	1

Equivalently (see [8,13]), PWK can be obtained out of (propositional) classical logic (CL) imposing the following syntactical restriction:

$$\Gamma \vdash_{\mathrm{PWK}} \varphi \iff \text{there is } \Delta \subseteq \Gamma \text{ s.t. } Var(\Delta) \subseteq Var(\varphi) \text{ and } \Delta \vdash_{\mathrm{CL}} \varphi,$$

J. Desharnais et al. (Eds.): RAMiCS 2018, LNCS 11194, pp. 166–183, 2018.
https://doi.org/10.1007/978-3-030-02149-8_11

where $Var(\varphi)$ is the set of variables really occurring in φ.

Involutive bisemilattices consist of a *regular* variety, namely one satisfying identities of the form $\varepsilon \approx \tau$, where $Var(\varepsilon) = Var(\tau)$. More precisely, involutive bisemilattices satisfy only the regular identities holding in Boolean algebras. Due to the general theory of regular varieties, which traces back to the pioneering work of Płonka [26], involutive bisemilattices can be represented as *Płonka sums* of Boolean algebras, that is, a sum over semilattice direct systems of Boolean algebras. Over the years, Płonka sums and (some) regular varieties have been studied in depth both from a purely algebraic perspective [3,17,18,21] and in connection with their topological duals [6,31,32]. The machinery of Płonka sums has also found useful applications in the study of the constraint satisfaction problem [4] and in database semantics [23,29]. Recently, thanks to the extension of this formalism to logical matrices [10,11], Płonka sums have turned out to play a useful role in the investigation of logics featuring the presence of a non-sensical, infectious truth-value. This family of logics – including PWK and Bochvar logic [5] – provides valuable formal instruments to model computer-programs affected by errors [14]. In this paper we exploit the Płonka sum representation for the purpose of counting the finite members of a specific subclass of involutive bisemilattices, whose representation consists of a linearly ordered semilattice. In particular, we provide an algorithm offering a solution to the fine spectrum problem [34] for the class of linearly ordered involutive bisemilattices. In order to achieve this goal, we use the categorical apparatus developed in [9]. We believe that the application of the above-mentioned algebraic methods allows us to develop algorithms that are more efficient than "brute-force" procedures. This is confirmed by the computational experiments. In particular, a comparison between the efficiency of the algorithm introduced in this paper and of Mace4 is briefly discussed in Sect. 5.

2 Preliminaries

A *semilattice* is an algebra $\mathbf{A} = \langle A, \vee \rangle$, where $\vee$ is a binary commutative, associative and idempotent operation. Given a semilattice $\mathbf{A}$ and $a, b \in A$, we set $a \leq b \iff a \vee b = b$. It is easy to see that $\leq$ is a partial order on A.

We briefly recall the category of semilattice direct systems, introduced in [6,9]. Intuitively, they consists of a specialization of direct (and inverse) systems of an arbitrary category, obtained by assuming the index set to be a semilattice instead of a (directed) pre-ordered set. For any unexplained notion in category theory, the reader is referred to [24].

Definition 1. *Let $\mathfrak{C}$ be an arbitrary category. A* semilattice direct system *in $\mathfrak{C}$ is a triple $\mathbb{X} = \langle X_i, p_{ii'}, I \rangle$ such that*

1. *I is a semilattice.*
2. *$\{X_i\}_{i \in I}$ forms an indexed family of objects in $\mathfrak{C}$ with disjoint universes;*
3. *$p_{ii'} : X_i \to X_{i'}$ is a morphism of $\mathfrak{C}$, for each pair $i \leqslant i'$ $(i, i' \in I)$, satisfying that p_{ii} is the identity in X_i and such that $i \leq i' \leq i''$ implies $p_{i'i''} \circ p_{ii'} = p_{ii''}$.*

We refer to I, X_i and $p_{ii'}$ as the index set, the terms and the transition morphisms, respectively, of the (semilattice direct) system.

A morphism between two semilattice direct systems $\mathbb{X} = \langle X_i, p_{ii'}, I\rangle$ and $\mathbb{Y} = \langle Y_j, p_{jj'}, J\rangle$ is a pair $(\varphi, \{f_i\}_{i\in I})\colon \mathbb{X} \to \mathbb{Y}$ such that

(i) $\varphi\colon I \to J$ is a semilattice homomorphism
(ii) $f_i\colon X_i \to Y_{\varphi(i)}$ is a morphism in $\mathfrak{C}$, making the following diagram commutative, for each $i, i' \in I$, $i \leq i'$.

$$\begin{array}{ccc} X_i & \xrightarrow{p_{ii'}} & X_{i'} \\ {\scriptstyle f_i}\downarrow & & \downarrow{\scriptstyle f_{i'}} \\ Y_{\varphi(i)} & \xrightarrow{q_{\varphi i \varphi i'}} & Y_{\varphi(i')} \end{array}$$

It is easy to check that the semilattice direct systems of a category $\mathfrak{C}$ form a category, which we denote by Sem-dir-$\mathfrak{C}$. Semilattice *inverse* systems of an arbitrary category are obtained analogously, by, intuitively, reversing the directions of transition morphisms (see [9] for precise details). Moreover, provided that two categories $\mathfrak{C}$ and $\mathfrak{D}$ are dually equivalent, the duality can be lifted to Sem-dir-$\mathfrak{C}$ and Sem-inv-$\mathfrak{D}$ (see [9, Remark 3.6]). The class of *involutive bisemilattices* has been introduced in [8] as the most suitable candidate to be the algebraic counterpart of the logic PWK.

Definition 2. An *involutive bisemilattice* is an algebra $\mathbf{B} = \langle B, \vee, \wedge, \neg, 0, 1\rangle$ of type $(2, 2, 1, 0, 0)$ satisfying:

I1. $x \vee x \approx x$;
I2. $x \vee y \approx y \vee x$;
I3. $x \vee (y \vee z) \approx (x \vee y) \vee z$;
I4. $\neg(\neg x) \approx x$;
I5. $x \wedge y \approx \neg(\neg x \vee \neg y)$;
I6. $x \wedge (\neg x \vee y) \approx x \wedge y$;
I7. $0 \vee x \approx x$;
I8. $1 \approx \neg 0$.

Involutive bisemilattices form an equational class denoted by $\mathcal{IBSL}$. Examples of involutive bisemilattices include any Boolean algebra, as well as any semilattice with zero. In the latter case, the two binary operations coincide and the unary operation is the identity. The variety $\mathcal{IBSL}$ is the *regularization*[1] of the variety $\mathcal{BA}$, of Boolean algebras (see [8,27]), i.e. $\mathcal{IBSL} \models \varepsilon \approx \tau$ if and only if $\mathcal{BA} \models \varepsilon \approx \tau$ and $Var(\varphi) = Var(\tau)$. Involutive bisemilattices can be connected to semilattice direct systems, in a way that we sketch. It is always possible to construct an algebra out of a semilattice direct system in an algebraic category. The construction we have in mind is called *Płonka sum* and is due to J. Płonka [26]. For standard information on Płonka sums we refer the reader to [28].

Definition 3. *Let* $\mathbb{A} = \langle \mathbf{A}_i, p_{ii'}, I\rangle$ *be a semilattice direct system of algebras of a fixed type* ν*. The* Płonka sum *over* $\mathbb{A}$ *is the algebra* $\mathcal{P}_l(\mathbb{A}) = \langle \bigsqcup_I A_i, g^{\mathcal{P}}\rangle$*, whose universe is the disjoint union of the algebras* $\mathbf{A}_i$ *and the operations* $g^{\mathcal{P}}$ *are*

[1] For the theory of regular varieties and regularizations we refer the reader to [28].

defined as follows: for every n-ary $g \in \nu$, *and* $a_1, \ldots, a_n \in \bigsqcup_I A_i$, *where* $n \geq 1$ *and* $a_r \in A_{i_r}$, *we set* $j = i_1 \vee \cdots \vee i_n$ *and define*[2]

$$g^{\mathcal{P}}(a_1, \ldots, a_n) = g^{\mathbf{A}_j}(p_{i_1 j}(a_1), \ldots, p_{i_n j}(a_n)).$$

Płonka sums provide a useful tool to represent algebras belonging to regular varieties. We recall here the representation theorem for involutive bisemilattices.

Involutive bisemilattices, as well as bisemilattices admit a representation as Płonka sums over a semilattice system of Boolean algebras. From [8, Thm. 46] we know that, if $\mathbb{A}$ is a semilattice direct system of Boolean algebras, then $\mathcal{P}_l(\mathbb{A})$ is an involutive bisemilattice, and if $\mathbf{B}$ is an involutive bisemilattice, then $\mathbf{B} \cong \mathcal{P}_l(\mathbb{A})$, where $\mathbb{A}$ is a semilattice direct system of Boolean algebras. The above facts can be strengthened to a full categorical equivalence.

Theorem 4 ([9, Thm. 4.5])**.** *The categories* $\mathfrak{IBSL}$ *and* Sem-dir-$\mathfrak{BA}$ *are equivalent.*

The equivalence is proved by the functors associating to each involutive bisemilattice the semilattice direct system of Boolean algebras corresponding to its Płonka sum representation. Conversely, to each semilattice direct system (of Boolean algebras), it is associated the Płonka sum. Upon considering Stone duality [33] between Boolean algebras and Stone spaces, $\mathfrak{SA}$ for short, namely compact totally disconnected Hausdorff topological spaces (see e.g. [20]), we have that

Theorem 5 ([9, Thm. 4.6])**.** *The categories* $\mathfrak{IBSL}$ *and* Sem-inv-$\mathfrak{SA}$ *are dually equivalent.*

For the purpose of the present work, we restrict Stone duality to the finite setting, which reduces to the well-known duality between the category of finite Boolean algebras and their homomorphisms $\mathfrak{FBA}$, and the category of finite sets and set-functions $\mathfrak{FS}$. The functor implementing such duality maps any Boolean algebra $\mathbf{A}$ to the set $\widehat{A}$ given by the atoms of $\mathbf{A}$. Thanks to the previous considerations, the problem of counting finite involutive bisemilattices coincides with counting semilattice direct systems (with finite index set) of finite Boolean algebras.

3 Linearly Ordered $\mathcal{IBSL}$

We confine our concern to a specific class of involutive bisemilattices, namely those ones whose corresponding direct system has a linearly ordered index set. We call this class *linearly ordered* involutive bisemilattices, $\mathcal{L}$-$\mathcal{IBSL}$ for short.

[2] In case ν contains constants, then, it is necessary to assume that I has a least element, see [27] for details.

Remark 6. The class $\mathcal{L}$-$\mathcal{IBSL}$ is closed under subalgebras and homomorphic images but not under products. Closure under subalgebras is obvious. For homomorphic images, it is enough to observe that any homomorphism between elements in $\mathcal{L}$-$\mathcal{IBSL}$ corresponds to a morphism between the equivalent semilattice direct systems, which, restricted to the index sets, is a homomorphism of semilattices. Moreover, any homomorphic image of a totally ordered semilattice is totally ordered. As regards products, it is clear that the product of (semilattice direct) systems whose index set is linearly order may, in general, be a system whose index set is not linearly ordered.

In the following part, we provide a first-order characterization of the class $\mathcal{L}$-$\mathcal{IBSL}$. Recall from [8] that, given an *involutive bisemilattice* $\mathbf{B}$, an element $a \in B$ is called *positive* if $a \vee \neg a = a$. We denote by $P(\mathbf{B})$ the set of positive elements of $\mathbf{B}$.

Remark 7. The set $P(\mathbf{B})$ of positive elements of an involutive bisemilattice $\mathbf{B}$ coincides with the set of the constants 1 of each Boolean algebra in the Płonka decomposition of $\mathbf{B}$. In other words, $\mathbf{B} \cong \mathcal{P}_l(\mathbb{A})$, with $\mathbb{A} = \langle \mathbf{A}_i, p_{ii'}, I\rangle$ a semilattice direct system of Boolean algebras, then $P(\mathbf{B}) = \bigcup_{i\in I}\{1_i\}$, where 1_i denotes the element 1 in the Boolean algebra A_i. Checking that 1_i is positive for each $i \in I$ is immediate. On the other hand, suppose that $a \in P(\mathbf{B})$, i.e. $a \vee \neg a = a$. Clearly, $a \in A_i$, for some $i \in I$, hence $\neg a \in A_i$. Therefore $a = a \vee^{\mathcal{P}_l} \neg a = a \vee^{\mathbf{A}_i} \neg a = 1_i$.

In the following result, we refer (with a slight abuse of notation) to the semilattice of indexes of a Płonka sum as $\langle I, \leq\rangle$ and to the semilattice formed by the positive elements of an involutive bisemilattice with respect to the reducts $\wedge$ ($\vee$, respectively) as $\langle P(\mathbf{B}), \wedge\rangle$ ($\langle P(\mathbf{B}), \vee\rangle$, respectively).

Proposition 8. *Let* $\mathbf{B} \in \mathcal{IBSL}$ *and let* $P(\mathbf{B})$ *be the set of positive elements. Then*

$$1.\ \langle P(\mathbf{B}), \wedge\rangle \cong \langle P(\mathbf{B}), \vee\rangle; \qquad 2.\ \langle P(\mathbf{B}), \vee\rangle \cong \langle I, \leq\rangle.$$

Proof. 1. The isomorphism is given by the identity map. We just check that, for any $a, b \in P(\mathbf{B})$, $a\wedge b = a\vee b$. In virtue of Remark 7, we can assume that $a = 1_i$ and $b = 1_j$, for some $i, j \in I$. Let $k = i\vee j$, then: $a\wedge b = 1_i \wedge^{\mathcal{P}_l} 1_j = p_{ik}(1_i) \wedge^{\mathbf{A}_k} p_{jk}(1_j) = 1_k \wedge^{\mathbf{A}_k} 1_k = 1_k = 1_k \vee^{\mathbf{A}_k} 1_k = p_{ik}(1_i) \vee_{\mathbf{A}_k} p_{jk}(1_j) = 1_i \vee^{\mathcal{P}_l} 1_j = a\vee b$.

2. We again assume the identification highlighted in Remark 7. Consider the map $f\colon I \to P(\mathbf{B})$, defined as $f(i) := 1_i$. The map is invertible (with inverse $g\colon P(\mathbf{B}) \to I$, $g(1_i) = i$). Moreover, we check that f is a homomorphism (of semilattices). Let $i, j \in I$ and $i\vee j = k$. Then $f(i\vee j) = f(k) = 1_k = 1_k \vee 1_k = p_{ik}(1_i) \vee p_{jk}(1_j) = 1_i \vee 1_j = f(i) \vee f(j)$. □

The above results shows that any consideration about the index set of the Płonka sum representation of an involutive bisemilattice can be expressed over the (partially ordered) set of its positive elements. This turns out to be convenient since the subset of positive elements is equationally definable. Observe, moreover that the two partial orders induced by the binary operations of an involutive bisemilattice (see [8] for details) coincide over the set of positive elements.

Corollary 9. *Let* $\mathbf{B} \in \mathcal{IBSL}$ *and* $\langle P(\mathbf{B}), \leq \rangle$ *the poset of its positive elements. The following are equivalent:*

1. $\mathbf{B} \in \mathcal{L}\text{-}\mathcal{IBSL}$; 2. *either* $x \leq y$ *or* $y \leq x$, *for any* $x, y \in P(\mathbf{B})$.

We provide a useful criteria to detect isomorphic copies of linearly ordered involutive bisemilattices.

Lemma 10. *Let* $\mathbb{A} = \langle \mathbf{A}_i, p_{ii'}, I \rangle$ *and* $\mathbb{B} = \langle \mathbf{B}_i, q_{ii'}, I \rangle$ *be two finite semilattice direct systems of Boolean algebras, with* I *linearly ordered and containing no trivial algebras. Then, the following statements are equivalent:*

1. $\mathbb{A} \cong \mathbb{B}$
2. $\mathbf{A}_i \cong \mathbf{B}_i$, *for every* $i \in I$, *and* $\mid \widehat{A}_{i'}/ker(\widehat{p}_{ii'}) \mid = \mid \widehat{B}_{i'}/ker(\widehat{q}_{ii'}) \mid$, *for every* $i < i'$.

Proof. ($\Rightarrow$) Assume that $\mathbb{A} \cong \mathbb{B}$ via an isomorphism (φ, f_i), for each $i \in I$. Since $\mathbb{A}$ and $\mathbb{B}$ share the same, linearly ordered, index set$I$, we necessarily have that $\varphi = id$. Moreover, for each $i \in I$, $\mathbf{A}_i \cong \mathbf{B}_i$ via the Boolean isomorphism f_i, and the following diagram on the left (we deliberately drop indexes from p and q to make notation less cumbersome) is commutative, for each $i < i'$, and in virtue of the duality established in Theorem 5 and the definition of inverse systems, the following diagram on the right is also commutative:

$$
\begin{array}{ccc}
\mathbf{A}_i & \xrightarrow{\ p\ } & \mathbf{A}_{i'} \\
{\scriptstyle f_i}\downarrow & & \downarrow{\scriptstyle f_{i'}} \\
\mathbf{B}_i & \xrightarrow{\ q\ } & \mathbf{B}'_i
\end{array}
\qquad\qquad
\begin{array}{ccc}
\widehat{A}_i & \xleftarrow{\ \widehat{p}\ } & \widehat{A}_{i'} \\
{\scriptstyle \widehat{f}_i^{-1}}\downarrow & & \downarrow{\scriptstyle \widehat{f}_{i'}^{-1}} \\
\widehat{B}_i & \xleftarrow[\ \widehat{q}\ ]{} & \widehat{B'}_i
\end{array}
$$

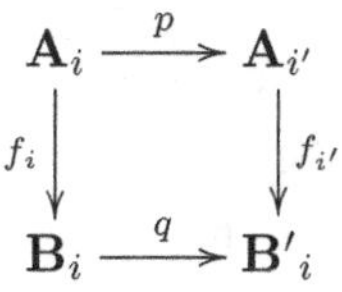

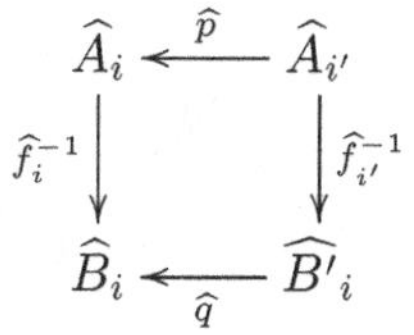

By the first isomorphism theorem (for sets), we have that there exist two (unique) embeddings $\psi\colon \widehat{A}_{i'}/ker(\widehat{p}\,) \to \widehat{A}_i$ and $\chi\colon \widehat{B}_{i'}/ker(\widehat{q}\,) \to \widehat{B}_i$ such that $\widehat{p} = \psi \circ \pi_{\widehat{A}_{i'}}$ and $\widehat{q} = \chi \circ \pi_{\widehat{B}_{i'}}$, where $\pi_{\widehat{A}_{i'}}$ and $\pi_{\widehat{B}_{i'}}$ indicate the natural projections onto the quotients $\widehat{A}_{i'}/ker(\widehat{p}\,)$, $\widehat{B}_{i'}/ker(\widehat{q}\,)$, respectively. The above diagram can therefore be split into the following:

$$
\begin{array}{ccc}
\widehat{A}_{i'} & \xrightarrow{\qquad \widehat{f}_{i'}^{-1} \qquad} & \widehat{B}_{i'} \\
{\scriptstyle \pi_{\widehat{A}_{i'}}}\downarrow & & \downarrow{\scriptstyle \pi_{\widehat{B}_{i'}}} \\
\widehat{A}_{i'}/ker(\widehat{(p)}) & \dashrightarrow^{\ \Phi\ } & \widehat{B}_{i'}/ker(\widehat{q}) \\
{\scriptstyle \psi}\downarrow & & \downarrow{\scriptstyle \chi} \\
\widehat{A}_i & \xrightarrow[\qquad \widehat{f}_i^{-1} \qquad]{} & \widehat{B}_i
\end{array}
$$

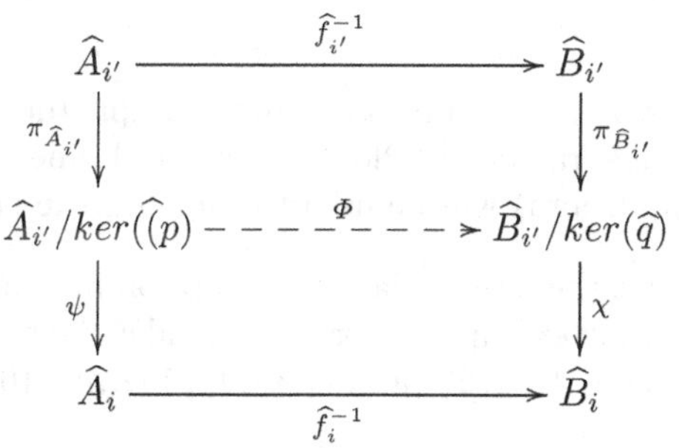

We define the map $\Phi\colon \widehat{A}_{i'}/ker(\widehat{p}\,) \to \widehat{B}_{i'}/ker(\widehat{q}\,)$ as $\Phi([a]_p) := [\widehat{f}_{i'}^{-1}(a)]_q$.

We claim that Φ is a bijection and this would conclude this part of the proof. In order to show the claim, let $[b] \in \widehat{B}_{i'}/ker(q)$, then $[b] = \pi_{\widehat{B}_{i'}}(b)$, for some $b \in \widehat{B}_{i'}$. By surjectivity of $\widehat{f}_{i'}^{-1}$, there exists an element $a \in \widehat{A}_{i'}$ such that $b = \widehat{f}_{i'}^{-1}(a)$. Therefore $[b]_q = [\widehat{f}_{i'}^{-1}(a)]_q = \Phi([a]_p)$, i.e. Φ is surjective. To show that Φ is also injective, assume $[a]_p \neq [b]_p$, i. e. $a \neq b$, with $a, b \in \widehat{A}_{i'}$. Suppose, in view of a contradiction, that $\Phi([a]_p) = \Phi([b]_p)$, i. e. $[\widehat{f}_{i'}^{-1}(a)]_q = [\widehat{f}_{i'}^{-1}(b)]_q$. By assumption and the fact that ψ is an embedding, we have that $\psi \circ \pi_{\widehat{A}_{i'}}(a) \neq \psi \circ \pi_{\widehat{A}_{i'}}(b)$, i.e. $p(a) \neq p(b)$, whence $\widehat{f}_i^{-1} \circ p(a) \neq \widehat{f}_i^{-1} \circ p(b)$, since $\widehat{f}_i^{-1}$ is a bijection. On the other hand, $q \circ \widehat{f}_{i'}^{-1}(a) = \chi \circ \pi_{\widehat{B}_{i'}} \circ \widehat{f}_{i'}^{-1}(a) = \chi \circ \pi_{\widehat{B}_{i'}} \circ \widehat{f}_{i'}^{-1}(b) = q \circ \widehat{f}_{i'}^{-1}(b)$, which is in contradiction with the commutativity of the above diagram. This shows our claim.

($\Leftarrow$) Assume that $\mathbf{A}_i \cong \mathbf{B}_i$, by the family of isomorphisms $f_i\colon \mathbf{A}_i \to \mathbf{B}_i$, for each $i \in I$, and that there exists a bijection $\Phi\colon \widehat{A}_{i'}/ker(\widehat{p}) \to \widehat{B}_{i'}/ker(\widehat{q})$. Then, it is easy to check that (id, f_i) gives the desired isomorphism. □

Example 11. The two linearly ordered involutive bisemilattices in the following picture (lines indicate orders in the Boolean components, dashed lines indicate Boolean homomorphisms) are isomorphic (this is an consequence of Lemma 10).

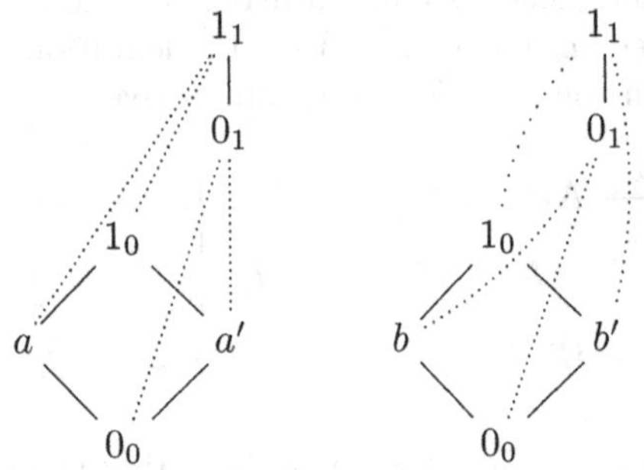

In order to exhibit a concrete isomorphism, observe that the two algebras are constructed using the very same index set (the two element lattice). An isomorphism is given by considering the identity map *id* on the lattice of indexes, the unique isomorphism on the two elements Boolean algebras and the Boolean isomorphism given by $\varphi(a) = \neg b$ (and $\varphi(\neg a) = b$) between the 4-elements Boolean algebras.

The graphical convention adopted in the above example (dotted lines for homomorphisms between algebras in the Płonka sum, black lines for the usual order relation in a Boolean algebra) will be used throughout the paper.

Remark 12. Notice that the non-triviality assumption in Lemma 10 is crucial as witnessed by the following example, where we consider the two semilattice direct systems $\langle\{\mathbf{A}_0, \mathbf{1}_1, \mathbf{1}_2\}, p_{ii'}, \{0, 1, 2\}\rangle$ and $\langle\{\mathbf{A}_0, \mathbf{1}_1, \mathbf{1}_2\}, q_{ii'}, \{0, 1, 2\}\rangle$.

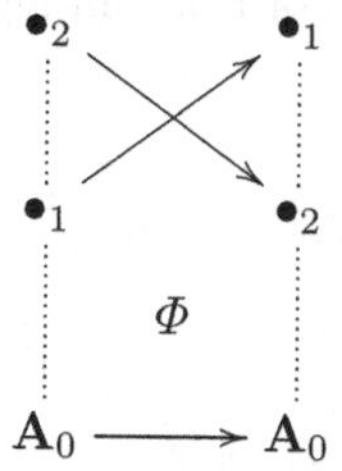

The map Φ depicted between the two systems is not an isomorphism (as it is clearly not an isomorphism on the index set), however $1_2/ker(\widehat{p}_{12}) = 1_2 = 1_2/ker(\widehat{q}_{12})$.

We denote by $\mathbb{A}_{p_{i-1,i}}$ the family $\{\langle \mathbf{A}_i, p_{i-1,i}, I\rangle\}_{i\in I}$ of all finite semilattice direct systems of Boolean algebras obtainable from the family of algebras $\{\mathbf{A}_i\}_{i\in I}$, indexed over the linearly ordered index set I. When clear from the context, we will write $\mathbb{A}_p$ instead of $\mathbb{A}_{p_{i-1,i}}$.

We are interested in the following question: what is the number of non-isomorphic elements in the family $\mathbb{A}_p$? This, in turn, will provide an answer to the question about how many linearly ordered involutive bisemilattices have the cardinality $\bigcup_{i\in I} \mid A_i \mid$, up to isomorphism.

In the light of Theorem 5 and Lemma 10, the answer to is provided by considering the dual semilattice inverse systems $\widehat{\mathbb{A}}_{\widehat{p}}$.

Lemma 13. *Let* $\mathbb{A}_{p_{01}} = \langle\{\mathbf{A}_0, \mathbf{A}_1\}, p, \{0 < 1\}\rangle$ *be a family of linearly ordered semilattice direct system of Boolean algebras. The number of non-isomorphic involutive bisemilattices obtained over* $\mathbb{A}_{p_{01}}$ *is the number of non-isomorphic involutive bisemilattices of cardinality* $\mid A_0 \mid + \mid A_1 \mid$ *and is equal to*

$$N(\mathbb{A}_{p_{01}}) := N(A_0, A_1) = \min(\mid \widehat{A}_0 \mid, \mid \widehat{A}_1 \mid),$$

where $\widehat{A}_0$ *(*$\widehat{A}_1$*, resp.) is the dual space of* $\mathbf{A}_0$ *(*$\mathbf{A}_1$*, resp.)*

Proof. At first observe that two elements in the family $\mathbb{A}_{p_{01}}$ differ only for the Boolean homomorphism from $\mathbf{A}_0$ to $\mathbf{A}_1$. Therefore, by Lemma 10, two involutive bisemilattices constructed over the system $\mathbb{A}_{p_{01}}$ are not isomorphic if and only if $\mid \widehat{A}_1/ker(\widehat{p}) \mid \neq \mid \widehat{A}_0/ker(\widehat{q}) \mid$ (where p and q are the Boolean homomorphisms). In other words, this means that the kernels of $\widehat{p}$ and $\widehat{q}$ generate two partitions, over $\mid \widehat{A}_1 \mid$, with a different number of equivalence classes. It is known that the number of equivalence classes partitioning a finite algebra $\mathbf{A}$ into a different number of blocks is equal to $\mid A \mid$. Therefore, if $\mid A_1 \mid \leq \mid A_0 \mid$ then $N(A_0, A_1) = \mid \widehat{A}_1 \mid$. Differently, since we only consider (the number of) partitions induced by (kernels of) maps from $\mid \widehat{A}_1 \mid$ to $\mid \widehat{A}_0 \mid$, we have that $N(A_0, A_1) = \mid \widehat{A}_0 \mid$ (Fig. 1). □

Remark 14. It is easily checked that the function $N(\mathbb{A}_{p_{01}})$, counting the number of involutive bisemilattices obtained over $\mathbb{A}_{p_{01}}$, can be generalized to the family

$\mathbb{A}_{p_{0m}}$ of semilattice direct systems of Boolean algebras $\langle\{A_0,\ldots,A_m\},p_{i-1,i},I\rangle$. More precisely,

$$N(\mathbb{A}_{p_{0m}}) = N(\mathbb{A}_{p_{01}})\cdot N(\mathbb{A}_{p_{12}})\cdot\ldots\cdot N(\mathbb{A}_{p_{m-1m}}) = \prod_{i=0}^{m-1} N(\mathbb{A}_{p_{ii+1}}).$$

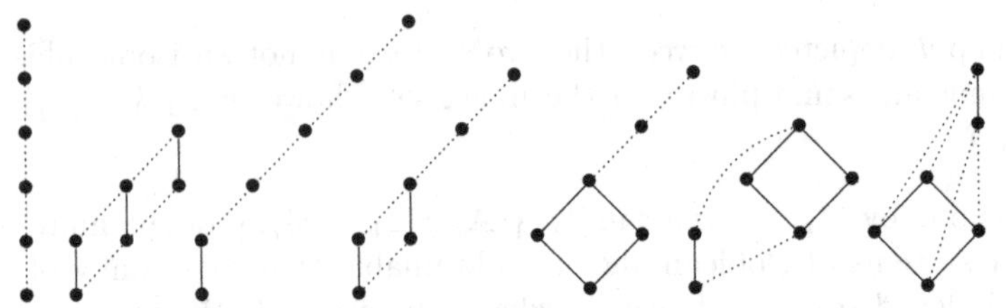

Fig. 1. The linearly ordered non-isomorphic IBSLs of cardinality 6.

4 Generating and Counting $\mathcal{L}$-$\mathcal{IBSL}$

The previous section provides sufficient and necessary conditions in order to identify the non-isomorphic $\mathcal{L}$-$\mathcal{IBSL}$ obtained over direct systems sharing both the index set and the family of Boolean algebras. In general, the elements of such a family of direct systems $\mathbb{A}_p$ differ at most with respect to the definition of their homomorphisms. It is easy to check that, given $\mathbb{X} = \langle \mathbf{A}_i, p_{ii'}, I\rangle$ and $\mathcal{P}_l(\mathbb{X}) \in \mathcal{L}$-$\mathcal{IBSL}$, we can always single out the family of direct systems $\mathbb{A}_p = \{\langle \mathbf{A}_i, p_{i-1,i}, I\rangle\}$. As this fact is central in the rest of the paper, we introduce the following definition

Definition 15. *Let* $\mathbb{X} = \langle \mathbf{A}_i, p_{ii'}, I\rangle$ *be a semilattice direct system of Boolean algebras and* $\mathbf{B} \cong \mathcal{P}_l(\mathbb{X}) \in \mathcal{L}$-$\mathcal{IBSL}$. *We define the* shape *of* $\mathbf{B}$ *as* $\mathbb{X}_{\mathbf{B}} = \langle \mathbf{A}_i, I\rangle$.

Clearly, two $\mathcal{L}$-$\mathcal{IBSL}$, $\mathbf{A} \cong \mathcal{P}_l(\mathbb{X}), \mathbf{B} \cong \mathcal{P}_l(\mathbb{Y})$ have the same shape if and only if the semilattice direct systems $\mathbb{X}$ and $\mathbb{Y}$ differ at most with respect to their homomorphisms, i.e. they belong to the same family $\mathbb{A}_p$. So, with this terminology at hand, Lemma 13 and Remark 14 tell us the number of non-isomorphic $\mathcal{L}$-$\mathcal{IBSL}$ of a fixed shape. Moreover, as a consequence of Lemma 10, two $\mathcal{L}$-$\mathcal{IBSL}$ with different shapes are non-isomorphic.

Therefore, in order to answer our question concerning the number of finite algebras in $\mathcal{L}$-$\mathcal{IBSL}$ it only remains to count, for a given $n \in \mathbb{N}$, the number of shapes that an $\mathcal{L}$-$\mathcal{IBSL}$ of order n can have. The present section is devoted to this issue.

Let $n \in \mathbb{N}^+$, and $e, m_i \in \mathbb{N}$ for $0 \le i \le e$. A *binary partition* of n is a decomposition of n into powers of two, that is

$$n = m_e \cdot 2^e + m_{e-1}\cdot 2^{e-1} + \cdots + m_0 \cdot 2^0. \tag{1}$$

Hence, the number $b(n)$ of binary partitions[3] of n is the number of solutions of (1). In general, binary partitions which differs by the orders of summands are considered identical.

Knowing that each finite linearly ordered $\mathcal{IBSL}$ $\mathbf{B}$ can be decomposed as a Płonka sum of Boolean algebras $\mathbf{A}_i$ whose direct system is indexed by a totally ordered set I, it follows that the cardinality n of $\mathbf{B}$ is always given by a solution of a binary partition (1) where $2^i = |\mathbf{A}_i|$ and $|I| = \sum_{i=0}^{e} m_i$.

The fact that binary partitions cannot differ only for the order of summands, together with (iii) in Definition 1 and Remark 14, implies that $b(n)$ cannot account for the number of shapes that an $\mathcal{L}$-$\mathcal{IBSL}$ of order n can assume. Indeed, given a certain shape $\langle \mathbf{A}_i, I\rangle$ of $\mathbf{B} \in \mathcal{L}$-$\mathcal{IBSL}$, every permutation ϕ over I that moves at least two indexes i, j such that $\mid \mathbf{A}_i \mid \neq 1, \mid \mathbf{A}_j \mid \neq 1$ defines a new shape $\langle \mathbf{A}_i, \phi(I)\rangle$. Notice that the condition about $\mid \mathbf{A}_i \mid \neq 1, \mid \mathbf{A}_j \mid \neq 1$ is justified by the fact that for any non-trivial Boolean algebra $\mathbf{A}$ there are no homomorphisms from a trivial Boolean algebra to $\mathbf{A}$. Hence, we can remove from the permutations $\phi(I)$ of I all the algebras whose cardinality is 2^0. The number of such algebras is expressed in (1) as m_0. Now, looking at $\langle \mathbf{A}_i, I\rangle$ and $\langle \mathbf{A}_i, \phi(I)\rangle$ as binary partitions, it is immediate to observe that they only differ for the order of summands.

Define $I^+ = I \setminus \{0\}$. For these reasons, we have to consider the *permutations with repetitions* of $|I| - m_0 = \sum_{i \in I^+} m_i$, knowing that each $\mathbf{A}_i$ is repeated m_i times. The number of such permutations is given by the *multinomial coefficient* [15]

$$pr(|I| - m_0) = \binom{|I| - m_0}{\{m_i\}_{i \in I^+}} = \frac{(\sum_{i \in I^+} m_i)!}{\prod_{i \in I^+} m_i!}, \tag{2}$$

We now start by introducing a routine procedure that generates all the binary partitions of n in a form that makes the ensuing computations easy to handle.

Definition 16. *A* sequence *is a list of pairs of the form* $s = (m_{e_1}, 2^{e_1}) \to \cdots \to (m_{e_p}, 2^{e_p})$. *We define its* presentation *as*

$$P(s) = (2_1^{e_1} \to \cdots \to 2_{m_{e_1}}^{e_1}) \to \cdots \to (2_1^{e_p} \to \cdots \to 2_{m_{e_p}}^{e_p}).$$

Given a set S of sequences, we denote by $P(S)$ the set of the presentations of sequences in S, that is $P(S) = \{P(s) \mid s \in S\}$.

Definition 17. *Given a natural number* $n \in \mathbb{N}^+$ *we define*

$$L(n) = \{(m_{e_1}, 2^{e_1}) \to (m_{e_2}, 2^{e_2}) \to \cdots \to (m_{e_k}, 2^{e_k}) \mid n = \sum_{i=1}^{k} (m_{e_i} \cdot 2^{e_i})\},$$

such that $e^1 > e^2 > \cdots > e^k$, *that is the set of sequences that give all decompositions of* n *into powers of two, such as in* (1).

[3] See sequence http://oeis.org/A018819 at *The On-Line Encyclopedia of Integer Sequences*, published electronically at https://oeis.org.

For any sequence $l(n) \in L(n)$*, define* $F(l(n)) = \{m_{e_i} \mid (m_{e_i}, 2^{e_i}) \in l(n)\}$ *as the set of the* multiplicities *in* $l(n)$*, and* $F^{+}(l(n)) = \{m_{e_i} \in F(l(n)) \mid e_i \neq 0\}$ *as the subset of* $F(l(n))$ *given by the* positive multiplicities *in* $l(n)$.

Observe that the above definition of multiplicity and positive multiplicity in $l(n)$ can be equally defined by looking at the presentation $P(l(n))$. Moreover, given a sequence $l(n) = (m_{e_1}, 2^{e_1}) \to \cdots \to (m_{e_p}, 2^{e_p})$ with positive factors $m_{e_1}, ..., m_{e_z}$, we denote by $P^{+}(l(n))$ the presentation of the sequence $(m_{e_1}, 2^{e_1}) \to ... \to (m_{e_z}, 2^{e_z})$.

Definition 18. *Let* $n, n_1, n_2 \in \mathbb{N}$ *and* $E(n) = \{2^e, \ldots, 2^0\}$ *the set of powers of two such that* $2^i \leq \log_2 n$*, for any* $i = 0, \ldots, e$*. The map* $d\colon \mathbb{N} \times \mathbb{N} \to E(n)$ *defined as*

$$d(n_1, n_2) := \begin{cases} max\{m \in E(n) : m < n_1, n_2\} & \text{if } n_1, n_2 \notin \{0, 1\} \\ 1 & \text{otherwise.} \end{cases}$$

is called the division map *of* (n_1, n_2) *with respect to* n.

By a forest we mean a disjoint union $\sqcup$ of trees.

Given two sequences $l(n) = (m_{e_1}, 2^{e_1}) \to \cdots \to (m_{e_k}, 2^{e_k})$, $l'(n) = (m_{f_1}, 2^{f_1}) \to \cdots \to (m_{f_h}, 2^{f_h})$ in $L(n)$, we say that they share a *common prefix* when for some $i \leq min(k, h)$, we have $e_j = f_j$, for $1 \leq j \leq i$. We write $l(n) = P \to s$ and $l'(n) = P \to s'$ to denote the fact that P is the common prefix of $l(n)$ and $l'(n)$.

Given a set of sequences S, we construct a forest $\Gamma(S)$ in the following way. Let l and l' be two chains in S, such that P is their longest common prefix, that is $l = P \to s$ and $l' = P \to s'$. Then, the tree $P \to (s \sqcup s')$ belongs to $\Gamma(S)$. For each sequence s in S there is a unique branch of $\Gamma(S)$ that is a unique copy of s, and every branch of $\Gamma(S)$ is a copy of a unique chain in S.

The pseudocode in Algorithm 1 introduces a couple of functions that, for any given $n \in \mathbb{N}^{+}$, produce the set $\Gamma(L(n))$, that is the set of trees whose branches are binary partitions of n expressed as sequences of Definition 17. Notice that, to better follow the construction of sequences of $\Gamma(L(n))$, in the pseudocode of Algorithm 1 we repeatedly use expressions like 2^e. Obviously, a real-world implementation of the algorithm does not need such a level of detail, and exponents e can be used instead.

Theorem 19. $\Gamma(L(n)) = \text{GenForest}(n)$.

Proof. $\supseteq$. This inclusion follows by direct inspection.

$\subseteq$. We prove this inclusion by an induction on n.

(B). $n = 1$. If $n = 1$ then $\Gamma(l(n)) = \{(1, 2^0)\}$. As $E(1) = \{2^0\}$ by Line 6 the algorithm calculates $\text{GenSeq}(1, 2^0)$ which by Line 13 has as output exactly the sequence $\{(1, 2^0)\}$.

(IND). Assume the statement holds for any $\{1, \ldots, n\}$ and consider a sequence $l(n+1) = (m_1, 2^{e_1}) \to ... \to (m_p, 2^{e_p}) \in \Gamma(l(n+1))$. Clearly $2^{e_1} \in E(n+1)$, then in $\text{GenSeq}(n+1, 2^{e_1})$ at Lines 15 we obtain $(n+1)/2^{e_1} = q$.

Algorithm 1

```
function GenForest(n)
    e = ⌊log_2 n⌋
    E = {2^e, 2^{e-1}, ..., 2^0}
    F empty list of trees
    for each 2^e in E do
        T_e = GenSeq(n, 2^e)                          ▷ T_e is a list of trees.
        add T_e to F
    end for
    return F
end function

function GenSeq(n,2^e)
    if e == 0 then
        return a tree with root (n, 2^0)
    end if
    q = n/2^e
    if q > 1 then
        for each i ∈ {1, ..., q} do
            create a tree P_i with root (i, 2^e)
            m = n − i · 2^e
            if m > 0 then
                if m = 2^x and x < e then
                    d_i = m
                else
                    d_i = d(m, 2^e)        ▷ d refers to the division map of Definition 18
                end if
                for each 2^j in {2^0, ..., d_i} do
                    T = GenSeq(n − i · 2^e, 2^j)      ▷ T is a list of trees.
                    for every t ∈ T, add t as a child of P_i
                end for
            end if
        end for
    end if
    if q == 1 then
        create a tree P_i with root (1, 2^e)
        r = n mod 2^e
        if r > 0 then
            T = GenForest(r)                          ▷ T is a list of trees.
            for every t ∈ T, add t as a child of P_i
        end if
    end if
    return P_i
end function
```

We have two cases (a) $m_1 = 1$ or (b) $m_1 \gneq 1$. In the first case (a), Lines 17–18 entails that $(1, 2^{e_1})$ is generated as root of the sequence in case $q > 1$. Similarly, when $q = 1$, Lines 34–35 generate $(1, 2^{e_1})$ as a root of the sequence. For the second case (b), we have $m_1 \in \{1, ..., q\}$ and by Line 18 the algorithm gives $(m_1, 2^{e_1})$ as a root of the sequence.

Now, clearly $l(n+1) \setminus (m_1, 2^{e_1}) \in \Gamma(l((n+1) - m_1 \cdot 2^{e_1}))$, i.e. $l(n+1) \setminus (m_1, 2^{e_1})$ is a sequence for $(n+1) - m_1 \cdot 2^{e_1}$. By induction hypothesis $l(n+1) \setminus (m_1, 2^{e_1}) \in$GenForest$((n+1) - m_1 \cdot 2^{e_1})$. To complete the proof we have to show that the second pair $(m_2, 2^{e_2})$ in $l(n+1)$ is generated as child of the root $(m_1, 2^{e_1})$ is some branch. In order to simplify the notation we fix $k = (n+1) - m_1 \cdot 2^{e_1}$.

We distinguish two cases:

(1). $m_2 = 1$. Then either (a) $2^{e_2} = k$ or (b) $2^{e_2} \in \{2^0, ..., d(2^1, k)\}$. In the first case (a) we have that k is a power of 2 strictly smaller than 2^1, and by Line 21,25,26,27, follows that the computation of GenSeq$(k, 2^0)$ returns $(1, k)$ as child of $(m_1, 2^{e_1})$ (by Lines 12–13). In the second case (b), by Lines 26,27 the algorithm computes GenSeq$(2^{e_2}, k)$. So, it is immediate to verify that for any $k/2^{e_2}$ computed at Line 15, by Lines 17,18 and 33,34 we obtain $(1, 2^{e_2})$ as child of $(m_1, 2^{e_1})$.

(2). $m_2 > 1$. This implies $2^{e_2} < 2^{e_2}$ and therefore $2^{e_2} \in \{2^0, ..., d(k, 2^{e_1})\}$. By Line 26 we have the computation of GenSeq$(k, 2^{e_2})$. Clearly $m_2 \leq k/2^{e_2}$ and this, together with the assumption $m_2 > 1$ implies $k/2^{e_2} > 1$. So, by Lines 17,18 $(m_2, 2^{e_2})$ is a child of $(m_1, 2^{e_1})$.

So, as $(m_2, 2^{e_2})$ is the root of the sequence $l(n+1) \setminus (m_1, 2^{e_1})$ and by induction hypothesis $l(n+1) \setminus (m_1, 2^{e_1})$ is generated by the algorithm, the fact that $(m_2, 2^{e_2})$ is generated as child of $(m_1, 2^{e_1})$ entails that $l(n+1)$ is generated by the algorithm, as desired. □

Example 20. For $n = 10$ the output of GenForest(10) is depicted in Fig. 2.

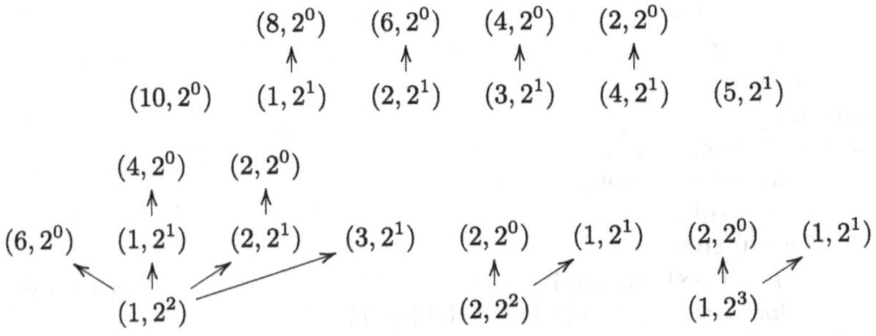

Fig. 2. The forest of sequences generated by GenForest(10), see Example 20.

Remark 21. It is worth noticing that, given a sequence $s = (m_{e_1}, 2^{e_1}) \to \cdots \to (m_{e_p}, 2^{e_p})$ such that $e_i \neq 0$ for any $e_1 \leq e_i < e_p$, its presentation $P(s) = (2_1^{e_1} \to \cdots \to 2_{m_{e_1}}^{e_1}) \to \cdots \to (2_1^{e_p} \to \cdots \to 2_{m_{e_p}}^{e_p})$ always describe a shape

$\mathbb{X}_{\mathbf{B}}$ of an $\mathcal{L}$-$\mathcal{IBSL}$ $\mathbf{B}$. More precisely, $\mathbb{X}_{\mathbf{B}} = \langle \mathbf{A}_i, I \rangle$ where $I = \{1_{e_1}, \ldots, m_{e_1}\} \cup \{1_{e_2}, \ldots, m_{e_2}\} \cup \cdots \cup \{1_{e_p}, \ldots, m_{e_p}\}$ and for each $k_{e_i} \in I$, $\mathbf{A}_{k_{e_i}}$ is a Boolean Algebra of order 2^{e_i}. Dually, given an $\mathcal{L}$-$\mathcal{IBSL}$ $\mathbf{B} \cong \mathcal{P}_l(\mathbb{X})$, its shape $\mathbb{X}_{\mathbf{B}}$ can always be described by the presentation of an appropriate sequence.

Notation. In what follows we adopt the following notation. Given $l(n) \in L(n)$, we denote by $\pi(l(n)) = \{s_1, \ldots, s_{c(l(n))}\}$ the set of sequences obtained by a permutation with repetition over $P^+(l(n))$.

Example 22. Let $l'(10) = (1, 2^2) \to (2, 2^1) \to (2, 2^0)$ be a sequence in $L(10)$ (compare with the branches in the biggest tree in Fig. 2). The only permutation in $P^+(l'(10))$ is $(2, 2^1) \to (1, 2^2) \to (2, 2^0)$.

Theorem 23. *Let $n \in \mathbb{N}^+$, $l(n) = (m_{e_1}, 2^{e_1}) \to \ldots \to (m_{e_q}, 2^{e_q}) \in L(n)$ with $k = m_{e_1} + \ldots + m_{e_q}$ and let also $m_{e_1}, \ldots, m_{e_z}$ be the members of $F^+(l(n))$. Then, there are*

$$c(l(n)) = \frac{(m_{e_1} + \ldots + m_{e_z})!}{m_{e_1}! \cdot \ldots \cdot m_{e_z}!}$$

shapes $\langle I_1, \mathbf{A}_i^1 \rangle, \ldots, \langle I_{c(l(n))}, \mathbf{A}_i^{c(l(n))} \rangle$ of finite $\mathcal{L}$-$\mathcal{IBSL}$ of order n such that, for each $1 \leq j \leq c(l(n))$, $\mid I_j \mid = k$ and $\mathbf{A}_i^j$ is a family of Boolean Algebras whose cardinalities correspond to the members of $P(l(n))$.

Proof. Observe that the above formula is an instance of (2), counting the permutations with repetitions of a set of cardinality $m_{e_1} + \ldots + m_{e_z}$ knowing that each $1 \leq i \leq z$ object is repeated m_{e_i} times. By Remark 21, each $s \in \pi(l(n))$ describes a shape of an $\mathcal{L}$-$\mathcal{IBSL}$ of cardinality n.

Now consider $s, s' \in \pi(l(n))$, with $s \neq s'$ and let $\leq_s, \leq_{s'}$ be the order of their presentations. The fact that $s \neq s'$ implies that for at least two elements $2_x^{e_i}, 2_{x'}^{e_i'} \in P^+(s), P^+(s')$ (with $e_i \neq e_i'$) it holds $2_x^{e_i} \leq_s 2_{x'}^{e_i'} \iff 2_{x'}^{e_i'} \leq_{s'} 2_x^{e_i}$. This, by applying Remark 21, proves that the shapes described by $P(s), P(s')$ are different. Finally, consider $\mathbb{X}_{\mathbf{B}} = \langle \mathbf{A}_i, I \rangle$ a shape of an $\mathcal{L}$-$\mathcal{IBSL}$ $\mathbf{B} \cong \mathcal{P}_l(\mathbb{X})$ with cardinality n such that $\mid I \mid = k$ and such that $\mathbf{A}_i$ is a family of Boolean algebras whose cardinalities are exactly the members of $P(l(n))$. By Remark 21, $\mathbb{X}_{\mathbf{B}}$ can be described by the presentation $P(r)$ of an appropriate sequence r. Moreover, the fact that the cardinalities of the algebras in the family $\mathbf{A}_i$ are all and only the members of $P(l(n))$, implies that r can be obtained from $l(n)$ by permuting at least one element $2_x^{e_i} \in P^+(l(n))$ with another $2_{x'}^{e_i'} \in P^+(l(n))$ such that $e_i \neq e_i'$. By construction, this implies $r \in \pi(l(n))$, as desired. □

Corollary 24. *Let $l_1(n), \ldots, l_p(n)$ be the elements of $L(n)$. Then the number of shapes of $\mathcal{L}$-$\mathcal{IBSL}$ of order n is equal to $\sum_{i=1}^{p} c(l_i(n))$.*

Theorem 25. *The number of all the non-isomorphic $\mathcal{L}$-$\mathcal{IBSL}$ of cardinality n is given by* L-IBSL(n).

Algorithm 2

```
1: function L-IBSL(n)
2:     F = GenForest(n)
3:     for each tree T in F do
4:         for each branch B in T do        ▷ Notice that B is a list of couples (i, 2^e)
5:             B' = B without all the couples (i, 2^0)
6:             for each permutation P of B' do
7:                 for each (i, 2^{e_i}) → (j, 2^{e_j}) in P do
8:                     t = t×N(2^{e_i}, 2^{e_j})
9:                 end for
10:            end for
11:        end for
12:    end for
13:    return t
14: end function
```

Proof. Let B be one of the branches of a tree in F, as computed in Line 2. By Theorem 19, B is exactly a sequence $l(n) \in L(n)$. Each one of the $c(l(n))$ permutations (see Theorem 23) of $l(n)$ are computed in Line 6. Consider $s \in \pi(l(n))$. In the light of Lemma 13, for every pair $2^{e_i}, 2^{e_j}$ such that $2^{e_i} \rightarrow 2^{e_j} \in s$ the number $\prod_{i,j} N(2^{e_i}, 2^{e_j})$ gives us all the non-isomorphic $\mathcal{L}$-$\mathcal{IBSL}$ of the shape described by s, as computed in Line 8. □

5 Conclusions

The following results have been obtained on a GNU/Linux Debian 4.9.82-1 system with an Intel Core i7-5500U CPU and 8 GB of RAM[4].

To study the effectiveness of our algorithm, we have used Mace4 [25] to compute the number of finite linearly ordered IBSLs, relying on the first-order theory provided in Sect. 3. We point out that using this FO theory, in the LIBSL given by the Płonka sum of n trivial Boolean algebras we have $0 = 1$. Mace4 assumes $0 \neq 1$, and hence such type of LIBSLs are not generated by it. To obtain Płonka sums of trivial Boolean algebras, we have to replace 1 for another constants in the FO theory.

Mace4 produces in reasonable time the algebraic structures of cardinality up to 11. For cardinality 12, Mace4 exits reaching its internal time limit. After running the `isofilter` program associated to Mace4, we obtain a file containing the non-isomorphic LIBLs of cardinality $2 \leq n \leq 11$ generated by Mace4.

The procedure introduced in the previous section has been implemented in Python and has been used to compute the number of all the non-isomorphic LIBSLs of cardinality $1 \leq n \leq 23$, the results are reported in Table 2. For cardinality $n = 24$, the script uses too much RAM and was automatically killed by the system. To measure the running times of both experiments we have used

[4] The Python implementation, the Mace4 input and output files can be downloaded from: https://homes.di.unimi.it/~valota/code/libsl.zip.

Table 1. The running times of our experiments as calculated by the tool `time`. The second column reports the total time used by the Python implementation of our algorithm to count all the non-isomorphic LIBSLs of cardinality $1 \leq n \leq 23$. The third column reports the time used by Mace4 to generate the first-order models of cardinality $2 \leq n \leq 11$. The fourth column reports the time used by `interpformat` to transform the Mace4 models in a format useful for `isofilter`. The fifth column reports the time used by `isofilter` to produce a file with all the non-isomorphic LIBSLs of cardinality $2 \leq n \leq 11$.

Running times	Algorithm 2	Mace4	`interpformat`	`isofilter`
`real`	0m1.331s	1m1.115s	0m40.891s	0m43.611s
`user`	0m1.176s	1m0.008s	0m40.484s	0m43.524s
`sys`	0m0.156s	0m0.908s	0m0.232s	0m0.068s

the Debian GNU/Linux command-line tool `time`, the results are summarized in Table 1.

Comparing these running times, it is clear that also a non-optimized implementation of our algorithm is more efficient than the brute-force approach of Mace4, for counting purposes. We should point out, however, that Mace4 generates full models with tables for each operation, whereas our algorithm only produces shapes of LIBSLs and then it performs counting computations. As reported above, our Python script is memory consuming. Hence, it would be interesting to improve our implementation with a better memory management, and with additional options to generate also the algebraic structure of the counted LIBSLs. Moreover, computational complexity study and asymptotic analysis of our algorithm seems to be within reach, allowing us to establish upper and lower bounds on the number of LIBSLs of cardinality n. These research directions are outside the scope of this paper, and are left as future work.

Table 2. The numbers of L-IBSL with n elements for $1 \leq n \leq 23$.

n	1	2	3	4	5	6	7	8	9	10	11	12	13	14	15	16	17	18	19	20	21	22	23
L-IBSL(n)	1	2	2	4	4	7	7	14	14	26	26	52	52	99	99	199	199	386	386	772	772	1508	1508

The set of cardinalities of the finite members of a variety of algebras $\mathcal{V}$, goes by the name of *fine spectrum* of $\mathcal{V}$ and it has been introduced by Taylor in [34]. According to Quackenbush, when dealing with ordered structures, "the fine spectrum problem is usually hopeless" [30]. In this note we have introduced a procedure to count a specific subclass, namely linearly ordered IBSL: this represents a first step to solve the fine spectrum problem for the variety $\mathcal{IBSL}$. As our approach relies on the algebraic representation theorem, it can possibly be extended to different subclasses of $\mathcal{IBSL}$. An option is considering, for instance, the quasi-variety of $\mathcal{IBSL}$ whose maps, in the Płonka representation, are injective (this class appears to be useful in the study of probability measures

[7]). Since the index set of the Płonka sum representation of a finite IBSL is a finite semilattice, and finite semilattices coincides with finite lattices, our next step is to improve our approach with the algorithm to generates finite lattices established in [19]. Finally, we notice that our approach is heavily grounded on the duality of Theorem 5. Hence, it appears that spectra problems are easier to handle when restated in dual terms. Indeed, in literature one can find several duality-based solutions to the *free spectrum problem* (counting the number of k-generated free algebras) for varieties related to many-valued logics[5] [2,12], and very recently this dual approach has been used to compute the fine spectrum of the variety of prelinear Heyting algebras [35].

Acknowledgments. The authors wish to thank the anonymous referees for their helpful comments.

References

1. Aguzzoli, S., Bova, S., Gerla, B.: Free algebras and functional representation for fuzzy logics. In: Cintula, P., Hájek, P., Noguera, C. (eds.) Handbook of Mathematical Fuzzy Logic. Volume 2, volume 38 of Studies in Logic. Mathematical Logic and Foundation, pp. 713–791. College Publications (2011)
2. Aguzzoli, S., Busaniche, M., Marra, V.: Spectral duality for finitely generated nilpotent minimum algebras, with applications. J. Logic Comput. **17**(4), 749–765 (2007)
3. Balbes, R.: A representation theorem for distributive quasilattices. Fundamenta Mathematicae **68**, 207–214 (1970)
4. Bergman, C., Failing, D.: Commutative idempotent groupoids and the constraint satisfaction problem. Algebra Universalis **73**(3), 391–417 (2015)
5. Bochvar, D.: On a three-valued calculus and its application in the analysis of the paradoxes of the extended functional calculus. Mathematicheskii Sbornik **4**, 287–308 (1938)
6. Bonzio, S.: Dualities for Płonka sums. Logica Universalis (2018). https://doi.org/10.1007/s11787-018-0209-4
7. Bonzio, S., Flaminio, T., Loi, A.: States over Płonka sums of Boolean algebras (2018, in preparation)
8. Bonzio, S., Gil-Férez, J., Paoli, F., Peruzzi, L.: On Paraconsistent Weak Kleene Logic: axiomatization and algebraic analysis. Stud. Logica **105**(2), 253–297 (2017)
9. Bonzio, S., Loi, A., Peruzzi, L.: A duality for involutive bisemilattices. Stud. Logica (2018). https://doi.org/10.1007/s11225-018-9801-0
10. Bonzio, S., Moraschini, T., Pra Baldi, M.: Logics of left variable inclusion and Płonka sums of matrices (2018, Submitted manuscript)
11. Bonzio, S., Pra Baldi, M.: Containment logics and Płonka sums of matrices (2018, Submitted manuscript)
12. Bova, S., Valota, D.: Finitely generated RDP-algebras: spectral duality, finite coproducts and logical properties. J. Logic Comput. **22**(3), 417–450 (2012). http://dx.doi.org/10.1093/logcom/EXR006

[5] For a complete account on Lindenbaum algebras representations in many-valued logic setting, we refer the interested reader to [1].

13. Ciuni, R., Carrara, M.: Characterizing logical consequence in paraconsistent weak kleene. In: Felline, L., Ledda, A., Paoli, F., Rossanese, E. (eds.) New Developments in Logic and the Philosophy of Science. College, London (2016)
14. Ferguson, T.: A computational interpretation of conceptivism. J. Appl. Non-Classical Logics **24**(4), 333–367 (2014)
15. Graham, R., Knuth, D., Patashnik, O.: Concrete Mathematics. Addison-Wesley, Boston (1989)
16. Halldén, S.: The Logic of Nonsense. Lundequista Bokhandeln, Uppsala (1949)
17. Harding, J., Romanowska, A.B.: Varieties of Birkhoff systems: part I. Order **34**(1), 45–68 (2017)
18. Harding, J., Romanowska, A.B.: Varieties of Birkhoff Systems: Part II. Order **34**(1), 69–89 (2017)
19. Heitzig, J., Reinhold, J.: Counting finite lattices. Algebra Universalis **48**(1), 43–53 (2002)
20. Johnstone, P.T.: Stone Spaces. Cambridge Studies in Advanced Mathematics, vol. 3. Cambridge University Press, Cambridge (1982)
21. Kalman, J.: Subdirect decomposition of distributive quasilattices. Fundamenta Mathematicae **2**(71), 161–163 (1971)
22. Kleene, S.: Introduction to Metamathematics. North Holland, Amsterdam (1952)
23. Libkin, L.: Aspects of Partial Information in Databases. Ph.D Thesis, University of Pennsylvania (1994)
24. Mac Lane, S.: Categories for the Working Mathematician. GTM, vol. 5. Springer, New York (1978). https://doi.org/10.1007/978-1-4757-4721-8
25. McCune, W.: Prover9 and mace4, 2005–2010. http://www.cs.unm.edu/~mccune/prover9/
26. Płonka, J.: On a method of construction of abstract algebras. Fundamenta Mathematicae **61**(2), 183–189 (1967)
27. Płonka, J.: On the sum of a direct system of universal algebras with nullary polynomials. Algebra Universalis **19**(2), 197–207 (1984)
28. Płonka, J., Romanowska, A.: Semilattice sums. Universal Algebra and Quasigroup Theory, pp. 123–158 (1992)
29. Puhlmann, H.: The snack powerdomain for database semantics. In: Borzyszkowski, A.M., Sokołowski, S. (eds.) MFCS 1993. LNCS, vol. 711, pp. 650–659. Springer, Heidelberg (1993). https://doi.org/10.1007/3-540-57182-5_56
30. Quackenbush, R.W.: Enumeration in classes of ordered structures. In: Rival, I. (ed.) Ordered Sets. NATO Advanced Study Institute Series (Series C – Mathematical and Physical Sciences), vol. 83, pp. 523–554. Springer, Dordrecht (1982). https://doi.org/10.1007/978-94-009-7798-3_17
31. Romanowska, A., Smith, J.: Semilattice-based dualities. Stud. Logica **56**(1/2), 225–261 (1996)
32. Romanowska, A., Smith, J.: Duality for semilattice representations. J. Pure Appl. Algebra **115**(3), 289–308 (1997)
33. Stone, M.: Applications of the theory of boolean rings to general topology. Trans. Am. Math. Soc. **41**, 375–481 (1937)
34. Taylor, W.: The fine spectrum of a variety. Algebra Universalis **5**(1), 263–303 (1975)
35. Valota, D.: Spectra of Gödel Algebras (2018, Submitted manuscript)

MIX ⋆-Autonomous Quantales and the Continuous Weak Order

Maria João Gouveia[1] and Luigi Santocanale[2(✉)]

[1] Universidade de Lisboa, 1749-016 Lisboa, Portugal
mjgouveia@fc.ul.pt
[2] LIS, CNRS UMR 7020, Aix-Marseille Université, Marseille, France
luigi.santocanale@lis-lab.fr

Abstract. The set of permutations on a finite set can be given a lattice structure (known as the weak Bruhat order). The lattice structure is generalized to the set of words on a fixed alphabet $\Sigma = \{x, y, z, \dots\}$, where each letter has a fixed number of occurrences (these lattices are known as multinomial lattices and, in dimension 2, as lattices of lattice paths). By interpreting the letters $x, y, z, \dots$ as axes, these words can be interpreted as discrete increasing paths on a grid of a d-dimensional cube, where $d = \text{card}(\Sigma)$.

We show in this paper how to extend this order to images of continuous monotone paths from the unit interval to a d-dimensional cube. The key tool used to realize this construction is the quantale $\mathsf{L}_\vee(\mathbb{I})$ of join-continuous functions from the unit interval to itself; the construction relies on a few algebraic properties of this quantale: it is $\star$-autonomous and it satisfies the mix rule.

We begin developing a structural theory of these lattices by characterizing join-irreducible elements, and by proving these lattices are generated from their join-irreducible elements under infinite joins.

1 Introduction

Combinatorial objects (trees, permutations, discrete paths, ...) are pervasive in mathematics and computer science; often these combinatorial objects can be organised into some ordered collection in such a way that the underlying order is a lattice.

Building on our previous work on lattices of binary trees (known as Tamari lattices or associahedra) and lattices of permutations (known as weak Bruhat orders or permutohedra) as well as on related constructions [6,21–26], we have been led to ask whether these constructions can still be performed when the underlying combinatorial objects are replaced with geometric ones.

More precisely we have investigated the following problem. Multinomial lattices [3] generalize permutohedra in a natural way. Elements of a multinomial lattice are words on a finite totally ordered alphabet $\Sigma = \{x, y, z \dots\}$ with a fixed number of occurrences of each letter. The order is obtained as the reflexive and transitive closure of the binary relation $\prec$ defined by $wabu \prec wbau$, whenever $a, b \in \Sigma$ and $a < b$ (if we consider

M. J. Gouveia—Partially supported by FCT under grant SFRH/BSAB/128039/2016
L. Santocanale—Partially supported by the TICAMORE project ANR-16-CE91-0002-01.

J. Desharnais et al. (Eds.): RAMiCS 2018, LNCS 11194, pp. 184–201, 2018.
https://doi.org/10.1007/978-3-030-02149-8_12

words with exactly one occurrence of each letter, then we have a permutohedron). Now these words can be given a geometrical interpretation as discrete increasing paths in some Euclidean cube of dimension $d = \mathrm{card}(\Sigma)$, so the weak order can be thought of as a way of organising these paths into a lattice structure. When Σ contains only two letters, then these lattices are also known as lattices of (lattice) paths [9] and we did not hesitate in [21] to call the multinomial lattices "lattices of paths in higher dimensions". The question that we raised is therefore whether the weak order can be extended from discrete paths to continuous increasing paths.

We already presented at the conference TACL 2011 the following result, positively answering this question.

Proposition. Let $d \geq 2$. Images of increasing continuous paths from $\mathbf{0}$ to $\mathbf{1}$ in $\mathbb{R}^d$ can be given the structure of a lattice; moreover, all the permutohedra and all the multinomial lattices can be embedded into one of these lattices while respecting the dimension d.

We called this lattice the *continuous weak order*. The proof of this result was complicated by the many computations arising from the structure of the reals and from analysis. We recently discovered a cleaner proof of the above statement where all these computations are uniformly derived from a few algebraic properties. The algebra we need to consider is the one of the quantale $\mathsf{L}_\vee(\mathbb{I})$ of join-continuous functions from the unit interval to itself. This is a ⋆-autonomous quantale, see [2], and moreover it satisfies the mix rule, see [7]. The construction of the continuous weak order is actually an instance of a general construction of a lattice $\mathsf{L}_d(Q)$ from a ⋆-autonomous quantale Q satisfying the mix rule. When $Q = \mathbf{2}$ (the two-element Boolean algebra) this construction yields the usual weak Bruhat order; when $Q = \mathsf{L}_\vee(\mathbb{I})$, this construction yields the continuous weak order. Thus, the step we took is actually an instance of moving to a different set of (non-commutative, in our case) truth values, as notably suggested in [17]. What we found extremely surprising is that many deep geometric notions (continuous monotone path, maximal chains, ...) might be characterised via this simple move and using the algebra of quantales.

Let us state our first main result. Let $\langle Q, 1, \otimes, {}^\star \rangle$ be a ⋆-autonomous quantale (or a residuated lattice), denote by 0 and $\oplus$ the dual monoidal operations. Q is not supposed to be commutative, but we assume that it is cyclic ($x^\star = x \multimap 0 = 0 \multimapinv x$, for each $x \in Q$) and that it satisfies the MIX rule ($x \otimes y \leq x \oplus y$, for each $x, y \in Q$). Let $d \geq 2$, $[d]_2 := \{\, (i, j) \mid 1 \leq i < j \leq d \,\}$ and consider the product $Q^{[d]_2}$. Say that a tuple $f \in Q^{[d]_2}$ is *closed* if $f_{i,j} \otimes f_{j,k} \leq f_{i,k}$, and that it is *open* if $f_{i,k} \leq f_{i,j} \oplus f_{j,k}$; say that f is *clopen* if it is closed and open.

Theorem. The set of clopen tuples of $Q^{[d]_2}$, with the pointwise ordering, is a lattice.

The above lattice is the one we denoted $\mathsf{L}_d(Q)$. The second main result we aim to present relates the algebraic setting to the analytic one:

Theorem. Clopen tuples of $\mathsf{L}_\vee(\mathbb{I})^{[d]_2}$ bijectively correspond to images of monotonically increasing continuous functions $\mathsf{p} : \mathbb{I} \to \mathbb{I}^d$ such that $\mathsf{p}(0) = \mathbf{0}$ and $\mathsf{p}(1) = \mathbf{1}$.

The results presented in this paper undoubtedly have a mathematical nature, yet our motivations for developing these ideas originate from various researches in computer science that we recall next.

Directed homotopy [13] was developed to understand behavioural equivalence of concurrent processes. Monotonically increasing paths might be seen as behaviours of distributed processes whose local state variable can only increase. The relationship between directed homotopies and lattice congruences (in lattices of lattice paths) was already pinpointed in [21]. In that paper we did not push further these ideas, mainly because the mathematical theory of a continuous weak order was not yet available.

In *discrete geometry* discrete paths (that is, words on the alphabet $\{ x, y, \ldots \}$) are used to approximate continuous lines. In dimension 2, Christoffel words [4] are well-established approximations of a straight segment from $(0,0)$ to some point (n,m). The lattice theoretic nature of this kind of approximation is apparent from the fact that Christoffel words can equivalently be defined as images of the identity/diagonal via the right/left adjoints to the canonical embedding of the binomial lattice $\mathsf{L}(n,m)$ into the lattice $\mathsf{L}_{\vee}(\mathbb{I})$. For higher dimensions, there are multiple proposals on how to approximate a straight segment, see for example [1,5,10,19]. It is therefore tempting to give a lattice theoretic notion of approximation by replacing the binomial lattices with the multinomial lattices and the lattice $\mathsf{L}_{\vee}(\mathbb{I})$ with the lattice $\mathsf{L}(\mathbb{I}^d)$. The structural theory of the lattices $\mathsf{L}(\mathbb{I}^d)$ already identifies difficulties in defining such a notion of approximation. For $d \geq 3$, the lattice $\mathsf{L}(\mathbb{I}^d)$ is no longer the Dedekind-MacNeille completion of the sublattice of discrete paths whose steps are on rational points—this is the colimit of the canonical embeddings of the multinomial lattices into $\mathsf{L}(\mathbb{I}^d)$; defining approximations naively via right/left adjoints of these canonical embeddings is bound to be unsatisfying. This does not necessarily mean that we should discard lattice theory as an approach to discrete geometry; for example, we expect that notions of approximation that take into consideration the degree of generation of $\mathsf{L}(\mathbb{I}^d)$ from multinomial lattices will be more robust.

The paper is organized as follows. We recall in Sect. 2 some facts on join-continuous (or meet-continuous) functions and adjoints. Section 3 describes the construction of the lattice $\mathsf{L}_d(Q)$, for an integer $d \geq 2$ a lattice and a mix $\star$-autonomous quantale Q. In Sect. 4 we show that the quantale $\mathsf{L}_{\vee}(\mathbb{I})$ of continuous functions from the unit interval to itself is a mix $\star$-autonomous quantale, thus giving rise to a lattice $\mathsf{L}_d(\mathsf{L}_{\vee}(\mathbb{I}))$ (we shall denote this lattice $\mathsf{L}(\mathbb{I}^d)$, to ease reading). In the following sections we formally instantiate our geometrical intuitions. Section 5 introduces the crucial notion of path and discusses its equivalent characterizations. In Sect. 6 we shows that paths in dimension 2 are in bijection with elements of the quantale $\mathsf{L}_{\vee}(\mathbb{I})$. In Sect. 7 we argue that paths in higher dimensions bijectively correspond to clopen tuples of the lattice $\mathsf{L}_{\vee}(\mathbb{I})^{[d]_2}$. In Sect. 8 we discuss some structural properties of the lattices $\mathsf{L}_{\vee}(\mathbb{I})$. We add concluding remarks in the final section.

2 Elementary Facts on Join-Continuous Functions

Throughout this paper, $[d]$ shall denote the set $\{ 1, \ldots, d \}$ while we let $[d]_2 := \{ (i, j) \mid 1 \leq i < j \leq d \}$.

Let P and Q be complete posets; a function $f : P \to Q$ is *join-continuous* (resp., *meet-continuous*) if

$$f(\bigvee X) = \bigvee_{x \in X} f(x), \quad (\text{resp., } f(\bigwedge X) = \bigwedge_{x \in X} f(x)), \tag{1}$$

for every $X \subseteq P$ such that $\bigvee X$ (resp., $\bigwedge X$) exists. Recall that $\bot_P := \bigvee \emptyset$ (resp., $\top_P := \bigwedge \emptyset$) is the least (resp., greatest) element of P. Note that if f is join-continuous (resp., meet-continuous) then f is monotone and $f(\bot_P) = \bot_Q$ (resp., $f(\top_P) = \top_Q$). Let f be as above; a map $g : Q \to P$ is *left adjoint* to f if $g(q) \leq p$ holds if and only if $q \leq f(p)$ holds, for each $p \in P$ and $q \in Q$; it is *right adjoint* to f if $f(p) \leq q$ is equivalent to $p \leq g(q)$, for each $p \in P$ and $q \in Q$. Notice that there is at most one function g that is left adjoint (resp., right adjoint) to f; we write this relation by $g = f_\ell$ (resp., $g = f_\rho$). Clearly, when f has a right adjoint, then $f = (g_\rho)_\ell$, and a similar formula holds when f has a left adjoint. We shall often use the following fact:

Lemma 1. *If $f : P \to Q$ is monotone and P and Q are two complete posets, then the following are equivalent:*

1. *f is join-continuous (resp., meet-continuous),*
2. *f has a right adjoint (resp., left adjoint).*

If f is join-continuous (resp., meet-continuous), then we have

$$f_\rho(q) = \bigvee\{\, p \in P \mid f(p) \leq q \,\} \qquad (\text{resp., } f_\ell(q) = \bigwedge\{\, p \in P \mid q \leq f(p) \,\}),$$
$$\text{for each } q \in Q.$$

Moreover, if f is surjective, then these formulas can be strengthened so to substitute inclusions with equalities:

$$f_\rho(q) = \bigvee\{\, p \in P \mid f(p) = q \,\} \qquad (\text{resp., } f_\ell(q) = \bigwedge\{\, p \in P \mid q = f(p) \,\}), \tag{2}$$
$$\text{for each } q \in Q.$$

The set of monotone functions from P to Q can be ordered point-wise: $f \leq g$ if $f(p) \leq g(p)$, for each $p \in P$. Suppose now that f and g both have right adjoints; let us argue that $f \leq g$ implies $g_\rho \leq f_\rho$: for each $q \in Q$, the relation $g_\rho(q) \leq f_\rho(q)$ is obtained by transposing $f(g_\rho(q)) \leq g(g_\rho(q)) \leq q$, where the inclusion $g(g_\rho(q)) \leq q$ is the counit of the adjunction. Similarly, if f and g both have left adjoints, then $f \leq g$ implies $g_\ell \leq f_\ell$.

3 Lattices from Mix ⋆-Autonomous Quantales

A *⋆-autonomous quantale* is a tuple $\mathcal{Q} = \langle Q, 1, \otimes, 0, \oplus, (-)^\star \rangle$ where Q is a complete lattice, $\otimes$ is a monoid operation on Q that distributes over arbitrary joins, $(-)^\star : Q^{op} \to Q$ is an order reversing involution of Q, and $(0, \oplus)$ is a second monoid structure on Q which is dual to $(1, \otimes)$. This means that

$$0 = 1^\star \quad \text{and} \quad f \oplus g = (g^\star \otimes f^\star)^\star.$$

Last but not least, the following relation holds:

$$f \otimes g \leq h \quad \text{iff} \quad f \leq h \oplus g^{\star} \quad \text{iff} \quad g \leq f^{\star} \oplus h.$$

Let us mention that we could have also defined a $\star$-autonomous quantale as a residuated (bounded) lattice $\langle Q, \bot, \vee, \top, \wedge, 1, \otimes, \multimap, \circ\!\!- \rangle$ such that Q is complete and comes with a cyclic dualizing element 0. The latter condition means that, for each $x \in Q$, $x \multimap 0 = 0 \circ\!\!- x$ and, letting $x^{\star} := x \multimap 0$, $x^{\star\star} = x$. This sort of algebraic structure is also called *(pseudo) $\star$-autonomous lattice* or *involutive residuated lattice*, see e.g. [8,18,28].

Example 2. Boolean algebras are the $\star$-autonomous quantales such that $\wedge = \otimes$ and $\vee = \oplus$. For a further example consider the following structure on the ordered set $\{-1 < 0 < 1\}$:

$\otimes$	-1	0	1
-1	-1	-1	-1
0	-1	0	1
1	-1	1	1

$\oplus$	-1	0	1
-1	-1	-1	1
0	-1	0	1
1	1	1	1

	$\star$
-1	1
0	0
1	-1

Together with the lattice structure on the chain, this structure yields a $\star$-autonomous quantale, known as the Sugihara monoid on the three-element chain, see e.g. [11].

We presented in [25] several ways on how to generalize the standard construction of the permutohedron (aka the weak Bruhat order). We give here a new one. Given a $\star$-autonomous quantale Q, we consider the product $Q^{[d]_2} := \prod_{1 \leq i < j \leq d} Q$. Observe that, as a product, $Q^{[d]_2}$ has itself the structure of a quantale, the structure being computed coordinate-wise. We shall say that a tuple $f = \langle f_{i,j} \mid 1 \leq i < j \leq d \rangle$ is *closed* (resp., *open*) if

$$f_{i,j} \otimes f_{j,k} \leq f_{i,k} \qquad (\textit{resp.},\ f_{i,k} \leq f_{i,j} \oplus f_{j,k}).$$

Clearly, closed tuples are closed under arbitrary meets and open tuples are closed under arbitrary joins. Observe that f is closed if and only if $f^{\star} = \langle (f_{\sigma(j),\sigma(i)})^{\star} \mid 1 \leq i < j \leq d \rangle$ is open, where for $i \in [d]$, $\sigma(i) = d - i + 1$. Thus, the correspondence sending f to $f^{\star}$ is an anti-isomorphism of $Q^{[d]_2}$, sending closed tuples to open ones, and vice versa. We shall be interested in tuples $f \in Q^{[d]_2}$ that are *clopen*, that is, they are at the same time closed and open.

For $(i, j) \in [d]_2$, a subdivision of the interval $[i, j]$ is a sequence of the form $i = \ell_0 < \ell_1 < \ldots \ell_{k-1} < \ell_k = j$ with $\ell_i \in [d]$, for $i = 1, \ldots, k$. We shall use $S_{i,j}$ for the set of subdivisions of the interval $[i, j]$. As closed tuples are closed under arbitrary meets, for each $f \in Q^{[d]_2}$ there exists a least tuple $\overline{f}$ such that $f \leq \overline{f}$ and $\overline{f}$ is closed; this tuple is easily computed as follows:

$$\overline{f}_{i,j} = \bigvee_{i<\ell_1<\ldots\ell_{k-1}<j \in S_{i,j}} f_{i,\ell_1} \otimes f_{\ell_1,\ell_2} \otimes \ldots \otimes f_{\ell_{k-1},j}.$$

Similarly and dually, if we set

$$f^{\circ}_{i,j} := \bigwedge_{i<\ell_1<\ldots\ell_{k-1}<j \in S_{i,j}} f_{i,\ell_1} \oplus f_{\ell_1,\ell_2} \oplus \ldots \oplus f_{\ell_{k-1},j}.$$

then f° is the greatest open tuple f° below f.

Proposition 3. *Suppose that, for each $f \in Q^{[d]_2}$, $\overline{(\overline{f})^{\circ}} = (\overline{f})^{\circ}$. Then, for each $f \in Q^{[d]_2}$, $(\overline{f^{\circ}})^{\circ} = (\overline{f^{\circ}})$ as well. The set of clopen tuples is then a lattice.*

Proof. The first statement is a consequence of the duality sending $f \in Q^{[d]_2}$ to $f^{\star} \in Q^{[d]_2}$. Now the relation $\overline{(\overline{f})^{\circ}} = (\overline{f})^{\circ}$ amounts to saying that the interior of any closed f is again closed. The other relation amounts to saying that the closure of an open is open. For a family $\{ f_i \mid i \in I \}$, with each f_i clopen, define then

$$\bigvee_{\mathsf{L}_d(Q)} \{ f_i \mid i \in I \} := \overline{\bigvee_{Q^{[d]_2}} \{ f_i \mid i \in I \}}, \qquad \bigwedge_{\mathsf{L}_d(Q)} \{ f_i \mid i \in I \} := (\bigwedge_{Q^{[d]_2}} \{ f_i \mid i \in I \})^{\circ},$$

and remark that, by our assumptions, the expressions on the right of the equalities denote clopen tuples. It is easily verified then these are the joins and meets, respectively, among clopen tuples. □

Lemma 4. *Consider the following inequalities:*

$$(\alpha \oplus \beta) \otimes (\gamma \oplus \delta) \leq \alpha \oplus (\beta \otimes \gamma) \oplus \delta \tag{3}$$

$$\alpha \otimes \beta \leq \alpha \oplus \beta. \tag{4}$$

Then (3) *is valid and* (4) *is equivalent to* $0 \leq 1$.

The inequation (4) is known as the *mix rule.* We say that a ★-autonomous quantale Q is a mix ★-autonomous quantale if the mix rule holds in Q.

Theorem 5. *If Q is a mix ★-autonomous quantale and $f \in Q^{[d]_2}$ is closed, then so is f°. Consequently, the set of clopen tuples of $Q^{[d]_2}$ is a lattice.*

Proof. Let $i, j, k \in [d]$ with $i < j < k$. We need to show that

$$f^{\circ}_{i,j} \otimes f^{\circ}_{j,k} \leq f_{i,\ell_1} \oplus \ldots \oplus f_{\ell_{n-1},k}$$

whenever $i < \ell_1 < \ldots \ell_{n-1} < k \in S_{i,k}$. This is achieved as follows. Let $u \in \{ 0, 1, \ldots, n-1 \}$ such that $j \in [\ell_u, \ell_{u+1})$ and put

$$\alpha := f_{i,\ell_1} \oplus \ldots \oplus f_{\ell_u}, \qquad \delta := f_{\ell_{u+1}} \oplus \ldots \oplus f_{\ell_{n-1},k}$$

$$\beta := f_{\ell_u,j} \qquad \gamma := f_{j,\ell_{u+1}}.$$

We let in the above definition $f_{\ell_u,j} := 0$ when $j = \ell_u$. Then

$$\begin{aligned} f^{\circ}_{i,j} \otimes f^{\circ}_{j,k} &\leq (\alpha \oplus \beta) \otimes (\gamma \oplus \delta), && \text{by definition of } f^{\circ}_{i,j} \text{ and } f^{\circ}_{j,k}, \\ &\leq \alpha \oplus (\beta \otimes \gamma) \oplus \delta, && \text{by the inequation (3)}, \\ &\leq \alpha \oplus f_{\ell_u,\ell_{u+1}} \oplus \delta, && \text{since } f \text{ is closed}, \end{aligned}$$

(or, when $j = \ell_u$, by using $\beta = 0 \leq 1$ and $\gamma = f_{\ell_u,\ell_{u+1}}$)

$$= f_{i,\ell_1} \oplus \ldots \oplus f_{\ell_{n-1},k}.$$

The last statement of the theorem is an immediate consequence of Proposition 3. □

Definition 6. *For Q a mix $\star$-autonomous quantale, $\mathsf{L}_d(Q)$ shall denote the lattice of clopen tuples of $Q^{[d]_2}$.*

Example 7. Suppose $Q = \mathbf{2}$, the Boolean algebra with two elements $0, 1$. Identify a tuple $\chi \in \mathbf{2}^{[d]_2}$ with the characteristic map of a subset S_χ of $\{(i, j) \mid 1 \leq i < j \leq d\}$. Think of this subset as a relation. Then χ is clopen if both S_χ and its complement in $\{(i, j) \mid 1 \leq i < j \leq d\}$ are transitive relations. These subsets are in bijection with permutations of the set $[d]$, see [6]; the lattice $\mathsf{L}_d(\mathbf{2})$ is therefore isomorphic to the well-known permutohedron, aka the weak Bruhat order. On the other hand, if Q is the Sugihara monoid on the three-element chain described in Example 2, then the lattice of clopen tuples is isomorphic to the lattice of pseudo-permutations, see [16,25].

Remark 8. For a fixed integer d the definition of the lattice $\mathsf{L}_d(Q)$ relies only on the algebraic structure of Q. This allows to say that the construction $\mathsf{L}_d(-)$ is functorial: if $f : Q_0 \to Q_1$ is a $\star$-autonomous quantale homomorphism, then we shall have a lattice homomorphism $\mathsf{L}_d(f) : \mathsf{L}_d(Q_0) \to \mathsf{L}_d(Q_1)$ (it might be also argued that if f is injective, then so is $\mathsf{L}_d(f)$). It also means that we can interpret the theories of the lattices $\mathsf{L}_d(Q)$ in the theory of the quantale Q. For example, if the equational theory of a quantale Q is decidable, then the equational theory of the lattice $\mathsf{L}_d(Q)$ is decidable as well.

4 The Mix $\star$-Autonomous Quantale $\mathsf{L}_\vee(\mathbb{I})$

In this paper $\mathbb{I}$ shall denote the unit interval of the reals, that is $\mathbb{I} := [0, 1]$. We use $\mathsf{L}_\vee(\mathbb{I})$ for the set of join-continuous functions from $\mathbb{I}$ to itself. Notice that a monotone function $f : \mathbb{I} \to \mathbb{I}$ is join-continuous if and only if

$$f(x) = \bigvee_{y<x,\, y\in\mathbb{I}\cap\mathbb{Q}} f(y), \tag{5}$$

see Proposition 2.1, Chap. II of [12]. As the category of complete lattices and join-continuous functions is a symmetric monoidal closed category, for every complete lattice L the set of join-continuous functions from Q to itself is a monoid object in that category, that is, a quantale, see [14,20]. Thus, we have:

Lemma 9. *Composition induces a quantale structure on $\mathsf{L}_\vee(\mathbb{I})$.*

Let now $\mathsf{L}_\wedge(\mathbb{I})$ denote the collection of meet-continuous functions from $\mathbb{I}$ to itself. By duality, we obtain:

Lemma 10. *Composition induces a dual quantale structure on $\mathsf{L}_\wedge(\mathbb{I})$.*

With the next set of observations we shall see $\mathsf{L}_\vee(\mathbb{I})$ and $\mathsf{L}_\wedge(\mathbb{I})$ are order isomorphic. For a monotone function $f : \mathbb{I} \to \mathbb{I}$, define

$$f^\wedge(x) = \bigwedge_{x<x'} f(x'), \qquad f^\vee(x) = \bigvee_{x'<x} f(x').$$

Lemma 11. *If $x < y$, then $f^\wedge(x) \leq f^\vee(y)$.*

Proof. Pick $z \in \mathbb{I}$ such that $x < z < y$ and observe then that $f^{\wedge}(x) \leq f(z) \leq f^{\vee}(y)$. □

Proposition 12. *$f^{\wedge}$ is the least meet-continuous function above f and $f^{\vee}$ is the greatest join-continuous function below f. The relations $f^{\vee\wedge} = f^{\wedge}$ and $f^{\wedge\vee} = f^{\vee}$ hold and, consequently, the operations $(\cdot)^{\vee} : \mathsf{L}_{\wedge}(\mathbb{I}) \to \mathsf{L}_{\vee}(\mathbb{I})$ and $(\cdot)^{\wedge} : \mathsf{L}_{\vee}(\mathbb{I}) \to \mathsf{L}_{\wedge}(\mathbb{I})$ are inverse order preserving bijections.*

Proof. We prove only one statement. Let us show that $f^{\wedge}$ is meet-continuous; to this goal, we use Eq. (5):

$$\bigwedge_{x<t} f^{\wedge}(t) = \bigwedge_{x<t} \bigwedge_{t<t'} f(t') = \bigwedge_{x<t} f(t') = f^{\wedge}(x).$$

We observe next that $f \leq f^{\wedge}$, as if $x < t$, then $f(x) \leq f(t)$. This implies that if $g \in \mathsf{L}_{\wedge}(\mathbb{I})$ and $f^{\wedge} \leq g$, then $f \leq f^{\wedge} \leq g$. Conversely, if $g \in \mathsf{L}_{\wedge}(\mathbb{I})$ and $f \leq g$, then

$$f^{\wedge}(x) = \bigwedge_{x<t} f(t) \leq \bigwedge_{x<t} g(t) = g(x).$$

Let us prove the last sentence. Clearly, both maps are order preserving. Let us show that $f^{\vee\wedge} = f^{\wedge}$ whenever f is order preserving. We have $f^{\vee\wedge} \leq f^{\wedge}$, since $f^{\vee} \leq f$ and $(-)^{\wedge}$ is order preserves the pointwise ordering. For the converse inclusion, recall from the previous lemma that if $x < y$, then $f^{\wedge}(x) \leq f^{\vee}(y)$, so

$$f^{\wedge}(x) \leq \bigwedge_{x<y} f^{\vee}(y) = f^{\vee\wedge}(x),$$

for each $x \in \mathbb{I}$. Finally, to see that $(-)^{\wedge}$ and $(-)^{\vee}$ are inverse to each other, observe that of $f \in \mathsf{L}_{\wedge}(\mathbb{I})$, then $f^{\vee\wedge} = f^{\wedge} = f$. The equality $f^{\wedge\vee} = f$ for $f \in \mathsf{L}_{\vee}(\mathbb{I})$ is derived similarly. □

Recall that if $f \in \mathsf{L}_{\vee}(\mathbb{I})$ (resp., $g \in \mathsf{L}_{\wedge}(\mathbb{I})$), then $f_{\rho} \in \mathsf{L}_{\wedge}(\mathbb{I})$ (resp., $f_{\ell} \in \mathsf{L}_{\vee}(\mathbb{I})$) denotes the right adjoint of f (resp., left adjoint of g). The following relation is the key observation to uncover the ★-autonomous quantale structure on $\mathsf{L}_{\vee}(\mathbb{I})$.

Lemma 13. *For each $f \in \mathsf{L}_{\vee}(\mathbb{I})$, the relation $(f_{\rho})^{\vee} = (f^{\wedge})_{\ell}$ holds.*

Proof. Let $f \in \mathsf{L}_{\vee}(\mathbb{I})$; we shall argue that $x \leq f^{\wedge}(y)$ if and only if $(f_{\rho})^{\vee}(x) \leq y$, for each $x, y \in \mathbb{I}$.

We begin by proving that $x \leq f^{\wedge}(y)$ implies that $(f_{\rho})^{\vee}(x) \leq y$. Suppose $x \leq f^{\wedge}(y)$ so, for each z with $y < z$, we have $x \leq f(z)$. Suppose that $(f_{\rho})^{\vee}(x) \not\leq y$, thus there exists $w < x$ such that $f_{\rho}(w) \not\leq y$. Then $y < f_{\rho}(w)$, so $x \leq f(f_{\rho}(w)) \leq w$, contradicting $w < x$. Therefore, $(f_{\rho})^{\vee}(x) \leq y$.

Dually, we can argue that if $g \in \mathsf{L}_{\wedge}(\mathbb{I})$, then $g^{\vee}(x) \leq y$ implies $x \leq (g_{\ell})^{\wedge}(y)$. Letting $g := f_{\rho}$ in this statement we obtain the converse implication: $(f_{\rho})^{\vee}(x) \leq y$ implies $x \leq ((f_{\rho})_{\ell})^{\wedge}(y) = f^{\wedge}(y)$. □

For $f, g \in \mathsf{L}_{\vee}(\mathbb{I})$, let us define

$$f \otimes g := g \circ f, \qquad f \oplus g := (g^{\wedge} \circ f^{\wedge})^{\vee}, \qquad f^{\star} = (f_{\rho})^{\vee} = (f^{\wedge})_{\ell}.$$

Proposition 14. *The tuple* $\langle \mathsf{L}_\vee(\mathbb{I}), id, \otimes, id, \oplus, (-)^\star \rangle$ *is a mix* $\star$*-autonomous quantale.*

Proof. The correspondence $(\cdot)^\star$ is order reversing as it is the composition of an order reversing function with a monotone one; by Lemma 13, it is an involution:

$$f^{\star\star} = (((f_\rho)^\wedge)^\vee)_\ell = (f_\rho)_\ell = f.$$

To verify that

$$(f \otimes g)^\star = g^\star \oplus f^\star \tag{6}$$

holds, for any $f, g \in \mathsf{L}_\vee(\mathbb{I})$, we compute as follows:

$$\begin{aligned} g^\star \oplus f^\star &= ((f^\star)^\wedge \circ (g^\star)^\wedge)^\vee \\ &= (f_\rho^{\vee\wedge} \circ g_\rho^{\vee\wedge})^\vee = (f_\rho \circ g_\rho)^\vee = (g \circ f)_\rho^\vee = (g \circ f)^\star = (f \otimes g)^\star. \end{aligned}$$

We verify next that, for any $f, g, h \in \mathsf{L}_\vee(\mathbb{I})$,

$$f \otimes g \leq h \quad \text{iff} \quad f \leq h \oplus g^\star. \tag{7}$$

Notice that $h \oplus g^\star = ((g^\star)^\wedge \circ h^\wedge)^\vee = (g_\rho^{\vee\wedge} \circ h^\wedge)^\vee = (g_\rho \circ h^\wedge)^\vee$, so

$$\begin{aligned} f \leq h \oplus g^\star \quad &\text{iff} \quad f \leq (g_\rho \circ h^\wedge)^\vee, && \text{by the equality just established,} \\ &\text{iff} \quad f \leq g_\rho \circ h^\wedge, && \text{by Proposition 2,} \\ &\text{iff} \quad g \circ f \leq h^\wedge, && \text{since } g(x) \leq y \text{ iff } x \leq g_\rho(y), \\ &\text{iff} \quad f \otimes g = g \circ f \leq h^{\wedge\vee} = h, && \text{using again Proposition 2.} \end{aligned}$$

It is an immediate algebraic consequence of (6) and (7) that $f \otimes g \leq h$ is equivalent to $g \leq f^\star \oplus h$, for any $f, g, h \in \mathsf{L}_\vee(\mathbb{I})$. Namely, we have

$$\begin{aligned} f \otimes g \leq h \quad &\text{iff} \quad f \leq h \oplus g^\star \\ &\text{iff} \quad (h \oplus g^\star)^\star \leq f^\star \\ &\text{iff} \quad g \otimes h^\star = g^{\star\star} \otimes h^\star \leq f^\star \\ &\text{iff} \quad g \leq f^\star \oplus h^{\star\star} = f^\star \oplus h. \end{aligned}$$

Finally, recall that the identity *id* is both join-continuous and meet-continuous and therefore $id^\wedge = id$. Then it is easily seen that *id* is both a unit for $\otimes$ and for $\oplus$. As seen in Lemma 4, this implies that $\mathsf{L}_\vee(\mathbb{I})$ satisfies the mix rule. □

5 Paths

Let in the following $d \geq 2$ be a fixed integer; we shall use $\mathbb{I}^d$ to denote the d-fold product of $\mathbb{I}$ with itself. That is, $\mathbb{I}^d$ is the usual geometric cube in dimension d. Let us recall that $\mathbb{I}^d$, as a product of the poset $\mathbb{I}$, has itself the structure of a poset (the order being coordinate-wise) which, moreover, is complete.

Definition 15. *A path in $\mathbb{I}^d$ is a chain $C \subseteq \mathbb{I}^d$ with the following properties:*

1. *if $X \subseteq C$, then $\bigwedge X \in C$ and $\bigvee X \in C$,*
2. *C is dense as an ordered set: if $x, y \in C$ and $x < y$, then $x < z < y$ for some $z \in C$.*

We have given a working definition of the notion of path in $\mathbb{I}^d$, as a totally ordered dense sub-complete-lattice of $\mathbb{I}^d$. The next theorem state the equivalence among several properties, each of which could be taken as a definition of the notion of path.

Theorem 16. *Let $d \geq 2$ and let $C \subseteq \mathbb{I}^d$. The following conditions are then equivalent:*

1. *C is a path as defined in Definition 15;*
2. *C is a maximal chain of the poset $\mathbb{I}^d$;*
3. *There exists a monotone (increasing) topologically continuous map $\mathsf{p} : \mathbb{I} \to \mathbb{I}^d$ such that $\mathsf{p}(0) = \mathbf{0}$, $\mathsf{p}(1) = \mathbf{1}$, whose image is C.*

6 Paths in Dimension 2

We give next a further characterization of the notion of path, valid in dimension 2. The principal result of this Section, Theorem 20, states that paths in dimension 2 are (up to isomorphism) just elements of the quantale $\mathsf{L}_\vee(\mathbb{I})$.

For a monotone function $f : \mathbb{I} \to \mathbb{I}$ define $C_f \subseteq \mathbb{I}^2$ by the formula

$$C_f := \bigcup_{x \in \mathbb{I}} \{x\} \times [f^\vee(x), f^\wedge(x)]. \tag{8}$$

Notice that, by Proposition 12, $C_f = C_{f^\vee} = C_{f^\wedge}$.

Proposition 17. *C_f is a path in $\mathbb{I}^2$.*

Proof. We prove first that C_f, with the product ordering induced from $\mathbb{I}^2$, is a linear order. To this goal, we shall argue that, for $(x, y), (x', y') \in C_f$, we have $(x, y) < (x', y')$ iff either $x < x'$ or $x = x'$ and $y < y'$. That is, C_f is a lexicographic product of linear orders, whence a linear order. Let us suppose that one of these two conditions holds: a) $x < x'$, b) $x = x'$ and $y < y'$. If a), then $f^\wedge(x) \leq f^\vee(x')$. Considering that $y \in [f^\vee(x), f^\wedge(x)]$ and $y' \in [f^\vee(x'), f^\wedge(x')]$ we deduce $y \leq y'$. This proves that $(x, y) < (x', y')$ in the product ordering. If b) then we also have $(x, y) < (x', y')$ in the product ordering. The converse implication, $(x, y) < (x', y')$ implies $x < x'$ or $x = x'$ and $y < y'$, trivially holds.

We argue next that C_f is closed under joins from $\mathbb{I}^2$. Let (x_i, y_i) be a collection of elements in C_f, we aim to show that $(\bigvee x_i, \bigvee y_i) \in C_f$, i.e. $\bigvee y_i \in [f^\vee(\bigvee x_i), f^\wedge(\bigvee x_i)]$. Clearly, as $y_i \leq f^\wedge(x_i)$, then $\bigvee y_i \leq \bigvee f^\wedge(x_i) \leq f^\wedge(\bigvee x_i)$. Next, $f^\vee(x_i) \leq y_i$, whence $f^\vee(\bigvee x_i) = \bigvee f^\vee(x_i) \leq \bigvee y_i$. By a dual argument, we have that $(\bigwedge x_i, \bigwedge y_i) \in C_f$.

Finally, we show that C_f is dense; to this goal let $(x, y), (x', y') \in C_f$ be such that $(x, y) < (x', y')$. If $x < x'$ then we can find a z with $x < z < x'$; of course, $(z, f(z)) \in C_f$ and, but the previous characterisation of the order, $(x, y) < (z, f(z)) < (x', y')$ holds. If $x = x'$ then $y < y'$ and we can find a w with $y < w < y'$; as $w \in [y, y'] \subseteq [f^\vee(x), f^\wedge(x)]$, then $(x, w) \in C_f$; clearly, we have then $(x, y) < (x, w) < (x, y') = (x', y')$. □

For C a path in $\mathbb{I}^2$, define

$$f_C^-(x) := \bigwedge\{ y \mid (x,y) \in C \}, \qquad f_C^+(x) := \bigvee\{ y \mid (x,y) \in C \}. \tag{9}$$

Recall that a path $C \subseteq \mathbb{I}^2$ comes with bi-continuous surjective projections $\pi_1, \pi_2 : C \to \mathbb{I}$. Observe that the following relations hold:

$$f_C^- = \pi_2 \circ (\pi_1)_\ell, \qquad f_C^+ = \pi_2 \circ (\pi_1)_\rho. \tag{10}$$

Indeed, we have

$$\begin{aligned} \pi_2((\pi_1)_\ell(x)) &= \pi_2(\bigwedge\{ (x',y) \in C \mid x = x' \}), \qquad \text{using equation (2)} \\ &= \bigwedge \pi_2(\{ (x',y) \in C \mid x = x' \}) = \bigwedge\{ y \mid (x,y) \in C \}. \end{aligned}$$

The other expression for f^+ is derived similarly. In particular, the expressions in (10) show that $f^- \in \mathsf{L}_\vee(\mathbb{I})$ and $f^+ \in \mathsf{L}_\wedge(\mathbb{I})$.

Lemma 18. *We have*

$$f_C^- = (f_C^+)^\vee, \quad f_C^+ = (f_C^-)^\wedge, \quad \textit{and} \quad C = C_{f_C^+} = C_{f_C^-}.$$

Proof. Let us firstly argue that $(x,y) \in C$ if and only if $f_C^-(x) \leq y \leq f_C^+(y)$. The direction from left to right is obvious. Conversely, it is easily verified that if $f_C^-(x) \leq y \leq f_C^+(y)$, then the pair (x,y) is comparable with all the elements of C; then, since C is a maximal chain, necessarily $(x,y) \in C$.

Therefore, let us argue that $f_C^+ = (f_C^-)^\wedge$; we do this by showing that f_C^+ is the least meet-continuous function above f_C^-. We have $f_C^-(x) \leq f_C^+(x)$ for each $x \in \mathbb{I}$ since the fiber sets $\pi_1^{-1}(x) = \{ (x',y) \in C \mid x' = x \}$ are non empty. Suppose now that $f_C^- \leq g \in \mathsf{L}_\wedge(\mathbb{I})$. In order to prove that $f_C^+ \leq g$ it will be enough to prove that $f_C^+(x) \leq g(x')$ whenever $x < x'$. Observe that if $x < x'$ then $f_C^+(x) \leq f_C^-(x')$: this is because if $(x,y), (x',y') \in C$, then $x < x'$ and C a chain imply $y \leq y'$. We deduce therefore $f_C^+(x) \leq f_C^-(x') \leq g(x')$. The relation $f_C^- = (f_C^+)^\vee$ is proved similarly. □

Lemma 19. *Let $f : \mathbb{I} \to \mathbb{I}$ be monotone and consider the path C_f. Then $f^\vee = f_{C_f}^-$ and $f^\wedge = f_{C_f}^+$.*

Proof. For a monotone $f : \mathbb{I} \to \mathbb{I}$ define $f' : \mathbb{I} \to C_f$ by $f' := \langle id_\mathbb{I}, f \rangle$, so $f = \pi_2 \circ f'$. Recall that $f_{C_f}^- = \pi_2 \circ (\pi_1)_\ell$. Therefore, in order to prove the relation $f^\vee = f_{C_f}^- = \pi_2 \circ (\pi_1)_\ell$ it shall be enough to prove that $\langle id, f^\vee \rangle$ is left adjoint to the first projection (that is, we prove that $\langle id, f^\vee \rangle = (\pi_1)_\ell$, from which it follows that $f^\vee = \pi_1 \circ \langle id, f^\vee \rangle = \pi_2 \circ (\pi_1)_\ell$). This amounts to verify that, for $x \in \mathbb{I}$ and $(x',y) \in C_f$ we have $x \leq \pi_1(x',y)$ if and only if $(x, f^\vee(x)) \leq (x',y)$. To achieve this goal, the only non trivial observation is that if $x \leq x'$, then $f^\vee(x) \leq f^\vee(x') \leq y$. The relation $f^\wedge = \pi_2 \circ (\pi_1)_\rho$ is proved similarly. □

Theorem 20. *There is a bijective correspondence between the following data:*

1. *paths in* $\mathbb{I}^2$,
2. *join-continuous functions in* $\mathsf{L}_\vee(\mathbb{I})$,
3. *meet-continuous functions in* $\mathsf{L}_\wedge(\mathbb{I})$.

Proof. According to Lemmas 18 and 19, the correspondence sending a path C to $f_C^- \in \mathsf{L}_\vee(\mathbb{I})$ has the mapping sending f to C_f as an inverse. Similarly, the correspondence $C \mapsto f_C^+ \in \mathsf{L}_\wedge(\mathbb{I})$ has $f \mapsto C_f$ as inverse. □

7 Paths in Higher Dimensions

We show in this Section that paths in dimension d, as defined in Sect. 5, are in bijective correspondence with clopen tuples of $\mathsf{L}_\vee(\mathbb{I})^{[d]_2}$, as defined in Sect. 3; therefore, as established in that Section, there is a lattice $\mathsf{L}_d(\mathsf{L}_\vee(\mathbb{I}))$ whose underlying set can be identified with the set of paths in dimension d.

Let $f \in \mathsf{L}_\vee(\mathbb{I})^{[d]_2}$, so $f = \{ f_{i,j} \mid 1 \leq i < j \leq d \}$. We define then, for $1 \leq i < j \leq d$,

$$f_{j,i} := (f_{i,j})^\star = ((f_{i,j})_\rho)^\vee .$$

Moreover, for $i \in [d]$, we let $f_{i,i} := id$.

Definition 21. *We say that a tuple* $f \in \mathsf{L}_\vee(\mathbb{I})^{[d]_2}$ *is* compatible *if* $f_{j,k} \circ f_{i,j} \leq f_{i,k}$, *for each triple of elements* $i, j, k \in [d]$.

Lemma 22. *A tuple is compatible if and only if it is clopen.*

Proof. For $i < j < k$, compatibility yields $f_{i,j} \otimes f_{j,k} \leq f_{i,k}$ (closedness) and $f_{k,j} \otimes f_{j,i} \leq f_{k,i}$ which in turn is equivalent to $f_{i,k} \leq f_{i,j} \oplus f_{j,k}$ (openness).

Conversely, suppose that f is clopen. Say that the pattern (ijk) is satisfied by f if $f_{i,j} \otimes f_{i,j} \leq f_{i,k}$. If $\text{card}(\{ i, j, k \}) \leq 2$, then f satisfies the pattern (ijk) if $i = j$ or $j = k$, since then $f_{i,j} = id$ or $f_{j,k} = id$. If $i = k$, then $f_{i,j} \otimes f_{j,i} \leq id$ is equivalent to $f_{i,j} \leq id \oplus f_{i,j}$. Suppose therefore that $\text{card}(\{ i, j, k \}) = 3$.

By assumption, f satisfies (ijk) and (kji) whenever $i < j < k$. Then it is possible to argue that all the patterns on the set $\{ i, j, k \}$ are satisfied by observing that if (ijk) is satisfied, then (jki) is satisfied as well: from $f_{i,j} \otimes f_{j,k} \leq f_{i,k}$, derive $f_{i,j} \leq f_{i,k} \oplus f_{k,j}$ and then $f_{j,k} \otimes f_{k,i} \leq f_{j,i}$. □

Remark 23. Let $f \in \mathsf{L}_\vee(\mathbb{I})^{[d]_2}$ and suppose that, for some $i, j, k \in [d]$, with $i < j < k$, $f_{i,k} = f_{i,k} \circ f_{i,j}$. That is, we have $f_{i,k} = f_{i,j} \otimes f_{j,k}$ and, using the mix rule, we derive $f_{i,k} \leq f_{i,j} \oplus f_{j,k}$. Dually, a relation of the form $f_{i,k}^\wedge = f_{j,k}^\wedge \circ f_{i,j}^\wedge$ is equivalent to $f_{i,k} = f_{i,j} \oplus f_{j,k}$ and implies $f_{i,j} \otimes f_{j,k} \leq f_{i,k}$.

Remark 24. Lemma 22 shows that a clopen tuple of $\mathsf{L}_\vee(\mathbb{I})^{[d]_2}$ can be extended in a unique way to a *skew* enrichment of the set $[n]$ over $\mathsf{L}_\vee(\mathbb{I})$, see [17,27]. Dually, a clopen tuple gives rise to a unique skew metric on the set $[n]$ with values in $\mathsf{L}_\vee(\mathbb{I})$. For a skew enrichment (or metric) we mean, here, that the law $f_{j,i} = f_{i,j}^\star$ holds; this law, which replaces the more usual requirement that a metric is symmetric, has been considered e.g. in [15].

If $C \subseteq \mathbb{I}^d$ is a path, then we shall use $\pi_i : C \to \mathbb{I}$ to denote the projection onto the i-th coordinate. Then $\pi_{i,j} := \langle \pi_i, \pi_j \rangle : C \to \mathbb{I} \times \mathbb{I}$.

Definition 25. *For a path C in $\mathbb{I}^d$, let us define $v(C) \in \mathsf{L}_\vee(\mathbb{I})^{[d]_2}$ by the formula:*

$$v(C)_{i,j} := \pi_j \circ (\pi_i)_\ell, \quad (i, j) \in [d]_2. \tag{11}$$

Remark 26. An explicit formula for $v(C)_{i,j}(x)$ is as follows:

$$v(C)_{i,j}(x) = \bigwedge \{ \pi_j(y) \in C \mid \pi_i(y) = x \}. \tag{12}$$

Let $C_{i,j}$ be the image of C via the projection $\pi_{i,j}$. Then $C_{i,j}$ is a path, since it is the image of a bi-continuous function from $\mathbb{I}$ to $\mathbb{I} \times \mathbb{I}$. Some simple diagram chasing (or the formula in (12)) shows that $v(C)_{i,j} = f^-_{C_{i,j}}$ as defined in (9).

Definition 27. *For a compatible $f \in \mathsf{L}_\vee(\mathbb{I})^{[d]_2}$, define*

$$C_f := \{ (x_1, \ldots, x_d) \mid f_{i,j}(x_i) \leq x_j, \text{ for all } i, j \in [d] \}.$$

Remark 28. Notice that the condition $f_{i,j}(x) \leq y$ is equivalent (by definition of $f_{i,j}$ or $f_{j,i}$) to the condition $x \leq f^{\wedge}_{j,i}(y)$. Thus, there are in principle many different ways to define C_f; in particular, when $d = 2$ (so any tuple $\mathsf{L}_\vee(\mathbb{I})^{[d]_2}$ is compatible), the definition given above is equivalent to the one given in (8).

Proposition 29. *C_f is a path.*

The proposition is an immediate consequence of the following Lemmas 30, 31 and 33.

Lemma 30. *C_f is a total order.*

Proof. Let $x, y \in C_f$ and suppose that $x \not\leq y$, so there exists $i \in [d]$ such that $x_i \not\leq y_i$. W.l.o.g. we can suppose that $i = 1$, so $y_1 < x_1$ and then, for $i > 1$, we have $f^{\wedge}_{1,i}(y_1) \leq f_{1,i}(x_1)$, whence $y_i \leq f^{\wedge}_{1,i}(y_1) \leq f_{1,i}(x_1) \leq x_1$. This shows that $y < x$. □

Lemma 31. *C_f is closed under arbitrary meets and joins.*

Proof. Let $\{ x^\ell \mid \ell \in I \}$ be a family of tuples in C_f. For all $i, j \in [d]$ and $\ell \in I$, we have $f_{i,j}(\bigwedge_{\ell \in I} x^\ell_i) \leq f_{i,j}(x^\ell_i) \leq x^\ell_j$, and therefore $f_{i,j}(\bigwedge_{\ell \in I} x^\ell_i) \leq \bigwedge_{\ell \in I} x^\ell_j$. Since meets in $\mathbb{I}^d$ are computed coordinate-wise, this shows that C_f is closed under arbitrary meets. Similarly, $f_{i,j}(x^\ell_i) \leq \bigvee_{\ell \in I} x^\ell_j$ and

$$f_{i,j}(\bigvee_{\ell \in I} x^\ell_i) = \bigvee_{\ell \in I} f_{i,j}(x^\ell_i) \leq \bigvee_{\ell \in I} x^\ell_j,$$

so C_f is also closed under arbitrary joins. □

Lemma 32. *Let $f \in \mathsf{L}_\vee(\mathbb{I})^{[d]_2}$ be compatible. Let $i_0 \in [d]$ and $x_0 \in \mathbb{I}$; define $x \in \mathbb{I}^d$ by setting $x_i := f_{i_0,i}(x_0)$ for each $i \in [d]$. Then $x \in C_f$ and $x = \bigwedge \{ y \in C_f \mid \pi_{i_0}(y) = x_0 \}$.*

Proof. Since f is compatible, $f_{i,j} \circ f_{i_0,i} \leq f_{i_0,j}$, for each $i, j \in [d]$, so

$$f_{i,j}(x_i) = f_{i,j}(f_{i_0,i}(x_0)) \leq f_{i_0,j}(x_0) = x_j.$$

Therefore, $x \in C_f$. Observe that since $f_{i_0,i_0} = id$, we have $x_{i_0} = x_0$ and x so defined is such that $\pi_{i_0}(x) = x_0$. On the other hand, if $y \in C_f$ and $x_0 \leq \pi_{i_0}(y) = y_{i_0}$, then $x_i = f_{i,i_0}(x_0) \leq f_{i,i_0}(y_{i_0}) \leq y_i$, for all $i \in [d]$. Thus $x = \bigwedge\{ y \in C_f \mid \pi_{i_0}(y) = x_0 \}$. □

Lemma 33. *C_f is dense.*

Proof. Let $x, y \in C_f$ and suppose that $x < y$, so there exists $i_0 \in [d]$ such that $x_{i_0} < y_{i_0}$. Pick $z_0 \in \mathbb{I}$ such that $x_{i_0} < z_0 < y_{i_0}$ and define $z \in C_f$ as in Lemma 32, $z_i := f_{i_0,i}(z_0)$, for all $i \in [d]$. We claim that $x_i \leq z_i \leq y_i$, for each $i \in [d]$. From this and $x_{i_0} < z_{i_0} < y_0$ it follows that $x < z < y$. Indeed, we have $z_i = f_{i_0,i}(z_0) \leq f_{i_0,i}(y_{i_0}) \leq y_i$. Moreover, $x_{i_0} < z_0$ implies $f^{\wedge}_{i_0,i}(x_{i_0}) \leq f_{i_0,i}(z_0)$; by Remark 28, we have $x_i \leq f^{\wedge}_{i_0,i}(x_{i_0})$. Therefore, we also have $x_i \leq f^{\wedge}_{i_0,i}(x_{i_0}) \leq f_{i_0,i}(z_0) = z_i$. □

Lemma 34. *If $f \in \mathsf{L}_{\vee}(\mathbb{I})^{[d]_2}$ is compatible, then $v(C_f) = f$.*

Proof. By Lemma 32, the correspondence sending x to $(f_{i,1}(x), \ldots, f_{d,1}(x))$ is left adjoint to the projection $\pi_i : C_f \to \mathbb{I}$. In turn, this gives that $v(C_f)_{i,j}(x) = \pi_j((\pi_i)_\ell(x)) = f_{i,j}(x)$, for any $i, j \in [d]$. It follows that $v(C_f) = f$. □

Lemma 35. *For C a path in $\mathbb{I}^d$, we have $C_{v(C)} = C$.*

Proof. Let us show that $C \subseteq C_{v(C)}$. Let $c \in C$; notice that for each $i, j \in [d]$, we have

$$v(C)_{i,j}(c_i) = \pi_j((\pi_i)_\ell(c_i)) = \pi_j((\pi_i)_\ell(\pi_i(c)) \leq \pi_j(c) = c_j,$$

so $c \in C_{v(C)}$. For the converse inclusion, notice that $C \subseteq C_{v(C)}$ implies $C = C_{v(C)}$, since every path is a maximal chain. □

Putting together Lemmas 34 and 35 we obtain:

Theorem 36. *The correspondences, sending a path C in $\mathbb{I}^d$ to the tuple $v(C)$, and a compatible tuple f to the path C_f, are inverse bijections.*

8 Structure of the Lattices $\mathsf{L}(\mathbb{I}^d)$

As final remarks, we present and discuss some structural properties of the lattices $\mathsf{L}(\mathbb{I}^d)$.

Recall that an element p of a lattice L is *join-prime* if, for any *finite* family $\{ x_i \mid i \in I \}$, $p \leq \bigvee_{i \in I} x_i$ implies $p \leq x_i$, for some $i \in I$. A *completely join-prime* element is defined similarly, by considering arbitrary families in place of finite ones. An element p of a lattice L is *join-irreducible* if, for any *finite* family $\{ x_i \mid i \in I \}$, $p = \bigvee_{i \in I} x_i$ implies $p = x_i$, for some $i \in I$; *completely join-irreducible* elements are defined similarly, by considering arbitrary families. If p is join-prime, then it is also join-irreducible, and the two notions coincide on distributive lattices.

Join-prime elements of $\mathsf{L}_\vee(\mathbb{I})$. We begin by describing the join-prime elements of $\mathsf{L}_\vee(\mathbb{I})$; this lattice being distributive, join-prime and join-irreducible elements coincide. For $x, y \in \mathbb{I}$, let us put

$$e_{x,y}(t) := \begin{cases} 0, & 0 \leq t \leq x, \\ y\,,, & x < t, \end{cases} \qquad E_{x,y}(t) := \begin{cases} 0, & 0 \leq t < x, \\ y, & x \leq t < 1, \\ 1, & t = 1. \end{cases}$$

so $e_{x,y} \in \mathsf{L}_\vee(\mathbb{I})$, $E_{x,y} \in \mathsf{L}_\wedge(\mathbb{I})$ and $E_{x,y} = e^{\wedge}_{x,y}$. We call a function of the form $e_{x,y}$ a *one step function*. Notice that if $x = 1$ or $y = 0$, then $e_{x,y}$ is the constant function that takes 0 as its unique value; said otherwise, $e_{x,y} = \bot$. We say that $e_{x,y}$ is a *prime one step function* if $x < 1$ and $0 < y$; we say that $e_{x,y}$ is *rational* if $x, y \in \mathbb{I} \cap \mathbb{Q}$.

Proposition 37. *Prime one step functions are exactly the join-prime elements of* $\mathsf{L}_\vee(\mathbb{I})$.

There are no completely join-prime elements in $\mathsf{L}_\vee(\mathbb{I})$. Yet we have:

Proposition 38. *Every element of* $\mathsf{L}_\vee(\mathbb{I})$ *is a join of rational one step functions.*

Meet-irreducible elements are easily characterized using duality; they belong to the join-semilattice generated by the join-prime elements. Using duality, the following proposition is derived.

Proposition 39. $\mathsf{L}_\vee(\mathbb{I})$ *is the Dedekind-MacNeille completion of the sublattice generated by the rational one step functions.*

Join-irreducible elements of $\mathsf{L}(\mathbb{I}^d)$. Let now $d \geq 3$ be fixed. The lattice $\mathsf{L}(\mathbb{I}^d)$ is no more distributive; we characterize therefore its join-irreducible elements. We associate to a vector $p \in \mathbb{I}^d$ the tuple $e_p \in \mathsf{L}_\vee(\mathbb{I})^{[d]_2}$ defined as follows:

$$e_p := \langle\, e_{p_i,p_j} \mid (i,j) \in [d]_2 \,\rangle.$$

Proposition 40. *The elements of the form* $e_p \in \mathsf{L}_\vee(\mathbb{I})^{[d]_2}$ *are clopen and they are exactly the join-irreducible elements of* $\mathsf{L}(\mathbb{I}^d)$ *(whenever* $e_p \neq \bot$*). Every element of* $\mathsf{L}(\mathbb{I}^d)$ *is the join of the join-irreducible elements below it.*

As before $\mathsf{L}(\mathbb{I}^d)$ is the Dedekind-MacNeille completion of its sublattice generated by the join-irreducible elements. Yet, it is no longer true that every element of $\mathsf{L}(\mathbb{I}^d)$ is a join of elements of the form e_p with all the p_i rational and therefore $\mathsf{L}(\mathbb{I}^d)$ is not the Dedekind-MacNeille completion of its sublattice generated by this kind of elements.

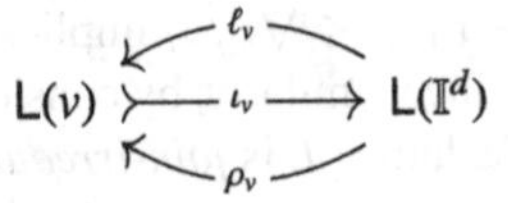

Let us explain the significance of the previous observations. For each vector $v \in \mathbb{N}^d$ there is an embedding ι_v of the multinomial lattice $\mathsf{L}(v)$ (see [3,25]) into $\mathsf{L}(\mathbb{I}^d)$, as in the diagram on the right, where ℓ_v and ρ_v are, respectively, the left and right adjoint to ι_v. These embeddings form a directed diagram whose colomit can be identified with the sublattice of $\mathsf{L}(\mathbb{I}^d)$ generated by the elements e_p with all the p_i, $i \in [d]$, rational. The fact $\mathsf{L}(\mathbb{I}^d)$ is not the Dedekind-MacNeille completion of this sublattice means that, while we can still define approximations of elements of $\mathsf{L}(\mathbb{I}^d)$ in the multinomial lattices via adjoints, these approximations do not converge to what they are meant to approximate. For example, we could define $\mathtt{appr}_v(f) := \ell_v(f)$ and yet have $\bigvee_{v \in \mathbb{N}^d} \iota_v(\mathtt{appr}_v(f)) < f$. On the other hand, it is possible to prove that every meet-irreducible element is an infinite join of join-irreducible elements arising from a rational point. Therefore we can state:

Proposition 41. *Every element of* $\mathsf{L}(\mathbb{I}^d)$ *is a meet of joins (and a join of meets) of elements in the sublattice of* $\mathsf{L}(\mathbb{I}^d)$ *generated by the* e_p *such that* p_i *is rational for each* $i \in [d]$.

Whether the last proposition is the key to use the lattices $\mathsf{L}(\mathbb{I}^d)$ as well as the multinomial lattices for higher dimensional approximations in discrete geometry is an open problem that we shall tackle in future research.

9 Conclusions

In this paper we have shown how to extend the lattice structure on a set of discrete paths (known as a multinomial lattice, or weak Bruhat order, if the words coding these paths are permutations) to a lattice structure on the set of (images of) continuous paths from $\mathbb{I}$, the unit interval of the reals, to the cube $\mathbb{I}^d$, for some $d \geq 2$.

By studying the structure of these lattices, called here $\mathsf{L}(\mathbb{I}^d)$, we have been able to identify an intrinsic difficulty in defining discrete approximations of lines in dimensions $d \geq 3$ (a problem that motivated us to develop this research). This stems from the fact that $\mathsf{L}(\mathbb{I}^d)$ is no longer (when $d \geq 3$) generated by its sublattice of discrete paths as a Dedekind-Macneille completion. Proposition 41 exactly describes how the lattice $\mathsf{L}(\mathbb{I}^d)$ is generated from discrete paths and might be the key to use the lattices $\mathsf{L}(\mathbb{I}^d)$ as well as the multinomial lattices for defining higher dimensional approximations of lines. We shall tackle this problem in future research.

As a byproduct, our paper also pinpoints that various generalizations of permutohedra crucially rely on the algebraic (but also logical) notion of mix ★-autonomous quantale. Every such quantale yields an infinite family of lattices indexed by positive integers. While the definition of these lattices becomes straightforward by means of the algebra, it turns out that the elements of these lattices are (as far as observed up to now) in bijective correspondence either with interesting combinatorial objects (permutations, pseudo-permutations) or with geometric ones (continuous paths, as seen in this paper). These intriguing correspondences suggest the existence of a deep connection between combinatorics/geometry and logic. Future research shall unravel these phenomena. A first step, already under way for the Sugihara monoids on a chain, shall systematically identify the combinatorial objects arising from a given mix ★-autonomous quantale Q.

References

1. Andres, E.: Discrete linear objects in dimension n: the standard model. Graph. Model. **65**(1), 92–111 (2003)
2. Barr, M.: ★-Autonomous Categories. LNM, vol. 752. Springer, Heidelberg (1979)

3. Bennett, M.K., Birkhoff, G.: Two families of Newman lattices. Algebra Universalis **32**(1), 115–144 (1994)
4. Berstel, J., Lauve, A., Reutenauer, C., Saliola, F.V.: Combinatorics on Words. CRM Monograph Series, vol. 27. American Mathematical Society, Providence (2009)
5. Berthé, V., Labbé, S.: An arithmetic and combinatorial approach to three-dimensional discrete lines. In: Debled-Rennesson, I., Domenjoud, E., Kerautret, B., Even, P. (eds.) DGCI 2011. LNCS, vol. 6607, pp. 47–58. Springer, Heidelberg (2011). https://doi.org/10.1007/978-3-642-19867-0_4
6. Caspard, N., Santocanale, L., Wehrung, F.: Algebraic and combinatorial aspects of permutohedra. In: Grätzer, G., Wehrung, F. (eds.) Lattice Theory: Special Topics and Applications, vol. 2, pp. 215–286. Birkhäuser, Cham (2016)
7. Cockett, J., Seely, R.: Proof theory for full intuitionistic linear logic, bilinear logic, and mix categories. Theory Appl. Categ. **3**, 85–131 (1997)
8. Emanovský, P., Rachůnek, J.: A non commutative generalization of ⋆-autonomous lattices. Czechoslov. Math. J. **58**(3), 725–740 (2008)
9. Ferrari, L., Pinzani, R.: Lattices of lattice paths. J. Stat. Plan. Inference **135**(1), 77–92 (2005)
10. Feschet, F., Reveillès, J.-P.: A generic approach for n-dimensional digital lines. In: Kuba, A., Nyúl, L.G., Palágyi, K. (eds.) DGCI 2006. LNCS, vol. 4245, pp. 29–40. Springer, Heidelberg (2006). https://doi.org/10.1007/11907350_3
11. Galatos, N., Raftery, J.: A category equivalence for odd sugihara monoids and its applications. J. Stat. Plan. Inference **216**(10), 2177–2192 (2012)
12. Gierz, G., Hofmann, K.H., Keimel, K., Lawson, J.D., Mislove, M., Scott, D.S.: A Compendium of Continuous Lattices. Springer, Heidelberg (1980). https://doi.org/10.1007/978-3-642-67678-9
13. Goubault, E.: Some geometric perspectives in concurrency theory. Homol. Homotopy Appl. **5**(2), 95–136 (2003)
14. Joyal, A., Tierney, M.: An extension of the Galois theory of Grothendieck. Mem. Am. Math. Soc. **51**(309) (1984)
15. Kabil, M., Pouzet, M., Rosenberg, I.G.: Free monoids and generalized metric spaces. Eur. J. Comb. (2018, in press)
16. Krob, D., Latapy, M. Novelli, J.-C., Phan, H.D., Schwer, S.: Pseudo-permutations I: first combinatorial and lattice properties. In: FPSAC 2001 (2001)
17. Lawvere, F.W.: Metric spaces, generalized logic and closed categories. Rendiconti del Seminario Matematico e Fisico di Milano **XLIII**, 135–166 (1973)
18. Paoli, F.: ⋆-autonomous lattices. Studia Logica **79**(2), 283–304 (2005)
19. Provençal, X., Vuillon, L.: Discrete segments of $\mathbf{Z}^3$ constructed by synchronization of words. Discret. Appl. Math. **183**, 102–117 (2015)
20. Rosenthal, K.: Quantales and Their Applications. Pitman Research Notes in Mathematics Series. Longman Scientific & Technical, New York (1990)
21. Santocanale, L.: On the join depenency relation in multinomial lattices. Order **24**(3), 155–179 (2007)
22. Santocanale, L., Wehrung, F.: Sublattices of associahedra and permutohedra. Adv. Appl. Math. **51**(3), 419–445 (2013)
23. Santocanale, L., Wehrung, F.: The extended permutohedron on a transitive binary relation. Eur. J. Comb. **42**, 179–206 (2014)
24. Santocanale, L., Wehrung, F.: Lattices of regular closed subsets of closure spaces. Int. J. Algebr. Comput. **24**(7), 969–1030 (2014)
25. Santocanale, L., Wehrung, F.: Generalizations of the permutohedron. In: Grätzer, G., Wehrung, F. (eds.) Lattice Theory: Special Topics and Applications, vol. 2, pp. 287–397. Birkhäuser, Cham (2016)

26. Santocanale, L., Wehrung, F.: The equational theory of the weak Bruhat order on finite symmetric groups. J. Eur. Math. Soc. **20**(8), 1959–2003 (2018)
27. Stubbe, I.: An introduction to quantaloid-enriched categories. Fuzzy Sets Syst. **256**, 95–116 (2014)
28. Tsinakis, C., Wille, A.M.: Minimal varieties of involutive residuated lattices. Studia Logica **83**(1), 407–423 (2006)

Reasoning About Computations and Programs

Calculational Verification of Reactive Programs with Reactive Relations and Kleene Algebra

Simon Foster(✉), Kangfeng Ye, Ana Cavalcanti, and Jim Woodcock

University of York, York, UK
simon.foster@york.ac.uk

Abstract. Reactive programs are ubiquitous in modern applications, and so verification is highly desirable. We present a verification strategy for reactive programs with a large or infinite state space utilising algebraic laws for reactive relations. We define novel operators to characterise interactions and state updates, and an associated equational theory. With this we can calculate a reactive program's denotational semantics, and thereby facilitate automated proof. Of note is our reasoning support for iterative programs with reactive invariants, which is supported by Kleene algebra. We illustrate our strategy by verifying a reactive buffer. Our laws and strategy are mechanised in Isabelle/UTP, which provides soundness guarantees, and practical verification support.

1 Introduction

Reactive programming [1,2] is a paradigm that enables effective description of software systems that exhibit both internal sequential behaviour and event-driven interaction with a concurrent party. Reactive programs are ubiquitous in safety-critical systems, and typically have a very large or infinite state space. Though model checking is an invaluable verification technique, it exhibits inherent limitations with state explosion and infinite-state systems that can be overcome by supplementing it with theorem proving.

Previously [3], we have shown how *reactive contracts* support automated proof. They follow the design-by-contract paradigm [4], where programs are accompanied by pre- and postconditions. Reactive programs are often non-terminating and so we also capture intermediate behaviours, where the program has not terminated, but is quiescent and offers opportunities to interact. Our contracts are triples, $[P_1 \vdash P_2 \mid P_3]$, where P_1 is the precondition, P_3 the postcondition, and P_2 the "pericondition". P_2 characterises the quiescent observations in terms of the interaction history, and the events enabled at that point.

Reactive contracts describe communication and state updates, so P_1, P_2, and P_3 can refer to both a trace history of events and internal program variables. They are, therefore, called "reactive relations": like relations that model sequential programs, they can refer to variables before (x) and later (x') in execution, but also the interaction trace (tt), in both intermediate and final observations.

J. Desharnais et al. (Eds.): RAMiCS 2018, LNCS 11194, pp. 205–224, 2018.
https://doi.org/10.1007/978-3-030-02149-8_13

Verification using contracts employs refinement ($\sqsubseteq$), which requires that an implementation weakens the precondition, and strengthens both the peri- and postcondition when the precondition holds. We employ the "programs-as-predicates" approach [5], where the implementation (Q) is itself denoted as a composition of contracts. Thus, a verification problem, $[P_1 \vdash P_2 \mid P_3] \sqsubseteq Q$, can be solved by calculating a program $[Q_1 \vdash Q_2 \mid Q_3] = Q$, and then discharging three proof obligations: (1) $Q_1 \sqsubseteq P_1$; (2) $P_2 \sqsubseteq (Q_2 \wedge P_1)$; and (3) $P_3 \sqsubseteq (Q_3 \wedge P_1)$. These can be further decomposed, using relational calculus, to produce verification conditions. In [3] we employ this strategy in an Isabelle/HOL tactic.

For reactive programs of a significant size, these relations are complex, and so the resulting proof obligations are difficult to discharge using relational calculus. We need, first, abstract patterns so that the relations can be simplified. This necessitates bespoke operators that allow us to concisely formulate the different kinds of observation. Second, we need calculational laws to handle iterative programs, which are only partly handled in our previous work [3].

In this paper we present a novel calculus for description, composition, and simplification of reactive relations in the stateful failures-divergences model [6,7]. We characterise conditions, external interactions, and state updates. An equational theory allows us to reduce pre-, peri-, and postconditions to compositions of these atoms using operators of Kleene algebra [8] (KA) and utilise KA proof techniques. Our theory is characterised in the Unifying Theories of Programming [6,9] (UTP) framework. For that, we identify a class of UTP theories that induce KAs, and utilise it in derivation of calculational laws for iteration. We use our UTP mechanisation, called Isabelle/UTP [10], to implement an automated verification approach for infinite-state systems with rich data structures.

The paper is structured as follows. Section 2 outlines preliminary material. Section 3 identifies a class of UTP theories that induce KAs, and applies this for calculation of iterative contracts. Section 4 specialises reactive relations with new atomic operators to capture stateful failures-divergences, and derives their equational theory. Section 5 extends this with support for calculating external choices. Section 6 completes the theoretical picture with while loops and reactive invariants. Section 7 demonstrates the resulting proof strategy in a small verification. Section 8 outlines related work and concludes. All our theorems have been mechanically verified in Isabelle/UTP[1] [10–13].

2 Preliminaries

Kleene Algebras [8] (KA) characterise sequential and iterative behaviour in nondeterministic programs using a signature $(K, +, 0, \cdot, 1, ^*)$, where $+$ is a choice operator with unit 0, and $\cdot$ a composition operator, with unit 1. Kleene closure P^* denotes iteration of P using $\cdot$ zero or more times. We consider the class of weak Kleene algebras [14], which build on weak dioids.

[1] All proofs can be found in the cited series of Isabelle/HOL reports. For historical reasons, we use the syntax $\mathbf{R}_s(P \vdash Q \diamond R)$ in our mechanisation for a contract $[P \vdash Q \mid R]$, which builds on Hoare and He's original syntax for the theory of designs [6].

Definition 2.1. *A weak dioid is a structure* $(K, +, 0, \cdot, 1)$ *such that* $(S, +, 0)$ *is an idempotent and commutative monoid;* $(S, \cdot, 1)$ *is a monoid;* $\cdot$ *left- and right-distributes over* $+$*; and* 0 *is a left annihilator for* $\cdot$.

The 0 operator represents miraculous behaviour. It is a left annihilator of composition, but not a right annihilator as this often does not hold for programs. K is partially ordered by $x \leq y \triangleq (x + y = y)$, which is defined in terms of $+$, and has least element 0. A weak KA extends this with the behaviour of the star.

Definition 2.2. *A weak Kleene algebra is a structure* $(K, +, 0, \cdot, 1, ^*)$ *such that*

1. $(K, +, 0, \cdot, 1)$ *is a weak dioid*
2. $1 + x \cdot x^* \leq x^*$
3. $z + x \cdot y \leq y \Rightarrow x^* \cdot z \leq y$
4. $z + y \cdot x \leq y \Rightarrow z \cdot x^* \leq y$

Various enrichments and specialisations of these axioms exist; for a complete survey see [8]. For our purposes, these axioms alone suffice. From this base, a number of useful identities can be derived, some of which are listed below.

Theorem 2.3. $x^{**} = x^* \quad x^* = 1 + x \cdot x^* \quad (x + y)^* = (x^* \cdot y^*)^* \quad x \cdot x^* = x^* \cdot x$

UTP [6,9] uses the "programs-as-predicate" approach to encode denotational semantics and facilitate reasoning about programs. It uses the alphabetised relational calculus, which combines predicate calculus operators like disjunction ($\vee$), complement ($\neg$), and quantification ($\exists\, x \bullet P(x)$), with relation algebra, to denote programs as binary relations between initial variables (x) and their subsequent values (x'). The set of relations *Rel* is partially ordered by refinement $\sqsubseteq$ (refined-by), denoting universally closed reverse implication, where ***false*** refines every relation. Relational composition ($\mathbin{;}$) denotes sequential composition with identity II. We summarise the algebraic properties of relations below.

Theorem 2.4. $(Rel, \sqsupseteq, \textbf{\textit{false}}, \mathbin{;}, \mathit{II})$ *is a Boolean quantale [15], so that:*

1. $(Rel, \sqsubseteq)$ *is a complete lattice, with infimum* $\bigvee$*, supremum* $\bigwedge$*, greatest element* ***false****, least element* ***true****, and weakest (least) fixed-point operator* $\mu\, F$*;*
2. $(Rel, \vee, \textbf{\textit{false}}, \wedge, \textbf{\textit{true}}, \neg)$ *is a Boolean algebra;*
3. $(Rel, \mathbin{;}, \mathit{II})$ *is a monoid with* ***false*** *as left and right annihilator;*
4. $\mathbin{;}$ *distributes over* $\bigvee$ *from the left and right.*

We often use $\bigsqcap_{i \in I} P(i)$ to denote an indexed disjunction over I, which intuitively refers to a nondeterministic choice. Note that the partial order $\leq$ of the Boolean quantale is $\sqsupseteq$, and so our lattice operators are inverted: for example, $\bigvee$ is the infimum with respect to $\sqsubseteq$, and $\mu\, F$ is the least fixed-point.

Relations can directly express sequential programs, whilst enrichment to characterise more advanced paradigms—such as object orientation [16], real-time [17], and concurrency [6]—can be achieved using UTP theories. A UTP theory is characterised as the set of fixed-points of a function $\mathbf{H} : Rel \rightarrow Rel$, called a healthiness condition. If P is a fixed-point of $\mathbf{H}$ it is said to be $\mathbf{H}$-healthy, and the set of healthy relations is $[\![\mathbf{H}]\!] \triangleq \{P \mid \mathbf{H}(P) = P\}$. In UTP, it is desirable that $\boldsymbol{H}$ is idempotent and monotonic so that $[\![\boldsymbol{H}]\!]$ forms a complete lattice under $\sqsubseteq$, and thus reasoning about both nondeterminism and recursion is possible.

Theory engineering and verification of programs using UTP is supported by Isabelle/UTP [10], which provides a shallow embedding of the relational calculus on top of Isabelle/HOL, and various approaches to automated proof. In this paper, we use a UTP theory to characterise reactive programs.

Reactive Programs. Whilst sequential programs determine the relationship between an initial and final state, reactive programs also pause during execution to interact with the environment. For example, the CSP [9,18] and *Circus* [7] languages can model networks of concurrent processes that communicate using shared channels. Reactive behaviour is described using primitives such as event prefix $a \rightarrow P$, which awaits event a and then enables P; conditional guard, $b \,\&\, P$, which enables P when b is true; external choice $P \,\Box\, Q$, where the environment resolves the choice by communicating an initial event of P or Q; and iteration $\textbf{while}\; b\; \textbf{do}\; P$. Channels can carry data, and so events can take the form of an input ($c?x$) or output ($c!v$). *Circus* processes also have local state variables that can be assigned ($x := v$). We exemplify *Circus* with an unbounded buffer.

Example 2.5. In the *Buffer* process below, variable $bf : \operatorname{seq} \mathbb{N}$ records the elements, and channels $inp(n : \mathbb{N})$ and $outp(n : \mathbb{N})$ represent inputs and outputs.

$$Buffer \triangleq bf := \langle\rangle \,;\, \left(\begin{array}{l} \textbf{while}\; true\; \textbf{do} \\ \left(\begin{array}{l} inp?v \rightarrow bf := bf \frown \langle v \rangle \\ \Box\, (\#bf > 0) \,\&\, out!(head(bf)) \rightarrow bf := tail(bf) \end{array} \right) \end{array} \right)$$

Variable bf is set to the empty sequence $\langle\rangle$, and then a non-terminating loop describes the main behaviour. Its body repeatedly allows the environment to either provide a value v over inp, followed by which bf is extended, or else, if the buffer is non-empty, receive the value at the head, and then bf is contracted. □

The semantics of such programs can be captured using reactive contracts [3]:

$$[\, P_1(\mathit{tt}, \mathit{st}, r) \vdash P_2(\mathit{tt}, \mathit{st}, r, r') \mid P_3(\mathit{tt}, \mathit{st}, \mathit{st}', r, r') \,]$$

Here, $P_{1 \cdots 3}$ are reactive relations that respectively encode, (1) the precondition in terms of the initial state and permissible traces; (2) permissible intermediate interactions with respect to an initial state; and (3) final states following execution. Pericondition P_2 and postcondition P_3 are both within the "guarantee" part of the underlying design contract, and so must be strengthened by refinement; see Appendix A and [3] for details. P_2 does not refer to intermediate state variables since they are concealed when a program is quiescent.

Variable tt refers to the trace, and $\mathit{st}, \mathit{st}' : \Sigma$ to the state, for state space Σ. Traces are equipped with operators for the empty trace $\langle\rangle$, concatenation $tt_1 \frown tt_2$, prefix $tt_1 \leq tt_2$, and difference $tt_1 - tt_2$, which removes a prefix tt_2 from tt_1. Technically, tt is not a relational variable, but an expression $\mathit{tt} \triangleq tr' - tr$ where tr, tr', as usual in UTP, encode the trace relationally [6]. Nevertheless, due to our previous results [10,19], tt can be treated as a variable. Here, traces are modelled as finite sequences, $tt : \text{seq}\, Event$, for some event set, though other models are also admitted [19]. Events can be parametric, written $a.x$, where a is a channel and x is the data. Moreover, the relations can encode additional semantic data, such as refusals, using variables r, r'. Our theory, therefore, provides an extensible denotational semantic model for reactive and concurrent languages.

To exemplify, we consider the event prefix and assignment operators from *Circus*, which require that we add variable $ref' : \mathbb{P}(Event)$ to record refusals.

$$a \rightarrow \textbf{Skip} \triangleq \left[\, \textbf{true}_r \vdash tt = \langle\rangle \wedge a \notin ref' \mid tt = \langle a \rangle \wedge st' = st \,\right]$$

$$x := v \triangleq \left[\, \textbf{true}_r \vdash \textbf{false} \mid st' = st \oplus \{x \mapsto v\} \wedge tt = \langle\rangle \,\right]$$

Prefix has a true precondition, indicated using the reactive relation $\textbf{true}_r$, since the environment cannot cause errors. In the pericondition, no events have occurred ($tt = \langle\rangle$), but a is not being refused. In the postcondition, the trace is extended by a, and the state is unchanged. Assignment also has a true precondition, but a false pericondition since it terminates without interaction. The postcondition updates the state, and leaves the trace unchanged.

Reactive relations and contracts are characterised by healthiness conditions **RR** and **NSRD**, respectively, which we have previously described [3], and reproduce in Appendix A. **NSRD** specialises the theory of reactive designs [7,9] to *normal stateful reactive designs* [3]. Both $[\![\textbf{RR}]\!]$ and $[\![\textbf{NSRD}]\!]$ are closed under sequential composition, and have units $I\!I_r$ and $I\!I_R$, respectively. Both also form complete lattices under $\sqsubseteq$, with top elements **false** and $\textbf{Miracle} = [\textbf{true}_r \vdash \textbf{false} \mid \textbf{false}]$, respectively. $\textbf{Chaos} = [\textbf{false} \vdash \textbf{false} \mid \textbf{false}]$, the least deterministic contract, is the bottom of the reactive contract lattice. We define the conditional operator $P \lhd b \rhd Q \triangleq ((b \wedge P) \vee (\neg b \wedge Q))$, where b is a condition on unprimed state variables, which can be used for both reactive relations and contracts. We then define the state test operator $[b]_r^\top \triangleq I\!I_r \lhd b \rhd \textbf{false}$.

Contracts can be composed using relational calculus. The following identities [3,12] show how this entails composition of the underlying pre-, peri-, and postconditions for $\sqcap$ and $⨾$, and also demonstrates closure under these operators.

Theorem 2.6 (Reactive Contract Composition).

$$\textstyle\bigsqcap_{i \in I} [P(i) \vdash Q(i) \mid R(i)] = \left[\, \bigwedge_{i \in I} P(i) \vdash \bigvee_{i \in I} Q(i) \mid \bigvee_{i \in I} R(i) \,\right] \qquad (1)$$

$$[P_1 \vdash P_2 \mid P_3] ⨾ [Q_1 \vdash Q_2 \mid Q_3] = [P_1 \wedge (P_3 \;\textbf{wp}_r\; Q_1) \vdash P_2 \vee (P_3 ⨾ Q_2) \mid P_3 ⨾ Q_3\,] \qquad (2)$$

Nondeterministic choice requires all preconditions, and asserts that one of the peri- and postcondition pairs hold. For sequential composition, the precondition assumes that P_1 holds, and that P_3 fulfils Q_1. The latter is formulated using a reactive weakest precondition ($\textbf{wp}_r$), which obeys standard laws [20] such as:

$$\left(\textstyle\bigvee_{i \in I} P(i)\right) \textbf{wp}_r\, R = \textstyle\bigwedge_{i \in I} P(i) \;\textbf{wp}_r\; R \qquad (P ⨾ Q) \;\textbf{wp}_r\; R = P \;\textbf{wp}_r (Q \;\textbf{wp}_r\; R)$$

In the pericondition, either the first contract is intermediate (P_2), or else it terminated (P_3) and then following this the second is intermediate (Q_2). In the postcondition the contracts have both terminated in sequence ($P_3 ⨾ Q_3$).

With these and related theorems [10], we can calculate contracts of reactive programs. Verification, then, can be performed by proving a refinement between two reactive contracts, a strategy we have mechanised in the Isabelle/UTP tactics rdes-refine and rdes-eq [10]. The question remains, though, how to reason about the underlying compositions of reactive relations for the pre-, peri-, and postconditions. For example, consider the action $(a \rightarrow \textbf{\textit{Skip}}) \mathbin{;} x := v$. For its postcondition, we must simplify $(\mathit{tt} = \langle a \rangle \wedge \mathit{st}' = \mathit{st}) \mathbin{;} (\mathit{st}' = \mathit{st} \oplus \{x \mapsto v\} \wedge \mathit{tt} = \langle\rangle)$. In order to simplify its precondition, we also need to consider reactive weakest preconditions. Without such simplifications, reactive relations can grow very quickly and hamper proof. Finally, of particular importance is the handling of iterative reactive relations. We address these issues in this paper.

3 Linking UTP and Kleene Algebra

In this section, we characterise properties of a UTP theory sufficient to identify a KA, and use this to obtain theorems for iterative contracts. We observe that UTP relations form a KA $(\mathit{Rel}, \sqcap, \mathbin{;}, ^*, \mathit{I\!I})$, where $P^* \triangleq (\nu X \bullet \mathit{I\!I} \sqcap P \mathbin{;} X)$. We have proved this definition equivalent to the power form: $P^* = (\bigsqcap_{i \in \mathbb{N}} P^i)$ where P^n iterates sequential composition n times.

Typically, UTP theories, like $[\![\textbf{\textit{NSRD}}]\!]$, share the operators for choice ($\sqcap$) and composition ($\mathbin{;}$), only redefining them when absolutely necessary. Formally, given a UTP theory defined by a healthiness condition $\textbf{\textit{H}}$, the set of healthy relations $[\![\textbf{\textit{H}}]\!]$ is closed under $\sqcap$ and $\mathbin{;}$. This has the major advantage that a large body of laws is directly applicable from the relational calculus. The ubiquity of $\sqcap$, in particular, can be characterised through the subset of continuous UTP theories, where $\textbf{\textit{H}}$ distributes through arbitrary non-empty infima, that is,

$$\textbf{\textit{H}}\left(\bigsqcap_{i \in I} P(i)\right) = \bigsqcap_{i \in I} \textbf{\textit{H}}(P(i)) \text{ provided } I \neq \emptyset.$$

Monotonicity of $\textbf{\textit{H}}$ follows from continuity, and so such theories induce a complete lattice. Continuous UTP theories include designs [6,14], CSP, and *Circus* [7]. A further consequence of continuity is that the relational weakest fixed-point operator $\mu X \bullet F(X)$ constructs healthy relations when $F : \mathit{Rel} \rightarrow [\![\textbf{\textit{H}}]\!]$.

Though these theories share infima and weakest fixed-points, they do not, in general, share $\top$ and $\bot$ elements, which is why the infima are non-empty in the above continuity property. Rather, we have a top element $\top_{\textbf{\textit{H}}} \triangleq \textbf{\textit{H}}(\textbf{\textit{false}})$ and a bottom element $\bot_{\textbf{\textit{H}}} \triangleq \textbf{\textit{H}}(\textbf{\textit{true}})$ [3]. The theories also do not share the relational identity $\mathit{I\!I}$, but typically define a bespoke identity $\mathit{I\!I}_{\textbf{\textit{H}}}$, which means that $[\![\textbf{\textit{H}}]\!]$ is not closed under the relational Kleene star. However, $[\![\textbf{\textit{H}}]\!]$ is closed under the related Kleene plus $P^+ \triangleq P \mathbin{;} P^*$ since it is equivalent to $(\bigsqcap_{i \in \mathbb{N}} P^{i+1})$, which iterates P one or more times. Thus, we can obtain a theory Kleene star with the definition $P^{\star} \triangleq \mathit{I\!I}_{\textbf{\textit{H}}} \sqcap P^+$, under which $\textbf{\textit{H}}$ is indeed closed. We, therefore, define the following criteria for a UTP theory.

Definition 3.1. *A Kleene UTP theory* $(\textbf{\textit{H}}, \mathit{I\!I}_{\textbf{\textit{H}}})$ *satisfies the following conditions: (1)* $\textbf{\textit{H}}$ *is idempotent and continuous; (2)* $\textbf{\textit{H}}$ *is closed under sequential composition;*

(3) identity $\mathit{II}_{\mathbf{H}}$ *is* $\mathbf{H}$*-healthy; (4)* $\mathit{II}_{\mathbf{H}} ⨾ P = P ⨾ \mathit{II}_{\mathbf{H}} = P$*, when* P *is* $\mathbf{H}$*-healthy; (5)* $\top_{\mathbf{H}} ⨾ P = \top_{\mathbf{H}}$*, when* P *is* $\mathbf{H}$*-healthy.*

From these properties, we can prove the following theorem.

Theorem 3.2. *If* $(\mathbf{H}, \mathit{II}_{\mathbf{H}})$ *is a Kleene UTP theory, then* $(\llbracket \mathbf{H} \rrbracket, \sqcap, \top_{\mathbf{H}}, ⨾, \mathit{II}_{\mathbf{H}}, ^*)$ *forms a weak Kleene algebra.*

Proof. We prove this in Isabelle/UTP by lifting of laws from the Isabelle/HOL KA hierarchy [21,22]. For details see [11].

All the identities of Theorem 2.3 hold in a Kleene UTP theory, thus providing reasoning capabilities for iterative programs. In particular, we can show that $(\llbracket \mathbf{NSRD} \rrbracket, \sqcap, \mathbf{Miracle}, ⨾, \mathit{II}_{\mathbf{R}}, ^*)$ and $(\llbracket \mathbf{RR} \rrbracket, \sqcap, \mathbf{false}, ⨾, \mathit{II}_{r}, ^*)$ both form weak KAs. Moreover, we can now also show how to calculate an iterative contract [12].

Theorem 3.3 (Reactive Contract Iteration).

$$[\, P \vdash Q \mid R \,]^* = [\, R^* \ \mathbf{wp}_r \ P \vdash R^* ⨾ Q \mid R^* \,]$$

Note that the outer and inner star are different operators. The precondition states that R must not violate P after any number of iterations. The pericondition has R iterated followed by Q holding, since the final observation is intermediate. The postcondition simply iterates R. Thus we have the basis for calculating and reasoning about iterative contracts.

4 Reactive Relations of Stateful Failures-Divergences

Here, we specialise our contract theory to incorporate failure traces, which are used in CSP, ***Circus***, and related languages [23]. We define atomic operators to describe the underlying reactive relations, and the associated equational theory to expand and simplify compositions arising from Theorems 2.6 and 3.3, and thus support automated reasoning. We consider external choice separately (Sect. 5).

Healthiness condition $\mathbf{NCSP} \triangleq \mathbf{NSRD} \circ \mathbf{CSP3} \circ \mathbf{CSP4}$ characterises the stateful failures-divergences model [6,7,9]. **CSP3** and **CSP4** ensure the refusal sets are well-formed [6,9]: ref' can only be mentioned in the pericondition (see also Appendix A). **NCSP**, like **NSRD**, is continuous and has **Skip**, defined below, as a left and right unit. Thus, $(\llbracket \mathbf{NCSP} \rrbracket, \sqcap, \mathbf{Miracle}, ⨾, \mathbf{Skip}, ^*)$ forms a Kleene algebra. An **NCSP** contract has the following specialised form [13].

$$[\, P(\mathit{tt}, \mathit{st}) \vdash Q(\mathit{tt}, \mathit{st}, ref') \mid R(\mathit{tt}, \mathit{st}, \mathit{st}') \,]$$

The underlying reactive relations capture a portion of the stateful failures-divergences. P captures the initial states and traces that do not induce divergence, that is, unpredictable behaviour like **Chaos**. Q captures the stateful failures of a program: the set of events not being refused (ref') having performed trace tt, starting in state st. R captures the terminated behaviours, where a final state is observed but no refusals. We describe the pattern of the underlying reactive relations using the following constructs.

Definition 4.1 (Reactive Relational Operators).

$$\mathcal{I}[b(\mathit{st}), t(\mathit{st})] \triangleq \mathbf{RR}(b(\mathit{st}) \wedge t(\mathit{st}) \leq \mathit{tt}) \tag{3}$$

$$\mathcal{E}[b(\mathit{st}), t(\mathit{st}), E(\mathit{st})] \triangleq \mathbf{RR}(b(\mathit{st}) \wedge \mathit{tt} = t(\mathit{st}) \wedge (\forall e \in E(\mathit{st}) \bullet e \notin \mathit{ref}')) \tag{4}$$

$$\Phi[b(\mathit{st}), \sigma, t(\mathit{st})] \triangleq \mathbf{RR}(b(\mathit{st}) \wedge \mathit{st}' = \sigma(\mathit{st}) \wedge \mathit{tt} = t(\mathit{st})) \tag{5}$$

In this definition, we utilise expressions b, t, and E that refer only to the variables by which they are parametrised. Namely, $b(\mathit{st}) : \mathbb{B}$ is a condition on st, $t(\mathit{st}) : \mathrm{seq}\, \mathit{Event}$ is a trace expression that describes a possible event sequence in terms of st, and $E(\mathit{st}) : \mathbb{P}\, \mathit{Event}$ is an expression that describes a set of events. Following [24], we describe state updates with substitutions $\sigma : \Sigma \to \Sigma$. We use $(\!| x \mapsto v |\!)$ to denote a substitution, which is the identity for every variable, except that v is assigned to x. Substitutions can also be applied to contracts and relations using operator $\sigma \dagger P$, and then $Q[v/x] \triangleq (\!| x \mapsto v |\!) \dagger Q$. This operator obeys similar laws to syntactic substitution, though it is a semantic operator [10].

$\mathcal{I}[b(\mathit{st}), t(\mathit{st})]$ is a specification of initial behaviour used in preconditions. It states that initially the state satisfies condition b, and t is a prefix of the overall trace. $\mathcal{E}[b(\mathit{st}), t(\mathit{st}), E(\mathit{st})]$ is used in periconditions to specify quiescent observations, and corresponds to a failure trace. It specifies that the state variables initially satisfy b, the interaction described by t has occurred, and finally we reach a quiescent phase where none of the events in E are being refused. $\Phi[b(\mathit{st}), \sigma, t(\mathit{st})]$ is used to encode final terminated observations in the postcondition. It specifies that the initial state satisfies b, the state update σ is applied, and the interaction t has occurred.

These operators are all deterministic, in the sense that they describe a single interaction and state-update history. There is no need for explicit nondeterminism here, as this is achieved using $\bigvee$. These operators allow us to concisely specify the basic operators of our theory as given below.

Definition 4.2 (Basic Reactive Operators).

$$\langle \sigma \rangle_{\mathbf{c}} \triangleq [\, \mathbf{true}_r \vdash \mathbf{false} \mid \Phi[\mathit{true}, \sigma, \langle\rangle]\,] \tag{6}$$

$$\mathbf{Do}(a) \triangleq [\, \mathbf{true}_r \vdash \mathcal{E}[\mathit{true}, \langle\rangle, \{a\}] \mid \Phi[\mathit{true}, \mathit{id}, \langle a \rangle]\,] \tag{7}$$

$$\mathbf{Stop} \triangleq [\, \mathbf{true}_r \vdash \mathcal{E}[\mathit{true}, \langle\rangle, \emptyset] \mid \mathbf{false}] \tag{8}$$

Generalised assignment $\langle \sigma \rangle_{\mathbf{c}}$ is again inspired by [24]. It has a $\mathbf{true}_r$ precondition and a ***false*** pericondition: it has no intermediate observations. The postcondition states that for any initial state (*true*), the state is updated using σ, and no events are produced ($\langle\rangle$). A singleton assignment $x := v$ can be expressed using a state update $(\!| x \mapsto v |\!)$. We define $\mathbf{Skip} \triangleq \langle \mathit{id} \rangle_{\mathbf{c}}$, which leaves all variables unchanged.

$\mathbf{Do}(a)$ encodes an event action. Its pericondition states that no event has occurred, and a is accepted. Its postcondition extends the trace by a, leaving the state unchanged. We can denote *Circus* event prefix $a \to P$ as $\mathbf{Do}(a) \,;\, P$.

Finally, ***Stop*** represents a deadlock: its pericondition states the trace is unchanged and no events are being accepted. The postcondition is false as there is no way to terminate. A ***Circus*** guard g & P can be denoted as $(P \lhd g \rhd \textbf{\textit{Stop}})$, which behaves as P when g is true, and otherwise deadlocks.

To calculate contractual semantics, we need laws to reduce pre-, peri-, and postconditions. These need to cater for various composition cases of operators $\sqcap$, $\fatsemi$, and $\Box$. So, we prove [13] the following composition laws for $\mathcal{E}$ and Φ.

Theorem 4.3 (Reactive Relational Compositions).

$$[b]_r^\top \fatsemi P = \Phi[b, id, \langle\rangle] \fatsemi P \tag{9}$$

$$\Phi[b_1, \sigma_1, t_1] \fatsemi \Phi[b_2, \sigma_2, t_2] = \Phi[b_1 \wedge \sigma_1 \dagger b_2, \sigma_2 \circ \sigma_1, t_1 \frown \sigma_1 \dagger t_2] \tag{10}$$

$$\Phi[b_1, \sigma_1, t_1] \fatsemi \mathcal{E}[b_2, t_2, E] = \mathcal{E}[b_1 \wedge \sigma_1 \dagger b_2, t_1 \frown \sigma_1 \dagger t_2, \sigma_1 \dagger E] \tag{11}$$

$$\Phi[b_1, \sigma_1, t_1] \lhd c \rhd \Phi[b_2, \sigma_2, t_2] = \Phi[b_1 \lhd c \rhd b_2, \sigma_1 \lhd c \rhd \sigma_2, t_1 \lhd c \rhd t_2] \tag{12}$$

$$\mathcal{E}[b_1, t_1, E_1] \lhd c \rhd \mathcal{E}[b_2, t_2, E_2] = \mathcal{E}[b_1 \lhd c \rhd b_2, t_1 \lhd c \rhd t_2, E_1 \lhd c \rhd E_2] \tag{13}$$

$$\left(\bigwedge_{i \in I} \mathcal{E}[b(i), t, E(i)]\right) = \mathcal{E}\left[\bigwedge_{i \in I} b(i), t, \bigcup_{i \in I} E(i)\right] \tag{14}$$

Law (9) states that a precomposed test can be expressed using Φ. (10) states that the composition of two terminated observations results in the conjunction of the state conditions, composition of the state updates, and concatenation of the traces. It is necessary to apply the initial state update σ_1 as a substitution to both the second state condition (s_2) and the trace expression (t_2). (11) is similar, but accounts for the enabled events rather than state updates. (10) and (11) are required because of Theorem 2.6-2, which sequentially composes a pericondition with a postcondition, and a postcondition with a postcondition. (12) and (13) show how conditional distributes through the operators. Finally, (14) shows that a conjunction of intermediate observations with a common trace takes the conjunction of the state conditions, and the union of the enabled events. It is needed for external choice, which conjoins the periconditions (see Sect. 5).

In order to calculate preconditions, we need to consider the weakest precondition operator. Theorem 2.6-2 requires that, in a sequential composition $P \fatsemi Q$, we need to show that the postcondition of contract P satisfies the precondition of contract Q. Theorem 4.3 explains how to eliminate most composition operators in a contract's postcondition, but not in general $\vee$. Postconditions are, therefore, typically expressed as disjunctions of the Φ operator, and so it suffices to calculate its weakest precondition using the theorem below.

Theorem 4.4. $\Phi[s, \sigma, t]\ \textbf{\textit{wp}}_r\ P = (\mathcal{I}[s, t] \Rightarrow (\sigma \dagger P)[tt - t/tt])$

In order for $\Phi[s, \sigma, t]$ to satisfy reactive condition P, whenever we start in the state satisfying s and the trace t has been performed, P must hold on the remainder of the trace ($tt - t$), and with the state update σ applied. We can now use these laws, along with Theorem 2.6, to calculate the semantics of processes, and to prove equality and refinement conjectures, as we illustrate below.

Example 4.5. We show that $(x:=1 \,;\, \mathbf{Do}(a.x) \,;\, x := x+2) = (\mathbf{Do}(a.1) \,;\, x:=3)$. By applying Definition 4.2 and Theorems 2.6 (2), 4.3, 4.4, both sides reduce to $[\, \mathbf{true}_r \vdash \mathcal{E}[true, \langle\rangle, \{a.1\}] \mid \Phi[true, \{x \mapsto 3\}, \langle a.1\rangle]\,]$, which has a single quiescent state, waiting for event $a.1$, and a single final state, where $a.1$ has occurred and state variable x has been updated to 3. We calculate the left-hand side below.

$$
\begin{aligned}
&(x:=1 \,;\, \mathbf{Do}(a.x) \,;\, x := x+2) \\
&= \begin{pmatrix} [\, \mathbf{true}_r \vdash \mathbf{false} \mid \Phi[true, (\!| x \mapsto 1 |\!), \langle\rangle]\,] \,; \\ [\, \mathbf{true}_r \vdash \mathcal{E}[true, \langle\rangle, \{a.x\}] \mid \Phi[true, id, \langle a.x\rangle]\,] \,; \\ [\, \mathbf{true}_r \vdash \mathbf{false} \mid \Phi[true, (\!| x \mapsto x+1 |\!), \langle\rangle]\,] \end{pmatrix} && [\text{Definition 4.2}] \\
&= \left[\mathbf{true}_r \vdash \begin{array}{l} \Phi[true, (\!| x \mapsto 1 |\!), \langle\rangle] \,; \\ \mathcal{E}[true, \langle\rangle, \{a.x\}] \end{array} \middle| \begin{array}{l} \Phi[true, (\!| x \mapsto 1 |\!), \langle\rangle] \,; \\ \Phi[true, id, \langle a.x\rangle] \,; \\ \Phi[true, (\!| x \mapsto x+2 |\!), \langle\rangle] \end{array} \right] && \begin{bmatrix} \text{Theorem 2.6,} \\ \text{Theorem 4.4} \end{bmatrix} \\
&= \left[\mathbf{true}_r \vdash \mathcal{E}[true, \langle\rangle[1/x], \{a.x\}[1/x]] \middle| \begin{array}{l} \Phi[true, (\!| x \mapsto 1 |\!), \langle a.1\rangle] \,; \\ \Phi[true, (\!| x \mapsto x+2 |\!), \langle\rangle] \end{array} \right] && [\text{Theorem 4.3}] \\
&= [\, \mathbf{true}_r \vdash \mathcal{E}[true, \langle\rangle, \{a.1\}] \mid \Phi[true, \{x \mapsto 3\}, \langle a.1\rangle]\,]
\end{aligned}
$$

□

Similarly, we can use our theorems, with the help of our mechanised proof strategy in rdes-eq, to prove a number of general laws [13].

Theorem 4.6 (Stateful Failures-Divergences Laws).

$$\langle\sigma\rangle_c \,;\, [P_1 \vdash P_2 \mid P_3] = [\sigma \dagger P_1 \vdash \sigma \dagger P_2 \mid \sigma \dagger P_3] \tag{15}$$

$$\langle\sigma\rangle_c \,;\, \mathbf{Do}(e) = \mathbf{Do}(\sigma \dagger e) \,;\, \langle\sigma\rangle_c \tag{16}$$

$$\langle\sigma\rangle_c \,;\, \langle\rho\rangle_c = \langle\rho \circ \sigma\rangle_c \tag{17}$$

$$\mathbf{Stop} \,;\, P = \mathbf{Stop} \tag{18}$$

Law (15) shows how assignment distributes substitutions through a contract. (16) and (17) are consequences of (15). (18) shows that deadlock is a left annihilator.

5 External Choice and Productivity

In this section we consider reasoning about programs with external choice, and characterise the important subclass of productive contracts [3], which are also essential in verifying recursive and iterative reactive programs.

An external choice $P \,\square\, Q$ resolves whenever either P or Q engages in an event or terminates. Thus, its semantics requires that we filter observations with a non-empty trace. We introduce healthiness condition $\mathbf{R4}(P) \triangleq (P \wedge tt > \langle\rangle)$, whose fixed points strictly increase the trace, and its dual $\mathbf{R5}(P) \triangleq (P \wedge tt = \langle\rangle)$ where the trace is unchanged. We use these to define indexed external choice.

Definition 5.1 (Indexed External Choice).

$$\Box\, i \in I \bullet [\, P_1(i) \vdash P_2(i) \mid P_3(i) \,] \triangleq$$
$$\left[\, \bigwedge_{i \in I} P_1(i) \vdash \left(\bigwedge_{i \in I} \mathbf{R5}(P_2(i))\right) \vee \left(\bigvee_{i \in I} \mathbf{R4}(P_2(i))\right) \mid \bigvee_{i \in I} P_3(i) \,\right]$$

This enhances the binary definition [6,7], and recasts our definition in [3] for calculation. Like nondeterministic choice, the precondition requires that all constituent preconditions are satisfied. In the pericondition **R4** and **R5** filter all observations. We take the conjunction of all **R5** behaviours: no event has occurred, and all branches are offering to communicate. We take the disjunction of all **R4** behaviours: an event occurred, and the choice is resolved. In the postcondition the choice is resolved, either by communication or termination, and so we take the disjunction of all constituent postconditions. Since unbounded choice is supported, we can denote indexed input prefix for any size of input domain A:

$$a?x\!:\!A \rightarrow P(x) \;\triangleq\; \Box\, x \in A \bullet a.x \rightarrow P(x)$$

We next show how **R4** and **R5** filter the various reactive relational operators, which can be applied to reason about contracts involving external choice.

Theorem 5.2 (Trace Filtering).

$$\mathbf{R4}\left(\bigvee_{i \in I} P(i)\right) = \bigvee_{i \in I} \mathbf{R4}(P(i)) \qquad \mathbf{R5}\left(\bigvee_{i \in I} P(i)\right) = \bigvee_{i \in I} \mathbf{R5}(P(i))$$
$$\mathbf{R4}(\Phi[s, \sigma, \langle\rangle]) = \mathbf{false} \qquad \mathbf{R5}(\mathcal{E}[s, \langle\rangle, E]) = \mathcal{E}[s, \langle\rangle, E]$$
$$\mathbf{R4}(\Phi[s, \sigma, \langle a, ...\rangle]) = \Phi[s, \sigma, \langle a, ...\rangle] \qquad \mathbf{R5}(\mathcal{E}[s, \langle a, ...\rangle, E]) = \mathbf{false}$$

Both operators distribute through $\bigvee$. Relations that produce an empty trace yield ***false*** under **R4** and are unchanged under **R5**. Relations that produce a non-empty trace yield ***false*** for **R5**, and are unchanged under **R4**. We can now filter the behaviours that do and do not resolve the choice, as exemplified below.

Example 5.3. Consider the calculation of $a \rightarrow b \rightarrow \mathbf{Skip} \;\Box\; c \rightarrow \mathbf{Skip}$. The left branch has two quiescent observations, one waiting for a, and one for b having performed a: its pericondition is $\mathcal{E}[\mathit{true}, \langle\rangle, \{a\}] \vee \mathcal{E}[\mathit{true}, \langle a \rangle, \{b\}]$. Application of **R5** to this will yield the first disjunct, since the trace has not increased, and **R4** will yield the second disjunct. For the right branch there is one quiescent observation, $\mathcal{E}[\mathit{true}, \langle\rangle, \{c\}]$, which contributes an empty trace and is **R5** only. The overall pericondition is $(\mathcal{E}[\mathit{true}, \langle\rangle, \{a\}] \wedge \mathcal{E}[\mathit{true}, \langle\rangle, \{c\}]) \vee \mathcal{E}[\mathit{true}, \langle a \rangle, \{b\}]$, which is simply $\mathcal{E}[\mathit{true}, \langle\rangle, \{a, c\}] \vee \mathcal{E}[\mathit{true}, \langle a \rangle, \{b\}]$. □

By calculation we can now prove that $([\![\mathbf{NCSP}]\!], \Box, \mathbf{Stop})$ forms a commutative and idempotent monoid, and ***Chaos***, the divergent program, is its annihilator. Sequential composition also distributes from the left and right through external choice, but only when the choice branches are productive [3].

Definition 5.4. *A contract* $[\, P_1 \vdash P_2 \mid P_3 \,]$ *is productive when* P_3 *is* ***R4*** *healthy.*

A productive contract is one that, whenever it terminates, strictly increases the trace. For example $a \to \textbf{\textsf{Skip}}$ is productive, but **Skip** is not. Constructs that do not terminate, like **Chaos**, are also productive. The imposition of **R4** ensures that only final observations that increase the trace, or are **false**, are admitted.

We define healthiness condition **PCSP**, which extends **NCSP** with productivity. We also define **ICSP**, which formalises instantaneous contracts where the postcondition is **R5** healthy and the pericondition is **false**. For example, both **Skip** and $x := v$ are **ICSP** healthy as they do not contribute to the trace and have no intermediate observations. This allows us to prove the following laws.

Theorem 5.5 (External Choice Distributivity).

$$(\Box\, i \in I \bullet P(i)) \,\text{;}\, Q = \Box\, i \in I \bullet (P(i) \,\text{;}\, Q) \quad [\textit{if}, \forall i \in I, P(i) \textit{ is } \textbf{\textsf{PCSP}} \textit{ healthy}]$$
$$P \,\text{;}\, (\Box\, i \in I \bullet Q(i)) = \Box\, i \in I \bullet (P \,\text{;}\, Q(i)) \quad [\textit{if } P \textit{ is } \textbf{\textsf{ICSP}} \textit{ healthy}]$$

The first law follows because every $P(i)$, being productive, must resolve the choice before terminating, and thus it is not possible to reach Q before this occurs. It generalises the standard guarded choice distribution law for CSP [6, p. 211]. The second law follows for the converse reason: since P cannot resolve the choice with any of its behaviour, it is safe to execute it first.

Productivity also forms an important criterion for guarded recursion that we utilise in Sect. 6 to calculate fixed points. **PCSP** is closed under several operators.

Theorem 5.6. (1) ***Miracle*, *Stop*, *Do***(a) *are* ***PCSP***; (2) $P \,\text{;}\, Q$ *is* ***PCSP*** *if either* P *or* Q *is* ***PCSP***; (3) $\Box\, i \in I \bullet P(i)$ *is* ***PCSP*** *if, for all* $i \in I$, $P(i)$ *is* ***PCSP***.

Calculation of external choice is now supported, and a notion of productivity defined. In the next section we use the latter for calculation of while-loops.

6 While Loops and Reactive Invariants

In this section we complete our verification strategy by adding support for iteration. Iterative programs can be constructed using the reactive while loop.

$$b \circledast P \triangleq (\mu X \bullet P \,\text{;}\, X \lhd b \rhd \textbf{\textsf{Skip}}).$$

We use the weakest fixed-point so that an infinite loop with no observable activity corresponds to the divergent action **Chaos**, rather than **Miracle**. For example, we can show that $(\mathit{true} \circledast x := x+1) = \textbf{\textsf{Chaos}}$. The *true* condition is not a problem in this context because, unlike its imperative cousin, the reactive while loop pauses for interaction with its environment during execution, and therefore infinite executions are observable and therefore potentially useful.

In order to reason about such behaviour, we need additional calculational laws. A fixed-point $(\mu X \bullet F(X))$ is guarded provided at least one event is contributed to the trace by F prior to it reaching X. For instance, $\mu X \bullet a \to X$ is guarded, but $\mu X \bullet y := 1 \,\text{;}\, X$ is not. Hoare and He's theorem [6, Theorem 8.1.13, p. 206] states that if F is guarded, then there is a unique fixed-point and hence $(\mu X \bullet F(X)) = (\nu X \bullet F(X))$. Then, provided F is continuous, we can invoke Kleene's fixed-point theorem to calculate νF. Our previous result [3] shows that if P is productive, then $\lambda X \bullet P \,;\, X$ is guarded, and so we can calculate its fixed-point. We now generalise this for the function above.

Theorem 6.1. *If P is productive, then $(\mu X \bullet P \,;\, X \lhd b \rhd \textbf{Skip})$ is guarded.*

Proof. In addition to our previous theorem [3], we use the following properties:

- If X is not mentioned in P then $\lambda X \bullet P$ is guarded;
- If F and G are both guarded, then $\lambda X \bullet F(X) \lhd b \rhd G(X)$ is guarded. □

This allows us to convert the fixed-point into an iterative form. In particular, we can prove the following theorem that expresses it in terms of Kleene star.

Theorem 6.2. *If P is **PCSP** healthy then $b \circledast P = ([b]_r^\top \,;\, P)^* \,;\, [\neg b]_r^\top$.*

This theorem is similar to the usual imperative definition [21,22]. P is executed multiple times when b is true initially, but each run concludes when b is false. However, due to the embedding of reactive behaviour, there is more going on than meets the eye; the next theorem shows how to calculate an iterative contract.

Theorem 6.3. *If R is **R4** healthy then*

$$b \circledast [P \vdash Q \mid R] = [\,([b]_r^\top ; R)^* \ \mathbf{wp}_r\ (b \Rightarrow P) \vdash ([b]_r^\top \,;\, R)^* ; [b]_r^\top ; Q \mid ([b]_r^\top ; R)^* ; [\neg b]_r^\top\,]$$

The precondition requires that any number of R iterations, where b is initially true, satisfies P. This ensures that the contract does not violate its own precondition from one iteration to the next. The pericondition states that intermediate observations have R executing several times, with b true, and following this b remains true and the contract is quiescent (Q). The postcondition is similar, but after several iterations, b becomes false and the loop terminates, which is the standard relational form of a while loop.

Theorem 6.3 can be utilised to prove a refinement introduction law for the reactive while loop. This employs "reactive invariant" relations, which describe how both the trace and state variables are permitted to evolve.

Theorem 6.4. *$[I_1 \vdash I_2 \mid I_3] \sqsubseteq b \circledast [Q_1 \vdash Q_2 \mid Q_3]$ provided that:*

1. *the assumption is weakened $(([b]_r^\top \,;\, Q_3)^* \ \mathbf{wp}_r\ (b \Rightarrow Q_1) \sqsubseteq I_1)$;*
2. *when b holds, Q_2 establishes the I_2 pericondition invariant $(I_2 \sqsubseteq ([b]_r^\top \,;\, Q_2))$ and, Q_3 maintains it $(I_2 \sqsubseteq [b]_r^\top \,;\, Q_3 \,;\, I_2)$;*
3. *postcondition invariant I_3 is established when b is false $(I_3 \sqsubseteq [\neg b]_r^\top)$ and Q_3 establishes it when b is true $(I_3 \sqsubseteq [b]_r^\top \,;\, Q_3 \,;\, I_3)$.*

Proof. By application of refinement introduction, with Theorems 2.2-3 and 6.3.

Theorem 6.4 shows the conditions under which an iterated reactive contract satisfies an invariant contract $[I_1 \vdash I_2 \mid I_3]$. Relations I_2 and I_3 are reactive invariants that must hold in quiescent and final observations, respectively. Both can refer to **st** and **tt**, I_2 can additionally refer to ref', and I_3 to **st**$'$. Combined with the results from Sects. 4 and 5, this result provides the basis for a proof strategy for iterative reactive programs that we now exemplify.

7 Verification Strategy for Reactive Programs

Our collected results give rise to an automated verification strategy for iterative reactive programs, whereby we (1) calculate the contract of a reactive program, (2) use our equational theory to simplify the underlying reactive relations, (3) identify suitable invariants for reactive while loops, and (4) finally prove refinements using relational calculus. Although the underlying relations can be quite complex, our equational theory from Sects. 4 and 5, aided by the Isabelle/HOL simplifier, can be used to rapidly reduce them to more compact forms amenable to automated proof. In this section we illustrate this strategy with the buffer in Example 2.5. We prove two properties: (1) deadlock freedom, and (2) that the order of values produced is the same as those consumed.

We first calculate the contract of the main loop in the *Buffer* process and then use this to calculate the overall contract for the iterative behaviour.

Theorem 7.1 (Loop Body). *The body of the loop is* $[\,\textbf{true}_r \vdash B_2 \mid B_3\,]$ *where*

$$B_2 = \mathcal{E}\left[true, \langle\rangle, \textstyle\bigcup_{v\in\mathbb{N}} \{inp.v\} \cup (\{out.head(bf)\} \lhd 0 < \#bf \rhd \emptyset)\right]$$

$$B_3 = \begin{pmatrix} \left(\bigvee_{v\in\mathbb{N}} \Phi[true, \{bf \mapsto bf \frown \langle v\rangle\}, \langle inp.v\rangle]\right) \vee \\ \Phi[0 < \#bf, \{bf \mapsto tail(bf)\}, \langle out.head(bf)\rangle] \end{pmatrix}$$

The $\textbf{true}_r$ precondition implies no divergence. The pericondition states that every input event is enabled, and the output event is enabled if the buffer is non-empty. The postcondition contains two possible final observations: (1) an input event occurred and the buffer variable was extended; or (2) provided the buffer was non-empty initially, then the buffer's head is output and *bf* is contracted.

Proof. To exemplify, we calculate the left-hand side of the choice, employing Theorems 2.6, 4.2, 4.3, and 5.2. The entire calculation is automated in Isabelle/UTP.

$$\begin{aligned}
& inp?v \to bf := bf \frown \langle v\rangle \\
&= \Box\, v \in \mathbb{N} \bullet \textbf{Do}(inp.v) \,;\, bf := bf \frown \langle v\rangle && \text{[Definitions]} \\
&= \Box\, v \in \mathbb{N} \bullet \begin{pmatrix} [\,\textbf{true}_r \vdash \mathcal{E}[true, \langle\rangle, \{inp.v\}] \mid \Phi[true, id, \langle inp.v\rangle]\,] \,; \\ [\,\textbf{true}_r \vdash \textbf{false} \mid \Phi[true, \llparenthesis bf \mapsto bf \frown \langle v\rangle \rrparenthesis, \langle\rangle]\,] \end{pmatrix} && \text{[4.2]} \\
&= \Box\, v \in \mathbb{N} \bullet \left[\, \textbf{true}_r \,\middle|\!\!-\, \begin{matrix} \mathcal{E}[true, \langle\rangle, \{inp.v\}] \\ \vee\ \textbf{false} \end{matrix} \,\middle|\, \begin{matrix} \Phi[true, id, \langle inp.v\rangle] \,; \\ \Phi[true, \llparenthesis bf \mapsto bf \frown \langle v\rangle \rrparenthesis, \langle\rangle] \end{matrix} \right] && \text{[2.6, 4.4]} \\
&= \Box\, v \in \mathbb{N} \bullet [\,\textbf{true}_r \vdash \mathcal{E}[true, \langle\rangle, \{inp.v\}] \mid \Phi[true, \llparenthesis bf \mapsto bf \frown \langle v\rangle \rrparenthesis, \langle inp.v\rangle]\,] && \text{[4.3]} \\
&= \left[\, \textbf{true}_r \,\middle|\!\!-\, \mathcal{E}\left[true, \langle\rangle, \bigcup_{v\in\mathbb{N}} \{inp.v\} \right] \,\middle|\, \bigvee_{v\in\mathbb{N}} \Phi[true, \llparenthesis bf \mapsto bf \frown \langle v\rangle \rrparenthesis, \langle inp.v\rangle] \right] && \text{[5.1, 5.2]}
\end{aligned}$$

Though this calculation seems complicated, in practice it is fully automated and thus a user need not be concerned with these minute calculational details, but can rather focus on finding suitable reactive invariants. □

Then, by Theorem 6.3 we can calculate the overall behaviour of the buffer.

$$Buffer = [\,\textbf{\textsf{true}}_r \vdash \Phi[true, \{bf \mapsto \langle\rangle\}, \langle\rangle] \,;\, B_3^* \,;\, B_2 \mid \textbf{\textsf{false}}\,]$$

This is a non-terminating contract where every quiescent behaviour begins with an empty buffer, performs some sequence of buffer inputs and outputs accompanied by state updates (B_3^*), and is finally offering the relevant input and output events (B_2). We can now employ Theorem 6.4 to verify the buffer. First, we tackle deadlock freedom, which can be proved using the following refinement.

Theorem 7.2 (Deadlock Freedom).

$$\left[\textbf{\textsf{true}}_r \,\middle\vdash\, \bigvee_{s,t,E,e} \mathcal{E}[s, t, \{e\} \cup E] \,\middle|\, \textbf{\textsf{true}}_r\right] \sqsubseteq Buffer$$

Since only quiescent observations can deadlock, we only constrain the pericondition. It characterises observations where at least one event e is being accepted: there is no deadlock. This theorem can be discharged automatically in 1.8 s on an Intel i7-4790 desktop machine. We next tackle the second property.

Theorem 7.3 (Buffer Order Property). *The sequence of items output is a prefix of those that were previously input. This can be formally expressed as*

$$[\,\textbf{\textsf{true}}_r \vdash outps(tt) \le inps(tt) \mid \textbf{\textsf{true}}_r\,] \sqsubseteq Buffer$$

where $inps(t), outps(t) : \mathrm{seq}\,\mathbb{N}$ *extract the sequence of input and output elements from the trace* t*, respectively. The postcondition is left unconstrained as Buffer does not terminate.*

Proof. First, we identify the reactive invariant $I \triangleq outps(tt) \le bf ^\frown inps(tt)$, and show that $[\textbf{\textsf{true}}_r \vdash I \mid \textbf{\textsf{true}}_r] \sqsubseteq \textbf{\textsf{true}}_r \circledast [\textbf{\textsf{true}}_r \vdash B_2 \mid B_3]$. By Theorem 6.4 it suffices to show case (2), that is $I \sqsubseteq B_2$ and $I \sqsubseteq B_3 \,;\, I$, as the other two cases are vacuous. These two properties can be discharged by relational calculus. Second, we prove that $[\textbf{\textsf{true}}_r \vdash outps(tt) \le inps(tt) \mid \textbf{\textsf{true}}_r] \sqsubseteq bf := \langle\rangle \,;\, [\textbf{\textsf{true}}_r \vdash I \mid \textbf{\textsf{true}}_r]$. This holds, by Theorem 4.6-15, since $I[\langle\rangle/bf] = outps(tt) \le inps(tt)$. Thus, the overall theorem holds by monotonicity of $;$ and transitivity of $\sqsubseteq$. The proof is semi-automatic—since we have to manually apply induction with Theorem 6.4—with the individual proof steps taking 2.2 s in total. □

8 Conclusion

We have demonstrated an effective verification strategy for reactive programs employing reactive relations and Kleene algebra. Our theorems and verification tool can be found in our theory repository[2], together with companion proofs.

Related work includes the works of Struth *et al.* on verification of imperative programs [21,22] using Kleene algebra for verification-condition generation,

[2] Isabelle/UTP: https://github.com/isabelle-utp/utp-main.

which our work heavily draws upon. Automated proof support for the failures-divergences model was previously provided by the CSP-Prover tool [25], which can be used to verify infinite-state systems in CSP. Our work is different both in its contractual semantics, and also in our explicit handling of state, which allows us to express variable assignments.

Our work also lies within the "design-by-contract" field [4]. The refinement calculus of reactive systems [26] is a language based on property transformers containing trace information. Like our work, they support reactive systems that are non-deterministic, non-input-receptive, and infinite state. The main differences are our handling of state variables, the basis in relational calculus, and our failures-divergences semantics. Nevertheless, our contract framework [3] can be linked to those results, and we expect to derive an assume-guarantee calculus.

In future work, we will extend our calculation strategy to parallel composition. We aim to apply it to more substantial examples, and are currently using it to build a prototype tactic for verifying robotic controllers [27]. In this direction, our semantics and techniques will be also be extended to cater for real-time, probabilistic, and hybrid computational behaviours [19].

Acknowledgments. This research is funded by the RoboCalc project (https://www.cs.york.ac.uk/circus/RoboCalc/), EPSRC grant EP/M025756/1.

A UTP Theory Definitions

In this appendix, we summarise our theory of reactive design contracts. The definitions are all mechanised in accompanying Isabelle/HOL reports [12,13].

A.1 Observational Variables

We declare two sets $\mathcal{T}$ and Σ that denote the sets of traces and state spaces, respectively, and operators $^\frown : \mathcal{T} \to \mathcal{T} \to \mathcal{T}$ and $\varepsilon : \mathcal{T}$. We require that $(\mathcal{T}, ^\frown, \varepsilon)$ forms a trace algebra [19], which is a form of cancellative monoid. Example models include $(\mathbb{N}, +, 0)$ and $(\text{seq } A, ^\frown, \langle\rangle)$. We declare the following observational variables that are used in both our UTP theories:

$ok, ok' : \mathbb{B}$ – indicate divergence in the prior and present relation;
$wait, wait' : \mathbb{B}$ – indicate quiescence in the prior and present relation;
$\mathsf{st}, \mathsf{st}' : \Sigma$ – the initial and final state;
$tr, tr' : \mathcal{T}$ – the trace of the prior and present relation.

Since the theory is extensible, we also allow further observational variables to be added, which are denoted by the variables r and r'.

A.2 Healthiness Conditions

We first describe the healthiness conditions of reactive relations.

Definition A.1 (Reactive Relation Healthiness Conditions).

$$\begin{aligned} \mathbf{R1}(P) &\triangleq P \wedge tr \le tr' \\ \mathbf{R2}_c(P) &\triangleq P[\varepsilon, tr' - tr/tr, tr'] \lhd tr \le tr' \rhd P \\ \mathbf{RR}(P) &\triangleq \exists (ok, ok', wait, wait') \bullet \mathbf{R1}(\mathbf{R2}_c(P)) \end{aligned}$$

RR healthy relations do not refer to ok or $wait$ and have a well-formed trace associated with them. The latter is ensured by the reactive process healthiness conditions [6,9,19], **R1** and $\mathbf{R2}_c$, which justify the existence of the trace pseudo variable $tt \triangleq tr' - tr$. **RR** is closed under relational calculus operators ***false***, $\vee$, $\wedge$, and $;$, but not ***true***, $\neg$, $\Rightarrow$, or $I\!I$. We therefore define healthy versions below.

Definition A.2 (Reactive Relation Operators).

$$\begin{aligned} \mathbf{true}_r &\triangleq \mathbf{R1}(true) \\ \neg_r P &\triangleq \mathbf{R1}(\neg P) \\ P \Rightarrow_r Q &\triangleq \neg_r P \vee Q \end{aligned} \qquad \begin{aligned} P \ \mathbf{wp}_r \ Q &\triangleq \neg_r (P \mathbin{;} (\neg_r Q)) \\ I\!I_r &\triangleq (tr' = tr \wedge st' = st \wedge r' = r) \end{aligned}$$

We define a reactive complement $\neg_r P$, reactive implication $P \Rightarrow_r Q$, and reactive true $\mathbf{true}_r$, which with the other connectives give rise to a Boolean algebra [3]. We also define the reactive skip $I\!I_r$, which is the unit of $;$, and the reactive weakest precondition operator $\mathbf{wp}_r$. The latter is similar to the standard UTP definition of weakest precondition [6], but uses the reactive complement.

We next define the healthiness conditions of reactive contracts.

Definition A.3 (Reactive Designs Healthiness Conditions).

$$\begin{aligned} \mathbf{R3}_h(P) &\triangleq I\!I_R \lhd wait \rhd P \\ \mathbf{RD1}(P) &\triangleq ok \Rightarrow_r P \\ \mathbf{RD2}(P) &\triangleq P \mathbin{;} \mathbf{J} \\ \mathbf{RD3}(P) &\triangleq P \mathbin{;} I\!I_R \end{aligned} \qquad \begin{aligned} I\!I_R &\triangleq \mathbf{RD1}((\exists st \bullet I\!I_r) \lhd wait \rhd I\!I_r) \\ \mathbf{R}_s &\triangleq \mathbf{R1} \circ \mathbf{R2}_c \circ \mathbf{R3}_h \\ \mathbf{SRD}(P) &\triangleq \mathbf{RD1} \circ \mathbf{RD2} \circ \mathbf{R}_s \\ \mathbf{NSRD}(P) &\triangleq \mathbf{RD1} \circ \mathbf{RD3} \circ \mathbf{R}_s \end{aligned}$$

$\mathbf{R3}_h$ states that if the predecessor is waiting then a reactive design behaves like $I\!I_R$, the reactive design identity. **RD1** is analagous to **H1** from the theory of designs [6,9], and introduces divergent behaviour: if the predecessor is divergent ($\neg ok$), then a reactive design behaves like $\mathbf{true}_r$ meaning that the only observation is that the trace is extended. **RD2** is identical to **H2** from the theory of designs [6,9]. **RD3** states that $I\!I_R$ is a right unit of sequential composition. $\mathbf{R}_s$ composes the reactive healthiness conditions and $\mathbf{R3}_h$. We then finally have the healthiness conditions for reactive designs: **SRD** for "stateful reactive designs", and **NSRD** for"normal stateful reactive designs".

Next we define the reactive contract operator.

Definition A.4 (Reactive Contracts).

$$\begin{aligned} P \vdash Q &\triangleq (ok \wedge P) \Rightarrow (ok' \wedge Q) \\ P \diamond Q &\triangleq P \lhd wait' \rhd Q \\ [P \vdash Q \mid R] &\triangleq \mathbf{R}_s(P \vdash Q \diamond R) \end{aligned}$$

A reactive contract is a "reactive design" [7,9]. We construct a UTP design [6] using the design turnstile operator, $P \vdash Q$, and then apply $\mathbf{R}_s$ to the resulting construction. The postcondition of the underlying design is split into two cases for $wait'$ and $\neg wait'$, which indicate whether the observation is quiescent, and correspond to the peri- or postcondition.

Finally, we define the healthiness conditions that specialise our theory to stateful-failure reactive designs.

Definition A.5 (Stateful-Failure Healthiness Conditions).

$$\begin{aligned} \mathbf{Skip} &\triangleq [\, \mathbf{true}_r \vdash \mathbf{false} \mid tt = \langle\rangle \wedge st' = st] \\ \mathbf{CSP3}(P) &\triangleq \mathbf{Skip} \,;\, P \\ \mathbf{CSP4}(P) &\triangleq P \,;\, \mathbf{Skip} \\ \mathbf{NCSP}(P) &\triangleq \mathbf{NSRD} \circ \mathbf{CSP3} \circ \mathbf{CSP4} \end{aligned}$$

$\mathbf{Skip}$ is similar to $I\!I_R$, but does not refer to ref in the postcondition. If P is $\mathbf{CSP3}$ healthy then it cannot refer to ref. If P is $\mathbf{CSP4}$ healthy then the postcondition cannot refer to ref', but the pericondition can: refusals are only observable when P is quiescent [9,18].

References

1. Harel, D., Pnueli, A.: On the development of reactive systems. In: Apt, K.R. (ed.) Logics and Models of Concurrent Systems. NATO ASI Series (Series F: Computer and Systems Sciences), vol. 13, pp. 477–498. Springer, Heidelberg (1985). https://doi.org/10.1007/978-3-642-82453-1_17
2. Bainomugisha, E., Carreton, A.L., Cutsem, T.V., Mostinckx, S., De Meuter, W.: A survey on reactive programming. ACM Comput. Surv. **45**(4), 34 pages (2013). Article No. 52
3. Foster, S., Cavalcanti, A., Canham, S., Woodcock, J., Zeyda, F.: Unifying theories of reactive design contracts. Submitted to Theoretical Computer Science, December 2017. Preprint: https://arxiv.org/abs/1712.10233
4. Meyer, B.: Applying "design by contract". IEEE Comput. **25**(10), 40–51 (1992)
5. Hehner, E.C.R.: A Practical Theory of Programming. Monographs in Computer Science. Springer, New York (1993). https://doi.org/10.1007/978-1-4419-8596-5
6. Hoare, C.A.R., He, J.: Unifying Theories of Programming. Prentice-Hall, Upper Saddle River (1998)

7. Oliveira, M., Cavalcanti, A., Woodcock, J.: A UTP semantics for Circus. Form. Asp. Comput. **21**, 3–32 (2009)
8. Kozen, D.: On Kleene algebras and closed semirings. In: Rovan, B. (ed.) MFCS 1990. LNCS, vol. 452, pp. 26–47. Springer, Heidelberg (1990). https://doi.org/10.1007/BFb0029594
9. Cavalcanti, A., Woodcock, J.: A tutorial introduction to CSP in *Unifying Theories of Programming*. In: Cavalcanti, A., Sampaio, A., Woodcock, J. (eds.) PSSE 2004. LNCS, vol. 3167, pp. 220–268. Springer, Heidelberg (2006). https://doi.org/10.1007/11889229_6
10. Foster, S., Zeyda, F., Woodcock, J.: Unifying heterogeneous state-spaces with lenses. In: Sampaio, A., Wang, F. (eds.) ICTAC 2016. LNCS, vol. 9965, pp. 295–314. Springer, Cham (2016). https://doi.org/10.1007/978-3-319-46750-4_17
11. Foster, S.: Kleene algebra in Unifying Theories of Programming. Technical report, University of York (2018). http://eprints.whiterose.ac.uk/129359/
12. Foster, S., et al.: Reactive designs in Isabelle/UTP. Technical report, University of York (2018). http://eprints.whiterose.ac.uk/129386/
13. Foster, S., et al.: Stateful-failure reactive designs in Isabelle/UTP. Technical report, University of York (2018). http://eprints.whiterose.ac.uk/129768/
14. Guttman, W., Möller, B.: Normal design algebra. J. Log. Algebr. Program. **79**(2), 144–173 (2010)
15. Möller, B., Höfner, P., Struth, G.: Quantales and temporal logics. In: Johnson, M., Vene, V. (eds.) AMAST 2006. LNCS, vol. 4019, pp. 263–277. Springer, Heidelberg (2006). https://doi.org/10.1007/11784180_21
16. Santos, T., Cavalcanti, A., Sampaio, A.: Object-orientation in the UTP. In: Dunne, S., Stoddart, B. (eds.) UTP 2006. LNCS, vol. 4010, pp. 18–37. Springer, Heidelberg (2006). https://doi.org/10.1007/11768173_2
17. Sherif, A., Cavalcanti, A., He, J., Sampaio, A.: A process algebraic framework for specification and validation of real-time systems. Form. Asp. Comput. **22**(2), 153–191 (2010)
18. Hoare, C.A.R.: Communicating Sequential Processes. Prentice-Hall, Upper Saddle River (1985)
19. Foster, S., Cavalcanti, A., Woodcock, J., Zeyda, F.: Unifying theories of time with generalised reactive processes. Inf. Process. Lett. **135**, 47–52 (2018)
20. Dijkstra, E.W.: Guarded commands, nondeterminacy and formal derivation of programs. Commun. ACM **18**(8), 453–457 (1975)
21. Armstrong, A., Gomes, V., Struth, G.: Building program construction and verification tools from algebraic principles. Form. Asp. Comput. **28**(2), 265–293 (2015)
22. Gomes, V.B.F., Struth, G.: Modal Kleene algebra applied to program correctness. In: Fitzgerald, J., Heitmeyer, C., Gnesi, S., Philippou, A. (eds.) FM 2016. LNCS, vol. 9995, pp. 310–325. Springer, Cham (2016). https://doi.org/10.1007/978-3-319-48989-6_19
23. Zhan, N., Kang, E.Y., Liu, Z.: Component publications and compositions. In: Butterfield, A. (ed.) UTP 2008. LNCS, vol. 5713, pp. 238–257. Springer, Heidelberg (2010). https://doi.org/10.1007/978-3-642-14521-6_14
24. Back, R.-J., Wright, J.: Refinement Calculus: A Systematic Introduction. Texts in Computer Science, 1st edn. Springer, New York (1998). https://doi.org/10.1007/978-1-4612-1674-2
25. Isobe, Y., Roggenbach, M.: CSP-Prover: a proof tool for the verification of scalable concurrent systems. J. Comput. Softw. Jpn. Soc. Softw. Sci. Technol. **25**(4), 85–92 (2008)

26. Preoteasa, V., Dragomir, I., Tripakis, S.: Refinement calculus of reactive systems. In: International Conference on Embedded Systems (EMSOFT). IEEE, October 2014
27. Miyazawa, A., Ribieiro, P., Li, W., Cavalcanti, A., Timmis, J.: Automatic property checking of robotic applications. In: International Conference on Intelligent Robots and Systems (IROS), pp. 3869–3876. IEEE (2017)

Verifying Hybrid Systems with Modal Kleene Algebra

Jonathan Julián Huerta y Munive(✉) and Georg Struth

Department of Computer Science, University of Sheffield, Sheffield, UK
{jjhuertaymunive1,g.struth}@sheffield.ac.uk

Abstract. Modal Kleene algebras have been used for building verification components for imperative programs with Isabelle/HOL. We integrate this approach with recent Isabelle components for ordinary differential equations to build two verification components for hybrid systems, where continuous phase space dynamics complement the discrete dynamics on program stores. We demonstrate their feasibility by deriving the most important domain-specific rules of differential dynamic logic and discussing four simple examples.

1 Introduction

Hybrid systems integrate continuous dynamics and discrete control. Their verification is increasingly important, yet notoriously difficult: applications range from the control of chemical plants, finance and traffic systems to the coordination of autonomous vehicles or robots, the optimisation of mechanical systems and biosystems engineering [1,29]. Mathematically, hybrid systems verification requires integrating dynamical systems, often modelled by ordinary differential equations (ODEs), into formalisms for their description and analysis, such as automata [15] or modal logics [9,28]. A prominent approach is differential dynamic logic $\mathsf{d}\mathcal{L}$ [28], an extension of dynamic logic [14] to hybrid programs with domain-specific inference rules for ODEs and their solutions. It is well supported by the KeYmaera X tool [10] and has already proved its worth in numerous case studies [21,24,28].

Mathematical components for dynamical systems have recently been formalised in proof assistants such as Isabelle/HOL [25]. An impressive theory stack—from topology and measure theory to analysis [17] and ODEs [18–20]—is ready for use. In real-word software verification, such proof assistants play an important role already [6,22,23], and verification components based on modal Kleene algebra (MKA) [7,11–13], which subsumes dynamic logic, exist in Isabelle as well. Yet an integration of the two theory stacks for $\mathsf{d}\mathcal{L}$-style hybrid program verification is missing.

Such an integration seems natural and interesting for several reasons. Firstly, the resulting components for $\mathsf{d}\mathcal{L}$-style hybrid program verification would be correct by construction relative to the proof assistant's small trustworthy core. The existence and uniqueness of solutions to ODEs, for instance, which is required

J. Desharnais et al. (Eds.): RAMiCS 2018, LNCS 11194, pp. 225–243, 2018.
https://doi.org/10.1007/978-3-030-02149-8_14

in the d$\mathcal{L}$ semantics, could be formally certified, in particular when these are computed by external tools. Secondly, such components could be engineered and used in open modular ways; limited only by the expressivity of the proof assistant's logic, the versatility of its mathematical components and the power of its provers and solvers. Specific components for particular classes of ODEs or decidable assertion languages, for instance for real-closed fields, could be developed and verified within the general framework. Finally, the integration would benefit from active communities that are driving the development of d$\mathcal{L}$ and a wide range of mathematical components forward.

Our main contribution is the proof of concept for this integration: a formal reconstruction of d$\mathcal{L}$ within MKA and Isabelle/HOL, resulting in prototypical d$\mathcal{L}$-based verification components for this general purpose proof assistant.

More detailed technical contributions are: (i) a relational flow model of MKA, suitable for hybrid systems, and its formalisation in Isabelle; (ii) a predicate transformer semantics for hybrid programs using this model, and a simple generic d$\mathcal{L}$-style Isabelle verification component based on equational reasoning with weakest liberal preconditions; (iii) an alternative MKA-based verification component using d$\mathcal{L}$-style differential invariants instead of flows of ODEs; (iv) the derivation of domain-specific d$\mathcal{L}$-style inference rules for these components within the relational flow semantics.

This is our approach in a nutshell: MKA provides variants of the laws of propositional dynamic logic for reasoning equationally about while programs without assignments, hence an algebraic predicate transformer semantics. Extant MKA-based verification components [12] capture the program store dynamics of assignments within the relational program store model of MKA. In Isabelle, a modular instantiation of MKA to this concrete semantics is supported by type polymorphism. Hence one can simply replace such discrete program stores with dynamical ones for hybrid programs; essentially with phase spaces. One can use evolution statements, which describe vector fields or systems of ODEs, in addition to assignment statements to model hybrid programs. The relational hybrid program semantics then uses flows (solutions of ODEs, provided they exist) for "updating" phase spaces together with update functions for assignments. This sets up an algebraic variant of d$\mathcal{L}$ as an instance of MKA.

Our first Isabelle verification component asks users—and ultimately external solvers—to supply flows for ODEs and certify them internally. MKA and Isabelle's ODE components are then used for manipulating predicate transformers over phase spaces and verifying correctness specifications for hybrid programs by equational reasoning. Alternatively to manipulating explicit solutions, users can provide invariants for ODEs in verification proofs, as with d$\mathcal{L}$, in our second verification component. Both are shown at work on a series of simple examples.

Our Isabelle components form a main contribution of this paper. They confirm the feasibility of the algebraic approach to d$\mathcal{L}$. The main statements in this article have been formalised (cf. Appendix C). The complete formalisation, including a PDF proof document, is available online[1].

[1] https://github.com/yonoteam/CPSVerification.

2 Modal Kleene Algebra

A *Kleene algebra* is a structure $(K, +, \cdot, 0, 1, ^*)$ such that $(K, +, \cdot, 0, 1)$ is a *dioid*, a semiring in which $+$ is idempotent. The Kleene star $^* : K \to K$ satisfies the unfold and induction axioms

$$1 + \alpha \cdot \alpha^* \leq \alpha^*, \qquad \gamma + \alpha \cdot \beta \leq \beta \to \alpha^* \cdot \gamma \leq \beta$$

and their opposites, where the order of multiplication is swapped. These axioms use the semilattice order, which is given by $\alpha \leq \beta \leftrightarrow \alpha + \beta = \beta$ on any dioid. The class of Kleene algebras is closed under opposition. The complete axioms for the algebras discussed in this article are listed in Appendix A.

Elements of K can represent programs: $+$ models nondeterministic choice, $\cdot$ sequential composition and * finite iteration; 0 models the abortive and 1 the ineffective program. Formally, programs are interpreted as binary relations over a state space S. In fact, the set $\mathcal{P}(S \times S)$ of all binary relations over S forms a *relation Kleene algebra* under $\cup$, relational composition ;, the empty relation $\emptyset$, the identity relation Id_S and the reflexive-transitive closure operation *. More realistic program semantics require modelling tests and assertions as well.

A *modal Kleene algebra* (MKA) [7] is a Kleene algebra K expanded by an *antidomain operation* $ad : K \to K$ and an *antirange operation* $ar : K \to K$. Antidomain is axiomatised by

$$ad\,\alpha \cdot \alpha = 0, \qquad ad\,\alpha + ad^2\,\alpha = 1, \qquad ad\,(\alpha \cdot \beta) \leq ad\,(\alpha \cdot ad^2\,\beta).$$

Antirange is antidomain in the opposite Kleene algebra. The operations interact via the axioms $ad^2 \circ ar^2 = ar^2$ and $ar^2 \circ ad^2 = ad^2$.

Intuitively, $ad\,\alpha$ models those states from which program α cannot be executed; the domain $d\,\alpha = ad^2\,\alpha$ of α models those states from which it can be. Dually, $ar\,\alpha$ models those states in which α cannot terminate; the range $r\,\alpha = ar^2\,\alpha$ models those states in which it can. The set $K_{ad} = \{\alpha \in K \mid ad\,\alpha = \alpha\}$ forms a boolean subalgebra of K bounded by 0 and 1. In K_{ad}, which is the same as K_{ar}, $+$ corresponds to join and $\cdot$ to meet; ad and ar correspond to complementation. Elements $p, q, r \in K_{ad}$ can therefore model tests and assertions. In particular, $p \cdot \alpha$ and $\alpha \cdot p$ model the domain and range restriction of α to states satisfying p. In relation MKAs, $ad\,R = \{(s, s) \mid \neg\exists s'.\ (s, s') \in R\}$. Hence $ad\,R \subseteq Id_S$ for all R. The class MKA is once again closed under opposition.

3 Modal Kleene Algebra and Dynamic Logic

On the one hand, MKA yields an algebraic semantics for simple while languages:

$$\alpha; \beta = \alpha \cdot \beta, \qquad \textbf{if } p \textbf{ then } \alpha \textbf{ else } \beta = p \cdot \alpha + \bar{p} \cdot \beta, \qquad \textbf{while } p \textbf{ do } \alpha = (p \cdot \alpha)^* \cdot \bar{p},$$

writing $\bar{p}$ instead of $ad\,p$. On the other hand, the forward modal operators of propositional dynamic logic (PDL) can be defined as

$$|\alpha\rangle p = d\,(\alpha \cdot p), \qquad |\alpha]p = ad\,(\alpha \cdot ad\,p);$$

backward ones are obtained by opposition: $\langle\alpha|p = r\,(\alpha{\cdot}p)$, and $[\alpha|p = ar\,(\alpha{\cdot}ar\,p)$. This is consistent with the relational (Kripke) semantics of PDL [7,12], where $|\alpha\rangle p$ models those states from which executing α may lead into states where p holds; and $|\alpha]p$ describes those states from which α must lead to states satisfying p. In particular, algebraic variants of the PDL axioms are derivable as theorems of MKA, so that the latter can be seen as an algebraic relative of the former.

It is well known from PDL that $|\alpha] : K_{ad} \to K_{ad}$ is a backward conjunctive predicate transformer on K_{ad}. In fact, $|\alpha]q$ models the weakest liberal precondition (wlp) of program α and postcondition q. The other modalities yield dual transformers that are less interesting for this article.

Predicate transformers are useful for specifying program correctness conditions and for verification condition generation. The identity

$$p \le |\alpha]q$$

encodes the standard partial correctness specification for programs: if program α is executed from states where precondition p holds, and if it terminates, then postcondition q holds in the states where it does. One can thus calculate $|\alpha]q$ backwards over the program structure from q and check that the result is above p. Calculating $|\alpha]q$ for straight-line programs is completely equational; yet loops require invariants. To this end we use while loops with annotations: **while** p **inv** i **do** α = **while** p **do** α and calculate wlps recursively over the program structure as follows [12]. For all $p, q, i, t \in K_{ad}$ and $x, y \in K$,

$$|\alpha;\beta]q = |\alpha]|\beta]q,$$
$$|\mathbf{if}\ p\ \mathbf{then}\ \alpha\ \mathbf{else}\ \beta]q = (\bar{p} + |\alpha]q)(p + |\beta]q),$$
$$p \le i \wedge i\bar{t} \le q \wedge it \le |\alpha]i \to p \le |\mathbf{while}\ t\ \mathbf{inv}\ i\ \mathbf{do}\ \alpha]q.$$

As binary relations form MKAs, the approach is consistent with the relational program semantics and the predicate transformer semantics obtained from it.

4 Integrating Discrete Program State Dynamics

Two important ingredients for program semantics and verification are still missing: a concrete relational semantics of the program store and program assignments, and rules for calculating wlps for these basic commands. To prepare for hybrid programs (see Appendix A for a syntax) we discuss stores and assignments in terms of discrete dynamical systems over state spaces.

Formally, a *dynamical system* [3,30] is an *action* of a monoid $(M, \star, e)$ on a set or state space S, that is, a monoid morphism $\varphi : M \to S \to S$ into the *transformation monoid* $(S^S, \circ, id_S)$ on the function space S^S. Thus $\varphi\,(m{\star}n) = (\varphi\,m)\circ(\varphi\,n)$ and $\varphi\,e = id_S$. The first action axiom captures the inherent determinism of dynamical systems. Conversely, each transformation monoid $(S^S, \circ, id_S)$ determines a monoid action in which the action $\varphi : S^S \to S \to S$ is function application.

States of simple while programs are functions $s : V \to E$ from program variables in V to values in E. State spaces for such discrete dynamical systems are function spaces E^V. An update function $f_a : V \to (E^V \to E) \to E^V \to E^V$ for assignment commands can be defined as $f_a\, v\, e\, s = s[v \mapsto (e\, s)]$, where $f[a \mapsto b]$ updates function $f : A \to B$ in argument $a \in A$ by value $b \in B$ ($f[a \mapsto b]\, x$ is b if $x = a$ and $f\, x$ otherwise). The "expression" $e : E^V \to E$ is evaluated in state s to $e\, s$. Yet with a shallow embedding of data domains in Isabelle, e is, by its type, a function. The functions $f_a\, v\, e$ generate a transformation monoid, hence a monoid action of type $(E^V)^{(E^V)} \to E^V \to E^V$ on $(E^V)^{(E^V)}$. They also connect the concrete program state semantics with the wlp semantics used for verification condition generation.

The relational semantics of assignment statements can hence be defined as

$$(v := e) = \{(s, f_a\, v\, e\, s) \mid s \in E^V\}, \tag{1}$$

and the wlp for assignments be derived in the relational program semantics as

$$|v := e]\lceil Q\rceil = \lceil \lambda s.\ Q(s[v \mapsto (e\, s)])\rceil,$$

where $\lceil - \rceil$ embeds predicates into relations: $\lceil P\rceil = \{(s, s) \mid P\, s\}$.

Adding this wlp law for assignments to those for the program structure suffices for generating data-level verification conditions for while programs. The development also yields a hybrid encoding of dynamic logic, where the propositional part is captured algebraically and the rest in set theory.

We have already implemented verification components based on MKA in Isabelle/HOL [12,13]. Apart from the uses outlined, they also support verification based on Hoare logic, verification of simple recursive programs, verification in the context of total correctness, verification by symbolic execution, as well as program construction, transformation and refinement.

5 Integrating Continuous Program State Dynamics

Hybrid programs require continuous dynamics as well as discrete ones. Actions φ are now of type $\mathbb{R} \to S \to S$. They are usually flows of systems of differential equations [3,30] over a phase space S. In general, for each $s \in S$, the functions $\varphi_s : \mathbb{R} \to S$ given by $\varphi_s = \lambda t.\ \varphi\, t\, s$ are maximal smooth integral curves generated by vector fields f that assign tangent vectors to points of some manifold S, varying smoothly from point to point in t, and starting in (or passing through) point s. The tangent vector of $\varphi_s\, t$ at any point in S must then coincide with the value of f at that point. The set $\gamma\, s = \{\varphi\, t\, s \mid t \geq 0\}$ is the (positive) *orbit* of φ through s. We write γ^φ to indicate the dependency of orbits on flows.

Example 5.1 (Vector field of fluid). Velocity vectors describe the motion of a fluid in an (open) subset of $\mathbb{R}^3$. They form a vector field that associates a velocity with each particle at each point of space. The integral curves are the trajectories, orbits or evolutions of fluid particles through time, and $\varphi_s\, t$ describes the movement of a particle from initial position $s \in \mathbb{R}^3$ at time $t = 0$. □

Typical phase spaces for continuous dynamical systems are open subsets of $\mathbb{R}^n$. System models are typically differential equations. Our approach is currently limited to autonomous systems of ordinary differential equations (ODEs) [30]

$$x'\,t = f\,(x\,t),$$

where $x : \mathbb{R} \to X$, for any open $X \subseteq \mathbb{R}^n$, and Lipschitz continuous vector fields $f \in C^k(X, \mathbb{R}^n)$ for some $k \geq 1$. The Picard-Lindelöf Theorem guarantees that such systems have unique solutions for initial value problems $x'\,t = f\,(x\,t)$, $x\,t_0 = x_0$ within certain closed intervals $I(x_0)$ [30]. These solutions are integral curves $\varphi_{x_0}\,t$. We currently allow only functions for which $I(x_0) = \mathbb{R}$. The general case is left for future work. It is already supported by Isabelle.

Hybrid programs, by contrast, require phase spaces $\mathbb{R}^V$, where V is a countable set of variables [28]. This leads to some subtleties in our model, which are due to Isabelle's strongly typed setting. As hybrid programs change only finitely many variables, there is always a subspace of $\mathbb{R}^V$ isomorphic to some $\mathbb{R}^n$ on which the program acts and where existence and uniqueness theorems apply, and an orthogonal subspace, where nothing changes.

We are now ready to define an analogue of the assignment semantics from Sect. 4 for hybrid programs and derive an equation for the corresponding wlp operator. First we replace assignment statements in the definiendum of (1) by *evolution statements* for vector fields of the form

$$x_1' = f_1, \ldots, x_n' = f_n,$$

or, more briefly, $x' = f$, where $x_i' = f_i$ is formally a pair $(x_i, f_i) : V \times (\mathbb{R}^V \to \mathbb{R})$. These, and the usual assignments, form the basic commands of hybrid programs. By analogy, we wish to replace the definiens of (1) by the flow $\varphi : \mathbb{R} \to \mathbb{R}^V \to \mathbb{R}^V$ for $x' = f$. Yet this requires checking its existence and uniqueness: a functional dependency of φ on f and x. For $X = \{x_1, \ldots, x_n\}$, we define the predicates

$$\begin{aligned}
\mathsf{Act}_1\,\varphi\,s &\Leftrightarrow \forall t_1, t_2 \geq 0.\ \varphi\,(t_1 + t_2)\,s = \varphi\,t_1\,(\varphi\,t_2\,s),\\
\mathsf{Act}_2\,\varphi\,s &\Leftrightarrow \lambda x \in V.\ \varphi\,0\,s\,x = s\,x,\\
\mathsf{Diff}\,\varphi\,x\,f\,s &\Leftrightarrow \forall t \geq 0, 1 \leq i \leq n.\ (\lambda \tau.\, \varphi\,\tau\,s\,x_i)'\,t = f_i\,(\varphi\,t\,s),\\
\mathsf{Id}\,\varphi\,s &\Leftrightarrow \forall t \geq 0, y \in V \setminus X.\ \varphi\,t\,s\,y = s\,y,\\
\mathsf{Flow}_{\exists}\,\varphi\,x\,f\,s &\Leftrightarrow \mathsf{Act}_1\,\varphi\,s \wedge \mathsf{Act}_2\,\varphi\,s \wedge \mathsf{Diff}\,\varphi\,x\,f\,s \wedge \mathsf{Id}\,\varphi\,s,\\
\mathsf{Flow}\,\varphi\,x\,f\,s &\Leftrightarrow \mathsf{Flow}_{\exists}\,\varphi\,x\,f\,s \wedge \forall \psi.\ \mathsf{Flow}_{\exists}\,\psi\,x\,f\,s \to \varphi = \psi.
\end{aligned}$$

Act_1 and Act_2 check that φ is actually a flow; in addition, Act_2 checks that the flow solves the initial value problem for $x' = f$ at point s. Diff checks that the coordinates of the derivative of φ are equal to f_i, hence that the tangent vectors of the flow are equal to the vector field f at each point $\varphi\,s\,t$ along an integral curve. These three conditions are standard for dynamical systems [3,30]. The fourth condition, Id, is particular to hybrid programs with countably many program variables. It guarantees that all variables that do not occur primed in $x' = f$ remain unchanged during the evolution. Finally, verifying uniqueness in practice requires checking that f satisfies conditions such as those in Picard-Lindelöf's theorem. In fact, the resulting uniqueness proof subsumes Act_1.

Next we define the relational flow semantics of evolution statements as

$$(\forall s \in \mathbb{R}^V.\ \mathsf{Flow}\,\varphi\, x\, f\, s) \rightarrow (x' = f) = \{(s, \varphi\, t\, s) \mid s \in \mathbb{R}^V \wedge t \geq 0\}, \qquad (2)$$

by analogy with (1). Each initial state s is thus related to its orbit $\gamma^\varphi\, s$. The wlp for evolution statements can now be calculated in this semantics.

Proposition 5.2. *Let* $\varphi : \mathbb{R} \rightarrow \mathbb{R}^V \rightarrow \mathbb{R}^V$, $Q : \mathbb{R}^V \rightarrow \mathbb{B}$, $f_i : \mathbb{R}^V \rightarrow \mathbb{R}$ *and* $x_i \in V$. *Then*

$$(\forall s \in \mathbb{R}^V.\ \mathsf{Flow}\,\varphi\, x\, f\, s) \ \rightarrow\ |x' = f]\lceil Q\rceil = \lceil \lambda s.\ \forall t \geq 0.\ Q\,(\varphi\, t\, s)\rceil.$$

Alternatively, we can write $\lceil \lambda s.\ \forall t \geq 0.\ Q\, s[x_1 \mapsto \varphi_s\, t\, x_1, \ldots, x_n \mapsto \varphi_s\, t\, x_n]\rceil$ for the right-hand side of the wlp-identity, where $f[a_1 \mapsto b_1, \ldots, a_m \mapsto b_m]$ indicates simultaneous function update. By definition, and consistently with $\mathsf{d}\mathcal{L}$, postconditions thus hold along the orbit of each particular state s, including s itself. Proposition 5.2 implies a generic inference rule for proving correctness specifications of evolution statements in the style of Hoare logic.

Lemma 5.3. *For* $\varphi : \mathbb{R} \rightarrow \mathbb{R}^V \rightarrow \mathbb{R}^V$, $P, Q : \mathbb{R}^V \rightarrow \mathbb{B}$, $f_i : \mathbb{R}^V \rightarrow \mathbb{R}$ *and* $x_i \in V$,

$$\frac{\forall s.\ \mathsf{Flow}\,\varphi\, x\, f\, s \qquad \forall s.\ P\, s \rightarrow \forall t \geq 0.\ Q\,(\varphi_s\, t)}{\lceil P\rceil \subseteq |x' = f]\lceil Q\rceil}\ \mathit{solve}$$

This rule can be used freely with the Hoare logic within MKA for hybrid program verification in the relational flow semantics. Proposition 5.2, by contrast, augments the predicate transformer laws from Sect. 3. The set-up of these rules requires users to supply flows for evolution statements and then certify that these actually solve the corresponding ODEs. See Sect. 9 for examples.

Example 5.4 (Motion with constant velocity). The evolution statement $x' = \lambda\, s.\ s\, v$ represents the ODE $x' = v$, where v is a constant, and notation is overloaded. For each initial value x_0, the flow $\varphi_{x_0}\, t = v \cdot t + x_0$ solves this ODE for all $t \geq 0$ in the phase space $\mathbb{R}$. With hybrid programs, we consider the phase space $\mathbb{R}^V$, where only the variable x changes. The relational flow semantics for evolution statements therefore requires, for all $t \geq 0$ and all $s \in \mathbb{R}^V$, the flow

$$\varphi_s\, t\, y = \begin{cases} (s\, v) \cdot t + (s\, x), & \text{if } y = x, \\ s\, y, & \text{if } y \in V \setminus \{x\}. \end{cases}$$

For postcondition $\lambda s.\ s\, x > 1$ we can then calculate

$$\begin{aligned} |x' = \lambda s.\ s\, v]\lceil \lambda s.\ s\, x > 1\rceil &= \lceil \lambda s.\ \forall t \geq 0.\ (\lambda s.\ s\, x > 1)\, s[x \mapsto (s\, v) \cdot t + (s\, x)]\rceil \\ &= \lceil \lambda s.\ \forall t \geq 0.\ (s\, v) \cdot t + (s\, x) > 1\rceil. \end{aligned}$$

Hence the wlp holds precisely of those states $s \in \mathbb{R}^V$ where $(s\, v) \cdot t + (s\, x) > 1$ for all $t \geq 0$. An Isabelle verification is presented in Sect. 9. □

6 Guarded Evolutions

Applications of dynamical systems and hybrid programs often depend on boundary conditions and similar constraints. These are called *guards* in d$\mathcal{L}$, and they are restricted to first-order formulas of real arithmetic. A flow is supposed to satisfy a guard during an entire evolution, just like a postcondition. With Isabelle, we support arbitrary higher-order formulas or predicates as guards.

A *guarded evolution statement* [28] with guard G has the form

$$x_1' = f_1, \ldots, x_n' = f_n \,\&\, G, \qquad \text{or more briefly,} \quad x' = f \,\&\, G.$$

Its relational flow semantics is captured by the flow predicate

$$\mathsf{Flow}_G\, \varphi\, x\, f\, G\, s \Leftrightarrow \mathsf{Flow}\, \varphi\, x\, f\, s \wedge \forall t \geq 0.\ G\,(\varphi\, t\, s).$$

and (2) with Flow_G in place of Flow. We can proceed like in Sect. 5.

Proposition 6.1. *Let* $\varphi : \mathbb{R} \to \mathbb{R}^V \to \mathbb{R}^V$, $G, Q : \mathbb{R}^V \to \mathbb{B}$, $f_i : \mathbb{R}^V \to \mathbb{R}$ *and* $x_i \in V$. *Then*

$$(\forall s \in \mathbb{R}^V.\mathsf{Flow}_G\, \varphi\, x\, f\, G\, s) \to |x' = f \,\&\, G]\lceil Q \rceil = \lceil \lambda s. \forall t \geq 0. G\,(\varphi_s\, t) \to Q\,(\varphi_s\, t)\rceil.$$

Lemma 6.2. *Let* $\varphi : \mathbb{R} \to \mathbb{R}^V \to \mathbb{R}^V$, $P, G, Q : \mathbb{R}^V \to \mathbb{B}$, $f_i : \mathbb{R}^V \to \mathbb{R}$ *and* $x_i \in V$, *then*

$$\frac{\forall s.\ \mathsf{Flow}_G\, \varphi_s\, x\, f\, G\, s \qquad \forall s \forall t \geq 0.\ P\, s \wedge G\,(\varphi_s\, t) \to Q\,(\varphi_s\, t)}{\lceil P \rceil \subseteq |x' = f \,\&\, G]\lceil Q \rceil}\ solve_G$$

Finally, we can derive another d$\mathcal{L}$-style inference rule for guards.

Lemma 6.3. *The* differential weakening *rule is derivable for* $P, G, Q : \mathbb{R}^V \to \mathbb{B}$, $f_i : \mathbb{R}^V \to \mathbb{R}$ *and* $x_i \in V$.

$$\frac{\lceil G \rceil \subseteq \lceil Q \rceil}{\lceil P \rceil \subseteq |x' = f \,\&\, G]\lceil Q \rceil}\ dW$$

Soundness of (dW) is rather trivial: $|x' = f \,\&\, G]\lceil G \rceil = Id$ by evaluating the wlp. Thus $\lceil P \rceil \subseteq Id = |x' = f \,\&\, G]\lceil G \rceil \subseteq |x' = f \,\&\, G]\lceil Q \rceil$ by isotonicity of boxes in MKA and the assumption $\lceil G \rceil \subseteq \lceil Q \rceil$.

An example verification of a guarded evolution is presented in Sect. 9.

7 Differential Invariants

In the theory of dynamical systems, an *invariant set* for a flow φ is a set $X \subseteq \mathbb{R}^n$ such that $\gamma^\varphi\, x \subseteq X$ for all $x \in X$ [30]. Alternatively, such sets can be described as *differential invariants* of the action φ of the Lie group $\mathbb{R}$ on $\mathbb{R}^n$ [26], which is beyond the scope of this article. Intuitively, if an invariant X holds at the initial point x_0 of φ_{x_0}, then it holds at all other points of the orbit through x_0.

In particular, all orbits or unions of orbits are invariants. Invariants can sometimes be used instead of flows to solve differential equations. It is straightforward to transfer this to hybrid program verification. A *differential invariant* [27,28] for $x' = f \,\&\, G$ is a predicate I that satisfies the partial correctness specification

$$\lceil I \rceil \subseteq |x'{=}f \,\&\, G]\lceil I \rceil.$$

If I holds before the evolution, then it remains true during and after it. Due to the intentional limitations of solving differential equations with $\mathsf{d}\mathcal{L}$, differential invariants form an integral part of this logic. The general approach is mathematically involved. We merely illustrate it by a classical example (cf. [27]).

Example 7.1 (Circular movement of particle). The ODEs

$$x' = y, \qquad y' = -x$$

can be solved by linear combinations of trigonometric functions. Alternatively, all orbits are "governed" by the separable differential equation

$$\frac{dy}{dx} = \frac{y'}{x'} = -\frac{x}{y},$$

obtained by parametric derivation. Rewriting it as $ydy + xdx = 0$ and integrating both sides yields $I = x^2 + y^2 = r^2$ for some constant $r > 0$. This invariant describes the circular orbits—the *phase portrait*—of the ODEs. Checking that I is indeed a differential invariant in the sense of $\mathsf{d}\mathcal{L}$ requires reverting this calculation in the specification statement $\lceil I \rceil \subseteq |x' = y, y' = -x]\lceil I \rceil$. It can be rewritten as $\lceil I \rceil \subseteq |x' = y, y' = -x]\lceil 2xx' + 2yy' = 0 \rceil$ by using the well known fact that two continuous differentiable functions differ by a constant if and only if they have the same derivatives, thus putting the above differential equation in place of the invariant. Substituting $(2xx' + 2yy')[y/x', -x/y'] = 2xy - 2yx$ in the postcondition then yields $\lceil I \rceil \subseteq |x' = y, y' = -x]\lceil \lambda s.\ \mathit{True} \rceil$, which holds irrespective of $|x' = y, y' = -x]$. Evaluating this wlp is therefore unnecessary for proving the correctness statement. □

One of the most intriguing features of $\mathsf{d}\mathcal{L}$ is its supports for such calculations in a proof-theoretic setting. This is achieved through four domain-specific inference rules for differential invariants, which may circumvent solving ODEs: a differential invariant, a differential cut, a differential ghost and sometimes a differential effect rule. We derive the first two in our setting (we currently do not see any need for the other ones in our setting).

We extend our language to support substitutions of primed variables, as in Example 7.1. We use the language of *differential rings*: rings R equipped with an additive group morphism $(-)' : R \to R$ satisfying the Leibniz rule $(\theta \cdot \eta)' = \theta' \cdot \eta + \eta \cdot \theta'$. We use constant symbols and primed variables, thus require $c' = 0$. We also extend $\mathbb{R}^V$ to $\mathbb{R}^{V \cup V'}$, such that $V \cap V' = \emptyset$ and $x \in V \Leftrightarrow x' \in V'$.

Within this approach, we can easily extend flows to $\mathbb{R}^{V \cup V'}$ and the flow predicate to $\mathsf{Flow}_{GI}\, \varphi\, x\, f\, G\, s \Leftrightarrow \mathsf{Flow}_G\, \varphi\, x\, f\, G\, s \wedge \mathsf{Prime}_1\, \varphi\, s \wedge \mathsf{Prime}_2\, \varphi\, s$. Here,

$\mathsf{Prime}_1\, \varphi\, s \Leftrightarrow \forall t \geq 0, 1 \leq i \leq n.\, \varphi\, t\, s\, x_i' = f_i\, (\varphi\, t\, s)$ and furthermore $\mathsf{Prime}_2\, \varphi\, s \Leftrightarrow \forall t \geq 0, z \in V \setminus X.\, \varphi\, t\, s\, z' = 0$ impose "solution conditions" for primed variables on φ. The relational flow semantics of evolution statements in (2) can then be extended to $\mathbb{R}^{V \cup V'}$ and precondition Flow_{GI}; wlps for these statements can be calculated in a straightforward extension of Proposition 6.1.

Proposition 7.2. *Let* $\varphi : \mathbb{R} \to \mathbb{R}^{V \cup V'} \to \mathbb{R}^{V \cup V'}$, $G, Q : \mathbb{R}^{V \cup V'} \to \mathbb{B}$, $f_i : \mathbb{R}^{V \cup V'} \to \mathbb{R}$, $x_i \in V$ *and* $z_i \in V \setminus X$. *Then* $\forall s \in \mathbb{R}^V$. $\mathsf{Flow}_G\, \varphi\, x\, f\, G\, s$ *implies*

$$\begin{aligned} &|x' = f \,\&\, G]\lceil Q \rceil \\ &= \lceil \lambda s.\ \forall t \geq 0.\ G\,(\varphi_s\, t) \to Q\,(s[x \mapsto \varphi_s\, t\, x, x' \mapsto f\,(\varphi_s\, t), z' \mapsto 0])\rceil. \end{aligned}$$

Once again, we suppress indices to simplify notation. Rules for solving guarded evolution statements and the differential weakening rule (dW) can readily be extended to $\mathbb{R}^{V \cup V'}$, too. Yet $(solve_G)$ is not particularly interesting in this context: the whole purpose of invariants is to avoid this rule!

To derive the specific invariant rules of d$\mathcal{L}$ we follow d$\mathcal{L}$ in extending $(-)'$ from terms to atomic predicates $=$ and $\leq$ and to positive formulas over the language of differential rings:

$$(\theta = \eta)' = (\theta' = \eta'), \qquad (\theta < \eta)' = (\theta \leq \eta)' = (\theta' \leq \eta'),$$
$$(\phi \wedge \psi)' = (\phi \vee \psi)' = \phi' \wedge \psi'.$$

These derivatives are symbolic; they apply to syntactic objects, which explains the change in notation. We have created Isabelle datatypes for this syntax and use an interpretation function $[\![-]\!]$ to link it with the functions and predicates of the relational semantics.

Lemma 7.3. *The* differential invariant *rule is derivable, for* $G, I : \mathbb{R}^{V \cup V'} \to \mathbb{B}$, $f_i : \mathbb{R}^{V \cup V'} \to \mathbb{R}$, $x_i \in V$, $[\![\phi]\!] = I$, $[\![\theta]\!] = f$ *and where no variable in* ϕ *is primed.*

$$\frac{\forall s.\, G\, s \wedge (\forall z \neq x_i.\, s\, z' = 0) \to [\![\phi'[\theta/x']]\!]\, s}{\lceil I \rceil \subseteq |x'{=}f \,\&\, G]\lceil I \rceil}\ dI$$

Intuitively, (dI)—also called *differential induction* rule—embodies the substitutions from Example 7.1. The proof uses properties of substitutions and the mean value theorem.

Example 7.4 (Circular motion cont.). Consider again the ODEs $x' = y$ and $y' = -x$ from Example 7.1 and let $G = \lambda s.$ *True*. Applying (dI) backwards to the specification statement in Example 7.1 yields the hypothesis

$$\forall s.\, \forall z \notin \{x, y\}.\ s\, z' = 0 \to [\![(x^2 + y^2 = r^2)'[y/x', -x/y']]\!]\, s.$$

Symbolic differentiation and substitution application confirm its validity. □

The rule (dI) can of course handle inequalities, conjunctions, disjunctions and quantified formulas as well. Its conditions on z and z', however, are not needed in its d$\mathcal{L}$ counterpart. In addition, we can derive the following variant.

Corollary 7.5. *For* $P, G, I, Q : \mathbb{R}^{V \cup V'} \to \mathbb{B}$, $f_i : \mathbb{R}^{V \cup V'} \to \mathbb{R}$ *and* $x_i \in V$,

$$\frac{\lceil P \rceil \subseteq \lceil I \rceil \qquad \forall s.\, G\, s \wedge (\forall z \neq x_i.\, s\, z' = 0) \to [\![\phi'[\theta/x']]\!]\, s \qquad \lceil I \rceil \subseteq \lceil Q \rceil}{\lceil P \rceil \subseteq |x'{=}f \,\&\, G] \lceil Q \rceil}\; dI_2$$

where again $[\![\phi]\!] = I, [\![\theta]\!] = f$ *and no variable in* ϕ *is primed.*

This rule is designed to be combined with (dW) and the final d$\mathcal{L}$ rule we derive.

Lemma 7.6. *The* differential cut *rule is derivable for* $P, G, I, Q : \mathbb{R}^{V \cup V'} \to \mathbb{B}$, $f_i : \mathbb{R}^{V \cup V'} \to \mathbb{R}$ *and* $x_i \in V$:

$$\frac{\lceil P \rceil \subseteq |x'{=}f \,\&\, G] \lceil I \rceil \qquad \lceil P \rceil \subseteq |x'{=}f \,\&\, (\lambda\, s. G\, s \wedge I\, s)] \lceil Q \rceil}{\lceil P \rceil \subseteq |x'{=}f \,\&\, G] \lceil Q \rceil}\; dC$$

This rule is essential for introducing differential invariants into proofs. An example for its use in combination with (dI_2) and (dW) can be found in Sect. 9.

Soundness of (dC) can be explained as follows: By the right premise, G, I and Q hold along the evolution $x' = f$ on part of the phase space. By the left premise, G and I holds along $x' = f$ anyway. Hence G and Q hold along $x' = f$.

This completes the derivation of the inference rules d$\mathcal{L}$ in Isabelle/HOL—except for the differential effect and ghost rule. They allow the verification of hybrid programs with a general purpose proof assistant in the style of d$\mathcal{L}$. A recent soundness proof of a uniform substitution calculus for d$\mathcal{L}$ with Isabelle [5] includes a proof term checker for this specific calculus and hence KeYmaera X, but not a prima facie verification component.

8 Isabelle Components

Our verification components for hybrid programs in Isabelle are based on the integration of two theory stacks from the *Archive of Formal Proofs*: those for Kleene algebras, MKA and the corresponding verification components [2,11,13], and those related to analysis and ODEs [19]. The second stack, in particular, is very deep. It contains mathematical components for topology, measure theory, analysis and differential equations. Theory engineering and theorem proving at this theory depth can certainly be challenging, but our overall experience was very positive. Our new verification components on top of these stacks with >120 mathematical components currently consist of >1500 lines of Isabelle code with >100 lemmas. The online Isabelle proof document consists of 36 pages in PDF format. Proofs are usually highly interactive; they relate mainly to set-theoretic and higher-order properties of the relational flow model. At the moment, many of them are rather long and complex. Scope for simplifying models and proofs remains and our components should be considered as prototypes.

Our previous MKA verification components provide syntactic illusions for program specifications in the relational program store model. Specification statements can be written as *PRE* ⟨*precondition*⟩ ⟨*program*⟩ *POST* ⟨*postcondition*⟩.

The ODE components contain useful functions and definitions for vector fields and flows, for instance ((*flow solves-ode vector-field*) *interval ran*). This statement expresses that a given flow is a solution of a vector field where, in our setting, *interval* is $\mathbb{R}$ and *ran* an open subset of $\mathbb{R}^n$.

We encode program variables as strings. Hence phase spaces $\mathbb{R}^V$ have type *real store*, which is a synonym of *string* $\Rightarrow$ *real*. The relational flow semantics of evolution statements (2) is formalised as

$$\mathit{ODEsystem}\ \mathit{xfList}\ \mathit{with}\ G = \{(s,\ F\ t) \mid s\ t\ F.\ 0 \leq t \wedge \mathit{solvesStoreIVP}\ F\ \mathit{xfList}\ s\ G\},$$

where *solvesStoreIVP F xfList s G* is the Isabelle-analogue of $\mathsf{Flow}_{GI}\ \varphi\ x\ f\ s$ and the term *xfList* :: (*string* $\times$ ((*string* $\Rightarrow$ *real*) $\Rightarrow$ *real*)) *list* that of the system represented by x and f. This deviates from our mathematical presentation. The $\mathsf{Flow}_{\exists}$ constraint is included into the definiens, instead of acting as a hypothesis. In addition, uniqueness checks are currently postponed to inference rules. This is suitable for Isabelle as the resulting definition is total. The semantics evaluates to $\emptyset$ if the flow is not a solution to the corresponding evolution statement.

Our formalisation includes guards as well as primed variables, that is, the semantics and ($solves_{GI}$) rule uses the predicate Flow_{GI} defined in Sect. 7. This rule requires that users supply a flow. We require that its input has type (*real* $\Rightarrow$ *real store* $\Rightarrow$ *real*) *list* to make it easier for users to supply a specification with λ-abstractions (see Sect. 9). It is then transformed into a semantics with update function *(s,t,xfList,uInput)* $\mapsto$ *sol s*[*xfList*$\leftarrow$*uInput*] *t*, where

$$\mathit{sol}\ s[\mathit{xfList}\leftarrow\mathit{uInput}]\ t = \begin{cases} v_i\, t\, s, & \text{if } w = x_i,\\ (\lambda\, \tau.\, v_i\, \tau\, s)'\, t, & \text{if } w = x_i',\\ 0, & \text{if } w \in V' \setminus X',\\ s\, w, & \text{if } w \in V \setminus X, \end{cases}$$

and *uInput* is the list of user-inputs $v_1, \ldots, v_n$. Proof obligations for

$$\mathsf{Flow}_{GI}\ (\lambda t\, s. \mathit{sol}\ s[\mathit{xfList}\leftarrow\mathit{uInput}]\ t)\ x\ f\ s,$$

where we use mixed notation, are then generated automatically:

1. $G\,(\mathit{sol}\ s[\mathit{xfList}\leftarrow\mathit{uInput}]\ t)$,
2. $P\, s \rightarrow Q\,(\mathit{sol}\ s[\mathit{xfList}\leftarrow\mathit{uInput}]\ t)$,
3. $(\lambda\, \tau.\,(\mathit{sol}\ s[\mathit{xfList}\leftarrow\mathit{uInput}]\ \tau)\ x_i)' = (\lambda\, \tau.\, f_i\,(\mathit{sol}\ s[\mathit{xfList}\leftarrow\mathit{uInput}]\ \tau))$ in the interval $(0, t)$,
4. $(\lambda\, \tau.\, f\,(\mathit{sol}\ s[\mathit{xfList}\leftarrow\mathit{uInput}]\ \tau))$ is continuous in $[0, t]$,
5. $\forall \varphi.\, \mathsf{Solves}_{GI}\ \varphi\, x\, f\, s \rightarrow \forall \tau \in [0, t].\, f_i\,(\varphi\, \tau) = f_i\,(\mathit{sol}\ s[\mathit{xfList}\leftarrow\mathit{uInput}]\ \tau)$.

Condition (1) requires that the guard holds along the flow; (2) that the postcondition holds along the flow whenever the initial state satisfies the precondition; (3) that the derivative of the flow is equal to the vector field; (4) that the flow is continuous; and (5)that flows are uniquely defined. There are some additional simple conditions that specify well formedness of the verification statement, that the f_i do not include primed variables or that the lists *xfList* and *uInput* have the same length.

9 Verification Examples

The four simple examples in this section are adapted from the literature on KeYmaera X [28,29]. Their main purpose is to illustrate the main rules of d$\mathcal{L}$; thus demonstrating the feasibility of the approach. MKA rules for the control flow are already well tested [12].

Motion with Constant Velocity. First we consider the evolution of an object x that moves with constant positive velocity v away from another point y.

```
lemma PRE (λ s. s ''y'' < s ''x''  ∧ s ''v'' > 0)
    (ODEsystem [(''x'',(λ s. s ''v''))] with (λ s. True))
    POST (λ s. (s ''y'' < s ''x''))
  apply(rule-tac uInput=[λ t s. s ''v'' · t + s ''x''] in dSolve-toSolveUBC)
  prefer 11 subgoal by(simp add: wp-trafo vdiff-def add-strict-increasing2)
  apply(simp-all add: vdiff-def varDiffs-def, clarify)
  apply(rule-tac f'1=λt. st ''v'' and g'1=λt. 0 in derivative-intros(173))
  apply(rule-tac f'1=λt.0 and g'1=λt.1 in derivative-intros(176))
    subgoal by (auto intro: derivative-intros)
    subgoal by(clarify, rule continuous-intros)
  by(simp add: solvesStoreIVP-def vdiff-def varDiffs-def)
```

The double quotes indicate Isabelle strings; they represent program variables. The postcondition describes that the object x is always above y. The first line of the proof supplies the solution $\lambda t\ s.\ s\ ''v'' \cdot t + s\ ''x''$ to the evolution statement. It could have been computed by a computer algebra system or a similar solver. The second line solves the wlp by invoking Isabelle's simplifier. The remaining lines prove that Flow holds, that is, the conditions (1)–(5) in the previous section.

System Where the Guard Implies the Postcondition. Next we consider a system that is initially at position $x = 0$ with velocity $x' = x + 1$ and satisfies postcondition $x \geq 0$. It can be verified without providing a flow or using a differential invariant, because the guard and the postcondition are equal. Our first proof works directly with the wlp for guarded evolutions in Proposition 6.1.

```
lemma PRE (λ s. s ''x'' = 0)
    (ODEsystem [(''x'',(λ s. s ''x'' + 1))] with (λ s. s ''x'' ≥ 0))
    POST (λ s. s ''x'' ≥ 0)
  apply(clarify, simp add: p2r-def)
  apply(simp add: rel-ad-def rel-antidomain-kleene- algebra.addual.ars-r- def)
  apply(simp add: rel-antidomain-kleene- algebra.fbox-def)
```

```
apply(simp add: relcomp-def rel-ad-def guarDiffEqtn-def)
by(simp add: solvesStoreIVP-def, auto)
```

The lemmas mentioned in the apply statements show that Isabelle pulls together facts about MKAs, ODEs and the relational flow semantics. A fully automated proof can, however, be obtained simply by applying (dW) and then simplifying: **using** *dWeakening* **by** *simp*.

Circular Motion. Next we formalise the manual proof from Example 7.1 to illustrate the use of differential invariants in Sect. 7.

```
lemma PRE (λs. (s ''x'') · (s ''x'') + (s ''y'') · (s ''y'') − (s ''r'') · (s ''r'') = 0)
    (ODEsystem [(''x'',(λs. s ''y'')),(''y'',(λs. − s ''x''))] with G)
    POST (λs. (s ''x'') · (s ''x'') + (s ''y'') · (s ''y'') − (s ''r'') · (s ''r'') = 0)
  apply(rule-tac η= ''x'' ⊗ ''x'' ⊕ ''y'' ⊗ ''y'' ⊖ ''r'' ⊗ ''r''
    and uInput=[''y'', - ''x''] in dInvForTrms)
  apply(simp-all add: vdiff-def varDiffs-def)
  by(clarsimp, erule-tac x=''r'' in allE, simp)
```

In the first line of the proof, we supply the invariant η in the differential ring syntax developed, as explained in Sect. 8. We need to supply invariants as terms, in this example as $x^2+y^2-r^2$. It is translated into $x^2+y^2=r^2$ internally. Moreover, a term for the ODE system has to be given. This, however, should be automated, as this information has already been provided in another syntax in the *ODEsystem* declaration. The syntax with $\oplus$, $\ominus$ and $\odot$ simplifies the one used in our Isabelle theories. Isabelle's simplifiers can then deal with the remaining proof obligations, using some domain-specific definitions for differential rings. The guard G in the program is generic; it plays no role in the proof.

Single-Hop Ball. Finally, we use invariants in a two-modes example of a particle falling from an initial height H to the ground, and then changing its velocity upwardly by a factor of $0 \leq c \leq 1$ by using an assignment statement.

```
lemma single-hop-ball:
PRE (λ s. 0 ≤ s ''x'' ∧ s ''x'' = H ∧ s ''v'' = 0 ∧ s ''g'' > 0 ∧ 1 ≥ c ∧ c ≥ 0)
(((ODEsystem [(''x'', λ s. s ''v''),(''v'',λ s. − s ''g'')] with (λ s. 0 ≤ s ''x'')));
(IF (λ s. s ''x'' = 0) THEN (''v'' ::= (λ s. − c · s ''v'')) ELSE (''v'' ::= (λ s. s
''v'')) FI)) POST (λ s. 0 ≤ s ''x'' ∧ s ''x'' ≤ H)
apply(simp add: d-p2r, subgoal-tac rdom ⌈λs. 0 ≤ s ''x'' ∧ s ''x'' = H ∧ s ''v'' = 0
∧ 0 < s ''g'' ∧ c ≤ 1 ∧ 0 ≤ c⌉ ⊆ wp (ODEsystem [(''x'', λs. s ''v''), (''v'', λs. − s
''g'')] with (λs. 0 ≤ s ''x'') ) ⌈inf (sup (− (λs. s ''x'' = 0)) (λs. 0 ≤ s ''x'' ∧ s ''x''
≤ H)) (sup (λs. s ''x'' = 0) (λs. 0 ≤ s ''x'' ∧ s ''x'' ≤ H))⌉)
apply(simp add: d-p2r, rule-tac C = λ s. s ''g'' > 0 in dCut)
apply(rule-tac φ = (tC 0) ≺ (tV ''g'') and uInput=[tV ''v'', ⊖ tV ''g'']in dInvFinal)
apply(simp-all add: vdiff-def varDiffs-def, clarify, erule-tac x=''g'' in allE, simp)
apply(rule-tac C =λ s. s ''v'' ≤ 0 in dCut)
apply(rule-tac φ = (tV ''v'') ≼ (tC 0) and uInput=[tV ''v'', ⊖ tV ''g''] in dInvFinal)
apply(simp-all add: vdiff-def varDiffs-def)
```

apply(*rule-tac* $C = \lambda\ s.\ \ s\ ''x'' \leq\ \ H$ **in** *dCut*)
apply(*rule-tac* $\varphi = (t_V\ ''x'') \preceq (t_C\ H)$ **and** *uInput*=$[t_V\ ''v'', \ominus\ t_V\ ''g'']$**in** *dInvFinal*)
apply(*simp-all add*: *varDiffs-def vdiff-def*) **using** *dWeakening* **by** *simp*

The Isabelle syntax in our theory files is simplified by using $\preceq$. The proof consists of two blocks, in which differential invariants are provided through (dC) and then discharged with (dI_2). Using semantic notation, the invariants are $g > 0$, $v \leq 0$, and $x \leq H$. The proof is then finished by (dW).

10 Conclusion

We have constructed generic modular verification components à la d$\mathcal{L}$ for hybrid programs, based on MKA, and formalised with Isabelle. These include a relational flow semantics for hybrid programs for MKA, generic rules for solving and reasoning about ODEs in predicate transformer semantics, and soundness proofs for the most relevant inference rules of d$\mathcal{L}$ relative to this semantics.

Our soundness proofs complement those in [5], which focus on a uniform substitution calculus within a proof theory for d$\mathcal{L}$. We work entirely within the semantics of MKA and, by and large, use store updates in place of substitutions.

Our general goals and motivations differ from those of d$\mathcal{L}$ and KeYmaera X as well. There, a main concern is scalable hybrid program verification in practice. This is achieved via significant restrictions on the ODEs admitted and the assertion language used. We aim at an open experimental platform for mathematically simple, light-weight and effective integrations of continuous dynamics and discrete control into proof assistants, beyond d$\mathcal{L}$. MKA could, for instance, be reduced to Kleene algebra with tests over a relational flow semantics—and dynamic logic thus be reduced to Hoare logic. This semantics could also be combined with a duration calculus, for which we already have Isabelle support [8] or with other algebraic semantics [16]. Another upshot of MKA is its close relationship to refinement calculi based on predicate transformers [4]. Domain-specific refinement rules for hybrid systems seem another avenue worth exploring.

Our verification components are still prototypes, and there is much scope for making the development both more abstract and more efficient. In particular, more refined domains and intervals of existence for flows are needed. Even the simple examples in Sect. 9 currently require a significant amount of user interaction. Better simplification mechanisms and Isabelle tactics should be developed before more large scale case studies are undertaken.

Acknowledgements. This work was funded by a CONACYT scholarship. We are grateful to André Platzer for sharing his wisdom on d$\mathcal{L}$, and to Achim Brucker, Michael Herzberg, Andreas Lochbihler and Makarius Wenzel for Isabelle advice.

A Axioms and Definitions

Semiring. A *semiring* is a structure $(S, +, \cdot, 0, 1)$ that, for all $\alpha, \beta, \gamma \in S$ satisfies

$$\alpha \cdot (\beta \cdot \gamma) = (\alpha \cdot \beta) \cdot \gamma, \qquad 1 \cdot \alpha = \alpha, \qquad \alpha \cdot 1 = \alpha,$$
$$\alpha + (\beta + \gamma) = (\alpha + \beta) + \gamma, \qquad \alpha + \beta = \beta + \alpha, \qquad \alpha + 0 = \alpha, \qquad 0 + \alpha = \alpha,$$
$$\alpha \cdot (\beta + \gamma) = \alpha \cdot \beta + \alpha \cdot \gamma, \qquad (\alpha + \beta) \cdot \gamma = \alpha \cdot \gamma + \beta \cdot \gamma,$$
$$0 \cdot \alpha = 0, \qquad \alpha \cdot 0 = 0.$$

Dioid. A *dioid* is a semiring S that satisfies $\alpha + \alpha = \alpha$ for all $x \in S$.

Kleene Algebra. A *Kleene algebra* is a structure $(K, +, \cdot, 0, 1, ^*)$ such that $(K, +, \cdot, 0, 1)$ is a dioid and, for all $x, y, z \in K$,

$$1 + \alpha \cdot \alpha^* \leq \alpha^*, \qquad \gamma + \alpha \cdot \beta \leq \beta \rightarrow \alpha^* \cdot \gamma \leq \beta,$$
$$1 + \alpha^* \cdot \alpha \leq \alpha^*, \qquad \gamma + \beta \cdot \alpha \leq \beta \rightarrow \gamma \cdot \alpha^* \leq \beta.$$

Modal Kleene Algebra. A *modal Kleene algebra* is a tuple $(K, +, \cdot, 0, 1, ^*, ad, ar)$ such that $(K, +, \cdot, 0, 1, ^*)$ is a Kleene algebra and, for all $x, y \in K$,

$$ad\, \alpha \cdot \alpha = 0, \qquad ad\, \alpha + ad^2\, \alpha = 1, \qquad ad\, (\alpha \cdot \beta) \leq ad\, (\alpha \cdot ad^2\, \beta),$$
$$\alpha \cdot (ar\, \alpha) = 0, \qquad ar\, \alpha + ar^2\, \alpha = 1, \qquad ar\, (\alpha \cdot \beta) \leq ar\, (ar^2\, \alpha \cdot \beta),$$
$$ad^2\, (ar^2\, \alpha) = ar^2\, \alpha, \qquad ar^2\, (ad^2\, \alpha) = ad^2\, \alpha.$$

Syntax of Differential Rings. $\mathcal{R} ::= 0 \mid 1 \mid x \mid c \mid \mathcal{R} + \mathcal{R} \mid \mathcal{R} - \mathcal{R} \mid \mathcal{R} \cdot \mathcal{R} \mid \mathcal{R}'$, where x is drawn from a set of variables and c from a set of constants.

d$\mathcal{L}$-Style Syntax of Hybrid Programs

$$\mathcal{C} ::= x := e \mid x' = f \;\&\; G \mid \mathcal{C} ; \mathcal{C} \mid \textbf{if } P \textbf{ then } \mathcal{C} \textbf{ else } \mathcal{C} \mid \textbf{while } P \textbf{ inv } I \textbf{ do } \mathcal{C},$$

where x is a variable, e an expression, f a vector field expression, G a guard and P a test. However we do not work with an explicit syntax for hybrid programs in Isabelle. The entire development is within the relational flow semantics. We are only creating syntactic illusions within the semantics.

B Background on Differential Equations

Ordinary Differential Equations A system of n ordinary differential equations (ODEs) can be described by a vector field $f \in C(X, \mathbb{R}^n)$, where X is an open subset of $\mathbb{R}^{n+1}$. Its solutions are integral curves $x : I \rightarrow \mathbb{R}^n$ that is, differentiable functions that satisfy $x'\, t = f\,(t, x\, t)$, where $t \in I \subseteq \pi_1[X]$ (and π_1 is the standard first projection map). An initial value problem for this system of ODEs is specified with an initial condition $x\; t_0 = x_0$ where $x_0 \in X$ and $t_0 \in I$.

Autonomous Systems of ODEs. The vector fields described above are time dependent. Time independent vector fields satisfy $f \in C(X, \mathbb{R}^n)$ with $X \subseteq R^n$. The system of ODEs is then called *autonomous* [30] and its solutions satisfy the equation $x'\ t = f\ (x\ t)$ for $t \in I \subseteq \mathbb{R}$.

Lipschitz Continuity. A function $f : V \to W$ between normed vector spaces V and W is Lipschitz continuous on $X \subseteq V$ if there is a constant $k \geq 0$ such that $||f\, x - f\, y|| \leq k||x - y||$ holds for all $x, y \in X$. The function f is a contraction if it is Lipschitz continuous with $k < 1$.

Solutions to ODEs: Existence and Uniqueness. A Banach space is a complete normed vector space, that is, all Cauchy sequences converge. Let $B \neq \emptyset$ be a closed subset of a Banach space. Then, by Banach's fixpoint theorem, every contraction $f : B \to B$ has a unique fixpoint. This theorem is essential for proving the Picard-Lindelöf Theorem, which guarantees unique solutions for initial value problems. We state a special case for autonomous systems of ODEs.

Theorem B.1 (Picard-Lindelöf Theorem). *For every Lipschitz continuous vector field $f : X \subseteq \mathbb{R}^n \to \mathbb{R}^n$ and $x_0 \in X$ there exists a unique integral curve $x \in C^1(I(x_0))$ that satisfies the initial value problem for the autonomous system $x'\ t = f\,(x\,t)$ and $x\,t_0 = x_0$. $I(x_0)$ is the maximal interval of existence for x around t_0.*

Flow of an Autonomous System of ODEs. Let $f : X \subseteq \mathbb{R}^n \to X$ be a Lipschitz continuous vector field that admits, for each $x \in X$, a unique integral curve $\phi_x \in C^1(I(x))$ such that $\phi_x\ 0 = x$ by Picard-Lindelöf's theorem. The local flow $\varphi : T \to X \to X$ for f is defined by $\varphi\ t\ x = \phi_x\ t$, where $T = \bigcap_{x \in X} I(x)$.

For $\mathbb{R} = T$, the group action equations $\varphi\ 0 = id$ and $\varphi\ (s + t) = \varphi\ s \circ \varphi\ t$ are immediately derivable from this definition. The flow is thus indeed an action of the additive group $\mathbb{R}$ on X, and hence a dynamical system.

C Cross-References to Isabelle Proofs

The following table links the results in this article with facts from the Isabelle repository. The repository contains a readme file with further instructions.

Result in article	Result in Isabelle theories
Definition of evolution statements	subsumed by *guarDiffEqtn*
Definition of Flow	*solvesStoreIVP* (no uniqueness)
Proposition 5.2	subsumed by *dS*
Lemma 5.3	subsumed by *dSolve*
Definition of guarded evolution statement	*guarDiffEqtn*
Proposition 6.1	*dS*
Lemma 6.2	subsumed by *dSolve*,
Lemma 6.3	subsumed by *dWeakening*
Definition of differential ring language	**datatype** *trms*, **primrec** *rdiff*
Lemma 7.2	implied by *prelim-dSolve*
Rules for solving ODEs in $\mathbb{R}^{V \cup V'}$	*dSolve*
Differential weakening rule for $\mathbb{R}^{V \cup V'}$	*dWeakening*
Lemma 7.3	*dInv*
Corollary 7.5	*dInvFinal*
Lemma 7.6	*dCut*

References

1. Alur, R.: Formal verification of hybrid systems. In: EMSOFT 2011, pp. 273–278. ACM (2011)
2. Armstrong, A., Struth, G., Weber, T.: Kleene algebra. Archive of Formal Proofs (2013)
3. Arnol'd, V.I.: Ordinary Differential Equations. Springer, Heidelberg (1992)
4. Back, R., von Wright, J.: Refinement Calculus—A Systematic Introduction. Springer, New York (1998). https://doi.org/10.1007/978-1-4612-1674-2
5. Bohrer, B., Rahli, V., Vukotic, I., Völp, M., Platzer, A.: Formally verified differential dynamic logic. In: CPP 2017, pp. 208–221. ACM (2017)
6. Chlipala, A.: Certified Programming with Dependent Types–A Pragmatic Introduction to the Coq Proof Assistant. MIT Press (2013)
7. Desharnais, J., Struth, G.: Internal axioms for domain semirings. Sci. Comput. Program. **76**(3), 181–203 (2011)
8. Dongol, B., Hayes, I.J., Struth, G.: Relational convolution, generalised modalities and incidence algebras. CoRR, abs/1702.04603 (2017)
9. Fainekos, G.E. Kress-Gazit, H., Pappas, G.J.: Hybrid controllers for path planning: a temporal logic approach. In: IEEE Conference on Decision and Control, pp. 4885–4890 (2005)
10. Fulton, N., Mitsch, S., Quesel, J.-D., Völp, M., Platzer, A.: KeYmaera X: an axiomatic tactical theorem prover for hybrid systems. In: Felty, A.P., Middeldorp, A. (eds.) CADE 2015. LNCS (LNAI), vol. 9195, pp. 527–538. Springer, Cham (2015). https://doi.org/10.1007/978-3-319-21401-6_36
11. Gomes, V.B.F., Guttman, W., Höfner, P., Struth, G., Weber, T.: Kleene algebra with domain. Archive of Formal Proofs (2016)

12. Gomes, V.B.F., Struth, G.: Modal kleene algebra applied to program correctness. In: Fitzgerald, J., Heitmeyer, C., Gnesi, S., Philippou, A. (eds.) FM 2016. LNCS, vol. 9995, pp. 310–325. Springer, Cham (2016). https://doi.org/10.1007/978-3-319-48989-6_19
13. Gomes, V.B.F., Struth, G.: Program construction and verification components based on Kleene algebra. Archive of Formal Proofs (2016)
14. Harel, D., Kozen, D., Tiuryn, J.: Dynamic Logic. MIT Press (2000)
15. Henzinger, T.A.: The theory of hybrid automata. In: LICS 1996, pp. 278–292. IEEE Computer Society (1996)
16. Höfner, P., Möller, B.: An algebra of hybrid systems. J. Logic Algebraic Program. **78**(2), 74–97 (2009)
17. Hölzl, J., Immler, F., Huffman, B.: Type classes and filters for mathematical analysis in Isabelle/HOL. In: Blazy, S., Paulin-Mohring, C., Pichardie, D. (eds.) ITP 2013. LNCS, vol. 7998, pp. 279–294. Springer, Heidelberg (2013). https://doi.org/10.1007/978-3-642-39634-2_21
18. Immler, F., Hölzl, J.: Numerical analysis of ordinary differential equations in Isabelle/HOL. In: Beringer, L., Felty, A. (eds.) ITP 2012. LNCS, vol. 7406, pp. 377–392. Springer, Heidelberg (2012). https://doi.org/10.1007/978-3-642-32347-8_26
19. Immler, F., Hölzl, J.: Ordinary differential equations. Archive of Formal Proofs (2012)
20. Immler, F., Traut, C.: The flow of ODEs. In: Blanchette, J.C., Merz, S. (eds.) ITP 2016. LNCS, vol. 9807, pp. 184–199. Springer, Cham (2016). https://doi.org/10.1007/978-3-319-43144-4_12
21. Jeannin, J., et al.: A formally verified hybrid system for safe advisories in the next-generation airborne collision avoidance system. STTT **19**(6), 717–741 (2017)
22. Klein, G., et al.: Comprehensive formal verification of an OS microkernel. ACM Trans. Comput. Syst. **32**(1), 2:1–2:70 (2014)
23. Leroy, X.: Formal verification of a realistic compiler. CACM **52**(7), 107–115 (2009)
24. Loos, S.M., Platzer, A., Nistor, L.: Adaptive cruise control: hybrid, distributed, and now formally verified. In: Butler, M., Schulte, W. (eds.) FM 2011. LNCS, vol. 6664, pp. 42–56. Springer, Heidelberg (2011). https://doi.org/10.1007/978-3-642-21437-0_6
25. Nipkow, T., Wenzel, M., Paulson, L.C. (eds.): Isabelle/HOL. LNCS, vol. 2283. Springer, Heidelberg (2002). https://doi.org/10.1007/3-540-45949-9
26. Olver, P.J.: Applications of Lie Groups to Differential Equations. Springer, New York (1986). https://doi.org/10.1007/978-1-4684-0274-2
27. Platzer, A.: The structure of differential invariants and differential cut elimination. LMCS **8**(4), 1–38 (2008)
28. Platzer, A.: Logical Analysis of Hybrid Systems. Springer, Heidelberg (2010). https://doi.org/10.1007/978-3-642-14509-4
29. Quesel, J., Mitsch, S., Loos, S.M., Arechiga, N., Platzer, A.: How to model and prove hybrid systems with KeYmaera: a tutorial on safety. STTT **18**(1), 67–91 (2016)
30. Teschl, G.: Ordinary Differential Equations and Dynamical Systems. AMS (2012)

Algebraic Derivation of Until Rules and Application to Timer Verification

Jessica Ertel[1], Roland Glück[2(✉)], and Bernhard Möller[3]

[1] msg-life, Leinfelden-Echterdingen, Germany
ertel-jessica@hotmail.de
[2] German Aerospace Center, Augsburg, Germany
roland.glueck@dlr.de
[3] University of Augsburg, Augsburg, Germany
moeller@uni-augsburg.de

Abstract. Using correspondences between linear temporal logic and modal Kleene Algebra, we prove in an algebraic manner rules of linear temporal logic involving the until operator. These can be used to verify programmable logic controllers; as a case study we use a part of the control of pedestrian lights, verified with the interactive tool KIV.

1 Introduction

Overview. Semirings, Kleene Algebra and their algebraic relatives have proved to be a flexible tool for reasoning about a broad variety of topics such as graph problems, algorithms and transformations [12,14,23,24,27,30], energy problems [20], fuzzy logic and relations [31], software development and verification [38,41] and database theory [36,40]. Here we use and apply an approach from [17,39] which relates Modal Kleene Algebra (MKA) and Linear Temporal Logic (LTL). It shows a.o. that sets of LTL traces form an MKA and that the standard LTL operators can be represented as compositions of MKA operators. Along these lines we first prove algebraically a few new properties of the LTL *until* operator in MKA. Since we use the MKA formalization, we prove in fact much more general theorems which hold in all MKAs, not just the LTL variant mentioned above. But, of course, the results apply to LTL itself as well.

We apply these in the interactive verification of programmable logic controllers (PLCs). Encouraged by the results in [21] on this, we tackle as a considerably more difficult new task a substantial part of traffic light control systems in PLC, including a formalization of timers. Besides this, the paper deals substantially with temporal phenomena, in which the until operator is a big structuring help. These issues were not yet covered in [21]; the treatment is based on [19]. As verification tool we choose KIV [3], a rather uncommon interactive verifier, hosted at the university of Augsburg. Despite not being widely known, it succeeded in some verification competitions [7,8]. Moreover, the present work continues [21] which also used KIV.

J. Desharnais et al. (Eds.): RAMiCS 2018, LNCS 11194, pp. 244–262, 2018.
https://doi.org/10.1007/978-3-030-02149-8_15

Related Work. Recent approaches to PLC verification are simulation based [15], use data-flow analysis [28] or model checking [35,42]. Simulation based approaches and model checking (based on timed automata) in a naïve manner suffer from the same problem: when confronted with a timer they have to execute or check all timer values, see e.g. the handling of the variable `timer` in the specification `robot.smv` from [4]. There all possible timer values between 0 and 400 are evaluated, whereas only some of them are important for the verification of the system under consideration (note that this is not a property of the complexity of the system's description and formalization but rather of the employed verification mechanism). Verification of timed PLC programs is a very sparse field; e.g., [35] deals only with Boolean values, whereas [42] explicitely excludes timers. An interactive approach to PLC timer verification using COQ [1] is presented in [47]. This is closely related to our approach; however, it does not reason about until-properties as we will do here.

Our Contribution. The present paper introduces an interactive approach to the verification of timed programs, in particular to timed PLC programs. Based on an algebraic modeling and a timer formalization that takes only significant values of the timer into account, we circumvent the problems sketched above. This idea can be seen as a variant of the zone-graph technique (see [9,48]) which divides the set of possible clock values into equivalence classes and achieves a model considering only important clock values. Formalization in MKA and proofs were conducted in KIV [3] due to the preliminary work [21] and our desire to show the possibility of a purely algebraic approach. As a side effect, interactive verification has the potential to guide humans to better bug-fixing than model checking which only outputs faulty traces. Moreover, algebraic rules as derived in Sect. 2.3 can be deployed in a large context not restricted to a single particular formalization. Finally, algebraic reasoning is much more compact and needs much fewer steps than pointwise reasoning in the original formulation of LTL or Dynamic Logic, even though the latter is directly supported by KIV.

Structure. The paper is organized as follows: In Sects. 2.1 and 2.2 we recall the basics of MKA and the connection between MKA and LTL. Section 2.3 gives algebraic proofs of some important rules concerning the until operator. Section 3 adapts and substantially extends the earlier results on PLC verification from [21]. After a quick introduction to PLCs in Sect. 3.1 and their MKA modeling in Sect. 3.2, we show in Sect. 3.3 how to formalize a PLC timer in our framework. Section 3.4 ties all threads together in a case study verifying a central part of the control of pedestrian lights. Conclusion and outlook are given in Sect. 4.

2 Modal Kleene Algebra and Linear Temporal Logic

For this section we assume basic knowledge about lattice theory and semirings (e.g. [13,22,29]), and about temporal logic (e.g. [11,34]). As usual, we often omit the semiring multiplication sign for better readability. Furthermore, we use $\sum$ and $\prod$ for general finite sums and products in semirings.

2.1 Modal Kleene Algebra

As stated in Sect. 1, MKA is a by now well established subdiscipline of Algebraic Logic with numerous applications. Its several axiomatic variants have different advantages and disadvantages; hence we make precise which one we use.

First, a *Kleene algebra* [32] is a structure $(M, +, \cdot, 0, 1, {}^*)$ where $(M, +, \cdot, 0, 1)$ is an idempotent semiring with *natural order* $x \leq y \Leftrightarrow x + y = y$, and the *Kleene star* operator * satisfies the following axioms for all $x, y, z \in M$:

$$1 + xx^* \leq x^* \qquad 1 + x^*x \leq x^* \qquad \text{(right and left unfold)}$$
$$y + zx \leq z \Rightarrow yx^* \leq z \qquad y + xz \leq z \Rightarrow x^*y \leq z \qquad \text{(right and left induction)}$$

Star has many useful properties like *reflexivity*, *multiplicative idempotence* and *isotony*, i.e., $1 \leq x^*$, $x^* \cdot x^* = x^*$ and $x \leq y \Rightarrow x^* \leq y^*$ for all x, y.

Assume now an idempotent semiring $S = (M, +, \cdot, 0, 1)$ whose elements correspond to sets of possible transitions (e.g., relations) between states of some kind. In particular, 0 models the empty set of transitions. To get an algebraic representation for sets of states one introduces the notion of tests [26,33,37]. An element $p \in M$ is called a *test* if there exists an element $\neg p$ (the *complement* of p) such that $p + \neg p = 1$ and $p \cdot \neg p = 0 = \neg p \cdot p$ hold. Clearly, all tests p satisfy $p \leq 1$. In the case of relations, tests are subrelations of the identity relation and hence can indeed be viewed as representations of sets of states. The set $\mathsf{test}(S)$ of all tests of S forms a Boolean algebra with multiplication as infimum, addition as supremum and $\neg$ as complement operator. Moreover, 0 and 1 are the least and greatest tests, with 0 also representing the empty set of states. For that reason, in view of the formal semantics of LTL to come, we call a test p *valid*, in signs $\models p$, if $p = 1$ (equivalently, if $1 \leq p$). Finally, it is useful to define test implication by $p \to q =_{df} \neg p + q$. This satisfies the important *shunting* equivalence $p \cdot q \leq r \Leftrightarrow p \leq q \to r$, with $p \cdot q \leq r \Leftrightarrow p \leq \neg q + r$ as a consequence. Moreover, this implies $\models p \to q \Leftrightarrow p \leq q$.

In an idempotent semiring $S = (M, +, \cdot, 0, 1)$ one can axiomatize the *(forward) diamond* operator $|\ \rangle$ of type $M \times \mathsf{test}(S) \to \mathsf{test}(S)$ by the equivalence $|x\rangle p \leq q \Leftrightarrow_{df} \neg q x p \leq 0$ for all $x \in M$ and $p, q \in \mathsf{test}(S)$. In the case of existence, the operator is unique. The test $|x\rangle p$ represents the inverse image of p under x, i.e., the states that are related by x to at least one p-state. A backward diamond $\langle\ |$ representing the image operator can be defined symmetrically by $\langle x|p \leq q \Leftrightarrow_{df} px\neg q \leq 0$ for all $x \in M$ and $p, q \in \mathsf{test}(S)$.

The diamonds distribute over $+$ and hence are isotone in both arguments. Moreover, the *import/export law* $|px\rangle q = p(|x\rangle q)$ and its dual hold for all x and tests p, q. Finally, the diamonds of tests are characterized by $|p\rangle q = pq = \langle p|q$ (and hence, in particular, $|1\rangle q = q = \langle 1|q)$ for all $p, q, \in \mathsf{test}(S)$.

The structure $(M, +, \cdot, 0, 1, |\ \rangle, \langle\ |)$ is called a *modal semiring* if additionally the *modality condition* $|xy\rangle p = |x\rangle|y\rangle p$ and its dual hold for all x, y and tests p.

As the De Morgan dual of the diamonds we introduce the *boxes* by the equality $|x]p =_{df} \neg|x\rangle\neg p$ and its dual. These operators are isotone in their second but antitone in their first argument. Clearly, in a modal semiring we also have $|xy]p = |x]|y]p$ and its dual.

Finally, we call a structure $(M, +, \cdot, 0, 1, ^*, |\ \rangle, \langle\ |)$ a *Modal Kleene Algebra* (or briefly *MKA*) if $(M, +, \cdot, 0, 1, ^*)$ is a Kleene Algebra and $(M, +, \cdot, 0, 1, |\ \rangle, \langle\ |)$ forms a modal semiring. Every MKA satisfies the important *modal star unfold and induction rules* (and their right duals) for all x and tests p, q:

$$\begin{array}{ll} p + |x\rangle|x^*\rangle p \leq |x^*\rangle p\ , & |x^*]p \leq p \cdot |x]|x^*]p\ , \\ q \leq p \wedge |x\rangle p \leq p \Rightarrow |x^*\rangle q \leq p\ , & p \leq q \wedge p \leq |x]p \Rightarrow p \leq |x^*]q\ . \end{array} \tag{1}$$

MKAs are related to Dynamic Algebras (e.g. [25,43]); a decisive difference is that, via tests, MKAs allow nested modalities such as $|a \cdot |b\rangle p]q$ which restricts a to target states in an inverse image under a transition b. The relationship between Dynamic Algebras, MKAs and Test Algebras has been worked out in [18]. Variants of the modal operators are also present in [10] and the algebraic counterpart [45], which is a special case of MKA. But no temporal operators are treated there. Finally, we mention the framework in [46]; this is rather specialised, whereas we are interested in re-using the more general framework of MKAs. The details are not relevant to the present paper and hence omitted.

2.2 Modal Kleene Algebra and Linear Temporal Logic

The syntax of the language Ψ of LTL formulas over a set Φ of atomic propositions is given by the context-free grammar

$$\Psi ::= \bot \mid \Phi \mid \neg\Psi \mid \Psi \rightarrow \Psi \mid \Psi \wedge \Psi \mid \Psi \vee \Psi \mid \mathsf{O}\,\Psi \mid \Box\,\Psi \mid \Diamond\,\Psi \mid \Psi\ \mathsf{U}\ \Psi$$

where $\bot$ denotes falsity, $\rightarrow$ is logical implication and O and U are the *next-time* and *until* operators. We are well aware of the redundancies in this definition; they serve to make the presentation of the semantics smoother.

In [39] (refined in [16]) a correspondence between MKA and LTL was established. It uses an MKA $S = (M, +, \cdot, 0, 1, ^*, |\ \rangle, \langle\ |)$ and an element $a \in M$ that models a transition relation transforming a set of states into the set of their successors. Then to every LTL formula ψ one assigns as semantics a test $[\![\psi]\!] \in \mathsf{test}(S)$ that represents the states in which ψ holds. Strictly speaking, the semantic function should be parametrised with the transition element a in the form $[\![\psi]\!]_a$; we omit this for better readability.[1] Further explanations can be found in [16,39]. Since the algebraic semantics only uses the forward diamond and box, we omit the word "forward" in the sequel.

[1] This abstracts from the classical LTL semantics in terms of sets of infinite traces of program states. That concrete semantics is mirrored by a modal semiring in which the elements are relations between sets of traces and tests are sets of traces; states in the sense of the above wording are then single traces, not program states.

We assume that to every atomic proposition $\varphi \in \Phi$ a test $[\![\varphi]\!] \in \mathsf{test}(S)$ has been assigned as the semantics. Then the semantics of the remaining formulas is inductively defined as follows.

$$
\begin{array}{llll}
[\![\bot]\!] & = 0 & [\![\circ\psi]\!] & = |a\rangle[\![\psi]\!] \\
[\![\neg\psi]\!] & = \neg[\![\psi]\!] & [\![\psi_1 \,\mathsf{U}\, \psi_2]\!] & = |([\![\psi_1]\!] \cdot a)^*\rangle[\![\psi_2]\!] \\
[\![\psi_1 \to \psi_2]\!] & = [\![\psi_1]\!] \to [\![\psi_2]\!] & [\![\Diamond\,\psi]\!] & = |a^*\rangle[\![\psi]\!] \\
[\![\psi_1 \wedge \psi_2]\!] & = [\![\psi_1]\!] \cdot [\![\psi_2]\!] & [\![\Box\,\psi]\!] & = |a^*][\![\psi]\!] \\
[\![\psi_1 \vee \psi_2]\!] & = [\![\psi_1]\!] + [\![\psi_2]\!] & &
\end{array}
$$

The semantics of U can be understood as follows. The element $[\![\psi_1]\!] \cdot a$ models the restriction of the transition relation a to those starting states that satisfy ψ_1. Thus, a transition along $([\![\psi_1]\!]\cdot a)^*$ traverses only ψ_1-states. Hence $|([\![\psi_1]\!]\cdot a)^*\rangle[\![\psi_2]\!]$ characterizes those states from which ψ_2-states can be reached by traversing only ψ_1-states; this is a faithful representation of the informal U-semantics. Finally, $\Diamond\,\psi$ and $\Box\,\psi$ hold if ψ holds in some/all subsequent states.

Motivated by these definitions we introduce temporal operators on tests by

$$\circ p =_{df} |a\rangle p \quad p \,\mathsf{U}\, q =_{df} |(p \cdot a)^*\rangle q \quad \Diamond\, q =_{df} |a^*\rangle q \quad \Box\, q =_{df} |a^*]q \tag{2}$$

for transition element a and tests p, q. This allows LTL formulas for tests.

Formula ψ_1 *entails* formula ψ_2, in signs $\psi_1 \models \psi_2$, if $[\![\psi_1]\!] \leq [\![\psi_2]\!]$. Formula ψ is *valid*, in signs $\models \psi$, if $\top \models \psi$, where $\top = \neg\bot$ is the true formula with $[\![\top]\!] = 1$. Since 1 is the greatest test, this is equivalent to $[\![\psi]\!] = 1$. We provide a frequently used rule concerning the validity of $\to$-formulas: by the above remark, the semantics of $\to$ and the shunting equivalence we obtain

$$\models \psi_1 \to \psi_2 \Leftrightarrow 1 \leq [\![\psi_1 \to \psi_2]\!] \Leftrightarrow 1 \leq [\![\psi_1]\!] \to [\![\psi_2]\!] \Leftrightarrow [\![\psi_1]\!] \leq [\![\psi_2]\!] \Leftrightarrow \psi_1 \models \psi_2 \tag{3}$$

Given a transition system, the relation transforming a set of states into the set of their successor states is a total function from sets to sets. In MKA, this behavior of an abstract relation a can be enforced by the requirement $|a\rangle p = |a]p$ for all tests p [16,39]; we call an element with this property also a *total function.*

Using the above correspondences, we can prove rules from LTL in an algebraic way, which avoids reasoning about traces and single states.

As an example, we show $\models \psi \to \Diamond\,\psi$: by (3), the semantics of $\Diamond$ and isotony of the diamond together with $|1\rangle p = p$, we obtain

$$\models \psi \to \Diamond\,\psi \Leftrightarrow [\![\psi]\!] \leq [\![\Diamond\,\psi]\!] \Leftrightarrow [\![\psi]\!] \leq |a^*\rangle[\![\psi]\!] \Leftarrow 1 \leq a^* ,$$

which holds by the definition of star.

2.3 Investigating the Until Operator

In this section we show some useful properties of the LTL until operator using the correspondences with MKA from the previous subsection. All proofs were also done interactively with the KIV system (see [3]), based on the work from [21].

The whole KIV treatment can be found online at [6]; however, we include the proofs to give the reader an impression of the algebraic framework and to demonstrate the power of reasoning in MKA. KIV has also the ability to conduct automated reasoning using adjustable heuristics. Since formulating these is not an easy task, we mostly forwent this feature; exploring its power in our setting will be future work. However, our experience so far shows that the right adjustment of heuristics can help a lot.

It turns out that many proofs about the transition element a only need the weaker condition $|a\rangle p \leq |a]p$ for all tests p. Such an element is called *modally deterministic*. We will show a number of properties of U over such elements.

First, assume that φ implies $\diamond\psi$ and that for every state satisfying φ the (by determinacy unique) successor state satisfies $\varphi \vee \psi$. We will prove that then φ implies $\varphi \,\mathsf{U}\, \psi$. In LTL notation, if $\models \varphi \rightarrow \circ(\varphi \vee \psi)$ and $\models (\varphi \rightarrow \diamond\psi)$ then $\models \varphi \rightarrow \varphi \,\mathsf{U}\, \psi$[2]. To save notation, in the sequel we identify formulas with their semantic values. E.g., p, q will stand for the values $[\![\varphi]\!], [\![\psi]\!]$ of formulas φ, ψ. With this convention, a translation into MKA looks as follows (remember (2) and the correspondence of $\wedge/\vee$ with $\cdot/+$):

Lemma 1. *For a modally deterministic element a and tests p, q, if $\models p \rightarrow \diamond q$ and $\models p \rightarrow \circ(p+q)$ then $\models p \rightarrow (p \,\mathsf{U}\, q)$.*

Proof. Plugging in the definitions and using (3) transform the claim into $p \leq |a\rangle(p+q) \;\wedge\; p \leq |a^*\rangle q \Rightarrow p \leq |(p \cdot a)^*\rangle q$.
First, by idempotence of multiplication on tests and the second assumption $p \leq |a^*\rangle q$ we obtain $p = p \cdot p \leq p \cdot |a^*\rangle q$. So we are done if we can show $p \cdot |a^*\rangle q \leq |(p \cdot a)^*\rangle q$. By shunting and diamond star induction, introducing $r =_{df} \neg p + |(p \cdot a)^*\rangle q$, we obtain

$$p \cdot |a^*\rangle q \leq |(p \cdot a)^*\rangle q \;\Leftrightarrow\; |a^*\rangle q \leq \neg p + |(p \cdot a)^*\rangle q \;\Leftarrow\; q \leq r \wedge |a\rangle r \leq r$$

The first conjunct of the latter formula holds by $1 \leq (p \cdot a)^*$ and hence $q \leq |(p \cdot a)^*\rangle q \leq r$. For the second one we continue as follows:

$$\begin{array}{lll} & |a\rangle r \leq r & \\ \Leftrightarrow & p \cdot |a\rangle r \leq |(p \cdot a)^*\rangle q & \{\!|\text{ definition of } r \text{ and shunting back }|\!\} \\ \Leftrightarrow & p \cdot |a\rangle \neg p \leq s \wedge p \cdot |a\rangle s \leq s & \{\!|\text{ setting } s =_{df} |(p \cdot a)^*\rangle q, \text{ definition of } r, \\ & & \quad \text{distributivity of } |a\rangle \text{ and } \cdot, \text{ lattice algebra }|\!\} \end{array}$$

For the second conjunct we reason as follows:

$$\begin{array}{lll} & p \cdot |a\rangle s & \\ = & |p \cdot a\rangle s & \{\!|\text{ import/export }|\!\} \\ = & |p \cdot a\rangle |(p \cdot a)^*\rangle q & \{\!|\text{ definition of } s \;|\!\} \\ = & |p \cdot a \cdot (p \cdot a)^*\rangle q & \{\!|\text{ modality }|\!\} \\ \leq & |(p \cdot a)^*\rangle q & \{\!|\; xx^* \leq x^* \text{ by right star unfold, isotony of} \\ & & \quad \text{diamond }|\!\} \\ = & s & \{\!|\text{ definition of } s \;|\!\} \end{array}$$

[2] Note that this is not the same as $((\varphi \rightarrow \circ(\varphi \vee \psi)) \wedge (\varphi \rightarrow \diamond\psi)) \models \varphi \rightarrow \varphi \,\mathsf{U}\, \psi$, which does not hold.

The first conjunct is the place where the first assumption is used:

$$
\begin{array}{ll}
p \leq |a\rangle(p+q) & \\
\Leftrightarrow p \leq |a\rangle p + |a\rangle q & \{\!\!\{ \text{distributivity} \}\!\!\} \\
\Leftrightarrow p \cdot \neg|a\rangle p \leq |a\rangle q & \{\!\!\{ \text{shunting} \}\!\!\} \\
\Leftrightarrow p \cdot |a]\neg p \leq |a\rangle q & \{\!\!\{ \text{definition forward box,} \\
 & \quad \text{Boolean algebra} \}\!\!\} \\
\Rightarrow p \cdot |a\rangle\neg p \leq |a\rangle q & \{\!\!\{ \text{modal determinacy of } a \}\!\!\} \\
\Rightarrow p \cdot p \cdot |a\rangle\neg p \leq p \cdot |a\rangle q & \{\!\!\{ \text{isotony} \}\!\!\} \\
\Leftrightarrow p \cdot |a\rangle\neg p \leq |p \cdot a\rangle q & \{\!\!\{ \text{idempotence of test multiplication, import/export} \}\!\!\} \\
\Rightarrow p \cdot |a\rangle\neg p \leq |(p \cdot a)^*\rangle q & \{\!\!\{ \text{star unfold and isotony} \}\!\!\}
\end{array}
$$

□

Next we relate U with $\Box$.

Lemma 2. *Assume an MKA, a modally deterministic element a and tests p, q. Set $u =_{df} q \mathsf{U} p$ and assume $\models p \to \circ u$.*

i) $\models p \to \Box u$.
ii) If additionally $\models p \to q$ *then* $\models p \to \Box q$.

Proof. (i) The claim transforms into $p \leq |a^*]u$. So we are done if we can show $p \leq u$ and $u \leq |a^*]u$. The first conjunct holds by diamond star unfold (1). The second conjunct reduces by box star induction (1) to $u \leq u \wedge u \leq |a]u$, of which the first part holds trivially. The second part is, by determinacy of a, implied by $u \leq |a\rangle u$. To show that we calculate, using the definition of u with diamond star unfold, $q \leq 1$ with isotony of diamond and the assumption, $u = p + |q \cdot a\rangle u \leq p + |a\rangle u = |a\rangle u$.

(ii) This follows from Part (i) by isotony of box if we can show $u \leq q$. To this purpose, we reason as follows:

$$
\begin{array}{ll}
u \leq q & \\
\Leftrightarrow |(q \cdot a)^*\rangle p \leq q & \{\!\!\{ \text{definition of } u \}\!\!\} \\
\Leftarrow p \leq q \wedge |q \cdot a\rangle q \leq q & \{\!\!\{ \text{diamond star induction} \}\!\!\} \\
\Leftrightarrow \mathsf{TRUE} \wedge q \cdot |a\rangle q \leq q & \{\!\!\{ \text{assumption, import/export} \}\!\!\} \\
\Leftrightarrow \mathsf{TRUE} & \{\!\!\{ |a\rangle q \leq 1, \text{ isotony} \}\!\!\}
\end{array}
$$

□

The next lemma shows that $\models p \wedge \neg q \to \circ p$ implies $\models p \wedge \Diamond q \to p \mathsf{U} q$ and $\models (r \to \circ(p \wedge \Diamond q)) \wedge (p \wedge \neg q \to \circ p)$ implies $\models r \to \circ(p \mathsf{U} q)$.

Lemma 3. *In an MKA we have for all total functions a and tests p, q, r the following properties:*

(i) $p \cdot \neg q \leq |a\rangle p \Rightarrow p \cdot |a^*\rangle q \leq |(pa)^*\rangle q$
(ii) $r \leq |a\rangle(p|a^*\rangle q) \wedge p \cdot \neg q \leq |a\rangle p \Rightarrow r \leq |a\rangle(|(pa)^*\rangle q)$

Proof. (i) We reason as follows:

$$
\begin{array}{ll}
p|a^*\rangle q \leq |(pa)^*\rangle q & \\
\Leftrightarrow |a^*\rangle q \leq \neg p + |(pa)^*\rangle q & \{\!\!\{ \text{shunting} \}\!\!\} \\
\Leftarrow q + |a\rangle(\neg p + |(pa)^*\rangle q) \leq \neg p + |(pa)^*\rangle q & \{\!\!\{ \text{diamond induction} \}\!\!\}
\end{array}
$$

$$\begin{array}{ll} \Leftrightarrow q \leq \neg p + |(pa)^*\rangle q \wedge & \{\!\!\{ \text{lattice algebra} \}\!\!\} \\ \quad |a\rangle(\neg p + |(pa)^*\rangle q) \leq \neg p + |(pa)^*\rangle q & \end{array}$$

The first conjunct is shown easily: $q \leq |(pa)^*\rangle q$ holds due to $1 \leq (pa)^*$ and isotony of $|\ \rangle$. Now adding $\neg p$ cannot decrease the right hand side.
For the second conjunct we argue first as follows:

$$\begin{array}{ll} \quad |a\rangle(\neg p + |(pa)^*\rangle q) \leq \neg p + |(pa)^*\rangle q & \\ \Leftrightarrow p \cdot |a\rangle(\neg p + |(pa)^*\rangle q) \leq |(pa)^*\rangle q & \{\!\!\{ \text{shunting} \}\!\!\} \\ \Leftrightarrow p \cdot |a\rangle\neg p + p \cdot |a\rangle(|(pa)^*\rangle q) \leq |(pa)^*\rangle q & \{\!\!\{ \text{distributivity} \}\!\!\} \\ \Leftrightarrow p \cdot |a\rangle\neg p \leq |(pa)^*\rangle q \wedge & \{\!\!\{ \text{sum properties} \}\!\!\} \\ \quad p \cdot |a\rangle(|(pa)^*\rangle q) \leq |(pa)^*\rangle q & \end{array}$$

A shunted form of the second conjunct was already shown in the proof of Lemma 1. The first one follows from the assumption $p \cdot \neg q \leq |a\rangle p$ as follows:

$$\begin{array}{ll} \quad p \cdot \neg q \leq |a\rangle p & \\ \Leftrightarrow p \cdot \neg|a\rangle p \leq q & \{\!\!\{ \text{shunting, twice} \}\!\!\} \\ \Rightarrow p \cdot \neg|a\rangle p \leq |(pa)^*\rangle q & \{\!\!\{ 1 \leq x^*, \text{ diamond properties} \}\!\!\} \\ \Leftrightarrow p \cdot |a\rangle\neg p \leq |(pa)^*\rangle q & \{\!\!\{ \text{definition of box and } a \text{ being} \\ & \quad \text{total and deterministic} \}\!\!\} \end{array}$$

(ii) By Part (i) the second conjunct of the premiss of Part (ii) implies $p \cdot |a^*\rangle q \leq |(pa)^*\rangle q$ by Part (i). Isotony of $|\ \rangle$ yields $|a\rangle(p \cdot |a^*\rangle q) \leq |a\rangle(|(pa)^*\rangle q)$, and now the assumption $r \leq |a\rangle(p|a^*\rangle q)$ and transitivity of $\leq$ show the claim. □

3 Verifying Programmable Logic Controllers

We now apply our semantic foundations to a concrete verification task.

3.1 Basics of Programmable Logic Controllers

Programmable logic controllers (PLCs) are widely used for the control of robots, plants and mechanical devices. They work in a cyclic way: in each cycle they read values from inputs (which stem from the environment and may be, e.g., switch signals or sensor values) and internal variables (which serve for storing values during the execution); from these they compute new values of the internal and output variables (which are forwarded to the environment and may, e.g., start or stop a machine or control the speed of a motor). By default, the names of input and output variables start with `IN` and `OUT`, resp., whereas internal variables have the form `M`x or `M`$x.y$; here the latter form is used to access single bits. It is possible to use variable aliasing to improve readability. Standards for PLCs are defined in [2]; we follow closely the syntax of STEP7 (see [5]).

One of the most common notations for PLCs is provided by function block diagrams (FBD) which use rectangles to represent predefined functions, such as elementary Boolean gates. The inputs of such a rectangle or *block* are on its left side, the outputs on its right. For instance, a block corresponding to conjunction has an ampersand (&) at its top, whereas a disjunction is symbolized by >=1. The negation of an input or output variable is denoted by a small circle. Normally, no block can ever change the value of an input from the environment; but see Sect. 3.2 for an exception.

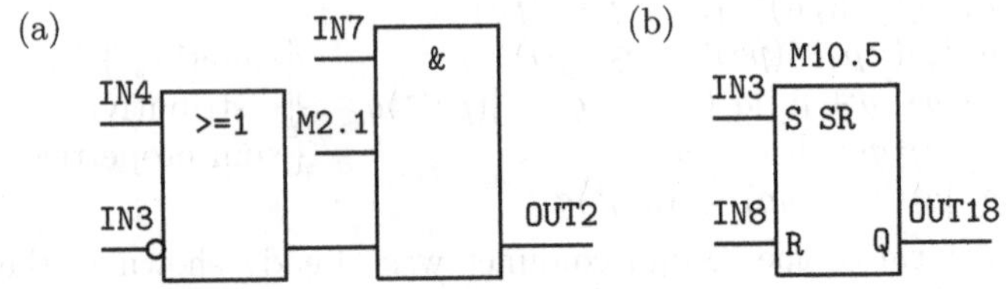

Fig. 1. Boolean functions (a) and an SR-Flip-flop (b) in FBD

More complex functions can be obtained by linking elementary rectangles, where the *evaluation order* is from left to right and from top to bottom. So the FBD in Fig. 1(a) computes the Boolean function (IN4 $\vee$ $\neg$IN3) $\wedge$ IN7 $\wedge$ M2.1 and returns the result on output OUT2.

Blocks for logical connectives lack the possibility of dynamic behavior and storing of values. A *flip-flop* is an elementary block with such abilities. Flip-flops have two inputs: one set and one reset input, marked by S and R in their FBDs. Moreover, they have an internal variable (called *marker*, in FBDs written above the top line) and an output Q which always has the same value as the marker. If the set input is TRUE and the reset input is FALSE then output and marker are set to TRUE. A FALSE-signal on the set input and a TRUE-signal on the reset input set output and marker to FALSE. If both the set and reset inputs receive a FALSE-signal then the values of output and marker remain unchanged. A set/reset conflict occurs if both the set and reset inputs are TRUE. There are two types of flip-flops, namely set-dominant and reset-dominant or RS- and SR-flip-flops, resp. Upon a set/reset conflict, an RS-flip-flop sets marker and output to FALSE, while an SR-flip-flop sets both to TRUE. Figure 1(b) shows the FBD of a reset-dominant flip-flop with IN3 on its set input, IN8 on its reset input, and output and internal markers OUT18 and M10.5 (which by the above conventions refers to bit 5 of variable M10), resp.

As the last elementary block we consider a simple assignment as shown in Fig. 2(a). It assigns the value of IN5 to the internal variable M20.3. Such blocks can be used for every data type (of course, the variables involved have to be of compatible types, as in other programming languages).

Although our formalization mostly works with Boolean values, we also need some blocks working with non-Boolean values. For instance, an important further concept is that of a counter, depicted in Fig. 2(b). It has two Boolean inputs CU and R, one integer input PV, one Boolean output Q and one integer output CV. A TRUE-signal on R resets the counter to zero. The counter value of the output CV is increased by one upon a positive edge (i.e., a change from FALSE to TRUE) on

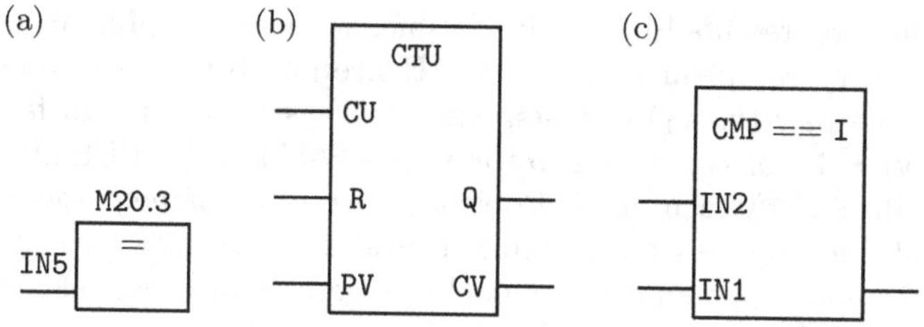

Fig. 2. Assignment, counter and comparator in FBD

the input CU. The input PV can be used for setting the counter to a desired value (so one can also reset the counter by feeding zero into it). Finally, the output Q returns the truth value of the comparison CV ≠ 0.

Almost self-explanatory, the FBD of Fig. 2(c) is a comparator which compares the numerical values of its inputs IN1 and IN2.

In order to obtain timed signals most PLCs offer the possibility of configuring the single bits of a specified internal byte as pulse generators with various frequencies. Often one chooses the byte M100 and assigns to the single bits frequencies as in Fig. 3. In the sequel, we will follow this convention.

Bit	M100.7	M100.6	M100.5	M100.4	M100.3	M100.2	M100.1	M100.0
Frequency	2 Hz	1.6 Hz	1 Hz	0.8 Hz	0.5 Hz	0.4 Hz	0.2 Hz	0.1 Hz

Fig. 3. Common frequencies of pulse generators

The signal corresponding to one such bit is a wave of rectangular pulses with the associated frequency. E.g., the signal corresponding to bit 100.5 is set alternately half a second to TRUE and half a second to FALSE (Fig. 3).

3.2 Modeling Function Block Diagrams in Modal Kleene Algebra

As already shown in [21], FBD programs can be translated into MKA expressions modeling their behavior. The representation uses a glassbox view of components, i.e., all connections and their names are visible. In a relational model then a state is a function from the set of all names to values, and a component (and hence even the whole program) a relation between states. Since by the PLC conventions evaluation follows the left-to-right top-to-bottom diagram order, one can describe a composite component as a linear sequence of relational compositions of elementary components. However, one can abstract from the relational view by associating with each elementary block an MKA element and considering as components only linear products of such elements.

Boolean connections and the values of the corresponding input, output and internal variables can be modeled by tests. However, following PLC conventions, each Boolean is represented by a pair of values with the coupling invariant that they always carry complementary values. Hence an abstract variable v is represented by the pair $(\mathtt{v}_0, \mathtt{v}_1)$ of tests, where always $\mathtt{v}_0 = \neg\mathtt{v}_1$. In fact, usually $\mathtt{v}_0 = 0$ and $\mathtt{v}_1 = 1$, corresponding to the values FALSE and TRUE of v, resp.

We formally specify each simple Boolean gate by a set of inequations involving the diamond operator. As an example, consider an OR-gate with inputs in1, in2 and in3 and output out1. To model this gate as an MKA element or, we characterize its behavior by the following inequations:

$$\mathtt{in1}_1 + \mathtt{in2}_1 + \mathtt{in3}_1 \leq |\mathtt{or}\rangle\mathtt{out1}_1 \qquad \mathtt{in1}_0 \cdot \mathtt{in2}_0 \cdot \mathtt{in3}_0 \leq |\mathtt{or}\rangle\mathtt{out1}_0$$

For simplicity, we do not treat negations as blocks but simply swap $\mathtt{v}_0$ and $\mathtt{v}_1$ for a variable v. Every gate gat is deterministic and total; so we require $|\mathtt{gat}\rangle p = |\mathtt{gat}]p$ for every test p. Hence, a quick calculation using shunting shows

$$q \leq |\mathtt{gat}\rangle p \wedge \neg q \leq |\mathtt{gat}\rangle\neg p \Leftrightarrow q = |\mathtt{gat}\rangle p \Leftrightarrow \neg q = |\mathtt{gat}\rangle\neg p$$

This is precisely the shape of the axioms for the OR-gate above.

Moreover, we have to ensure that a block at most modifies its output and internal variables. In the above example, we have to add the inequations $\mathtt{v}_1 \leq |\mathtt{or}\rangle\mathtt{v}_1$ and $\mathtt{v}_0 \leq |\mathtt{or}\rangle\mathtt{v}_0$ for all variables v except out1 as *tracking conditions* for v. The only exception is with the last-evaluated block of an FBD. To allow composition of an FBD with itself and hence also its star iteration, we have to allow that the input channels in the next execution cycle receive new values; so for the last block we drop the above condition for all input variables of the overall FBD. The same holds for output variables which are computed from scratch in every cycle. Note that internal variables have tracking conditions also at the last-evaluated gate because their value is stored for the next cycle. In our formalization, we even omitted the conditions for input variables that are not used later on in the FBD in order to keep the formalization as small as possible. An input variable that is used in a certain block B but does not appear as input of any block following it (in evaluation order) does not have to be tracked across the program after block B. Its new value will be determined by its input channel in the following execution cycle. For example, in the FBD from the left part of Fig. 1, there are neither the inequation $\mathtt{IN4}_0 \leq |\mathtt{or}\rangle\mathtt{IN4}_0$ nor $\mathtt{IN7}_1 \leq |\mathtt{and}\rangle\mathtt{IN7}_1$.

As stated, the overall behavior of a single program cycle can then be described by the product of all blocks in their evaluation order (which is basically a topological sorting corresponding to western reading conventions; for details see [2]). For example, if we consider the whole Fig. 1 as a PLC program and denote the blocks by or1, and1 and sr1, resp., it corresponds to the expression $\mathtt{or1}\cdot\mathtt{and1}\cdot\mathtt{sr1}$ (recall that negations are modeled by simply swapping $\mathtt{v}_0$ and $\mathtt{v}_1$). If not indicated otherwise, this product describing (a single execution cycle of) an FBD program is named cycle; it corresponds to the transition element a from Sect. 2.2, and we use O, U, $\Diamond$ and $\Box$ w.r.t. it. It is easy to see that total functionality of the single blocks propagates to the whole program. Following [21], one can give analogous formalizations for the other Boolean gates. In the next section we will deal with the other blocks we introduced in Sect. 3.1.

3.3 Formalization of Timers

A common mechanism for generating timed signals is shown in Fig. 4. There, a TRUE-value on `req` activates the flip-flop (we will discuss its reset input soon) whose output is conjoined with a timer signal of 1 Hz. As long as the output of the flip-flop equals TRUE the counter value will be increased every second by one and is stored in the internal variable M50. This behavior will persist as long as `res` does not become TRUE (even if `req` changes its value to FALSE). However, a TRUE-signal of `res` resets both the counter and the flip-flop (note that the flip-flop is reset dominant and that `res` acts as a resetter for both the flip-flop and the counter). The further behavior depends on the value of `req`.

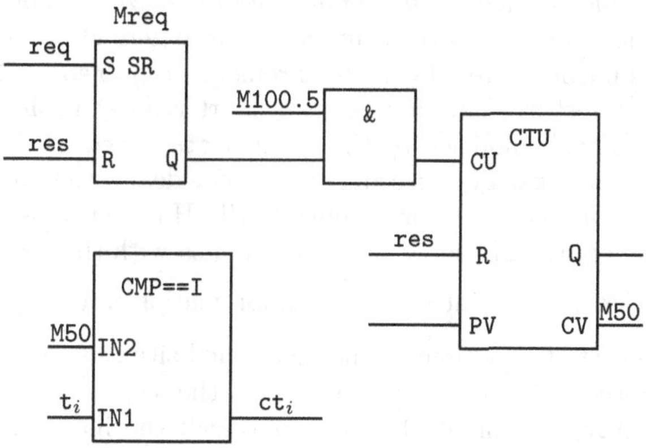

Fig. 4. Generating Time Signals in FBD

For further processing, one often wants to trigger some action at a certain time after starting the counter. In this case, the counter value is compared with the desired time; the Boolean output is used as trigger signal. This is shown exemplarily in the bottom part of Fig. 4: $\mathtt{ct}_i$ becomes TRUE when the output M50 of the counter equals the value of $\mathtt{t}_i$. Usually, for $\mathtt{t}_i$ one uses a constant value to start an action after a given time as we will do in the further course.

In general, one such counter can be associated with several comparators to enable a timed sequence of activations. In this context, one regularly has also a cut-off time `tcu` after which the timer should be reset to zero. This can be achieved easily by feeding the output of a suitable comparator to the counter's and flip-flop's reset inputs.

In the sequel, we will view the described *timer* component as a black box with inputs `req` and `res` as start and reset signals, and the comparator results as outputs. To formalize timers, we arrange the compared values (fed into the IN1 lines of the comparators) in increasing order as a sequence $\mathtt{t}_0$, $\mathtt{t}_1$, ..., $\mathtt{t}_n$, such that the final value $\mathtt{t}_n$ is used as reset input as described above, and denote the (Boolean) outputs of the respective comparators by $\mathtt{ct}_0$, $\mathtt{ct}_1$, ..., $\mathtt{ct}_n$.

For better readability, we use in the sequel the notation $\mathtt{ct}_{i,1}$ instead of $(\mathtt{ct}_i)_1$ for indicating a TRUE-value of $\mathtt{ct}_i$ and define $\mathtt{ct}_{i,0}$ analogously. In particular, $\mathtt{ct}_{i,1}$ corresponds to a state where the counter stands at $\mathtt{t_i}$. Then we require the following properties:

- **No simultaneity:** If the output of one timer comparator is TRUE then the others have to be FALSE. In MKA, this can be described by $\mathtt{ct}_{i,1} \leq \prod_{j \neq i} \mathtt{ct}_{j,0}$.
- **Order:** The timer comparators output TRUE according to the above ordering. This means, we have $\mathtt{ct}_{i,1} \leq \Diamond\, \mathtt{ct}_{i+1,1}$ for all $0 \leq i < n$.
- **Resetting:** A TRUE-value of $\mathtt{ct}_n$ resets the counter in the following cycle. Therefore, we have $\mathtt{ct}_{n,1} \leq \circ\mathtt{ct}_{0,1}$.
- **Resting:** We have to ensure that the counter does not start until it gets a request. This means that if the counter stands at zero (modeled by $\mathtt{ct}_{0,1}$) and there is no request then the counter stands at zero also in the subsequent state. To this purpose, we add the requirement $\mathtt{ct}_{0,1} \cdot \mathtt{req}_0 \leq \circ\, \mathtt{ct}_{0,1}$.
- **Starting:** If a resting counter receives a start request it should eventually output $\mathtt{ct}_{1,1}$. This is modeled by the formula $\mathtt{ct}_{0,1} \cdot \mathtt{req}_1 \leq \Diamond\, \mathtt{ct}_{1,1}$.
- **Intermediate states:** The preceding properties deal with situations in which at least the counter comparator outputs TRUE. However, most of the time all the outputs equal FALSE; so we have to deal also with this situation. To ease reading and writing, we introduce the abbreviation $\mathtt{nst} =_{df} \prod_{i=0}^{n} \mathtt{ct}_{i,0}$ (nst stands for 'no significant time'). The mentioned situation occurs if the value of M50 is between two consecutive values of the sequence $\mathtt{t_i}$; so we require $\mathtt{ct}_{i,1} \leq \mathtt{nst}\, \mathsf{U}\, \mathtt{ct}_{i+1,1}$ for all $0 \leq i < n$ (recall the modeling of the until-operator in Sect. 3.2). Note also that, by the resetting rule, $\mathtt{ct}_{n,1}$ is followed temporally immediately by $\mathtt{ct}_{0,1}$.
- **No other states:** The system is either in a situation where a counter comparator's output is TRUE or it is awaiting another counter comparator's output to become TRUE. In our framework, this reads

$$\models \mathtt{ct}_{n,1} \vee \bigvee_{i=0}^{n-1} (\mathtt{nst}\, \mathsf{U}\, \mathtt{ct}_{i,1}) \qquad \models \mathtt{ct}_{n,1} + \sum_{i=0}^{n-1} |(\mathtt{nst} \cdot \mathtt{cycle})^*\rangle \mathtt{ct}_{i,1} \quad (4)$$

The counter comparator outputs cannot be altered by any block except the last one in the evaluation ordering. This ensures that time does not change during the execution of one cycle but also may progress between two consecutive cycles. The modeling of this behavior is analogous to that given in Sect. 3.2.

3.4 A Case Study: Traffic Lights

As an example application we chose an FBD controlling the pedestrian lights of a traffic control signal. With the conventions of Sect. 3.3 it looks as in Fig. 5. If the button is pressed, the pedestrian lights should eventually become green for ten seconds. After a green phase of the pedestrian lights it should take at least nine seconds before the pedestrian lights can become green again (to respect car drivers). Also, there should be a time of three seconds for the car traffic lights to become yellow after pushing the request button. So, if one starts with red pedestrian lights and a timer at zero, a push should lead to green pedestrian lights after three seconds. Then, the lights should stay green for ten seconds, and a new such cycle can start only nine seconds later. Here, the variables have the following definitions and meanings:

- push is a Boolean input from the request button of pedestrian lights.
- gr is an internal Boolean variable whose value indicates whether the pedestrian lights are green (this value can be forwarded to different output signals; however, for our verification purposes it suffices to consider only gr itself).
- req and res are the start and reset inputs of a timer.
- c_0, c_3, c_{13} and c_{22} are the outputs of a timer, corresponding to values of zero, three, thirteen and twenty-two seconds, resp., after starting the timer.

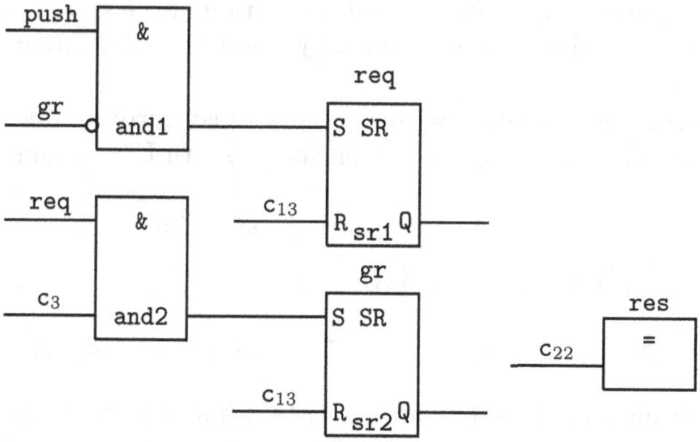

Fig. 5. Pedestrian Lights in Modified FBD

Let us take a short look at the functionality described in Fig. 5. A push of the request button has only an effect if the pedestrian lights are not yet green (this is the purpose of and1). If such a push manages to pass and1 then it activates the timer via the flip-flop sr1. The timer start is reset after thirteen seconds (note that this does not stop or reset the timer) to prevent a new activation of the timer after finishing the current cycle. and2 and sr2 set the lights to green after three seconds if req was set to TRUE by some preceding push of the request button, and reset the lights to red (implicitly by setting gr to FALSE) after thirteen seconds. Finally, the timer is reset after twenty-two seconds by the assignment of c_{22} to the timer reset signal res. A sample from the KIV formalization together with explanations can be found in the readme file of [6].

The program should fulfill some temporal properties concerning the interplay between the variable gr and the timer. We introduce three examples together with proof sketches in order to give the reader an impression how the rules from Sect. 2.3 can be used in our context.

- If the timer value is three and there is a request then the lights should be green in the following cycle. The LTL specification of this reads

$$\models \Box(\mathtt{req}_1 \wedge c_{3,1} \rightarrow \bigcirc \mathtt{gr}_1) \tag{5}$$

which can be rewritten in MKA as

$$\models |\mathtt{lig}^*](\mathtt{req}_1 \cdot c_{3,1} \rightarrow |\mathtt{lig}\rangle \mathtt{gr}_1) \tag{6}$$

Here we used `lig` (short for `lights`) instead of `cycle` as in Sect. 3.2 for the overall behavior of the system in one cycle. Concretely, we have $\mathtt{lig} = \mathtt{and1} \cdot \mathtt{sr1} \cdot \mathtt{and2} \cdot \mathtt{sr2}$ using the namings from Fig. 5 (the counter properties are added as axioms to the the formalization, so the assignment of c_{22} to `res` is covered by the associated `Resetting` rule). One may wonder why we stipulate additionally a request in the precondition. The reason is that it makes later reasoning a lot more convenient, and it is justified by the fact that under reasonable assumptions about the start condition of the system the timer can reach three only if there is an additional request (see also Equations (9)/(10) below).
- Clearly, we are not satisfied with the lights just turning green after three seconds, they should stay green for ten seconds. An LTL formula for this is

$$\models \Box(\mathtt{req}_1 \wedge c_{3,1} \rightarrow \bigcirc(\mathtt{gr}_1 \, \mathsf{U} \, c_{13,1})) \tag{7}$$

with an equivalent MKA characterization

$$\models |\mathtt{lig}^*](\mathtt{req}_1 \cdot c_{3,1} \rightarrow |\mathtt{lig}\rangle(|(\mathtt{gr}_1 \cdot \mathtt{lig})^*\rangle c_{13,1})) \tag{8}$$

- These two properties do not require a certain initial state of the system (the outermost operator is an always operator). However, a start in an inappropriate state can lead to undesired behavior. For example, one can show that $c_{0,1} \cdot \mathtt{gr}_1 \leq |\mathtt{lig}^*]c_{0,1} \cdot \mathtt{gr}_1$ holds, which means that the lights stay green all the time. This initial state contradicts the intention of the program, because the lights should turn green only after the timer reaches the value three, and they have to be set to red again before the timer is reset; so the state $c_{0,1} \cdot \mathtt{gr}_1$ would represent an inconsistency. A reasonable choice for the initial state is that the lights are red and the timer is zero. Starting in such a state and pushing the button while the lights are red, we want that from the next state we eventually reach a state with the following properties:
  - the lights are green,
  - the counter value is thirteen, and
  - in the next state the lights are red until the counter reaches 22.

  Note that this is essentially a statement only about the last cycle where the timer value is thirteen. We chose this example because it is well suited as an illustration of the application of our techniques. An LTL-formulation of this property is

$$\models \mathtt{gr}_0 \wedge c_{0,1} \rightarrow \Box(\mathtt{gr}_0 \wedge \mathtt{push}_1 \rightarrow \bigcirc(\Diamond\, \mathtt{gr}_1 \wedge c_{13,1} \wedge \bigcirc(\mathtt{gr}_0 \, \mathsf{U} \, c_{22,1}))) \tag{9}$$

and its translation into MKA reads

$$\begin{aligned}&\models \mathtt{gr}_0 \cdot \mathtt{c}_{0,1} \rightarrow \\ &\quad |\mathtt{lig}^*](\mathtt{gr}_0 \cdot \mathtt{push}_1 \rightarrow \\ &\qquad\qquad |\mathtt{lig}\rangle(|\mathtt{lig}^*\rangle \mathtt{gr}_1 \cdot \mathtt{c}_{13,1} \cdot |\mathtt{lig}\rangle|(\mathtt{gr}_0 \cdot \mathtt{lig})^*\rangle \mathtt{c}_{22,1}))\end{aligned} \tag{10}$$

Property (6) is rather easy to prove: First, we note that by (3) it suffices to show $\models \mathtt{req}_1 \cdot \mathtt{c}_{3,1} \rightarrow |\mathtt{lig}\rangle \mathtt{gr}_1$ due to the property $|x]1 = 1$ in all MKAs. Equivalently, we can show that $\mathtt{req}_1 \cdot \mathtt{c}_{3,1} \leq |\mathtt{lig}\rangle \mathtt{gr}_1$ holds. This is done easily by unfolding the definition of lig and iterated application of modality and isotony of the diamond with the aid of the characterization of the respective blocks.

Similarly, we can prove **Property** (8) by showing the inequation $\mathtt{req}_1 \cdot \mathtt{c}_{3,1} \leq |\mathtt{lig}\rangle|(\mathtt{gr}_1 \cdot \mathtt{lig})^*\rangle \mathtt{c}_{13,1}$ which is an example for the application of Lemma 3. This means we have to show $\mathtt{gr}_1 \cdot \neg \mathtt{c}_{13,1} \leq |\mathtt{lig}\rangle \mathtt{gr}_1$ and $\mathtt{req}_1 \cdot \mathtt{c}_{3,1} \leq |\mathtt{lig}\rangle(\mathtt{gr}_1 \cdot |\mathtt{lig}^*\rangle \mathtt{c}_{13,1})$. The first inequation can be shown analogously to the proof sketch of Equation (6). Due to total functionality, definition of the diamond and distributivity, the second one can be split up into $\mathtt{req}_1 \cdot \mathtt{c}_{3,1} \leq |\mathtt{lig}\rangle \mathtt{gr}_1$ and $\mathtt{req}_1 \cdot \mathtt{c}_{3,1} \leq |\mathtt{lig}\rangle|\mathtt{lig}^*\rangle \mathtt{c}_{13,1}$. The first inequation is already known from the proof of Equation (6). For the last one, we have the chain of inequalities $\mathtt{req}_1 \cdot \mathtt{c}_{3,1} \leq \mathtt{c}_{3,1} \leq |\mathtt{lig}\rangle|(\mathtt{nst} \cdot \mathtt{lig})^*\rangle \mathtt{c}_{13,1} \leq |\mathtt{lig}\rangle|\mathtt{lig}^*\rangle \mathtt{c}_{13,1}$ by isotony, timer properties and isotony of multiplication and diamond.

For **Property** (10) we will resort only to rough explanations and refer the reader for details to the full KIV project file [6]. As in the previous cases, we transform the claim into the equivalent inequation

$$\mathtt{gr}_0 \cdot \mathtt{c0}_1 \leq |\mathtt{lig}^*](\mathtt{gr}_0 \cdot \mathtt{push}_1 \rightarrow (|\mathtt{lig}^*\rangle \mathtt{gr}_1 \cdot \mathtt{c}_{13,1} \cdot |\mathtt{lig}\rangle|(\mathtt{gr}_0 \cdot \mathtt{lig})^*\rangle \mathtt{c}_{22,1}))$$

Here we can simplify the right side using the equality $\mathtt{gr}_1 \cdot \mathtt{c}_{13,1} = \mathtt{gr}_1 \cdot \mathtt{c}_{13,1} \cdot |\mathtt{lig}\rangle|(\mathtt{gr}_0 \cdot \mathtt{lig})^*\rangle \mathtt{c}_{22,1}$ (this follows from $\mathtt{gr}_1 \cdot \mathtt{c}_{13,1} \leq |\mathtt{lig}\rangle|(\mathtt{gr}_0 \cdot \mathtt{lig})^*\rangle \mathtt{c}_{22,1}$ which in turn can be shown using timer properties and Lemma 3) and reduce our task — after exploiting isotony — to showing $\mathtt{gr}_0 \cdot \mathtt{c}_{0,1} \leq |\mathtt{lig}^*](\mathtt{gr}_0 \cdot \mathtt{push}_1 \rightarrow (|\mathtt{lig}^*\rangle \mathtt{gr}_1 \cdot \mathtt{c}_{13,1}))$. Introducing the abbreviations $\mathtt{init} =_{df} \mathtt{gr}_0 \cdot \mathtt{c}_{0,1}$ and $\mathtt{result} =_{df} |\mathtt{lig}^*\rangle \mathtt{gr}_1 \cdot \mathtt{c}_{13,1}$ this goal reads $\mathtt{init} \leq |\mathtt{lig}^*](\mathtt{gr}_0 \cdot \mathtt{push}_1 \rightarrow \mathtt{result})$. Now we use the timer's no-other-state property (4) and replace the above inequation by seven of the type $\mathtt{init} \leq |\mathtt{lig}^*](\mathtt{gr}_0 \cdot \mathtt{push}_1 \cdot \mathtt{itm} \rightarrow \mathtt{result})$ where the intermediate value itm has the form $\mathtt{itm} = \mathtt{c}_{\mathtt{i},1}$ for $\mathtt{i} = 0, 3, 13, 22$ or $\mathtt{itm} = |(\mathtt{nst} \cdot \mathtt{lig})^*\rangle \mathtt{c}_{\mathtt{j},1}$ with $\mathtt{j} = 3, 13, 22$. Some of these cases can be handled by showing that the argument of the diamond evaluates to 1. For the remaining properties, a crucial point is the application of Lemmas 1 and 2 together with appropriate timer properties, isotony and MKA calculus.

Summing up our experiences, we can say that after some time of familiarization, proving in KIV became routine work without greater difficulties. An increasing amount of calculation rules and lemmata in MKA made it a pleasant task.

4 Conclusion and Outlook

After proving some useful LTL rules concerning the until-operator in MKA we applied them successfully to a considerable extension of the verification framework developed in [21].

Now that the theoretical foundations are laid, notably concerning the treatment of timing issues, the next step will be to tackle the verification of larger, more lifelike systems. A great help for this goal will be be the by now substantial body of reusable rules and lemmata we have accumulated. Other topics of future work concern the automated construction of input files, comparison with model checkers and extension of the approach to other PLC languages.

Another point is a formal proof of the properties from Sect. 3.3. In the present paper, these rules were inserted as axioms without further verification. A proof needs meta-knowledge about the natural numbers which has to be added in some way. Moreover, one has to assume and to model the condition that the execution time of one cycle of the PLC does not exceed the period of the used frequency generator. Otherwise, effects similar to the Nyquist-Shannon sampling theorem (see [44]) would destroy the functionality of Fig. 4.

Acknowledgement. We are grateful to the anonymous referees for their careful scrutiny and helpful remarks.

References

1. Coq. https://coq.inria.fr/. Accessed 7 July 2015
2. IEC61131. http://webstore.iec.ch/webstore/webstore.nsf/artnum/048541!opendocument. Accessed 20 Mar 2018
3. The KIV system. http://www.isse.uni-augsburg.de/en/software/kiv/. Accessed 20 Mar 2018
4. NuSMVExamples. http://nusmv.fbk.eu/examples/examples.html. Accessed 7 Aug 2018
5. Step7. http://w3.siemens.com/mcms/simatic-controller-software/en/step7/Pages/Default.aspx. Accessed 20 Mar 2018
6. Verification of pedestrian lights in MKA. http://rolandglueck.de/Downloads/Pedestrian_lights_verified.zip. Accessed 20 Mar 2018
7. VerifyThis 2015. http://verifythis2015.cost-ic0701.org/results. Accessed 8 Aug 2018
8. VerifyThis 2017. http://www.pm.inf.ethz.ch/research/verifythis/Archive/2017.html. Accessed 8 Aug 2018
9. Alur, R., Dill, D.L.: A theory of timed automata. Theor. Comput. Sci. **126**(2), 183–235 (1994)
10. Back, R.-J., von Wright, J.: Refinement Calculus - A Systematic Introduction. Graduate Texts in Computer Science. Springer, New York (1998)
11. Ben-Ari, M.: Mathematical Logic for Computer Science, 3rd edn. Springer, London (2012)
12. Berghammer, R., Stucke, I., Winter, M.: Using relation-algebraic means and tool support for investigating and computing bipartitions. J. Log. Algebr. Meth. Prog. **90**, 102–124 (2017)

13. Birkhoff, G.: Lattice Theory, 3rd edn. American Mathematical Society, Providence (1967)
14. Brunet, P., Pous, D., Stucke, I.: Cardinalities of finite relations in Coq. In: Blanchette, J.C., Merz, S. (eds.) ITP 2016. LNCS, vol. 9807, pp. 466–474. Springer, Cham (2016). https://doi.org/10.1007/978-3-319-43144-4_29
15. Carlsson, H., Svensson, B., Danielson, F., Lennartson, B.: Methods for reliable simulation-based PLC code verification. IEEE Trans. Ind. Inform. **8**(2), 267–278 (2012)
16. Desharnais, J., Möller, B.: Non-associative Kleene algebra and temporal logics. In: Höfner, P., Pous, D., Struth, G. (eds.) RAMICS 2017. LNCS, vol. 10226, pp. 93–108. Springer, Cham (2017). https://doi.org/10.1007/978-3-319-57418-9_6
17. Desharnais, J., Möller, B., Struth, G.: Modal Kleene algebra and applications - a survey. J. Relat. Methods Comput. Sci. **1**, 93–131 (2004)
18. Ehm, T., Möller, B., Struth, G.: Kleene modules. In: Berghammer, R., Möller, B., Struth, G. (eds.) RelMiCS 2003. LNCS, vol. 3051, pp. 112–123. Springer, Heidelberg (2004). https://doi.org/10.1007/978-3-540-24771-5_10
19. Ertel, J.: Verifikation von SPS-Programmen MIT Kleene Algebra. Master's thesis, Institut of Informatics, University of Augsburg (2017)
20. Ésik, Z., Fahrenberg, U., Legay, A., Quaas, K.: Kleene algebras and semimodules for energy problems. In: Van Hung, D., Ogawa, M. (eds.) ATVA 2013. LNCS, vol. 8172, pp. 102–117. Springer, Cham (2013). https://doi.org/10.1007/978-3-319-02444-8_9
21. Glück, R., Krebs, F.B.: Towards interactive verification of programmable logic controllers using modal Kleene algebra and KIV. In: Kahl, W., Winter, M., Oliveira, J.N. (eds.) RAMICS 2015. LNCS, vol. 9348, pp. 241–256. Springer, Cham (2015). https://doi.org/10.1007/978-3-319-24704-5_15
22. Gondran, M., Minoux, M.: Graphs, Dioids and Semirings. Springer, Heidelberg (2008)
23. Guttmann, W.: Stone relation algebras. In: Höfner, P., Pous, D., Struth, G. (eds.) RAMICS 2017. LNCS, vol. 10226, pp. 127–143. Springer, Cham (2017). https://doi.org/10.1007/978-3-319-57418-9_8
24. Höfner, P., Möller, B.: Dijkstra, Floyd and Warshall meet Kleene. Formal Asp. Comput. **24**(4–6), 459–476 (2012)
25. Hollenberg, M.: An equational axiomatization of dynamic negation and relational composition. J. Log. Lang. Inf. **6**(4), 381–401 (1997)
26. Hollenberg, M.: Equational axioms of test algebra. In: Nielsen, M., Thomas, W. (eds.) CSL 1997. LNCS, vol. 1414, pp. 295–310. Springer, Heidelberg (1998). https://doi.org/10.1007/BFb0028021
27. Jackson, M., McKenzie, R.: Interpreting graph colorability in finite semigroups. IJAC **16**(1), 119–140 (2006)
28. Jee, E., Yoo, J., Cha, S.D., Bae, D.-H.: A data flow-based structural testing technique for FBD programs. Inf. Softw. Technol. **51**(7), 1131–1139 (2009)
29. Jipsen, P., Rose, H.: Varieties of Lattices, 1st edn. Springer, Heidelberg (1992)
30. Kahl, W.: Graph transformation with symbolic attributes via monadic coalgebra homomorphisms. ECEASST **71**, 5.1–5.17 (2014)
31. Kawahara, Y., Furusawa, H.: An algebraic formalization of fuzzy relations. Fuzzy Sets Syst. **101**(1), 125–135 (1999)
32. Kozen, D.: A completeness theorem for Kleene algebras and the algebra of regular events. Inf. Comput. **110**(2), 366–390 (1994)
33. Kozen, D.: Kleene algebra with tests. ACM Trans. Prog. Lang. Syst. **19**(3), 427–443 (1997)

34. Kröger, F., Merz, S.: Temporal Logic and State Systems. Texts in Theoretical Computer Science. An EATCS Series. Springer, Heidelberg (2008)
35. Li, J., Qeriqi, A., Steffen, M., Yu, I.C.: Automatic translation from FBD-PLC-programs to NuSMV for model checking safety-critical control systems. In: NIK 2016. Bibsys Open Journal Systems, Norway (2016)
36. Litak, T., Mikulás, S., Hidders, J.: Relational lattices. In: Höfner, P., Jipsen, P., Kahl, W., Müller, M.E. (eds.) RAMICS 2014. LNCS, vol. 8428, pp. 327–343. Springer, Cham (2014). https://doi.org/10.1007/978-3-319-06251-8_20
37. Manes, E., Benson, D.: The inverse semigroup of a sum-ordered semiring. Semigroup Forum **31**, 129–152 (1985)
38. Michels, G., Joosten, S., van der Woude, J., Joosten, S.: Ampersand. In: de Swart, H. (ed.) RAMICS 2011. LNCS, vol. 6663, pp. 280–293. Springer, Heidelberg (2011). https://doi.org/10.1007/978-3-642-21070-9_21
39. Möller, B., Höfner, P., Struth, G.: Quantales and temporal logics. In: Johnson, M., Vene, V. (eds.) AMAST 2006. LNCS, vol. 4019, pp. 263–277. Springer, Heidelberg (2006). https://doi.org/10.1007/11784180_21
40. Möller, B., Roocks, P.: An algebra of database preferences. J. Log. Algebr. Meth. Program. **84**(3), 456–481 (2015)
41. Oliveira, J.N.: A relation-algebraic approach to the "Hoare logic" of functional dependencies. J. Log. Algebr. Meth. Prog. **83**(2), 249–262 (2014)
42. Pavlovic, O., Ehrich, H.-D.: Model checking PLC software written in function block diagram. In: ICST 2010, CEUR Workshop Proceedings. IEEE Computer Society (2010)
43. Pratt, V.: Dynamic algebras: examples, constructions, applications. Studia Logica **50**, 571–605 (1991)
44. Shannon, C.E.: Communication in the presence of noise. Proc. IRE **37**(1), 10–21 (1949)
45. Solin, K., von Wright, J.: Enabledness and termination in refinement algebra. Sci. Comput. Prog. **74**(8), 654–668 (2009)
46. von Karger, B.: Temporal algebra. Math. Struct. Comput. Sci. **8**(3), 277–320 (1998)
47. Wan, H., Chen, G., Song, X., Gu, M.: Formalization and verification of PLC timers in Coq. In: Ahamed, S.I., et al. (eds.): Proceedings of the COMPSAC 2009, pp. 315–323. IEEE Computer Society (2009)
48. Wimmer, S., Lammich, P.: Verified model checking of timed automata. In: Beyer, D., Huisman, M. (eds.) TACAS 2018. LNCS, vol. 10805, pp. 61–78. Springer, Cham (2018). https://doi.org/10.1007/978-3-319-89960-2_4

False Failure: Creating Failure Models for Separation Logic

Callum Bannister[1,2] and Peter Höfner[1,2](✉)

[1] Data61, CSIRO, Sydney, Australia
{Callum.Bannister,Peter.Hoefner}@data61.csiro.au
[2] Computer Science and Engineering,
University of New South Wales, Sydney, Australia

Abstract. Separation logic, an extension of Floyd-Hoare logic, finds countless applications in areas of program verification, but does not allow forward reasoning in the setting of total or generalised correctness. To support forward reasoning, separation logic needs to be equiped with a failure element. We present several ways on how to add such an element. We show that none of the 'obvious' extensions preserve all the algebraic properties desired. We develop more complicated models, satisfying the desired properties, and discuss their use for forward reasoning.

1 Introduction

Some of the most prominent methods for formal reasoning about the correctness of programs are Floyd-Hoare logic [13,14], Dijkstra's weakest-precondition calculus [10], and strongest postconditions [13]. The usefulness and importance of these approaches are undeniable and they have been used in the area of formal verification countless times.

However, a shortcoming of techniques based on Hoare logic is that they lack expressiveness for shared mutable data structures, such as structures where updatable fields can be referenced from more than one point. To overcome this deficiency, Reynolds, O'Hearn and others developed separation logic [27,30]. It extends Hoare logic by separating conjunctions, and adds assertions to express separation between memory regions, which allows for local reasoning by splitting memory into two halves: the part the program interacts with, and the part which remains untouched, called the *frame*. Later, O'Hearn extended this language to concurrent programs that work on shared mutable data structures [26].

It is known that sets of assertions, in combination with separating conjunction form a quantale [8]. Quantales [24,31], sometimes called standard Kleene algebras [6], or the more general concept of semirings have been used to derive algebraic characterisations for Hoare logic [19,22] and the wp-calculus of Dijkstra [23]. In the quantale of assertions all operations of separation logic, such as separating conjunction and separating implication are related via Galois connections and dualities [1,8]. Many useful theorems about separation logic follow 'for free' from the underlying algebraic theory.

J. Desharnais et al. (Eds.): RAMiCS 2018, LNCS 11194, pp. 263–279, 2018.
https://doi.org/10.1007/978-3-030-02149-8_16

Separation logic has been used for reasoning in weakest-precondition style [10], which proceeds backwards from a given postcondition and a given program by determining the weakest precondition [1]. It has also been used in the setting of forward reasoning, using strongest postconditions. However, in the latter setting only partial correctness can be considered, as separation logic does not contain formulae corresponding to the *failed* execution of programs. Both forward and backward reasoning in separation logic heavily rely on the Galois connections and dualities mentioned above.

In this paper we discuss how to extend the assertion quantale of separation logic to handle failed executions, bearing the application of forward reasoning in mind. The created models should maintain as much of the algebraic structure as possible, which would allow us to reuse knowledge from the original separation logic and from the algebraic meta-theory. We show that all *simple* models fail to satisfy all algebraic properties we hope for: some lose associativity of separating conjunction, while others maintain associativity but do not relate the operators via Galois connections. As our search for a good model is systematic, we conclude that there cannot be a simple, but powerful model for separation logic, which features failure and has 'nice' algebraic properties at the same time. Our final models, inspired by the construction of integers from natural numbers, achieve the desired properties. The cost is splitting separating implication from its dual. Although this looks like an acceptable trade-off, we conclude that the new model is not suitable for forward reasoning either as it leads to undesirable behaviour.

2 Algebraic Separation Logic

Assertions are a crucial ingredient in separation logic. They describe states consisting of a store and a heap, roughly corresponding to the state of local variables and dynamically-allocated objects. They are used as predicates to describe the contents of heaps and stores, and as pre- or postconditions of programs, as in Hoare logic. The semantics of the assertion language is given by the relation $s, h \models p$ of *satisfaction.*[1] Informally, $s, h \models p$ holds if the state (s, h) satisfies the assertion p. A *state* (s, h) contains a *store* $s : V \rightharpoonup \mathit{Values}$ – a partial function from the set V of all variables into a set of *Values*[2] – and a heap $h : \mathit{Addr} \rightharpoonup \mathit{Values}$, which maps an arbitrary set of (heap) addresses to values. An assertion p is called *valid* (or *pure*) iff p holds in every state and p is *satisfiable* if there exists a state (s, h) that satisfies p.

We denote the set of all heaps by *Heaps*, and the set of all states by *States*. The semantics of assertions is defined inductively as follows (e.g. [30]).

[1] We introduce its syntax implicitly; see e.g. [30] for an explicit definition.
[2] Often one assumes $\mathit{Values} = \mathbb{Z}$.

$$
\begin{array}{lll}
s,h \models b & \Leftrightarrow_{df} & b^S = \mathsf{true} \\
s,h \models \neg p & \Leftrightarrow_{df} & s,h \not\models p \\
s,h \models p \vee q & \Leftrightarrow_{df} & s,h \models p \quad \text{or} \quad s,h \models q \\
s,h \models \forall v.\ p & \Leftrightarrow_{df} & \forall x \in \mathit{Values} : \{(v,x)\} \mid s,h \models p \\
s,h \models \mathsf{emp} & \Leftrightarrow_{df} & h = \emptyset \\
s,h \models e_1 \mapsto e_2 & \Leftrightarrow_{df} & h = \{(e_1^S, e_2^S)\} \\
s,h \models p * q & \Leftrightarrow_{df} & \exists h_1, h_2 \in \mathit{Heaps}.\ \mathsf{dom}(h_1) \cap \mathsf{dom}(h_2) = \emptyset \text{ and} \\
 & & \quad h = h_1 \cup h_2 \text{ and } s, h_1 \models p \text{ and } s, h_2 \models q
\end{array}
$$

Here, b is a Boolean, and e_1 and e_2 are arbitrary expressions; p, q are assertions. The semantics e^s of an expression e with regards to a store s is straightforward. The domain of a relation modelling a partial function R is defined by $\mathsf{dom}(R) =_{df} \{x : \exists y.\ (x,y) \in R\}$, and the update function $\mid$ is defined by $f \mid g =_{df} f \cup \{(x,y) : (x,y) \in g \wedge x \notin \mathsf{dom}(f)\}$.

The first four clauses do not make any assumptions about the heap and are well known. The fifth clause defines the assertion emp, which ensures that the heap h is empty and does not contain any addressable cell. The assertion $e_1 \mapsto e_2$ characterises the heap of a state to contain exactly one cell at address e_1^s and value e_2^s. More complex heaps are built using *separating conjunction* $*$; it is a connective that ensures properties on disjoint regions of the underlying heap.

To create a denotational model one lifts the satisfaction-based semantics to a set-based one:

$$[\![p]\!] =_{df} \{(s,h) : s,h \models p\}.$$

In particular, $[\![\mathsf{false}]\!] = \emptyset$. This definition offers a set-based semantics [8].

$$
\begin{array}{ll}
[\![\neg p]\!] & = \{(s,h) : s,h \not\models p\} = \overline{[\![p]\!]} \\
[\![p \vee q]\!] & = [\![p]\!] \cup [\![q]\!] \\
[\![\forall v.\ p]\!] & = \bigcap_{x \in \mathit{Values}} \{(s,h) : ((v,x) \mid s, h) \in [\![p]\!]\} \\
[\![\mathsf{emp}]\!] & = \{(s,h) : h = \emptyset\} \\
[\![e_1 \mapsto e_2]\!] & = \{(s,h) : h = \{(e_1^s, e_2^s)\}\} \\
[\![p * q]\!] & = [\![p]\!] \circledast [\![q]\!]\,, \\
 & \quad \text{with } P \circledast Q =_{df} \{(s, h \cup h') : (s,h) \in P \wedge (s,h') \in Q \\
 & \qquad\qquad\qquad\qquad\qquad \wedge\ \mathsf{dom}(h) \cap \mathsf{dom}(h') = \emptyset\}
\end{array}
$$

Here, $\overline{}$ denotes set complementation with respect to the carrier set *States*. It has been shown that set union in combination with set complementation and lifted separating conjunction forms a useful algebraic structure.

A *quantale* [24,31] is a structure $(S, \leq, 0, \cdot, 1)$ where $(S, \leq)$ is a complete lattice and $(S, \cdot, 1)$ is a monoid where multiplication is completely disjunctive, i.e.,

$$a \cdot (\bigsqcup T) = \bigsqcup\{a \cdot x : x \in T\} \quad \text{and} \quad (\bigsqcup T) \cdot a = \bigsqcup\{x \cdot a : x \in T\},$$

for $T \subseteq S$ and $\bigsqcup$ denoting the supremum operator. The least element is 0.

The supremum of two elements a, b is denoted by $a \sqcup b$, and relates to the order by $a \sqcup b = b \Leftrightarrow a \leq b$. The definition implies that $\cdot$ is a *full annihilator* (strict), i.e., that $0 \cdot a = 0$ and $a \cdot 0 = 0$ for all $a \in S$. The notion of a quantale is equivalent to that of a *standard Kleene algebra* [6].

A quantale is *commutative* if $a \cdot b = b \cdot a$, for all $a, b \in S$; it is called *Boolean* if the underlying lattice is distributive and complemented, hence a Boolean algebra.

Classical examples are the algebra of Booleans $\mathsf{BA} = (\mathbb{B}, \Rightarrow, \mathsf{false}, \wedge, \mathsf{true})$, and binary relations $\mathtt{REL}_X = (\mathcal{P}(X \times X), \subseteq, \emptyset, ;, \mathrm{I})$, where ; denotes sequential composition, and $\mathrm{I} = \{(x, x) : x \in X\}$ is the identity relation.

We now relate sets of assertions of separation logic to algebra.

Theorem 1 [8]. The structure $\mathsf{AS} =_{df} (\mathcal{P}(States), \subseteq, [\![\mathsf{false}]\!], \uplus, [\![\mathsf{emp}]\!])$ of separation logic assertions is a commutative and Boolean quantale with $P \sqcup Q = P \cup Q$.

In its early days separation logic was based on intuitionistic logic [29]. Hence the underlying algebra, called BI algebra, was based on Heyting algebras, rather than Boolean algebras [28]. Ishtiaq and O'Hearn expanded BI algebras to capture the Boolean nature of contemporary separation logic [17]. Their approach, called Boolean BI algebra, does not require an underlying order. A more detailed comparison between these approaches is given in [7]. As we require an ordering for our verification technique – see below – we work in the setting of quantales.

Separation logic features three additional operators: separating implication (e.g. [30]), septraction [32], and separating coimplication [1].

$$
\begin{aligned}
s, h \models p \mathrel{-\!\!*} q &\Leftrightarrow_{df} \forall h_1 \in Heaps : (\mathsf{dom}(h_1) \cap \mathsf{dom}(h) = \emptyset \text{ and } s, h_1 \models p) \\
&\qquad \text{implies } s, h_1 \cup h \models q \\
s, h \models p \mathrel{-\!\circledast} q &\Leftrightarrow_{df} \exists h_1 \in Heaps : \mathsf{dom}(h_1) \cap \mathsf{dom}(h) = \emptyset \text{ and} \\
&\qquad s, h_1 \models p \text{ and } s, h \cup h_1 \models q \\
s, h \models p \mathrel{\leadsto\!\!*} q &\Leftrightarrow_{df} \forall h_1, h_2 \in Heaps : (\mathsf{dom}(h_1) \cap \mathsf{dom}(h_2) = \emptyset \text{ and} \\
&\qquad h = h_1 \cup h_2 \text{ and } s, h_1 \models p) \text{ implies } s, h_2 \models q
\end{aligned}
$$

A state (s, h) satisfies the *separating implication* $p \mathrel{-\!\!*} q$ if h ensures that whenever it is extended with a disjoint heap h_1 satisfying p then the combined heap $h \cup h_1$ satisfies q. *Septraction* ($\mathrel{-\!\circledast}$) denotes an existential version of separating implication, which quantifies over all subheaps h_1; it expresses that the heap can be extended with a state satisfying p, so that the extended state satisfies q. *Separating coimplication* ($\mathrel{\leadsto\!\!*}$) states that whenever there is a subheap h_1 satisfying p then the remaining heap satisfies q. It is straightforward to lift these operations to the algebra AS.

Algebraically these operators are related. In a quantale, the *left residual* $a \backslash b$ and the *right residual* a / b exist [2] and are defined by the Galois connections

$$x \le a \backslash b \Leftrightarrow_{df} a \cdot x \le b \quad \text{and} \quad x \le a / b \Leftrightarrow_{df} x \cdot b \le a.$$

The element $a \backslash b$ is a pseudo-inverse of multiplication and the greatest solution of the inequality $a \cdot x \le b$. Hence $a \backslash b$ can also be defined as $\bigsqcup\{x : a \cdot x \le b\}$. In case the underlying quantale is commutative both residuals coincide, i.e., $a \backslash b = b / a$. In a Boolean quantale, the *left detachment* $a \rfloor b$ and the *right detachment* $a \lfloor b$ are defined based on residuals.

$$a \rfloor b =_{df} \overline{a \backslash \overline{b}} \quad \text{and} \quad a \lfloor b =_{df} \overline{\overline{a} / b}$$

In BA residuals coincide with implication, and detachments with conjunction. In $\mathtt{REL}_X$, $R \backslash S = \overline{R^{\smile} ; \overline{S}}$ and $R \rfloor S = R^{\smile} ; S$, where $^{\smile}$ denotes the converse of a relation.

Theorem 2 [8]. In the algebra of assertions AS, separating implication is the residual of separating conjunction; septraction coincides with the detachment.

$$[\![p \mathrel{-\!\!*} q]\!] = [\![p]\!] \backslash [\![q]\!] = [\![q]\!] / [\![p]\!] \quad \text{and} \quad [\![p \mathrel{-\!\!\circledast} q]\!] = [\![p]\!] \rfloor [\![q]\!] = [\![q]\!] \lfloor [\![p]\!]$$

The algebraic theory of quantales is well established, and implies plenty of properties for separation logic. Examples are monotonicity properties of the operators, modus ponens, as well as the following exchange law.

$$\begin{aligned} a \cdot \overline{b} \le \overline{c} &\Leftrightarrow a \rfloor c \le b \\ [\![p]\!] \uplus \overline{[\![q]\!]} \subseteq \overline{[\![r]\!]} &\Leftrightarrow [\![p \mathrel{-\!\!\circledast} r]\!] \subseteq [\![q]\!] \\ (p * \neg q \Rightarrow \neg r) &\Leftrightarrow (p \mathrel{-\!\!\circledast} r \Rightarrow q) \end{aligned}$$

The first equivalence shows the law in the general setting of Boolean quantales, the second one in AS, and the last one the corresponding law in separation logic. It is standard that in algebraic settings the order coincides with implication. In this paper we use these representations interchangeably, depending on the current situation. Many more properties about quantales can be found in the literature; many useful properties about residuals and detachments are summarised in [21].

The exchange law implies another Galois connection based on detachments:

$$b \le a \rightarrow c \quad \Leftrightarrow \quad a \rfloor b \le c\,,$$

where $a \rightarrow c =_{df} \overline{a \cdot \overline{c}}$. As for residuals and detachments, a symmetric operator exists: $a \leftarrow c =_{df} \overline{\overline{a} \cdot c}$. In sum, any Boolean quantale features four operators (and their symmetric ones), which are related via dualities and Galois connections; Fig. 1 summarises the situation.

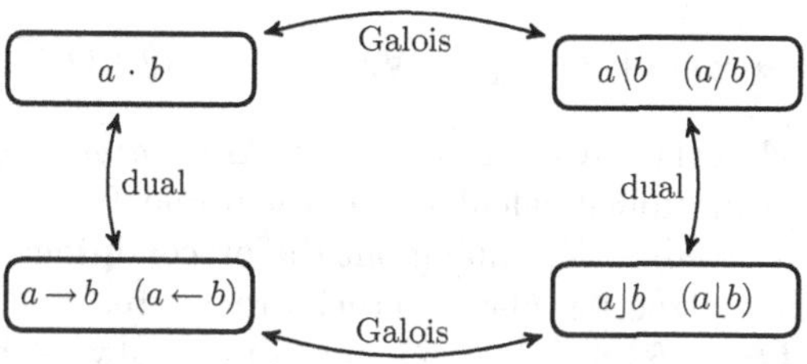

Fig. 1. Relationship between operators of quantales

Theorem 3 [1]. In AS, separating coimplication is upper adjoint of septraction.

$$[\![p \leadsto\!\!* \, q]\!] = [\![p]\!] \rightarrow [\![q]\!] = \overline{[\![p]\!] \uplus \overline{[\![q]\!]}} \quad \text{and} \quad [\![q]\!] \subseteq [\![p \leadsto\!\!* \, r]\!] \Leftrightarrow [\![p \mathrel{-\!\!\circledast} q]\!] \subseteq [\![r]\!]$$

3 Forward and Backward Reasoning in Separation Logic

The Galois connections are extremely useful when reasoning forwards and backwards in separation logic [1]. Since we aim at an extension of forward reasoning, including the possibility of failure, we briefly explain the method in this section.

Forward reasoning [13] proceeds forwards from a given precondition P and a given program m by calculating the *strongest postcondition* $\mathtt{sp}(P, m)$ such that $\{P\}\, m\, \{\mathtt{sp}(P, m)\}$ is a valid Hoare triple. *Backward reasoning* [10] proceeds the other way round and determines the *weakest precondition* $\mathtt{wp}(m, Q)$, for a given program m and postcondition Q.

Although backward reasoning is more common when it comes to verification efforts, there are several applications where forward reasoning is more convenient, such as for programs where the precondition is 'trivial', and the postcondition either too complex or unknown. Both reasoning techniques are well established for Hoare logic, and calculational reasoning is feasible. For example, the strongest postcondition for $\mathtt{sp}(P, m_1; m_2)$ for a sequential program equals $\mathtt{sp}(\mathtt{sp}(P, m_1), m_2)$.

Using separation logic in the context of forward and backward reasoning yield the problem commonly known as *frame calculation.* In separation logic any given specification $\{P\}\, m\, \{Q\}$, can be extended to $\{P * R\}\, m\, \{Q * R\}$, where R is a frame – a region of the memory that remains untouched during execution of m. Forward reasoning starts with a given precondition X, which then needs to be split into the actual precondition P and a frame R. That means for given X and P one has to find a frame R such that $X = P * R$. In large projects, frame calculations are usually challenging since X can be arbitrarily complex.

For forward reasoning the Galois connection between septraction and separating coimplication comes to aid.

$$\forall R.\, \{P \rightsquigarrow\!\!\ast\, R\}\, m\, \{Q * R\} \Leftrightarrow \forall X.\, \{X\}\, m\, \{Q * (P \mathbin{-\!\circledast} X)\}$$

The right hand side has the advantage that it works for *arbitrary* preconditions X and does not require an explicit calculation of the frame. Intuitively, it states that the postcondition is given by X, pulling out the precondition P and replacing it by Q. We note that septraction plays a crucial role here.

Specifications $\{P * R\}\, m\, \{Q * R\}$ can almost always be rewritten into $\{P \rightsquigarrow\!\!\ast R\}\, m\, \{Q * R\}$, especially if P is precise – preciseness ensures the existence of a unique subheap, i.e., $p \mapsto v$. A similar equivalence that can be used for backward reasoning exists, and is based on the Galois connection between $*$ and $-\!\!*$. Both techniques have been implemented in the theorem prover Isabelle/HOL [25] and are ready to be used for verification tasks [1].

4 Simple Models for Separation Logic

While backward reasoning works for both partial and total correctness, forward reasoning is more useful in the setting of partial correctness, where termination has to be proved separately. In total correctness, the existence of a postcondition implies termination, and therefore, it is not guaranteed that forward reasoning can proceed from a given precondition, limiting its application. We target *general correctness* [18], where the postcondition can handle both termination and nontermination. This is in contrast to partial correctness, where failure coincides with false, and total correctness where failure is unrepresentable.

We extend the assertion language by some notion of failure. We explore different ways to add failure to separation logic in this and the next section.

Ideally we can develop a failure model maintaining the algebraic relationships between the operators depicted in Fig. 1. In this section we present a series of models, all lacking at least one algebraic property. As our development is systematic, we conclude that no simple failure model exists.

A Single Failure Element

We start by extending our model by a single element $\bot$ indicating failure. Within separation logic that means we are looking at the set $\mathcal{P}(\mathit{States}) \cup \{\bot\}$; or in the abstract algebraic setting at $S \cup \{\bot\}$ as underlying set.

Two choices need to be made: extend multiplication, and integrate $\bot$ into the lattice of elements.

Model I: *Near-annihilation and largest element.* It seems reasonable to assume that failure cannot be erased by any non-zero element (in AS by any assertion different to $[\![\mathsf{false}]\!]$), i.e., $a \cdot \bot = \bot \cdot a = \bot$, for all $a \in (S \cup \{\bot\}) - \{0\}$. The interaction between $\bot$ and 0 ($[\![\mathsf{false}]\!]$ in AS) needs to be decided independently. Let us assume that zero cancels out failure: $0 \cdot \bot = \bot \cdot 0 = 0$. Moreover, let $\bot$ be the largest element in the underlying lattice: $a \leq \bot$, for all a.

If there are elements that cancel each other, i.e., $\exists a, b \in S.\, a \cdot b = 0$, we have

$$(a \cdot b) \cdot \bot = 0 \cdot \bot = 0 \quad \text{and} \quad a \cdot (b \cdot \bot) = a \cdot \bot = \bot.$$

Therefore either multiplication is not associative, and hence the underlying multiplicative structure $(S \cup \{\bot\}, \cdot, 1)$ is not even a monoid, or the algebra collapses: $\bot = 0 \leq a \leq \bot$. Separation logic features cancellative elements such as $(p \mapsto v) * (p \mapsto v) = \mathsf{false}$, where $p \mapsto v$ characterises a pointer p pointing to value v. Hence separating conjunction cannot be associative if a failure element with the above properties is present. ▲

Loosing associativity is not a priori bad (e.g. [9,11]), and we can still use models without associativity for reasoning in separation logic. We discuss this in Sect. 6. However, many useful properties are lost. For example the law $(a \cdot b) \rfloor c = a \rfloor (b \rfloor c)$, which holds in commutative quantales, does not hold without associativity. In separation logic this translates to $(p * q) -\circledast\, r = p -\circledast\, (q -\circledast\, r)$ and states that a heap can be 'removed' by removing subheaps consecutively.

Model II: *Near-annihilation and least element.* When using the same set up as in Model I, but forcing $\bot$ to be the least element, i.e., $\bot \leq a$, the problem with associativity stays, but the algebra does not collapse any longer when associativity is enforced – one does not have $\bot \leq a \leq \bot$. However, we get $\bot = 0$ when assuming associativity and the existence of elements that cancel each other. ▲

Model III: *Full annihilation and largest element.* Since near-annihilation of failure ($\bot \cdot a = a \cdot \bot = \bot$ for $a \neq 0$) yields problems with associativity, we now focus on situations where failure is a (full) annihilator: $a \cdot \bot = \bot \cdot a = \bot$, for all $a \in S \cup \{\bot\}$. As a consequence, 0 does not fully annihilate any longer, as we only have $a \cdot 0 = 0 \cdot a = 0$ for $a \neq \bot$. For this model we assume that $\bot$ is the largest element. While the resulting algebra is still a complete lattice[3], it does not form a quantale as multiplication is only positively distributive, i.e.,

$$T \neq \emptyset \Rightarrow (a \cdot (\bigsqcup T) = \bigsqcup\{a \cdot x : x \in T\} \text{ and } (\bigsqcup T) \cdot a = \bigsqcup\{x \cdot a : x \in T\}).$$

For $T = \emptyset$ we have $\bot \cdot \bigsqcup\emptyset = \bot \cdot 0 = \bot$ and $\bigsqcup\{a \cdot x : x \in \emptyset\} = \bigsqcup\emptyset = 0$. While one could still define residuals as suprema, the Galois connections between multiplication and residuals do not hold any longer. Since these connections are crucial for backward reasoning (see Sect. 3), this model is of no use for us. ▲

Model IV: *Full annihilation and least element.* Our final model featuring a single failure element defines failure as least element of the lattice and as full annihilator. By straightforward calculations it can be shown that this algebra forms indeed a quantale. In particular, multiplication is associative and completely distributive. Therefore this model is the first one which features an associative multiplication and establishes Galois connections. ▲

However, it is impossible to extend a Boolean quantale $(S, \leq, 0, \cdot, 1)$ to a Boolean quantale $(S \cup \{\bot\}, \sqsubseteq, \bot, \circ, 1)$ in this setting, if $\bot \notin S$: as Boolean algebras have size 2^n (for $n \in \mathbb{N}$) and $2^n + 1$ is not a power of 2, no complementation can be defined on the extended structure.

While this model is still a fine failure model for separation logic (without complementation), which can probably be used in many circumstances, it is not suitable for forward reasoning. As we want to use it in combination with reasoning in Hoare logic, we have to maintain fundamental aspects of this logic such, as the weakening rule

$$\frac{\{P\}\, m\, \{Q\} \qquad Q \leq Q'}{\{P\}\, m\, \{Q'\}}.$$

As mentioned above, $\leq$ coincides with logical implication and hence $Q \leq Q'$ describes the fact that the precondition Q' is weaker than Q. This rule in combination with the fact that $\bot$ is the least element yields false conclusions. Assume the program `set_ptr` $p\ v$, which assigns value v to pointer p. If the heap does not contain the pointer, for example when the heap is empty, then the program fails. We would like to have the Hoare triple $\{\mathtt{emp}\}$ `set_ptr` $p\ v\, \{\bot\}$ to be valid. Using the weakening rule we can replace the postcondition by any other, as $\bot$ is the least element. Using the weakening rule conclude that $\{\mathtt{emp}\}$ `set_ptr` $p\ v\, \{q \mapsto 7\}$ or $\{\mathtt{emp}\}$ `set_ptr` $p\ v\, \{\mathsf{true}\}$ hold. Both are invalid Hoare triples under general correctness.

Therefore, in a setting where separation logic is used for forward or backward reasoning the element representing false ($\emptyset = [\![\mathsf{false}]\!]$ in AS) needs to be the least element of any order to be used with Hoare logic.

[3] We omit straightforward proofs; most of them are available online (see Sect. 6).

The above four models conclude our failure models for separation logic featuring a single failure element; other models are not useful as it is not realistic that non-zero elements (proper heaps in AS) cancel out failure. It also does not seem plausible to have the failure element sitting at some place in the lattice that is not at the bottom or the top.

Sets of Failure and Non-failure Elements

We have shown that a failure element cannot be the least element w.r.t. the order underlying Hoare logic. Moreover, failure should be an annihilator in case elements exist that cancel each other out – otherwise associativity is lost.

Since a single failure yields severe shortcomings we now look at models based on subsets of $\mathit{States}' = \mathit{States} \cup \{\bot\}$. The intuition behind these models is that failure does not forget about the heap setting, but combines failure with possible heaps. It can be seen as introducing a flag indicating whether a calculation has failed or not; since there is only one flag, any calculation that *may* fail will have the failure flag set (non-failure executions may exist). This setting allows more flexibility when it comes to defining multiplication.

For the following two models we use the following separating conjunction.

$$P *_1 Q =_{df} ((P - \{\bot\}) \circledast (Q - \{\bot\})) \;\cup\; ((P \cup Q) \cap \{\bot\})$$

The first part calculates 'classical' separating conjunction on the non-failure parts of P and Q, and the second part adds the failure element in case either P or Q contains a failure. Using distributivity of $\cap$ over $\cup$, and introducing the shorthands $P_\bot$ for $P \cap \{\bot\}$ and $P^{\text{-}\bot}$ for $P - \{\bot\}$, the equation becomes

$$P *_1 Q \;=\; (P^{\text{-}\bot} \circledast Q^{\text{-}\bot}) \;\cup\; P_\bot \;\cup\; Q_\bot .$$

It is easy to see that $*_1$ is associative and commutative. Moreover, it is straightforward to prove that the set $\{\bot\}$ is an annihilator w.r.t. $*_1$, i.e., $P *_1 \{\bot\} = \{\bot\}$.

Model V: *Full annihilation with subset order.* One of the first models that comes into mind when creating a failure model for separation logic is the structure $(\mathcal{P}(\mathit{States} \cup \{\bot\}), \subseteq, \emptyset, *_1, [\![\mathsf{emp}]\!])$. Clearly, $(\mathcal{P}(\mathit{States} \cup \{\bot\}), \subseteq, \emptyset)$ is a complete, distributed and complemented lattice. Moreover, $(\mathcal{P}(\mathit{States} \cup \{\bot\}), *_1, [\![\mathsf{emp}]\!])$ is a monoid. Similar to Model III this algebra is only positively distributive. As before, residuals can be defined via the supremum operator, but do not establish the desired Galois connections. ▲

Model VI: *Full annihilation with more sophisticated order.* To turn Model V into a quantale we define a more sophisticated order.

$$\begin{aligned} P \sqsubseteq Q \;&\Leftrightarrow_{df}\; P^{\text{-}\bot} \subseteq Q^{\text{-}\bot} \;\wedge\; Q_\bot \subseteq P_\bot \\ &\Leftrightarrow\quad P^{\text{-}\bot} \subseteq Q^{\text{-}\bot} \;\wedge\; (\bot \in Q \Rightarrow \bot \in P) \end{aligned}$$

This order, illustrated in Fig. 2, is the subset-order on non-failure elements. It classifies a set containing $\bot$ worse than the same set without failure.

The structure $(\mathcal{P}(\mathit{States}'), \sqsubseteq, \{\bot\})$ is a complete lattice, the supremum operator coincides with $P \mathbin{[\!]} Q =_{df} (P^{\text{-}\bot} \cup Q^{\text{-}\bot}) \cup (Q_\bot \cap P_\bot)$, which is associative and commutative. The structure $(\mathcal{P}(States'), \sqsubseteq, \{\bot\}, *_1, [\![\mathsf{emp}]\!])$ forms a Boolean commutative quantale when using set-theoretic complementation over $States'$, denoted by$'$. In particular $(P \cup \{\bot\})' = \overline{P}$, with $\overline{}$ is complementation of $States$.

Since the largest element does not contain the failure element, this model suffers from the same problems as Model IV, and cannot be used in combination with forward reasoning. However, it is a decent extension of algebraic separation logic, and features all the desired algebraic properties. In particular, we can define the other three operators via Galois connections and duals. As this is an extension of separation logic, we use the symbols from separation logic rather than their algebraic counterparts (e.g. $-\circledast$ rather than $\rfloor$).

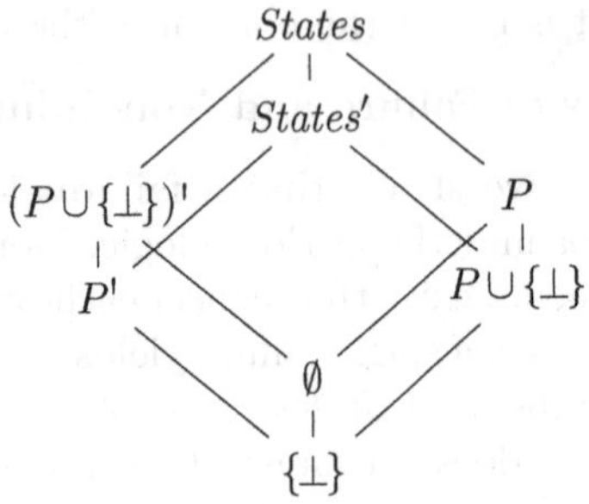

Fig. 2. Another order

$$
\begin{aligned}
P \mathrel{-\!\!*_1} Q &= (P^{\text{-}\bot} \mathrel{-\!\!*} Q^{\text{-}\bot}) \cup ((P')_\bot \cap Q_\bot)\\
P \mathrel{\leadsto\!\!*_1} Q &= (P^{\text{-}\bot} \mathrel{\leadsto\!\!*} Q^{\text{-}\bot}) \cup ((P')_\bot \cap Q_\bot)\\
P \mathrel{-\!\circledast_1} Q &= (P^{\text{-}\bot} \mathrel{-\!\circledast} Q^{\text{-}\bot}) \cup P_\bot \cup Q_\bot
\end{aligned}
$$

All these operations have the advantage that they are built based on the original operations of separation logic. In particular, the new and the original definitions behave identical on the non-failure elements (heaps). The second component defines the effect if either P or Q contain the error element. ▲

The above model, as well as all forthcoming ones, can be lifted to a more abstract level, based on a quantale $\mathcal{S} = (\mathcal{P}(S), \subseteq, \emptyset, \cdot, 1)$. As our motivation stems from separation logic, we stick with models based on AS.

Although the model cannot be used for forward reasoning, it is useful to show another important shortcoming: septraction as defined in Model VI cannot yield failure: $P \mathrel{-\!\circledast_1} Q$ contains an error only if either P or Q contains a failure element.

Consider the program `delete_ptr` p. It clears the allocated memory pointed to by p, and should fail if p does not exist. Since we would like to work with Hoare triples of the form

$$\{X\}\, \texttt{delete\_ptr}\ p\, \{p \mapsto \_ \mathrel{-\!\circledast_?} X\}$$

we require that $-\circledast_?$ can fail, even in the setting of non-failed X, where $-\circledast_?$ is an extension of $-\circledast$. More concretely, the modified septraction should imply

$$p \mapsto \_ \mathrel{-\!\circledast_?} \mathsf{emp} = \bot,$$

written as a formula of separation logic. A possible solution would be to modify $*_1$ or $-\circledast_1$ by generating a failure whenever separation logic would lead to false.

$$
\begin{aligned}
P *_2 Q &= \mathtt{F}(P^{\text{-}\bot} * Q^{\text{-}\bot}) \cup P_\bot \cup Q_\bot,\\
P \mathrel{-\!\circledast_2} Q &= \mathtt{F}(P^{\text{-}\bot} \mathrel{-\!\circledast} Q^{\text{-}\bot}) \cup P_\bot \cup Q_\bot,
\end{aligned}
$$

where function F is defined as $\mathtt{F}(X) =_{df} \{\bot\}$ if $X = \emptyset$ and $\mathtt{F}(X) =_{df} X$ otherwise. Both $*_2$ and $-\circledast_2$ can be used to create failure models. However, the only useful orderings we found for these algebras are either the subset-order or orders closely related to $\sqsubseteq$ of Model VI. As a consequence either multiplication is only positively distributive, or the set $\{\bot\}$ is the least element. As discussed, either setting is not ideal for forward reasoning.

As a side remark it is worth mentioning that in the commutative Boolean quantale $(\mathcal{P}(States'), \sqsubseteq, \{\bot\}, *_2, [\![\mathsf{emp}]\!])$ the operator $-\circledast_2$ is *not* the detachment, despite the symmetric definition.

Failure Flags for Every Single Heap

We turn to the finest sets of models, featuring a failure flag for every heap. In separation logic a heap would be of type $(Addr \rightharpoonup Values) \times \mathbb{B}$, where the Boolean flag indicates (non)failure: true indicates non-failure, and false failure.

In the set-theoretic setting where we consider sets of states, we model failure flags by pairs of sets. The first set lists all heaps stemming from non-failure calculations, and the second one lists all failed heaps: an element of the form $(P, \emptyset)$ indicates non-failure calculations.

Since we distinguish failed from non-failed heaps we do not need a separate failure element. Moreover, since the algebra is now built on $\mathcal{P}(States) \times \mathcal{P}(States)$, we can use standard algebraic constructions to create further models.

Model VII: *Cross-Product with component-wise multiplication.* We consider the structure $(\mathcal{P}(States) \times \mathcal{P}(States), \sqsubseteq, (\emptyset, \emptyset), \circ, ([\![\mathsf{emp}]\!], [\![\mathsf{emp}]\!]))$, where $\sqsubseteq$ and $\circ$ are the component-wise lifted order $\subseteq$ and multiplication $\circledast$, e.g.

$$(P_1, P_2) \circ (Q_1, Q_2) \;=_{df}\; (P_1 \circledast Q_1, P_2 \circledast Q_2).$$

Clearly, this model forms a Boolean commutative quantale, which is a proper extension of separation logic. However, since failed and non-failed heaps are kept separate, the derived operation of septraction (detachment) cannot produce failure heaps out of non-failure ones; a similar behaviour as Model VI. Using the same example as above, adapted to the model at hand, we get

$$([\![p \mapsto \_]\!], \emptyset) -\circledast_3 ([\![\mathsf{emp}]\!], \emptyset) \;=\; (\emptyset, \emptyset),$$

where $-\circledast_3$ is the component-wise lifted septraction. ▲

In general, there are multiple product constructions to create new algebras. The most common constructions, such as Model VII, form quantales again, which can be proved in an algebraic setting, e.g. by using general results from universal algebra.

However, all of these constructions keep the components more or less separate. Hence these models, although decent failure extensions, suffer the same shortcoming as Model VII. Some of those models, e.g. one that is identical to Model VII except that the order is given by $\subseteq \times \supseteq$, yield problems already discussed earlier – here, the problem occurring in reasoning in Hoare-logic style.

5 More Sophisticated Models

Since simple extensions do not work we consider two sophisticated models in this section. Instead of considering flags, we now split every heap into a set of actual configuration, and one of 'failed' configurations. Formally a heap is of type $Heaps_2 =_{df} (Addr \rightharpoonup Values) \times (Addr \rightharpoonup Values)$. The extended heap $(p \mapsto v, q \mapsto \_)$ states that a pointer p exists, pointing to value v, and that a pointer q is required, but does not exist. This idea is inspired by the construction of integers from natural numbers (e.g. [5]). Both models build on this idea of 'negative' heaps; the base set is $States_2 =_{df} \mathcal{P}(Stores \times Heaps_2)$. Instead of adding elements and operations on top of AS we change the underlying logic of AS.

As before the forthcoming models can be lifted to an abstract level, based on a quantale $(\mathcal{P}(S), \subseteq, \emptyset, \cdot, 1)$, but we stay within the setting of separation logic.

We aim at a failure model that allows non-failure elements to create failure, when used with septraction. Hence we have to develop operators that combine both parts of the pair of the underlying set of elements.

We define a new unary operator that eliminates those parts of a heap that are present and missing at the same time; again this is similar to the construction on top of natural numbers. Intuitively such heaps cannot exist:

$$\begin{aligned} {}^* : Heaps_2 &\rightarrow Heaps_2 \\ (h_1, h_2)^* &=_{df} (h_1|_{\overline{\mathsf{dom}(h_2)}}, h_2|_{\overline{\mathsf{dom}(h_1)}}), \end{aligned}$$

where the heaps are restricted to those parts that do not occur in both. This operator is lifted to $States_2$ by $P^* =_{df} \{(s, \mathtt{h}^*) : (s, \mathtt{h}) \in P\}$, where $\mathtt{h} \in Heaps_2$ is an extended heap.

Model VIII: *Septraction based on heap reduction.* The first model of this section modifies AS to take heap reduction * into account: in detail we consider the structure $(\mathcal{P}(States_2), \subseteq, \emptyset, \uplus_1, [\![\mathsf{emp}']\!])$, where $[\![\mathsf{emp}']\!] =_{df} \{(s, \emptyset, \emptyset)\}$ and $\uplus_1$ is an adapted version of $\uplus$:

$$\begin{aligned} P \uplus_1 Q =_{df} \{(s, h_1 \cup h_1', h_2 \cup h_2') : (s, h_1, h_2) \in P^* \wedge (s, h_1', h_2') \in Q^* \\ \wedge\ \mathsf{dom}(h_1) \cap \mathsf{dom}(h_1') = \mathsf{dom}(h_2) \cap \mathsf{dom}(h_2') = \emptyset\}. \end{aligned}$$

The structure forms a commutative quantale, with $(\mathcal{P}(States_2), \subseteq, \emptyset)$ being a complete and distributive lattice. Therefore the residuals of $\uplus_1$ exist, denoted by $-\!\!*_4$. We further define a new complementation operator $\sim$ by $\sim P =_{df} \overline{P^*}$. The idea of this negation is based on the negation of non-classical *relevance logics* (also called *relevant logics*) [12,20], where * is related to the logic's 'Routley Star'. We introduce separating coimplication and septraction as

$$P \sim\!\!*_4 Q =_{df} \sim(P *_4 \sim Q) \quad \text{and} \quad P -\!\circledast_4 Q =_{df} \sim(P -\!\!*_4 \sim Q).$$

Lifting the satisfaction-based semantics to a set-based one by $[\![p]\!] =_{df} \{(s, h_1, h_2) : s, h_1, h_2 \models p\}$, and using the shorthand $[\![h_1, h_2]\!]$ for $\{(s, h_1, h_2)\}$, we can derive useful laws such as

$$[\![p \mapsto v, \emptyset]\!] -\!\circledast_4 [\![\emptyset, \emptyset]\!] = [\![\emptyset, p \mapsto v]\!] \quad \text{and} \quad [\![p \mapsto v, \emptyset]\!] -\!\circledast_4 [\![p \mapsto v, \emptyset]\!] \supseteq [\![\mathsf{emp}']\!].$$

Both properties are desired and fit the intuition of septraction involving failure. The first property states that pulling out a resource ($p \mapsto v$) from the empty heap yields a negative heap; the second property states that one can pull out a resource if it exists. However, we can derive inequalities such as $[\![q \mapsto \_, q \mapsto \_]\!] \subseteq [\![p \mapsto v, \emptyset]\!] -\!\circledast_4 [\![p \mapsto v, \emptyset]\!]$, which seems to be weird: why should a statement that deletes a pointer p from a heap that only consists of p talk about another pointer q? Applying the operator * would remove such heap elements, but the current model has another severe disadvantage: $\leadsto\!\!*_4$ and $-\!\circledast_4$ do not form a Galois connection. This is not a surprise because $\sim$ is not a proper complement, e.g. $\sim(\sim P) \neq P$. ▲

Using standard set complementation turns $(\mathcal{P}(\mathit{States}_2), \subseteq, \emptyset)$ into a complemented lattice and the above structure into a Boolean quantale. However, it would not solve any of the problems mentioned before as it is similar to the original algebra AS, featuring two instead of one heap; but not interpreting the second heap as a negative one.

A similar model with the same problems as Model VIII is based on failure flags for every single heap (see Sect. 4): it considers the structure

$$(\mathcal{P}(\mathit{States}) \times \mathcal{P}(\mathit{States}), \subseteq \times \subseteq, (\emptyset, \emptyset), \uplus_2, ([\![\mathsf{emp}]\!], [\![\mathsf{emp}]\!])),$$

with $(P_1, P_2) \uplus_2 (Q_1, Q_2) =_{df} (P_1 \uplus P_2, Q_1 \uplus Q_2)$, and heap restriction operator $(P_1, P_2)^* =_{df} (P_1 - P_2, P_2 - P_1)$.

Model IX: *Septraction based on heap reduction under adapted heap addition.* As in Model VIII we consider the structure $(\mathcal{P}(\mathit{States}_2), \subseteq, \emptyset, \uplus_2, [\![\mathsf{emp}']\!])$, but this time we define multiplication as

$$P \uplus_2 Q =_{df} \{(s, \mathtt{h}^* \cup \mathtt{h}'^*) : (s, \mathtt{h}) \in P \wedge (s, \mathtt{h}') \in Q \wedge \mathsf{dom}(\mathtt{h}) \cap \mathsf{dom}(\mathtt{h}') = \emptyset\},$$

where $\cup$, $\cap$ and dom are defined component-wise on Heaps_2. As we changed the underlying heap addition, which now uses heap reduction *, we are not required to use a non-standard complement any longer. It is straightforward to show that this structure is indeed a commutative Boolean quantale; the proof is similar to the one of AS. As a consequence, residuals, detachments and duals exist; their characterisation is similar to the one of AS, presented in Sect. 2. At the level of septraction and separating coimplication they read as follows

$$\begin{aligned} s, \mathtt{h} \models p -\!\circledast_5\, q &\Leftrightarrow \exists \mathtt{h}_1 \in \mathit{Heaps}_2 : \mathsf{dom}(\mathtt{h}_1^*) \cap \mathsf{dom}(\mathtt{h}^*) = \emptyset \text{ and} \\ &\qquad s, \mathtt{h}_1 \models p \text{ and } s, \mathtt{h} \cup \mathtt{h}_1 \models q \\ s, \mathtt{h} \models p \leadsto\!\!*_5\, q &\Leftrightarrow \forall \mathtt{h}_1, \mathtt{h}_2 \in \mathit{Heaps} : (\mathsf{dom}(\mathtt{h}_1^*) \cap \mathsf{dom}(\mathtt{h}_2^*) = \emptyset \text{ and} \\ &\qquad \mathtt{h} = \mathtt{h}_1 \cup \mathtt{h}_2 \text{ and } s, \mathtt{h}_1 \models p) \text{ implies } s, \mathtt{h}_2 \models q. \end{aligned}$$

A big advantage of this model is that all the definitions are 'identical' to the ones of AS; even the logical formulas are identical. Moreover, it satisfies many useful properties such as

$$[\![p \mapsto v, \emptyset]\!] -\!\circledast_5 [\![p \mapsto v, \emptyset]\!] = [\![\mathsf{emp}']\!] \quad \text{and} \quad [\![p \mapsto v, \emptyset]\!] -\!\circledast_5 [\![\mathsf{emp}']\!] \subseteq [\![\emptyset, p \mapsto v]\!]$$

The first equation states that a subheap that exists can actually be pulled out; the second that pulling out a non-existing heap does indeed yield a failure – represented by its appearance in the second component.

As Galois connections are available we can use this model for forward reasoning (see Sect. 3). The connectives are related to Bi-Intuitionistic Boolean Bunched Implications (BiBBI) [3,4], which is a fragment of linear logic. Although the theories are slightly different, the proof theory developed for BiBBI should be usable for our model.

The model comes very close to the one we are looking for. However, it does not satisfy all the properties we desire. For example, we have

$$[\![p \mapsto v, \emptyset]\!] \mathbin{-\!\circledast_5} [\![\emptyset, p \mapsto v]\!] = \emptyset = [\![\mathsf{false}]\!].$$

This law states that pulling out a heap, which does not exist and which is already part of the negative heap, leads to the empty heap in both components. That means a non-existent heap cannot be pulled out twice. ▲

Clearly, the equation mentioned at the end of Model IX is not the intended behaviour we would expect from a model that can be used for forward reasoning in separation logic. The reason is that we use separating conjunction in both parts of the extended heap, on the 'positive' heap as well on the 'negative' heap. We suspect that relaxing the 'negative' heap structure could help. Rather than using a partial function $Addr \rightharpoonup Values$ one could use multisets. By doing so, we hope to achieve the following behaviour

$$(p \mapsto v, \emptyset) \mathbin{-\!\circledast_?} (\emptyset, p \mapsto v) = (\emptyset, \{p \mapsto v, p \mapsto v\}),$$

written in separation-logical style and stating that the resource $p \mapsto v$ is already missing twice. When using a multiset as negative heap, however, it is not possible to use set-theoretic complementation as negation. Part of our future work is to figure out all the details, e.g. how to integrate the heap reduction operator *.

6 Discussion and Conclusion

In this paper we have reported on our quest of extending separation logic in a way that it can be used for forward reasoning in non-partial contexts; in particular in the setting of total and general correctness. To support forward reasoning there, separation logic needs to be equipped with failure element(s). The problem with creating such a model is that we want it to be natural and intuitive when it comes to septraction, but at the same time we want to maintain the algebraic properties of the original separation logic.

In this paper we have focussed on the algebraic structure of separation logic, namely on quantales. We have developed a series of models that could potentially be used as failure models. It turns out that all extensions that we thought of being useful either loose algebraic properties such as dualities, Galois connections and/or associativity; or we cannot use the developed models for forward reasoning as the weakening rule of Hoare logic would not be valid any longer.

Although these models could be seen as negative results, we have decided to publish all the models nevertheless: (a) as we have shown that simple models cannot be used for extending separation logic, we want to share the experience to prevent other researchers from creating 'simple' models; (b) we want to give a justification why the models we are looking at the moment are complicated; and (c) we suspect that the developed models can also be used in other settings, such as Concurrent Kleene Algebra [15] or hybrid system analysis [16].

To present as many models as possible, we have decided to skip all the proofs as most of them are lengthy and boring, but not hard. Some of the proofs such as the proof that Model VI forms a commutative Boolean quantale have been mechanised in the interactive theorem prover Isabelle/HOL; these proofs can be found at http://hoefner-online.de/ramics18/.

We have mentioned in Sect. 3 that backward and forward reasoning for separation logic – the latter only for partial correctness – is supported by Isabelle/HOL. Although we have not found the perfect model yet, we have integrated some of the presented models in our Isabelle framework. In particular Model VIII, Model IX, as well as a similar (non-associative) one, which we have not listed. By doing so, we have figured out that the loss of associativity is not too bad. Non-associativity makes automation slightly more complicated, but certainly not impossible, as associativity still holds when no non-failure elements are involved. When performing forward reasoning, a single failed conjunct already indicates a problem of a program; hence a quick search for a failure element over a derived formula indicates failure. A link to our Isabelle framework can also be found at the above-mentioned webpage.

For future work we obviously will continue our search for an ideal model for our framework. If we prove that none exist, we will have to decide on which tradeoff suits our setting the most.

Acknowledgement. We are grateful to Gerwin Klein and Bernhard Möller for fruitful discussions and inspiring ideas. We also thank the anonymous referees for their valuable feedback.

References

1. Bannister, C., Höfner, P., Klein, G.: Backwards and forwards with separation logic. In: Avigad, J., Mahboubi, A. (eds.) ITP 2018. LNCS, vol. 10895, pp. 68–87. Springer, Cham (2018). https://doi.org/10.1007/978-3-319-94821-8_5
2. Birkhoff, G.: Lattice Theory, Colloquium Publications, vol. XXV. Annals of Mathematics Studies, 3rd edn. (1967)
3. Brotherston, J., Calcagno, C.: Classical BI: its semantics and proof theory. Log. Methods Comput. Sci. **6**(3) (2010)
4. Brotherston, J., Villard, J.: Sub-classical Boolean bunched logics and the meaning of par. In: Kreutzer, S. (ed.) Computer Science Logic (CSL 2015). Leibniz International Proceedings in Informatics (LIPIcs), vol. 41, pp. 325–342. Schloss Dagstuhl-Leibniz-Zentrum fuer Informatik (2015)
5. Campbell, H.E.: The Structure of Arithmetic. Appleton-Century-Crofts, New York (1970)
6. Conway, J.H.: Regular Algebra and Finite Machines. Chapman and Hall, London (1971)

7. Dang, H.H.: Algebraic calculi for separation logic. Ph.D. thesis, University of Augsburg, Germany (2014)
8. Dang, H.H., Höfner, P., Möller, B.: Algebraic separation logic. J. Logic Algebraic Program. **80**(6), 221–247 (2011)
9. Desharnais, J., Möller, B.: Non-associative Kleene algebra and temporal logics. In: Höfner, P., Pous, D., Struth, G. (eds.) RAMiCS 2017. LNCS, vol. 10226, pp. 93–108. Springer, Cham (2017). https://doi.org/10.1007/978-3-319-57418-9_6
10. Dijkstra, E.W.: A Discipline of Programming. Prentice Hall, Englewood Cliffs (1976)
11. Dongol, B., Hayes, I.J., Struth, G.: Relational convolution, generalised modalities and incidence algebras. arXiv:1702.04603 (2017)
12. Dunn, J.: Star and perp. Philos. Perspect. **7**, 331–357 (1993)
13. Floyd, R.W.: Assigning meanings to programs. Math. Aspects Comput. Sci. **19**, 19–32 (1967)
14. Hoare, C.A.R.: An axiomatic basis for computer programming. Commun. ACM **12**, 576–580 (1969)
15. Hoare, T., Möller, B., Struth, G., Wehrman, I.: Concurrent Kleene algebra and its foundations. J. Logic Algebraic Program. **80**(6), 266–296 (2011)
16. Höfner, P., Möller, B.: An algebra of hybrid systems. J. Logic Algebraic Program. **78**, 74–97 (2009)
17. Ishtiaq, S.S., O'Hearn, P.W.: BI as an assertion language for mutable data structures. SIGPLAN Not. **36**, 14–26 (2001)
18. Jacobs, D., Gries, D.: General correctness: a unification of partial and total correctness. Acta Inf. **22**(1), 67–83 (1985)
19. Kozen, D.: On Hoare logic and Kleene algebra with tests. ACM Trans. Comput. Logic **1**(1), 60–76 (2000)
20. Mares, E.: Relevance logic. In: Zalta, E.N. (ed.) The Stanford Encyclopedia of Philosophy. Metaphysics Research Lab, Stanford University, Spring 2014 edn. (2014)
21. Möller, B.: Residuals and detachments. Technical report 2005–20, Institut für Informatik, Universität Augsburg (2005)
22. Möller, B., Struth, G.: Algebras of modal operators and partial correctness. Theor. Comput. Sci. **351**(2), 221–239 (2006)
23. Möller, B., Struth, G.: wp Is wlp. In: MacCaull, W., Winter, M., Düntsch, I. (eds.) RelMiCS 2005. LNCS, vol. 3929, pp. 200–211. Springer, Heidelberg (2006). https://doi.org/10.1007/11734673_16
24. Mulvey, C.: &. In: Second Topology Conference (1986). Rendiconti del Circolo Matematico di Palermo **2**(12), 99–104
25. Nipkow, T., Wenzel, M., Paulson, L.C. (eds.): Isabelle/HOL: A Proof Assistant for Higher-Order Logic. LNCS, vol. 2283. Springer, Heidelberg (2002). https://doi.org/10.1007/3-540-45949-9
26. O'Hearn, P.: Resources, concurrency, and local reasoning. Theor. Comput. Sci. **375**, 271–307 (2007)
27. O'Hearn, P., Reynolds, J., Yang, H.: Local reasoning about programs that alter data structures. In: Fribourg, L. (ed.) CSL 2001. LNCS, vol. 2142, pp. 1–19. Springer, Heidelberg (2001). https://doi.org/10.1007/3-540-44802-0_1
28. Pym, D.: The Semantics and Proof Theory of the Logic of Bunched Implications. Kluwer Academic Publishers, Boston (2002)
29. Reynolds, J.C.: Intuitionistic reasoning about shared mutable data structure. In: Davies, J., Roscoe, B., Woodcock, J. (eds.) Millennial Perspectives in Computer Science, pp. 303–321. Palgrave (2000)

30. Reynolds, J.C.: An introduction to separation logic. In: Broy, M., Sitou, W., Hoare, T. (eds.) Engineering Methods and Tools for Software Safety and Security, NATO Science for Peace and Security Series - D: Information and Communication Security, vol. 22, pp. 285–310. IOS Press (2009)
31. Rosenthal, K.: Quantales and Their Applications. Pitman Research Notes in Mathematics Series, vol. 234 (1990)
32. Vafeiadis, V., Parkinson, M.: A marriage of rely/guarantee and separation logic. In: Caires, L., Vasconcelos, V.T. (eds.) CONCUR 2007. LNCS, vol. 4703, pp. 256–271. Springer, Heidelberg (2007). https://doi.org/10.1007/978-3-540-74407-8_18

Towards an Analysis of Dynamic Gossip in NetKAT

Malvin Gattinger[1] and Jana Wagemaker[2](✉)

[1] University of Groningen, Groningen, The Netherlands
malvin@w4eg.eu
[2] CWI, Amsterdam, The Netherlands
jana.wagemaker@cwi.nl

Abstract. In this paper we analyse the dynamic gossip problem using the algebraic network programming language NetKAT.

NetKAT is a language based on Kleene algebra with tests and describes packets travelling through networks. It has a sound and complete axiomatisation and an efficient coalgebraic decision procedure. Dynamic gossip studies how information spreads through a peer-to-peer network in which links are added dynamically.

In this paper we embed dynamic gossip into NetKAT. We show that a reinterpretation of NetKAT in which we keep track of the state of switches allows us to model Learn New Secrets, a well-studied protocol for dynamic gossip. We axiomatise this reinterpretation of NetKAT and show that it is sound and complete with respect to the packet-processing model, via a translation back to standard NetKAT.

Our main result is that many common decision problems about gossip graphs can be reduced to checks of NetKAT equivalences. We also implemented the reduction.

Keywords: Dynamic gossip · Kleene algebra with tests (KAT)
Network programming language · NetKAT
Peer-to-peer communication

1 Introduction

The dynamic gossip problem is a generalisation of the classic telephone problem, in which agents exchange secrets in phone calls with the goal to spread all secrets. In the dynamic setting, also phone numbers are exchanged, hence who can call whom constantly changes. More generally, dynamic gossip provides a formal model of any peer-to-peer setting in which information has to be spread or synchronised between multiple nodes.

As of now, no formal language and logic exists that captures dynamic gossip in a satisfactory way. There is existing work on gossip using formal languages which we will discuss in Sect. 7, but to our knowledge there is no sound and complete proof system to describe dynamic gossip axiomatically.

J. Desharnais et al. (Eds.): RAMiCS 2018, LNCS 11194, pp. 280–297, 2018.
https://doi.org/10.1007/978-3-030-02149-8_17

Anderson et al. in [1] and Foster et al. in [10], introduced NetKAT, a logic to describe packet-processing behaviour of networks. In this paper we will model dynamic gossip in NetKAT.

Our contributions are twofold. First, we show that we can simulate switch states in NetKAT without losing soundness and completeness, as long as different packets do not interact with each other. This observation is motivated by the application to dynamic gossip, but also applies to NetKAT in general. Second, we show that simulating switch states in NetKAT is a natural choice for the analysis of dynamic gossip protocols. We provide a way to compute all call sequences of the so-called Learn New Secrets (LNS) protocol by evaluating a NetKAT policy. Moreover, we reduce the decision problems whether LNS is weakly or strongly successful to a check of NetKAT equivalences. Given the complete axiomatisation and efficient decision procedure of NetKAT, any question about gossip graphs expressible as a NetKAT equivalence is decidable. More generally, this paper builds a bridge between the two areas of gossip protocols and network programming languages.

We proceed as follows. In Sect. 2 we summarise the main definitions of NetKAT. Then we present a reinterpretation of NetKAT simulating switch states in Sect. 3. In Sect. 4 we recapitulate the dynamic gossip problem and in Sect. 5 we show how to encode it in NetKAT, including proofs that these translations are sound. We describe an implementation of our methods in Sect. 6, discuss related work in Sect. 7 and conclude with ideas for future work in Sect. 8. Further details can also be found in the master thesis [20] on which this paper is based on.

2 Standard NetKAT

NetKAT was first presented in [1] and is a network programming language with a strong mathematical foundation. It extends Kleene Algebra with Tests (KAT) and relies on technical results from that field.

A Kleene Algebra with Tests is a Kleene algebra — the algebra of regular expressions — with a Boolean algebra. Its soundness and completeness with respect to relational models and language-theoretic models is proven in [16]. Formally, a Kleene algebra with tests (KAT) is a two-sorted algebraic structure $(K, B, +, \cdot, ^*, 0, 1, \neg)$ where $B \subseteq K$ and $\neg$ is a unary operator defined only on B such that

- $(K, +, \cdot, ^*, 0, 1)$ is a Kleene algebra,
- $(B, +, \cdot, 0, 1, \neg)$ is a Boolean algebra
- $(B, +, \cdot, 0, 1, \neg)$ is a subalgebra of $(K, +, \cdot, ^*, 0, 1)$

The equational theory of KAT is PSPACE-complete as shown in [16].

The Kleene algebra operators are $+$ for non-deterministic choice, $\cdot$ for sequential composition, the Kleene star * for finite iteration, 0 for fail and 1 for skip. Elements of the Boolean algebra are called tests, and on tests the $+$ and $\cdot$ operators behave as disjunction and conjunction respectively. The negation operator $\neg$ can only be applied to elements of the Boolean algebra.

The axioms of KAT are the axioms of Kleene algebra combined with the axioms of Boolean algebra. They can be found in [15].

NetKAT's main purpose is to describe how packets travel through a network, taking into account both the network topology and the individual switch behaviours, where a switch is a node in the network topology. To do so, NetKAT extends KAT with additional primitives for network behaviour and axioms governing those primitives. Syntactically, NetKAT expressions can be predicates and policies. Predicates are the constants true and false (1 and 0 respectively), tests ($f = n$), negation ($\neg a$), disjunction ($a+b$) and conjunction ($a \cdot b$). Policies are all predicates, modifications ($f \leftarrow n$), union ($p + q$), sequential composition ($p \cdot q$) and iteration (p^*). We assume finite NetKAT networks throughout this paper.

Definition 1. *The syntax of NetKAT is given by the following* predicates *a and* policies *p:*

$$a ::= 1 \mid 0 \mid f = n \mid a + b \mid a \cdot b \mid \neg a$$
$$p ::= a \mid f \leftarrow n \mid p + q \mid p \cdot q \mid p^*$$

where f ranges over some finite set of fields $f ::= f_1 \mid \cdots \mid f_k$ (including a switch field sw *and a port field* pt*) and n is a value from a finite domain.*

Example 1. The NetKAT expression $\mathsf{sw} = A \cdot \mathsf{pt} = 4 \cdot \mathsf{dst} \leftarrow H \cdot \mathsf{pt} \leftarrow 7$ can be read as "test whether the packet is located at port 4 of switch A and then set the destination to H and move the packet to port 7".

There exist multiple models that satisfy the NetKAT axioms, but we are mainly concerned with the packet-processing model. A packet pk is a tuple ($f_1 = v_1, \ldots, f_n = v_n$) which for each field f_k provides a value v_n from a finite domain. Among the fields, two are used for the location of the packet in the network: switch (sw) and port (pt). We write $\mathsf{pk}.f$ for the value in field f of pk and $\mathsf{pk}[f := n]$ for the packet obtained from pk by updating field f to n.

Standard NetKAT as presented in [1] also tracks the history of packets. It contains an operator *dup* to duplicate the current packet so that a copy of it is kept in the history. We do not need histories to model gossip in NetKAT, so we leave *dup* out here and simplify the semantics to work on packets instead of histories of packets.

We show the semantics of NetKAT in Fig. 1. The interpretation of a policy p is a function that maps packets to sets of packets: $[\![p]\!] : P \to 2^P$ where P is the set of all packets. A test or filter $f = n$ takes any input packet pk and outputs the singleton $\{\mathsf{pk}\}$ if field f of pk equals n, and $\varnothing$ otherwise. A modification $f \leftarrow n$ takes any input packet pk and yields the singleton set $\{\mathsf{pk}[f := n]\}$.

The $+$ is interpreted as a multicast operation: the outcome of the policy $(\mathsf{pt} \leftarrow 1 + \mathsf{pt} \leftarrow 2)$ are two copies of the input packet, one at port 1 and one at port 2. The policy $p \cdot q$ is the Kleisli composition of p and q which we define as $(f \bullet g)(x) = \bigcup\{g(y) \mid y \in f(x)\}$. To iterate sequential composition with * we define $F^0(h) := \{h\}$ and $F^{i+1}(h) := (F \bullet F^i)h$.

$$
\begin{aligned}
[\![1]\!](\mathsf{pk}) &:= \{\mathsf{pk}\} \\
[\![0]\!](\mathsf{pk}) &:= \varnothing \\
[\![\neg a]\!](\mathsf{pk}) &:= \{\mathsf{pk}\} \setminus [\![a]\!](\mathsf{pk}) \\
[\![f = n]\!](\mathsf{pk}) &:= \begin{cases} \{\mathsf{pk}\} & \text{if } \mathsf{pk}.f = n \\ \varnothing & \text{otherwise} \end{cases} \\
[\![f \leftarrow n]\!](\mathsf{pk}) &:= \{\mathsf{pk}[f := n]\} \\
[\![p + q]\!](\mathsf{pk}) &:= [\![p]\!] h \cup [\![q]\!] \mathsf{pk} \\
[\![p \cdot q]\!](\mathsf{pk}) &:= ([\![p]\!] \bullet [\![q]\!])(h) \\
[\![p^*]\!](\mathsf{pk}) &:= \bigcup_{i \in \mathbb{N}} [\![p]\!]^i(\mathsf{pk})
\end{aligned}
$$

Fig. 1. Semantics of NetKAT

$(f \leftarrow n \cdot f' \leftarrow n') \equiv (f' \leftarrow n' \cdot f \leftarrow n)$, if $f \neq f'$	MOD-MOD-COMM
$(f \leftarrow n \cdot f' = n') \equiv (f' = n' \cdot f \leftarrow n)$, if $f \neq f'$	MOD-FILTER-COMM
$(f \leftarrow n \cdot f = n) \equiv (f \leftarrow n)$	MOD-FILTER
$(f = n \cdot f \leftarrow n) \equiv (f = n)$	FILTER-MOD
$(f \leftarrow n \cdot f \leftarrow n') \equiv (f \leftarrow n')$	MOD-MOD
$(f = n \cdot f = n') \equiv 0$, if $n \neq n'$	CONTRA
$\sum_i (f = i) \equiv 1$	MATCH-ALL

Fig. 2. NetKAT axioms

The NetKAT network of switches and hosts is not used in the semantics. As shown in [1], any network topology can be incorporated into a policy. This is not important here, because we will use a totally connected NetKAT network.

The NetKAT axioms are the axioms of KAT together with additional axioms for the interactions between tests and modifications. We show these additional axioms in Fig. 2. The first one for instance tells us that when modifying two different fields, the order does not matter. The last axiom implies that the values of the fields are drawn from a finite domain. These axioms are sound and complete with respect to the packet-processing model of Fig. 1. For proofs, see [1].

3 Simulating Switch States in NetKAT

In standard NetKAT, a switch has to treat all incoming packets according to the same policy. For example, it cannot count how many packets of a certain type it has seen before. In this section we will demonstrate that we can reinterpret selected packet fields as switch states, which gives switches the ability to react differently, depending on previous packets. NetKAT can then express examples like dynamic gossip more naturally.

Standard NetKAT describes how packets travel through a network and it can describe the behaviour of multiple packets with the multicast interpretation of the + operator. However, each of these packets form their own network trace, and NetKAT does not allow any interaction between them.

Ideally, a global state would allow us to model interactions between packets via alterations of the global state. But this immediately yields questions about concurrency. Versions of Kleene algebra that treat concurrency can be found in [14] and [13], but to our knowledge a similar extension of KAT or NetKAT has not been found yet. For dynamic gossip as studied in [4] we do not actually need interaction between packets. It suffices to use a single packet, because in standard Dynamic Gossip no two phone calls can take place at the same time. Hence we can avoid concurrency questions by linking the global state of all switches to the current packet.

We have to change the intended meaning of the + operator in order to reinterpret selected fields as switch states. In standard NetKAT the + operator can be understood as multicasting. However, it is not realistic that a switch is in multiple states at the same time. Therefore, in NetKAT with switch states the + operator should be understood as non-deterministic choice.

For clarity, we introduce a new piece of syntax to separate the standard packet fields from the switch state fields.

Definition 2. *The syntax of NetKAT with switch states extends the one of NetKAT as follows. For f as in Definition 1, n a packet field value or a switch state from a finite domain and i a switch identifier:*

$$a ::= 1 \mid 0 \mid f = n \mid \mathsf{state}(i) = n \mid a + b \mid a \cdot b \mid \neg a$$
$$p ::= a \mid f \leftarrow n \mid \mathsf{state}(i) \leftarrow n \mid p + q \mid p \cdot q \mid p^*$$

We thus add two new operators: the state test $\mathsf{state}(i) = n$ and the state modification $\mathsf{state}(i) \leftarrow n$. They work similarly to normal tests and modifications but act specifically on the switch state fields.

The semantics of the new variant of NetKAT is shown in Fig. 3. In order to also highlight the reinterpretation of selected packet fields in the semantics, we use a *state vector* $\vec{s}$ that contains all the switch state fields. For the set of all state vectors we write $\vec{S}$. The interpretation of each policy p is thus a function $[\![p]\!]: P \times \vec{S} \to 2^{(P \times \vec{S})}$ where $P \times \vec{S}$ is the set of all tuples $\mathsf{ps} = (\mathsf{pk}, \vec{s})$ of a packet pk and a state vector $\vec{s}$. Note the similarity to Fig. 1: We only changed the underlying set from P to $P \times \vec{S}$ and added definitions for tests and assignments on the state vector.

Example 2. Consider policy $(\mathsf{sw} \leftarrow A \cdot \mathsf{state}(A) \leftarrow 1) + (\mathsf{sw} \leftarrow B \cdot \mathsf{state}(B) \leftarrow 2)$. Applied to $(\mathsf{pk}, \vec{s})$, this policy outputs $\{(\mathsf{pk}[\mathsf{sw} := A], \vec{s}[A := 1]), (\mathsf{pk}[\mathsf{sw} := B], \vec{s}[B := 2])\}$. Hence, state vectors denote the state of the network as a result of how the corresponding packet was processed by the policy.

Note that in Example 2 the packet first moves to a switch before modifying its state. The semantics of NetKAT with switch states in principle allows us to change the state of a switch without the packet actually being there at this moment. Such policies are unrealistic. We therefore restrict ourselves to policies that only modify the state of switches when they are there.

Definition 3 (Topology Respecting). *We say that a policy is* topology respecting *iff it is equivalent to its* localised version *which is obtained by replacing every* $\mathsf{state}(x) \leftarrow n$ *with* $\mathsf{sw} = x \cdot \mathsf{state}(x) \leftarrow n$.

The axioms of NetKAT with switch states are the same as for standard NetKAT, except that we add axioms for state tests and state modifications. These are exactly the same as those shown in Fig. 2 with $\mathsf{state}(i)$ replacing f.

It is easy to see that NetKAT with switch states can be translated back to standard NetKAT. This allows us to use the soundness and completeness of standard NetKAT to argue that NetKAT with switch states is sound and complete. Intuitively, the translation relies on the fact that the packet and state vector are always paired together. Hence we can simply map the state vector to extra fields w_i for each switch i.

Definition 4. *Let* $m\colon (P \times \vec{S}) \to P$ *be the translation defined by*

$$m\left(\{f_1 = v_1, \ldots, f_n = v_n\}, [s_1, \ldots, s_n]\right) := \left\{ \begin{matrix} f_1 = v_1, \ldots, f_n = v_n, \\ w_1 = s_1, \ldots, w_n = s_n \end{matrix} \right\}$$

and let m^{-1} *denote its inverse. Let* t *map a NetKAT expression with* state *to a standard NetKAT expression, proceeding by recursion for compositional policies and mapping policies concerning the state vector as follows:*

$$t(\mathsf{state}(i) = n) := w_i = n \qquad t(\mathsf{state}(i) \leftarrow n) := w_i \leftarrow n$$

For all other atomic policies p*, let* $t(p) := p$.

Lemma 1. *For every policy* p *and every packet with state* ps *we have:*

$$[\![p]\!](\mathsf{ps}) = \{m^{-1}(\mathsf{pk}) \mid \mathsf{pk} \in [\![t(p)]\!](m(\mathsf{ps}))\}$$

$$\begin{aligned}
[\![1]\!](\mathsf{ps}) &:= \{\mathsf{ps}\} \\
[\![0]\!](\mathsf{ps}) &:= \varnothing \\
[\![\neg a]\!](\mathsf{ps}) &:= \{\mathsf{ps}\} \setminus ([\![a]\!](\mathsf{ps})) \\
[\![f = n]\!]((\mathsf{pk}, \vec{s})) &:= \begin{cases} \{(\mathsf{pk}, \vec{s})\} & \text{if } \mathsf{pk}.f = n \\ \varnothing & \text{otherwise} \end{cases} \\
[\![f \leftarrow n]\!]((\mathsf{pk}, \vec{s})) &:= \{(\mathsf{pk}[f := n], \vec{s})\} \\
[\![p + q]\!](\mathsf{ps}) &:= [\![p]\!](\mathsf{ps}) \cup [\![q]\!](\mathsf{ps}) \\
[\![p \cdot q]\!](\mathsf{ps}) &:= ([\![p]\!] \bullet [\![q]\!])(\mathsf{ps}) \\
[\![p^*]\!](\mathsf{ps}) &:= \textstyle\bigcup_{i \in \mathbb{N}} [\![p]\!]^i(\mathsf{ps}) \\
[\![\mathsf{state}(i) = n]\!]((\mathsf{pk}, \vec{s})) &:= \begin{cases} \{(\mathsf{pk}, \vec{s})\} & \text{if } \vec{s}(i) = n \\ \varnothing & \text{otherwise} \end{cases} \\
[\![\mathsf{state}(i) \leftarrow n]\!]((\mathsf{pk}, \vec{s})) &:= \{(\mathsf{pk}, \vec{s}[i := n])\}
\end{aligned}$$

Fig. 3. Semantics of NetKAT with switch states

Lemma 2. *For every two policies p and q we have:*

1. $[\![p]\!] = [\![q]\!]$ *if and only if* $[\![t(p)]\!] = [\![t(q)]\!]$ *and*
2. $p \equiv q$ *if and only if* $t(p) \equiv t(q)$.

Both lemmas can be proven by induction on the structure of policies, but as these proofs do not contain any new insights we omit them here.

Theorem 1 (Soundness and Completeness). *For all policies p and q of* NetKAT *with switch states we have* $[\![p]\!] = [\![q]\!]$ *if and only if* $p \equiv q$.

This finishes our explanation of NetKAT with switch states. In Sect. 5 we will use it to give an intuitive formalisation of dynamic gossip.

4 Dynamic Gossip

Dynamic Gossip is an extension of the simpler gossip problem, also known as the telephone problem: A group of agents each has a secret. They can communicate via phone calls in which two agents exchange all the secrets they know. How many calls are needed, until all agents know all the secrets? This scenario was widely studied in the 1980s and a classic result is that for $n \geq 4$ agents $2n - 4$ phone calls are necessary and sufficient to distribute all secrets to everyone.

In the original setting every agent can call every other agent. Later studies removed this assumption and used a reachability graph to define which agents can communicate with each other. Different classes of graphs lead to different minimal numbers of calls. For a survey on classical static gossip, see [11].

Dynamic gossip from [7] is another variation of the problem: not only is there a reachability graph restricting who can call whom, but this graph is also manipulated when phone calls are made. Intuitively, the agents now also exchange phone numbers, in addition to the secrets. The dynamic gossip literature focuses on epistemic protocols [2,8], which can be executed by a group of agents without a central authority. The prime example of such a protocol is "Learn New Secrets", short LNS. It allows agent a to call agent b if and only if a knows the number of b but does not know the secret of b.

In the remainder of this section we briefly state the basic definitions and a main result about the dynamic gossip problem.

Definition 5 (Gossip Graph). *Given a finite set of agents A, a* gossip graph *G is a triple (A, N, S) where N and S are binary relations over A such that $I \subseteq S \subseteq N$ where I is the identity relation on A. We write N_a as an abbreviation for $\{b \mid (a, b) \in N\}$. We abbreviate $(a, b) \in N$ with Nab. An* initial gossip graph *is a gossip graph where $S = I$. The set of all initial gossip graphs is denoted by $\mathcal{G}$. A gossip graph is called* total *if $S = A \times A$.*

The relations model the basic knowledge of the agents. We say that agent a *knows the number* of b iff Nab and that a *knows the secret* of b iff Sab. Hence a *total* gossip graph is one in which everyone knows all secrets.

Definition 6 (Possible Call; Call Execution). *A* call *is an ordered pair of agents* $(a, b) \in (A \times A)$. *We usually write ab instead of* (a, b). *Given a gossip graph* $G = (A, N, S)$, *a call ab is* possible *iff* Nab. *Given a possible call ab,* G^{ab} *is the graph* (A', N', S') *such that* $A' := A$, $N'_a := N'_b := N_a \cup N_b$, $S'_a := S'_b := S_a \cup S_b$, *and* $N'_c := N_c$, $S'_c := S_c$ *for* $c \neq a, b$. *For a sequence of calls* $ab; cd; \ldots$ *from the set* $(A \times A)^*$ *we write* σ. *The empty sequence is* ϵ. *We extend the notation* G^{ab} *to sequences of calls:* $G^\epsilon := G$, $G^{\sigma;ab} := (G^\sigma)^{ab}$.

Definition 7 (LNS). *The* Learn New Secrets protocol *is defined as follows. Given a gossip graph* $G = (A, N, S)$, *the* set of LNS-allowed calls *is:*

$$lns(G) := \{(a, b) \in A \times A \mid Nab \text{ and not } Sab \text{ in } G\}$$

The extension of LNS *on G is the set of call sequences defined recursively by*

$$LNS(G) := \begin{cases} \{ab; \sigma \mid (a, b) \in lns(G) \text{ and } \sigma \in LNS(G^{ab})\} & \text{if } lns(G) \neq \varnothing \\ \{\epsilon\} & \text{otherwise} \end{cases}$$

Example 3. Consider an initial gossip graph G for three agents in which a knows the number of b, and b knows the number of c. Suppose that a calls b. We then obtain the gossip graph G^{ab} in which a and b know each other's secret and a now also knows the number of c. We can visualise the two graphs as follows, with dashed lines for N and solid lines for S:

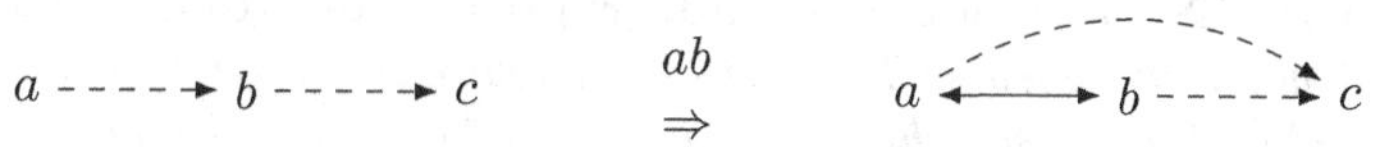

There are three LNS sequences on G, namely $ab; ac; bc$, $ab; bc; ac$ and $bc; ab$. The first two sequences are successful, i.e. they lead to a total graph in which everyone knows all three secrets. The shorter sequence $bc; ab$ is stuck: no more calls are allowed according to LNS but not everyone knows all the secrets yet.

The gossip graph in Example 3 is a case in which LNS is only weakly, but not strongly successful, which we define as follows.

Definition 8 (Weak and Strong Success). *Consider a gossip graph G.*

1. *We say that LNS is* weakly successful *on G if and only if there is a call sequence* $\sigma \in LNS(G)$ *such that* G^σ *is total.*
2. *We say that LNS is* strongly successful *on G if and only if for all call sequences* $\sigma \in LNS(G)$ *we have that* G^σ *is total.*

We conclude this section with a definition and theorem from [7].

Definition 9 (Sun Graphs). *An initial gossip graph G is a* sun graph *if and only if N is strongly connected on the restriction of G to non-terminal nodes (nodes with at least one out-going edge).*

Theorem 2 (Theorem 20 in [7]). *Let G be an initial gossip graph. LNS is strongly successful on G if and only if G is a sun graph.*

5 Dynamic Gossip in NetKAT

In this section we will discuss how to translate gossip to NetKAT. We will build a NetKAT network and input packet that together represent a gossip graph. Then we will construct a policy that performs the LNS protocol.

The switches in NetKAT will represent the gossiping agents, and the state of a switch will be the list of phone numbers and secrets this agent knows. We stress that the NetKAT network does *not* describe the gossip graph. Instead, we ensure that during the execution of LNS each agent can in principle communicate with every other agent by using a totally connected NetKAT network.

The numbers and secrets an agent knows are part of the local state of the corresponding switch. Instead of encoding the N and S relation into one integer field $\mathsf{state}(i)$ for each switch i, we simplify notation and use multiple NetKAT fields Nij and Sij with values 0 and 1 to describe the state of i. For example, the field $N12$ is part of the local state of switch 1 and $Nab = 1$ means that agent a knows the phone number of agent b. We will ensure that the gossip policies will be topology respecting in the sense of Definition 3 applied to all fields Nij and Sij belonging to switch i, instead of only one field $\mathsf{state}(i)$.

In addition to the Sij and Nij fields, the input packet will have fields for the location of the packet, sw and pt, and fields call_m for denoting what call took place in round m.

Definition 10 (NetKAT network and graph packet). *Consider a group A of $n := |A|$ gossiping agents. The NetKAT network $\mathcal{N}_n$ for n agents is a fully connected network of n switches, each of which has an additional port $home_i$ that is not connected to any other switch. Suppose we have a gossip graph $G = (A, N, S)$. The NetKAT input packet pk_G describes the S and N relations as follows. The call_m fields represent call rounds and we will explain them later.*

$$\begin{aligned} \mathsf{pk}_G := &\{\mathsf{sw} = 0, \mathsf{pt} = 0\} \cup \{\mathsf{call}_m = 0 \mid m \in \{0, \ldots, n(n-1)\}\} \\ &\cup \{Nij = 1 \mid (i,j) \in N\} \cup \{Nij = 0 \mid (i,j) \notin N\} \\ &\cup \{Sij = 1 \mid (i,j) \in S\} \cup \{Sij = 0 \mid (i,j) \notin S\} \end{aligned}$$

We will now describe a NetKAT policy that represents the LNS protocol. On an input packet representing a gossip graph $G = (A, N, S)$, this policy should output packets corresponding to call sequences σ and the Nij and Sij fields should describe N^σ and S^σ. We want an output for each $\sigma \in LNS(G)$.

The policy describing LNS starts by distributing the initial input packet to every agent's home port — a private port not connected to any other agent. This is necessary to ensure that any agent can make the first call.

The next part of the policy will describe all LNS-allowed calls. In particular, a call from a to b is defined in policy pol_{ab} and describes a packet moving from the caller to the callee and back. We first check that the LNS conditions are satisfied. If so, the call takes place by moving the packet back and forth and updating the knowledge of the agents when the packet is located at the corresponding switch. We also keep track of the call sequences by overwriting the field call_k in round k, up to round $n(n-1)$. This is a safe upper bound on the number of calls, because no call happens more than once in LNS. After a call has taken place, the resulting packet is again distributed to all home ports to ensure that any other agent can initiate the next call. Finally, we use the Kleene star to iterate the whole procedure.

Definition 11 (Gossip Policies). *Let A be the set of all agents and let $n := |A|$. Consider the NetKAT network $\mathcal{N}_n$ for n agents. For all $a, b \in A$, let $link_{ab}$ denote the port of agent a connected to agent b, and for all $i \in A$ let $home_i$ be the home port of agent i. For distributing the packet, we define a policy:*

$$pol_{\text{dstr}} := \sum_{i=1}^{n} (\mathsf{sw} \leftarrow i \cdot \mathsf{pt} \leftarrow home_i)$$

Suppose f is a field which takes values 0 and 1. We use the following NetKAT abbreviation for "if ... then ... " programs: $(\text{IF } f \text{ THEN } p) := (f = 1 \cdot p) + (f = 0)$. *For each call (a, b) we define a policy:*

$$\begin{aligned} pol_{ab} := \mathsf{sw} = a \ \cdot \ \mathsf{pt} = home_a \ \cdot \ Nab = 1 \ \cdot \ Sab = 0 \\ \cdot \ \mathsf{pt} \leftarrow link_{ab} \ \cdot \ \mathsf{sw} \leftarrow b \ \cdot \ \mathsf{pt} \leftarrow link_{ba} \\ \cdot \ \textstyle\prod_{x \in A} (\text{IF } Sax \text{ THEN } Sbx \leftarrow 1) \ \cdot \ \prod_{x \in A} (\text{IF } Nax \text{ THEN } Nbx \leftarrow 1) \\ \cdot \ \mathsf{sw} \leftarrow a \ \cdot \ \mathsf{pt} \leftarrow link_{ab} \\ \cdot \ \textstyle\prod_{x \in A} (\text{IF } Sbx \text{ THEN } Sax \leftarrow 1) \ \cdot \ \prod_{x \in A} (\text{IF } Nbx \text{ THEN } Nax \leftarrow 1) \\ \cdot \ \textstyle\sum_{k=1}^{n(n-1)} \left(\prod_{y<k} (\neg(\mathsf{call}_y = 0)) \ \cdot \ \mathsf{call}_k = 0 \ \cdot \ \mathsf{call}_k \leftarrow ab \right) \end{aligned}$$

To make any LNS call, we define the policy $pol_{\text{lns}} := \sum_{i \in A} \left(\sum_{j \in A \setminus \{i\}} pol_{ij} \right)$. *For the whole LNS protocol, let* $pol_{\text{LNS}} := (pol_{\text{dstr}} \cdot pol_{\text{lns}})^*$ *and for any sequences of LNS calls $\sigma = c_1; \ldots; c_k$, let* $pol_\sigma := pol_{\text{dstr}} \cdot pol_{c_1} \cdot \ldots \cdot pol_{\text{dstr}} \cdot pol_{c_k}$.

The pol_{LNS} policy only depends on the number of agents, and not on the gossip graph. It is topology respecting because we only update those N and S fields of agents that are emulated by the switches where the packet is at that moment of evaluation.

We now get the following correspondence between the original definition of G^σ and the result of applying pol_σ to pk_G.

Lemma 3. *Consider a gossip graph $G = (A, N, S)$, the corresponding pk_G, any LNS sequence of calls σ and the resulting graph $G^\sigma = (A, N^\sigma, S^\sigma)$. Then $[\![pol_\sigma]\!](\mathsf{pk}_G)$ describes G^σ, i.e. we have:*

$$N^\sigma = \{(a, b) \mid \forall \mathsf{pk} \in [\![pol_\sigma]\!](\mathsf{pk}_G) : \mathsf{pk}.Nab = 1\}$$

$$S^\sigma = \{(a, b) \mid \forall \mathsf{pk} \in [\![pol_\sigma]\!](\mathsf{pk}_G) : \mathsf{pk}.Sab = 1\}$$

In fact, pol_σ is such that the output is always exactly one packet.

Lemma 3 only talks about a single sequence, but we can lift it to the whole LNS protocol as follows. The LNS call sequences for a gossip graph G are the same call sequences as those generated by applying pol_{LNS} to pk_G and the distribution of knowledge in the output packets of pol_{LNS} is the same as in the gossip graphs resulting from the same call sequences (this latter fact follows directly from Lemma 3).

Let pol_{stop} be a policy checking whether the LNS protocol has finished, i.e. testing whether for every two agents i and j we have either $Sij = 1$ or $Nij = 0$: $pol_{\text{stop}} := \prod_{i,j \in A}(Sij = 1 + Nij = 0)$.

Theorem 3. *Definitions 7 and 11 agree with each other. Formally, for every gossip graph $G = (A, N, S) \in \mathcal{G}$ we have:*

$$LNS(G) = \{c_0; \ldots; c_k \in (A \times A)^* \mid \exists pk \in [\![pol_{\text{LNS}} \cdot pol_{\text{stop}}]\!](pk_G): \\ \forall m \leq k\colon pk.\mathsf{call}_m = c_m \textit{ and } pk.\mathsf{call}_{k+1} = 0\}$$

Proof. We first show $\subseteq$. Suppose an LNS call sequence σ is a result of the LNS protocol on gossip graph G. We now prove by induction on the length of σ that this call sequence is also happening through pol_{LNS}. In case σ is ϵ, it means no calls were allowed according to LNS for gossip graph G. The initial input pk_G to pol_{LNS} will resemble this, and thus no calls happen through pol_{LNS} either.

For the induction step, we use Lemma 3 as follows. Suppose we have a sequence $xy; \sigma \in LNS(G)$ and σ consists of k calls. Then by Definition 7, we know that $\sigma \in LNS(G^{xy})$. By the induction hypothesis we then know that there exists a packet $\mathsf{pk} \in [\![pol_{\text{LNS}} \cdot pol_{\text{stop}}]\!](\mathsf{pk}_{G^{xy}})$ such that for all $m \leq k - 1$ we have $\mathsf{pk.call}_m = c_m$. Moreover, $\mathsf{pk} \in [\![pol_\sigma]\!](\mathsf{pk}_{G^{xy}})$. From Lemma 3 we know that $N^{xy} = \{(a, b) \mid \forall \mathsf{pk} \in [\![pol_{xy}]\!](\mathsf{pk}_G) : \mathsf{pk}.Nab = 1\}$ and $S^{xy} = \{(a, b) \mid \forall \mathsf{pk} \in [\![pol_{xy}]\!](\mathsf{pk}_G) : \mathsf{pk}.Sab = 1\}$. Thus we can conclude that $\mathsf{pk}_{G^{xy}} \in [\![pol_{xy}]\!](\mathsf{pk}_G)$, as $\mathsf{pk}_{G^{xy}}$ is the packet resembling gossip graph G^{xy} and thus such that it corresponds to N^{xy} and S^{xy}. This gives us a $\mathsf{pk}' \in [\![pol_{xy;\sigma}]\!](\mathsf{pk}_G)$ where pk' is the same as packet pk except that pk' will have one more call field. The values of the call fields are such that they match $xy; \sigma$ in the sense that $c_0 = xy$ and $c_1; \ldots; c_k = \sigma$. Moreover, $xy; \sigma$ is a finished LNS sequence. Hence we have $\mathsf{pk}' \in [\![pol_{\text{LNS}} \cdot pol_{\text{stop}}]\!](\mathsf{pk}_G)$ such that for all $m \leq k$ we have $\mathsf{pk}'.\mathsf{call}_m = c_m$.

For the converse $\supseteq$, we again proceed by induction on the length of call sequences. For the base case, if pk_G matches graph G and no calls are allowed to take place by pol_{LNS}, then also $LNS(G)$ will be empty.

For the induction step, take any outcome $\mathsf{pk} \in [\![pol_{\text{LNS}} \cdot pol_{\text{stop}}]\!](\mathsf{pk}_G)$ that is the result of $k + 1$ iterations inside pol_{LNS}. Then we get a call sequence $c_0; \ldots; c_k$ such that for all $m \leq k$ we have that $\mathsf{pk.call}_m = c_m$. Let us say that $c_0 = xy$ and denote $c_0; \ldots; c_k$ with $xy; \sigma$. In other words, pk is the result of policy $pol_{xy;\sigma}$ applied to pk_G where $xy; \sigma$ is a finished LNS sequence on G. Similar to what we did before, we can conclude from Lemma 3 that $\mathsf{pk}_{G^{xy}} \in [\![pol_{xy}]\!](\mathsf{pk}_G)$. Hence, $\mathsf{pk}' \in [\![pol_\sigma]\!](\mathsf{pk}_{G_{xy}})$ where pk' is the same as pk except that pk' has one less call field as pk and the call fields of pk' correspond to σ which is a finished call sequence on G^{xy}. We then get that $\mathsf{pk}' \in [\![pol_{\text{LNS}} \cdot pol_{\text{stop}}]\!](\mathsf{pk}_{G^{xy}})$ such that there is a sequence $\sigma = c_0; \ldots; c_k$

where for all $m \leq k$ we have that $\mathsf{pk}'.\mathsf{call}_m = c_m$. By our induction hypothesis we can now conclude that $\sigma \in LNS(G^{xy})$. From Definition 7 we know that to show that $xy;\sigma \in LNS(G)$ we need to have that $(x,y) \in lns(G)$. This holds because $[\![pol_{xy}]\!](\mathsf{pk}_G)$ was nonempty and pk_G corresponds to G. □

This connection between the standard gossip definitions and our definitions in NetKAT might seem obvious to the reader — of course we defined the policies exactly to get this correspondence. Theorem 3 is still useful because it means that we can use NetKAT to compute all LNS call sequences and to check whether LNS is successful, with the following translation of Definition 8 to NetKAT.

Definition 12 (Graphs and Success in NetKAT). *Given a gossip graph G, we define a policy pol_G to check whether the current packet encodes G:*

$$pol_G := \begin{pmatrix} \prod\{Nij = 1 \mid Nij \; in \, G\} \cdot \prod\{Nij = 0 \mid \; not \, Nij \; in \, G\} \cdot \\ \prod\{Sij = 1 \mid Sij \; in \, G\} \cdot \prod\{Sij = 0 \mid \; not \; Sij \; in \, G\} \end{pmatrix}$$

We also define a policy to test whether LNS has been successful, i.e. whether the S relation encoded in a given packet is total: $pol_{\text{success}} := \prod_{i,j \in A}(Sij = 1)$.

Theorem 4. *The LNS protocol is weakly successful on gossip graph G if and only if the following NetKAT equivalence holds:*

$$pol_G \cdot pol_{\text{LNS}} \cdot pol_{\text{success}} \not\equiv 0$$

Similarly, we can reduce the check whether LNS is strongly successful to NetKAT as follows. See the appendix for proofs of Theorems 4 and 5.

Theorem 5. *The LNS protocol is strongly successful on gossip graph G if and only if the following NetKAT equivalence holds:*

$$pol_G \cdot pol_{\text{LNS}} \cdot pol_{\text{stop}} \equiv pol_G \cdot pol_{\text{LNS}} \cdot pol_{\text{success}}$$

We can also translate the notion of a sun graph from Definition 9 to NetKAT. For this we need to express that an agent has an N-path to every other agent.

Definition 13 (N-paths in NetKAT). *For any set of agents A, we define the following two policies:*

$$pol_{\text{num}} := \left(\sum_{x \in A} \sum_{y \in A \setminus \{x\}} \begin{pmatrix} \mathsf{sw} = x \cdot \mathsf{pt} = home_x \cdot Nxy = 1 \cdot \mathsf{pt} \leftarrow link_{xy} \\ \cdot \; \mathsf{sw} \leftarrow y \cdot \mathsf{pt} \leftarrow link_{yx} \cdot \mathsf{pt} \leftarrow home_y \end{pmatrix} \right)^*$$

$$pol_{Ni} := \prod_{j \in A \setminus \{i\}} (\mathsf{sw} = i \cdot \mathsf{pt} = home_i \cdot pol_{\text{num}} \cdot \mathsf{sw} = j \cdot \mathsf{sw} \leftarrow i \cdot \mathsf{pt} \leftarrow home_i)$$

Lemma 4. *Consider a gossip graph $G = (A, N, S)$. An agent $i \in A$ has an N-path in to every other agent if and only if we have $pol_G \cdot pol_{Ni} \not\equiv 0$.*

The policy pol_{num} distributes a packet following the N-relation between any two agents. We prepend and append it with location tests to check whether we can go from agent i to another agent j. As we want an N-path to every other agent in the network, we take a product over all agents. If this does not return the empty set, we know that agent i has an N-path to every agent in the network.

Theorem 6 (Sun Graphs in NetKAT). *A gossip graph G is a sun graph if and only if the following NetKAT equivalence holds:*

$$pol_G \cdot \prod_{i \in A} (\mathsf{sw} \leftarrow i \cdot \mathsf{pt} \leftarrow home_i \cdot (pol_{Ni} + pol_{\text{self}_i})) \not\equiv 0$$

where pol_{self_i} is a policy checking whether agent i only has its own number.

The proof is straightforward and can be found in the appendix.

Corollary 1. *For every gossip graph G, "LNS is strongly successful on G if and only if G is a sun" can be expressed and proven using NetKAT equivalences.*

Admittedly, this corollary is the easy direction: By Theorems 2 and 6 and the completeness of NetKAT, we know that for each G there *exists* a proof of Theorem 2 in NetKAT — see appendix. We leave the more interesting and challenging task as future work: to actually find these algebraic proofs. Moreover, it would be interesting to find a proof on the meta-level by working with schemata of NetKAT expressions instead of a specific G. This could yield a completely new proof of Theorem 2 or one can try to translate the proof from [7] to NetKAT.

6 Implementation

The embedding of dynamic gossip into NetKAT allows us to reduce decision problems about gossip graphs to NetKAT equivalence checking. We implemented the methods described in this paper in *Haskell*. The code is available at https://github.com/janawagemaker/GossipKATS and can be used in three ways.

1. Gossip-only: We provide a direct and explicit implementation of dynamic gossip, using custom data types in Haskell. A similar implementation is described in [20]. These methods do not use NetKAT at all and we only use them as a reference to check other methods for correctness.
2. Explicit-NetKAT: Also in Haskell, we implemented the packet-processing model for NetKAT, Definitions 10 and 11. We thus take a gossip graph G and generate the corresponding packet pk_G. We then apply the LNS policy on this packet. The result is a set of packets from which we can obtain $LNS(G)$ by reading the call_k fields.
3. Equivalence-NetKAT: If we are not interested in the call sequences, but only in whether LNS is successful on a given graph, then we can use Theorems 4 and 5. Given a gossip graph, we generate the NetKAT equivalence that holds iff LNS is weakly or strongly successful on it. We then pass this statement

to the implementation of [10], a coalgebraic decision procedure for NetKAT equivalence. The implementation we use is part of the general network programming framework *frenetic* available at https://github.com/frenetic-lang/frenetic.[1]

All three methods are fully automated, i.e. the user only needs to input an initial gossip graph. We also provide automated tests that randomly generate gossip graphs to verify that the methods agree.

Unfortunately, the third method is currently too slow for interesting examples and the implementation is mainly a proof of work. We hope to improve it in the near future. In particular, we plan to switch to a symbolic decision method as discussed in [18] and currently being developed by the authors of [19].

7 Related Work

The idea of NetKAT with additional state is not new: another version of it is discussed in [17]. However, the system developed there is no longer a KAT and hence cannot base a soundness and completeness proof on the corresponding proofs for KAT. Such a proof is also not given via a different route. Our system is much less expressive but still sound and complete. Also related to our work is the temporal version of NetKAT presented in [6]. The authors add temporal operators to inspect histories of packets. It also seems possible to embed the gossip problem into temporal NetKAT, but we expect this to be less intuitive.

Static and dynamic gossip has mainly been studied in the logic community, with a focus on how the primitive and higher-order knowledge of gossiping agents develops [3,12]. Some of these works also use formal languages and define logics for gossip. For example, in [4] action models of Dynamic Epistemic Logic are used to describe the effects of different gossip calls. This also yields an axiomatization via standard reduction axioms. However, the action models and axioms are of size exponential in the number of agents. This makes an axiomatization of gossip via action models impractical. Another language for the static gossip problem based on Propositional Dynamic Logic is studied in [9]. It is used to distinguish different variants of the gossip problem, but no axiomatization is given.

8 Conclusion

We have seen that switch states can be simulated in NetKAT without losing soundness and completeness. This reinterpretation of NetKAT provides a natural framework to describe dynamic gossip. We translated the LNS protocol to a NetKAT policy, and the definitions of strong/weak success and sun graphs. With these translations we can answer questions about gossip graphs by deciding NetKAT equivalences. By completeness we know that any question about

[1] To be precise, we use the `verification_and_felix` branch currently at commit `be47c929ed84904f9bdb81bf9765a0432db63069`. We would like to thank Steffen Smolka and Nate Foster for their help to get the decision method running.

gossip that is expressible in NetKAT can be decided this way. As mentioned above, we hope to find algebraic proofs and to improve the performance of our implementation in the future.

We also plan to model social influence and diffusion phenomena in NetKAT. They are often similar to dynamic gossip and have been modelled in dynamic epistemic logics, see for example [5]. A crucial difference however, is that social influence settings are, in contrast to dynamic gossip, not monotone: agents do not forget secrets or phone numbers, but they can change their behaviour and influence back and forth. Describing the fixpoints in these settings can thus be a challenge. We think that NetKAT is a suitable language to formalise such non-monotone phenomena, given that its fixpoints are allowed to exist of sets of packets describing multiple outcomes.

Acknowledgements. We received helpful feedback from Jan van Eijck, Tobias Kappé, Jurriaan Rot, Jan Rutten and the anonymous RAMiCS reviewers. The first author was affiliated with the ILLC at the University of Amsterdam during most of this work. The research of the second author is funded by the Dutch NWO project 612.001.210.

Appendix

Proof of Lemma 3

By induction on the call sequence σ. The base case follows instantly as $pol_\epsilon = 1$ outputs the same packet and thus the same gossip graph.

For the induction step we use that calling agents exchange everything they know and that this is encoded in the modifications done by pol_{ab}. As an example, consider the $\subseteq$ direction for N. By the induction hypothesis for σ we have

$$N^\sigma \subseteq \{(a,b) \mid \forall \mathsf{pk} \in [\![pol_\sigma]\!](\mathsf{pk}_G) : \mathsf{pk}.Nab = 1\}$$

and want to show for $\sigma; xy$ that

$$N^{\sigma;xy} \subseteq \{(a,b) \mid \forall \mathsf{pk} \in [\![pol_{\sigma;xy}]\!](\mathsf{pk}_G) : \mathsf{pk}.Nab = 1\}$$

Suppose $(a,b) \in N^{\sigma;xy}$. Either $(a,b) \in N^\sigma$, in which case we are done by the induction hypothesis because $\mathsf{pk}.Nab = 1$ is preserved by pol_{xy}, or $(a,b) \notin N^\sigma$. For the latter case, w.l.o.g. we can assume that $a = x$. Thus we know that $(y,b) \in N^\sigma$. From our induction hypothesis we can conclude that $\forall \mathsf{pk} \in [\![pol_\sigma]\!](\mathsf{pk}_G)$: $\mathsf{pk}.Nyb = 1$. Then in pol_{xy} the field Nxb gets set to 1. Hence we know that $\forall \mathsf{pk} \in [\![pol_{\sigma;xy}]\!](\mathsf{pk}_G) : \mathsf{pk}.Nxb = 1$. □

Proof of Theorem 4

Using soundness and completeness of NetKAT the given syntactic equivalence holds iff we semantically have $[\![pol_G \cdot pol_{\text{LNS}} \cdot pol_{\text{success}}]\!] \neq \varnothing$.

Suppose LNS is weakly successful on G. Let us consider input packet pk_G from Definition 10. We have $[\![pol_G]\!](\mathsf{pk}_G) = \{pk_G\}$ by definition. There is at least one call sequence $\sigma \in LNS(G)$ that is successful. By Theorem 3 we know that σ corresponds to an execution of pol_{LNS} on input pk_G and there is a packet pk such that $\mathsf{pk} \in [\![pol_{\text{LNS}} \cdot pol_{\text{stop}}]\!](\mathsf{pk}_G)$ and its fields call_m encode σ. Because σ is successful we also have $[\![pol_{\text{success}}]\!](\mathsf{pk}) = \{\mathsf{pk}\}$. Hence we can conclude $\mathsf{pk} \in [\![pol_G \cdot pol_{\text{LNS}} \cdot pol_{\text{success}}]\!](\mathsf{pk}_G)$ and thus that $[\![pol_G \cdot pol_{\text{LNS}} \cdot pol_{\text{success}}]\!] \neq \varnothing$.

The other direction is similar, so we omit the proof here. □

Proof of Theorem 5

By soundness and completeness of NetKAT the equivalence holds iff we have $[\![pol_G \cdot pol_{\text{LNS}} \cdot pol_{\text{stop}}]\!] = [\![pol_G \cdot pol_{\text{LNS}} \cdot pol_{\text{success}}]\!]$.

Suppose LNS is strongly successful on G. We immediately have $\supseteq$ because any packet passing the success check also passes the test that LNS has finished.

To show $\subseteq$, take any pk and pk' such that $\mathsf{pk}' \in [\![pol_G \cdot pol_{\text{LNS}} \cdot pol_{\text{stop}}]\!](\mathsf{pk})$. We then know that $\mathsf{pk} = \mathsf{pk}_G$ because pk passed the test pol_G. Hence $\mathsf{pk}' \in [\![pol_G \cdot pol_{\text{LNS}} \cdot pol_{\text{stop}}]\!](\mathsf{pk}_G)$. As $[\![pol_G]\!](\mathsf{pk}_G) = \{\mathsf{pk}_G\}$, we get that $\mathsf{pk}' \in [\![pol_{\text{LNS}} \cdot pol_{\text{stop}}]\!](\mathsf{pk}_G)$. From Theorem 3 we know that every output pk' of $[\![pol_{\text{LNS}} \cdot pol_{\text{stop}}]\!](\mathsf{pk}_G)$ corresponds to a call sequence $\sigma \in LNS(G)$. From the assumption that LNS is strongly successful on G we get that all of these σ are successful call sequences. We thus know that $[\![pol_{\text{success}}]\!](\mathsf{pk}') = \{\mathsf{pk}'\}$ and thereby $\mathsf{pk}' \in [\![pol_G \cdot pol_{\text{LNS}} \cdot pol_{\text{success}}]\!](\mathsf{pk})$.

The other direction is similar. □

Proof of Theorem 6

By soundness and completeness of NetKAT, the statement is equivalent to:

$$\left[\!\!\left[pol_G \cdot \prod_{i \in A} (\mathsf{sw} \leftarrow i \cdot \mathsf{pt} \leftarrow home_i \cdot (pol_{Ni} + pol_{\text{self}_i})) \right]\!\!\right] \neq \varnothing$$

$\Rightarrow$ Take the packet pk_G. We know that $[\![pol_G]\!](\mathsf{pk}_G) = \{\mathsf{pk}_G\}$. As G is a sun graph, we know that for each agent i either $[\![pol_{Ni}]\!](\mathsf{pk}_G) = \{\mathsf{pk}_G\}$ or $[\![pol_{\text{self}_i}]\!](\mathsf{pk}_G) = \{\mathsf{pk}_G\}$. Hence we have

$$\left[\!\!\left[\prod_{i \in A} (\mathsf{sw} \leftarrow i \cdot \mathsf{pt} \leftarrow home_i \cdot (pol_{Ni} + pol_{\text{self}_i})) \right]\!\!\right] (\mathsf{pk}_G) = \{\mathsf{pk}\}$$

for some packet pk. Thus we can conclude:

$$\mathsf{pk} \in \left[\!\!\left[pol_G \cdot \prod_{i \in A} (\mathsf{sw} \leftarrow i \cdot \mathsf{pt} \leftarrow home_i \cdot (pol_{Ni} + pol_{\text{self}_i})) \right]\!\!\right] (\mathsf{pk}_G)$$

$\Leftarrow$ is similar. □

Proof of Corollary 1

Fix some G. By Theorem 5 LNS is strongly successful on G if and only if

$$pol_G \cdot pol_{\text{LNS}} \cdot pol_{\text{stop}} \equiv pol_G \cdot pol_{\text{LNS}} \cdot pol_{\text{success}}$$

and from Theorem 6 that G is a sun graph if and only if

$$pol_G \cdot \prod_{i \in A} (\mathsf{sw} \leftarrow i \cdot \mathsf{pt} \leftarrow home_i \cdot (pol_{Ni} + pol_{\text{self}_i})) \not\equiv 0$$

Now suppose we have $pol_G \cdot pol_{\text{LNS}} \cdot pol_{\text{stop}} \equiv pol_G \cdot pol_{\text{LNS}} \cdot pol_{\text{success}}$. From Theorem 5 we then know that LNS is strongly successful on G. By Theorem 2 it follows that G is a sun graph. From Theorem 6 we know this is equivalent to

$$pol_G \cdot \prod_{i \in A} (\mathsf{sw} \leftarrow i \cdot \mathsf{pt} \leftarrow home_i \cdot (pol_{Ni} + pol_{\text{self}_i})) \not\equiv 0$$

The other direction is similar. □

References

1. Anderson, C.J., et al.: NetKAT: semantic foundations for networks. In: Proceedings of the 41st ACM SIGPLANSIGACT Symposium on Principles of Programming Languages, POPL 2014, pp. 113–126 (2014). ISBN: 978-1-4503-2544-8. Extended version https://hdl.handle.net/1813/34445. https://doi.org/10.1145/2535838.2535862
2. Apt, K.R., Grossi, D., van der Hoek, W.: Epistemic protocols for distributed gossiping. In: Proceedings of TARK 2015 (2015). Edited by Ramanujam. https://doi.org/10.4204/EPTCS.215.5
3. Apt, K.R., Wojtczak, D. Common knowledge in a logic of gossips. In: Proceedings of TARK 2017 (2017). Edited by Jérôme Lang. https://doi.org/10.4204/EPTCS.251.2
4. Attamah, M., van Ditmarsch, H., Grossi, D., van der Hoek, W.: Knowledge and gossip. In: Proceedings of the Twenty-First European Conference on Artificial Intelligence. Frontiers in Artificial Intelligence and Applications, pp. 21–26 (2014). ISBN: 978-1-61499-418-3. https://doi.org/10.3233/978-1-61499-419-0-21
5. Baltag, A., Christoff, Z., Rendsvig, R.K., Smets, S.: Dynamic epistemic logics of diffusion and prediction in social networks. In: Proceedings of the Twelfth Conference on Logic and the Foundations of Game and Decision Theory (2016). https://is.gd/DiffDEL
6. Beckett, R., Greenberg, M., Walker, D.: Temporal NetKAT. In: Proceedings of the 37th ACM SIGPLAN Conference on Programming Language Design and Implementation, PLDI 2016, pp. 386–401 (2016). ISBN: 978-1-4503-4261-2. https://doi.org/10.1145/2908080.2908108
7. van Ditmarsch, H., van Eijck, J., Pardo, P., Ramezanian, R., Schwarzentruber, F.: Dynamic gossip. Bulletin of the Iranian Mathematical Society, Sept 2018. ISSN: 1735-8515, https://doi.org/10.1007/s41980-018-0160-4

8. van Ditmarsch, H., van Eijck, J., Pardo, P., Ramezanian, R., Schwarzentruber, F.: Epistemic protocols for dynamic gossip. J. Appl. Log. **20**, 1–31 (2017). https://doi.org/10.1016/j.jal.2016.12.001
9. van Ditmarsch, H., Grossi, D., Herzig, A., van der Hoek, W., Kuijer, L.B.: Parameters for epistemic gossip problems. In: Proceedings of the Twelfth Conference on Logic and the Foundations of Game and Decision Theory (2016). https://is.gd/GosPar
10. Foster, N., Kozen, D., Milano, M., Silva, A., Thompson, L.: A coalgebraic decision procedure for NetKAT. In: Proceedings of the 42nd Annual ACM SIGPLAN-SIGACT Symposium on Principles of Programming Languages, POPL 2015, pp. 343–355 (2015). ISBN: 978-1-4503-3300-9. https://doi.org/10.1145/2676726.2677011
11. Hedetniemi, S.M., Hedetniemi, S.T., Liestman, A.L.: A survey of gossiping and broadcasting in communication networks. Networks **18**(4), 319–349 (1988). https://doi.org/10.1002/net.3230180406
12. Herzig, A., Maffre, F.: How to share knowledge by gossiping. AI Commun. **30**(1), 1–17 (2017). https://doi.org/10.3233/AIC-170723
13. Hoare, T., Möller, B., Struth, G., Wehrman, I.: Concurrent kleene algebra and its foundations. J. Log. Algebr. Program. **80**(6), 266–296 (2011). Relations and Kleene Algebras in Computer Science. ISSN: 1567–8326. https://doi.org/10.1016/j.jlap.2011.04.005
14. Kappé, T., Brunet, P., Silva, A., Zanasi, F.: Concurrent kleene algebra: free model and completeness. In: Ahmed, A. (ed.) Programming Languages and Systems (ESOP 2018), pp. 856–882 (2018). ISBN: 978-3-319-89884-1. https://doi.org/10.1007/978-3-319-89884-1_30
15. Kozen, D.: Kleene algebra with tests. ACM Trans. Program. Lang. Syst. **19**(3), 427–443 (1997). https://doi.org/10.1145/256167.256195
16. Kozen, D., Smith, F.: Kleene algebra with tests: completeness and decidability. In: Computer Science Logic, pp. 244–259 (1997). Edited by Dirk van Dalen and Marc Bezem. ISBN: 978-3-540-69201-0. https://doi.org/10.1007/3-540-63172-0_43
17. McClurg, J., Hojjat, H., Foster, N., Černý, P.: Event-driven network programming. ACM SIGPLAN Not. **51**(6), 369–385 (2016). https://doi.org/10.1145/2908080.2908097. PLDI 2016. ISSN: 0362–1340
18. Pous, D.: Symbolic algorithms for language equivalence and kleene algebra with tests. In: Proceedings of the 42nd Annual ACM SIGPLANSIGACT Symposium on Principles of Programming Languages, POPL 2015, pp. 357–368. ISBN: 978-1-4503-3300-9. https://doi.org/10.1145/2676726.2677007
19. Smolka, S. Eliopoulos, S., Foster, N., Guha, A.: A fast compiler for NetKAT. In: Proceedings of the 20th ACM SIGPLAN International Conference on Functional Programming, ICFP 2015, pp. 328–341 (2015). ISBN: 978-1-4503-3669-7. https://doi.org/10.1145/2784731.2784761
20. Wagemaker, J.: Gossip in NetKAT. Master's thesis, ILLC, University of Amsterdam (2017). https://eprints.illc.uva.nl/1552/

Coalgebraic Tools for Randomness-Conserving Protocols

Dexter Kozen(✉) and Matvey Soloviev

Cornell University, Ithaca, USA
{kozen,msoloviev}@cs.cornell.edu

Abstract. We propose a coalgebraic model for constructing and reasoning about state-based protocols that implement efficient reductions among random processes. We provide basic tools that allow efficient protocols to be constructed in a compositional way and analyzed in terms of the tradeoff between latency and loss of entropy. We show how to use these tools to construct various entropy-conserving reductions between processes.

1 Introduction

In low-level performance-critical computations—for instance, data-forwarding devices in packet-switched networks—it is often desirable to minimize local state in order to achieve high throughput. But if the situation requires access to a source of randomness, say to implement randomized routing or load-balancing protocols, it may be necessary to convert the output of the source to a form usable by the protocol. As randomness is a scarce resource to be conserved like any other, these conversions should be performed as efficiently as possible and with a minimum of machinery.

In this paper we propose a coalgebraic model for constructing and reasoning about state-based protocols that implement efficient reductions among random processes. Efficiency is measured by the ratio of entropy produced to entropy consumed. The efficiency cannot exceed the information-theoretic bound of unity, but it should be as close to unity as can be achieved with simple state-based devices. We provide basic tools that allow efficient protocols to be constructed in a compositional way and analyzed in terms of the tradeoff between latency and loss of entropy.

We use these tools to construct the following reductions between processes, where k is the latency parameter:

- d-uniform to c-uniform with loss $\Theta(k^{-1})$
- d-uniform to arbitrary rational with loss $\Theta(k^{-1})$
- d-uniform to arbitrary with loss $\Theta(k^{-1})$
- arbitrary to c-uniform with loss $\Theta(\log k/k)$
- $(1/r, (r-1)/r)$ to c-uniform with loss $\Theta(k^{-1})$.

Omitted proofs can be found in the full version of the paper [11].

J. Desharnais et al. (Eds.): RAMiCS 2018, LNCS 11194, pp. 298–313, 2018.
https://doi.org/10.1007/978-3-030-02149-8_18

1.1 Related Work

Since von Neumann's classic result showing how to simulate a fair coin with a coin of unknown bias, many authors have studied variants of this problem. Our work is heavily inspired by the work of Elias [8], who studies entropy-optimal generation of uniform distributions from known sources. The definition of conservation of entropy and a concept related to latency are defined there. Mossel, Peres, and Hillar [17] characterize the set of functions $f : (0,1) \to (0,1)$ for which it is possible to simulate an $f(p)$-biased coin with a p-biased coin when p is unknown. Peres [16] shows how to iterate von Neumann's procedure for producing a fair coin from a biased coin to approximate the entropy bound. Blum [2] shows how to extract a fair coin from a Markov chain. Pae and Loui [13,14] present several simulations for optimal conversions between discrete distributions, known and unknown. The main innovation in this paper is the coalgebraic model that allows compositional reasoning about such reductions.

There is also a large body of related work on extracting randomness from weak random sources (e.g. [7,12,18,19]). These models typically work with imperfect knowledge of the input source and provide only approximate guarantees on the quality of the output. Here we assume that the statistical properties of the input and output are known completely, and simulations must be exact.

2 Definitions

Informally, a **reduction** from a stochastic process X to another stochastic process Y is a deterministic protocol that consumes a finite or infinite stream of letters from an alphabet Σ and produces a finite or infinite stream of letters from another alphabet Γ. If the letters of the input stream are distributed as X, then the letters of the output stream should be distributed as Y. Of particular interest are reductions between **Bernoulli processes**, in which the letters of the input and output streams are independent and identically distributed according to distributions μ on Σ and ν on Γ, respectively. In this case, we say that procedure is a reduction from μ to ν.

To say that the protocol is **deterministic** means that the only source of randomness is the input stream. It makes sense to talk about the expected number of input letters read before halting or the probability that the first letter emitted is a, but any such statistical measurements are taken with respect to the distribution of the input stream.

There are several ways to formalize the notion of a reduction. One approach, following [16], is to model a reduction as a map $f : \Sigma^* \to \Gamma^*$ that is monotone with respect to the prefix relation on strings; that is, if $x, y \in \Sigma^*$ and x is a prefix of y, then $f(x)$ is a prefix of $f(y)$. Monotonicity implies that f can be extended uniquely by continuity to domain $\Sigma^* \cup \Sigma^\omega$ and range $\Gamma^* \cup \Gamma^\omega$. The map f would then constitute a reduction from the stochastic process $X = X_0X_1X_2\cdots$ to $f(X_0X_1X_2\cdots)$. To be a reduction from μ to ν, it must be that if the X_i are independent and identically distributed as μ, and if Y_i is the value of the ith letter of $f(X_0X_1X_2\cdots)$, then the Y_i are independent and identically distributed as ν.

In this paper we propose an alternative state-based approach in which protocols are modeled as coalgebras $\delta : S \times \Sigma \to S \times \Gamma^*$, where S is a (possibly infinite) set of **states**. This approach allows a more streamlined treatment of common programming constructions such as composition, which is perhaps more appealing from a programming perspective.

2.1 Protocols and Reductions

Let Σ, Γ be finite alphabets. Let Σ^* denote the set of finite words and Σ^ω the set of ω-words (streams) over Σ. We use $x, y, \ldots$ for elements of Σ^* and $\alpha, \beta, \ldots$ for elements of Σ^ω. The symbols $\preceq$ and $\prec$ denote the prefix and proper prefix relations, respectively.

If μ is a probability measure on Σ, we endow Σ^ω with the product measure in which each symbol is distributed as μ. The notation $\Pr(A)$ for an event A refers to this measure. The measurable sets of Σ^ω are the Borel sets of the Cantor space topology whose basic open sets are the **intervals** $\{\alpha \in \Sigma^\omega \mid x \prec \alpha\}$ for $x \in \Sigma^*$, and $\mu(\{\alpha \in \Sigma^\omega \mid x \prec \alpha\}) = \mu(x)$, where $\mu(a_1a_2\cdots a_n) = \mu(a_1)\mu(a_2)\cdots\mu(a_n)$.

A **protocol** is a coalgebra (S, δ) where $\delta : S \times \Sigma \to S \times \Gamma^*$. We can immediately extend δ to domain $S \times \Sigma^*$ by coinduction:

$$\delta(s, \varepsilon) = (s, \varepsilon)$$
$$\delta(s, ax) = \text{let } (t, y) = \delta(s, a) \text{ in let } (u, z) = \delta(t, x) \text{ in } (u, yz).$$

Since the two functions agree on $S \times \Sigma$, we use the same name. It follows that

$$\delta(s, xy) = \text{let } (t, z) = \delta(s, x) \text{ in let } (u, w) = \delta(t, y) \text{ in } (u, zw).$$

By a slight abuse, we define the **length** of the output as the length of its second component as a string in Γ^* and write $|\delta(s, x)|$ for $|z|$, where $\delta(s, x) = (t, z)$.

A protocol δ also induces a partial map $\delta^\omega : S \times \Sigma^\omega \rightharpoonup \Gamma^\omega$ by coinduction:

$$\delta^\omega(s, a\alpha) = \text{let } (t, z) = \delta(s, a) \text{ in } z \cdot \delta^\omega(t, \alpha).$$

It follows that

$$\delta^\omega(s, x\alpha) = \text{let } (t, z) = \delta(s, x) \text{ in } z \cdot \delta^\omega(t, \alpha).$$

Given $\alpha \in \Sigma^\omega$, this defines a unique infinite string in $\delta^\omega(s, \alpha) \in \Gamma^\omega$ except in the degenerate case in which only finitely many output letters are ever produced. A protocol is said to be **productive** (with respect to a given probability measure on input streams) if, starting in any state, an output symbol is produced within finite expected time. It follows that infinitely many output letters are produced with probability 1.

Now let ν be a probability measure on Γ. Endow Γ^ω with the product measure in which each symbol is distributed as ν. As with μ, define $\nu(a_1a_2\cdots a_n) = \nu(a_1)\nu(a_2)\cdots\nu(a_n)$ for $a_i \in \Gamma$. We say that a protocol (S, δ, s) with start state $s \in S$ is a **reduction from** μ *to* ν if for all $y \in \Gamma^*$,

$$\Pr(y \preceq \delta^\omega(s, \alpha)) = \nu(y), \tag{1}$$

where the probability Pr is with respect to the product measure μ on Σ^ω. This implies that the symbols of $\delta^\omega(s, \alpha)$ are independent and identically distributed as ν.

2.2 Restart Protocols

A **prefix code** is a subset $A \subseteq \Sigma^*$ such that every element of Σ^ω has at most one prefix in A. Thus the elements of a prefix code are $\preceq$-incomparable. A prefix code is **exhaustive** (with respect to a given probability measure on input streams) if $\Pr(\{\alpha \in \Sigma^\omega \text{has a prefix in} A\}) = 1$. By König's lemma, if every $\alpha \in \Sigma^\omega$ has a prefix in A, then A is finite.

A **restart protocol** is protocol (S, δ, s) of a special form determined by a function $f : A \rightharpoonup \Gamma^*$, where A is an exhaustive prefix code. Here s is a designated start state. Intuitively, starting in s, we read symbols of Σ from the input stream until encountering a string $x \in A$, output $f(x)$, then return to s and repeat. Note that we are not assuming A to be finite.

Formally, we can take the state space to be

$$S = \{u \in \Sigma^* \mid x \not\preceq u \text{ for any } x \in A\}$$

and define $\delta : S \times \Sigma \to S \times \Gamma^*$ by

$$\delta(u, a) = \begin{cases} (ua, \varepsilon), & ua \notin A, \\ (\varepsilon, z), & ua \in A \text{ and } f(ua) = z \end{cases}$$

with start state ε. Then for all $x \in A$, $\delta(\varepsilon, x) = (\varepsilon, f(x))$.

As with the more general protocols, we can extend to a function on streams, but here the definition takes a simpler form: for $x \in A$,

$$\delta^\omega(\varepsilon, x\alpha) = f(x) \cdot \delta^\omega(\varepsilon, \alpha), \quad x \in A, \ \alpha \in \Sigma^\omega.$$

A restart protocol is **positive recurrent** (with respect to a given probability measure on input streams) if, starting in the start state s, the probability of eventually returning to s is 1, and moreover the expected time before the next visit to s is finite. All finite-state restart protocols are positive recurrent, but infinite-state ones need not be.

2.3 Convergence

We will have the occasion to discuss the convergence of random variables. There are several notions of convergence in the literature, but for our purposes the most useful is **convergence in probability**. Let X and X_n, $n \geq 0$ be bounded nonnegative random variables. We say that the sequence X_n **converges to** X **in probability** and write $X_n \xrightarrow{\Pr} X$ if for all fixed $\delta > 0$,

$$\Pr(|X_n - X| > \delta) = o(1).$$

Let $\mathbb{E}(X)$ denote the expected value of X and $\mathbb{V}(X)$ its variance.

Lemma 1

(i) If $X_n \xrightarrow{\Pr} X$ and $X_n \xrightarrow{\Pr} Y$, then $X = Y$ with probability 1.
(ii) If $X_n \xrightarrow{\Pr} X$ and $Y_n \xrightarrow{\Pr} Y$, then $X_n + Y_n \xrightarrow{\Pr} X + Y$ and $X_n Y_n \xrightarrow{\Pr} XY$.
(iii) If $X_n \xrightarrow{\Pr} X$ and X is bounded away from 0, then $1/X_n \xrightarrow{\Pr} 1/X$.
(iv) If $\mathbb{V}(X_n) = o(1)$ and $\mathbb{E}(X_n) = e$ for all n, then $X_n \xrightarrow{\Pr} e$.

Proof Clause (iv) follows from the Chebyshev bound. Please see [11] for details. □

See [4,9] for a more thorough introduction.

2.4 Efficiency

The **efficiency** of a protocol is the long-term ratio of entropy production to entropy consumption. Formally, for a fixed protocol $\delta : S \times \Sigma \to S \times \Gamma^*$, $s \in S$, and $\alpha \in \Sigma^\omega$, define the random variable

$$E_n(\alpha) = \frac{|\delta(s, \alpha_n)|}{n} \cdot \frac{H(\nu)}{H(\mu)}, \tag{2}$$

where H is the Shannon entropy $H(p_1, \ldots, p_n) = -\sum_{i=1}^{n} p_i \log p_i$ (logarithms are base 2 if not otherwise annotated), μ and ν are the input and output distributions, respectively, and α_n is the prefix of α of length n. Intuitively, the Shannon entropy measures the number of fair coin flips the distribution is worth, and the random variable E_n measures the ratio of entropy production to consumption after n steps of δ starting in state s. Here $|\delta(s, \alpha_n)|\ H(\nu)$ (respectively, $nH(\mu)$) is the contribution along α to the production (respectively, consumption) of entropy in the first n steps. We write $E_n^{\delta,s}$ when we need to distinguish the E_n associated with different protocols and start states.

In most cases of interest, E_n converges in probability to a unique constant value independent of start state and history. When this occurs, we call this constant value the **efficiency** of the protocol δ and denote it by Eff_δ. Notationally, $E_n \xrightarrow{\Pr} \mathsf{Eff}_\delta$. One must be careful when analyzing infinite-state protocols: The efficiency is well-defined for finite-state protocols, but may not exist in general. For restart protocols, it is enough to measure the ratio for one iteration of the protocol.

In Sect. 3.2 we will give sufficient conditions for the existence of Eff_δ that is satisfied by all protocols considered in Sect. 4.

2.5 Latency

The **latency** of a protocol from a given state s is the expected consumption before producing at least one output symbol, starting from state s. This is proportional to the expected number of input letters consumed before emitting at least one symbol. The latency of a protocol is finite if and only if the protocol is productive. All positive recurrent restart protocols that emit at least one symbol are productive. We will often observe a tradeoff between latency and efficiency.

Suppose we iterate a positive recurrent restart protocol only until at least one output symbol is produced. That is, we start in the start state s and choose one string x in the prefix code randomly according to μ. If at least one output symbol is produced, we stop. Otherwise, we repeat the process. The sequence of iterations to produce at least one output symbol is called an **epoch**. The latency is the expected consumption during an epoch. If p is the probability of producing an output symbol in one iteration, then the sequence of iterations in a epoch forms a Bernoulli process with success probability p. The latency is thus $1/p$, the expected stopping time of the Bernoulli process, times the expected consumption in one iteration, which is finite due to the assumption that the protocol is positive recurrent.

3 Basic Results

Let $\delta : S \times \Sigma \to S \times \Gamma^*$ be a protocol. We can associate with each $y \in \Gamma^*$ and state $s \in S$ a prefix code in Σ^*, namely

$$\mathsf{pc}_\delta(s, y) = \{\text{minimal-length strings } x \in \Sigma^* \text{ such that } y \preceq \delta(s, x)\}.$$

The string y is generated as a prefix of the output if and only if exactly one $x \in \mathsf{pc}_\delta(s, y)$ is consumed as a prefix of the input. These events must occur with the same probability, so

$$\nu(y) = \Pr(y \prec \delta^\omega(s, \alpha)) = \mu(\mathsf{pc}_\delta(s, y)). \tag{3}$$

Note that $\mathsf{pc}_\delta(s, y)$ need not be finite.

Lemma 2. *If $A \subseteq \Gamma^*$ is a prefix code, then so is $\bigcup_{y \in A} \mathsf{pc}_\delta(s, y) \subseteq \Sigma^*$, and*

$$\nu(A) = \mu(\bigcup_{y \in A} \mathsf{pc}_\delta(s, y)).$$

If $A \subseteq \Gamma^*$ is exhaustive, then so is $\bigcup_{y \in A} \mathsf{pc}_\delta(s, y) \subseteq \Sigma^*$.

Proof. Please see [11]. □

Lemma 3

(i) The partial function $\delta^\omega(s,-) : \Sigma^\omega \rightharpoonup \Gamma^\omega$ is continuous, thus Borel measurable.
(ii) $\delta^\omega(s,\alpha)$ is almost surely infinite; that is, $\mu(\mathrm{dom}\,\delta^\omega(s,-)) = 1$.
(iii) The measure ν on Γ^ω is the push-forward measure $\nu = \mu \circ \delta^\omega(s,-)^{-1}$.

Proof Please see [11]. □

Lemma 4. *If δ is a reduction from μ to ν, then the random variables E_n defined in (2) are continuous and uniformly bounded by an absolute constant $R > 0$ depending only on μ and ν.*

Proof. Please see [11]. □

3.1 Composition

Protocols can be composed sequentially as follows. If

$$\delta_1 : S \times \Sigma \to S \times \Gamma^* \qquad\qquad \delta_2 : T \times \Gamma \to T \times \Delta^*,$$

then

$$(\delta_1 \,;\, \delta_2) : S \times T \times \Sigma \to S \times T \times \Delta^*$$
$$(\delta_1 \,;\, \delta_2)((s,t),a) = \text{let } (u,y) = \delta_1(s,a) \text{ in let } (v,z) = \delta_2(t,y) \text{ in } ((u,v),z).$$

Intuitively, we run δ_1 for one step and then run δ_2 on the output of δ_1. The following theorem shows that the map on infinite strings induced by the sequential composition of protocols is almost everywhere equal to the functional composition of the induced maps of the component protocols.

Theorem 1. *The partial maps $(\delta_1 \,;\, \delta_2)^\omega((s,t),-)$ and ${\delta^\omega}_2(t,{\delta^\omega}_1(s,-))$ of type $\Sigma^\omega \rightharpoonup \Delta^\omega$ are defined and agree on all but a μ-nullset.*

Proof. We restrict inputs to the subset of Σ^ω on which ${\delta^\omega}_1$ is defined and produces a string in Γ^ω on which ${\delta^\omega}_2$ is defined. These sets are of measure 1. To show that $(\delta_1 \,;\, \delta_2)^\omega((s,t),\alpha) = {\delta^\omega}_2(t,{\delta^\omega}_1(s,\alpha))$, we show that the binary relation

$$\beta R \gamma \;\Leftrightarrow\; \exists\alpha \in \Sigma^\omega\ \exists s \in S\ \exists t \in T\ \ \beta = (\delta_1 \,;\, \delta_2)^\omega((s,t),\alpha) \wedge \gamma = {\delta^\omega}_2(t,{\delta^\omega}_1(s,\alpha))$$

on Δ^ω is a bisimulation. Please see [11] for details. □

Corollary 1. *If $\delta_1(s,-)$ is a reduction from μ to ν and $\delta_2(t,-)$ is a reduction from ν to o, then $(\delta_1 \,;\, \delta_2)((s,t),-)$ is a reduction from μ to o.*

Proof. Please see [11]. □

Theorem 2. *If $\delta_1(s,-)$ is a reduction from μ to ν and $\delta_2(t,-)$ is a reduction from ν to o, and if Eff_{δ_1} and Eff_{δ_2} exist, then $\mathsf{Eff}_{\delta_1;\delta_2}$ exists and $\mathsf{Eff}_{\delta_1;\delta_2} = \mathsf{Eff}_{\delta_1} \cdot \mathsf{Eff}_{\delta_2}$.*

Proof. Please see [11]. □

In the worst case, the latency of compositions of protocols is also the product of their latencies: if the first protocol only outputs one character at a time, then the second protocol may have to wait the full latency of the first protocol for each of the characters it needs to read in order to emit a single one.

3.2 Serial Protocols

Consider a sequence $(S_0, \delta_0, s_0), (S_1, \delta_1, s_1), \ldots$ of positive recurrent restart protocols defined in terms of maps $f_k : A_k \to \Gamma^*$, where the A_k are exhaustive prefix codes, as described in Sect. 2.2. These protocols can be combined into a single **serial protocol** δ that executes one iteration of each δ_k, then goes on to the next. Formally, the states of δ are the disjoint union of the S_k, and δ is defined so that $\delta(s_k, x) = (s_{k+1}, f_k(x))$ for $x \in A_k$, and within S_k behaves like δ_k.

Let C_k and P_k be the number of input symbols consumed and produced, respectively, in one iteration of the component protocol δ_k starting from s_k. Let $e(n)$ be the index of the component protocol $\delta_{e(n)}$ in which the n-th step of the combined protocol occurs. These are random variables whose values depend on the input sequence $\alpha \in \Sigma^\omega$. Let $c_k = \mathbb{E}(C_k)$ and $p_k = \mathbb{E}(P_k)$.

To derive the efficiency of serial protocols, we need a form of the law of large numbers (see [4,9]). Unfortunately, the law of large numbers as usually formulated does not apply verbatim, as the random variables in question are bounded but not independent, or (under a different formulation) independent but not bounded. Our main result, Theorem 3 below, can be regarded as a specialized version of this result adapted to our needs.

Our version requires that the variances of certain random variables vanish in the limit. This holds under a mild condition (4) on the growth rate of m_n, the maximum consumption in the nth component protocol, and is true for all serial protocols considered in this paper. The condition (4) is satisfied by all finite serial protocols in which m_n is bounded, or $m_n = O(n)$ and c_n is unbounded.

Lemma 5. *Let $\mathbb{V}(X)$ denote the variance of X. Let $m_n = \max_{x \in A_k} |x| \cdot H(\mu)$. If*

$$m_n = o(\sum_{i=0}^{n-1} c_i), \tag{4}$$

then

$$\mathbb{V}(\frac{\sum_{i=0}^{n} C_i}{\sum_{i=0}^{n} c_i}) = o(1) \qquad \mathbb{V}(\frac{C_n}{\sum_{i=0}^{n-1} c_i}) = o(1). \tag{5}$$

If in addition $p_n = \Theta(c_n)$, then

$$\mathbb{V}(\frac{\sum_{i=0}^{n} P_i}{\sum_{i=0}^{n} p_i}) = o(1) \qquad \mathbb{V}(\frac{P_n}{\sum_{i=0}^{n-1} p_i}) = o(1). \tag{6}$$

Proof. Please see [11]. □

The following is our main theorem.

Theorem 3. *Let δ be a serial protocol with finite-state components $\delta_0, \delta_1, \ldots$ satisfying (4). If the limit $\ell = \lim_n \frac{\sum_{i=0}^{n} p_i}{\sum_{i=0}^{n} c_i}$ exists, then the efficiency of the serial protocol exists and is equal to ℓ.*

Proof. Please see [11]. □

4 Reductions

In this section we present a series of reductions between distributions of certain forms. Each example defines a sequence of positive recurrent restart protocols (Sect. 2.2) indexed by a latency parameter k with efficiency tending to 1. By Theorem 3, these can be combined in a serial protocol (Sect. 3.2) with asymptotically optimal efficiency, albeit at the cost of unbounded latency.

4.1 Uniform ⇒ Uniform

Let $c, d \geq 2$. In this section we construct a family of restart protocols with latency k mapping d-uniform streams to c-uniform streams with efficiency $1 - \Theta(k^{-1})$. The Shannon entropy of the input and output distributions are $\log d$ and $\log c$, respectively.

Let $m = \lfloor k \log_c d \rfloor$. Then $c^m \leq d^k < c^{m+1}$. It follows that

$$\frac{c^m}{d^k} = \Theta(1) \qquad 1 - \Theta(1) < \frac{m \log c}{k \log d} \leq 1. \tag{7}$$

Let the c-ary expansion of d^k be

$$d^k = \sum_{i=0}^{m} a_i c^i, \tag{8}$$

where $0 \leq a_i \leq c - 1$, $a_m \neq 0$.

The protocol P_k is defined as follows. Do k calls on the d-uniform distribution. For each $0 \leq i \leq m$, for $a_i c^i$ of the possible outcomes, emit a c-ary string of length i, every possible such string occurring exactly a_i times. For a_0 outcomes, nothing is emitted, and this is lost entropy, but this occurs with probability $a_0 d^{-k}$. After that, restart the protocol.

By elementary combinatorics,

$$\sum_{i=0}^{m-1} (m - i) a_i c^i \leq \sum_{i=0}^{m-1} (m - i)(c - 1) c^i = \frac{c(c^m - 1)}{c - 1} - m. \tag{9}$$

In each run of P_k, the expected number of c-ary digits produced is

$$\begin{aligned}
\sum_{i=0}^{m} i a_i c^i d^{-k} &= d^{-k}\Big(\sum_{i=0}^{m} m a_i c^i - \sum_{i=0}^{m} (m - i) a_i c^i\Big) \\
&\geq m - d^{-k}\Big(\frac{c(c^m - 1)}{c - 1} - m\Big) && \text{by (8) and (9)} \\
&= m - \Theta(1) && \text{by (7)},
\end{aligned}$$

thus the entropy production is at least $m \log c - \Theta(1)$. The number of d-ary digits consumed is k, thus the entropy consumption is $k \log d$. The efficiency is

$$\frac{m \log c - \Theta(1)}{k \log d} \geq 1 - \Theta(k^{-1}).$$

The output is uniformly distributed, as there are $\sum_{i=\ell}^{m} a_i c^i$ equal-probability outcomes that produce a string of length ℓ or greater, and each output letter a appears as the ℓth output letter in equally many strings of the same length, thus is output with equal probability.

4.2 Uniform $\Rightarrow$ Rational

Let $c, d \geq 2$. In this section, we will present a family of restart protocols D_k mapping d-uniform streams over Σ to streams over a c-symbol alphabet $\Gamma = \{1, \ldots, c\}$ with rational symbol probabilities with a common denominator d, e.g. $p_1 = a_1/d, \ldots, p_c = a_c/d$. Unlike the protocols in the previous section, here we emit a fixed number of symbols in each round while consuming a variable number of input symbols according to a particular prefix code $S \subseteq \Sigma^*$. The protocol D_k has latency at most $kH(p_1, \ldots, p_c)/\log d + 2$ and efficiency $1 - \Theta(k^{-1})$, exhibiting a similar tradeoff to the previous family.

To define D_k, we will construct an exhaustive prefix code S over the source alphabet, which will be partitioned into pairwise disjoint sets $S_y \subseteq \Sigma^*$ associated with each k-symbol output word $y \in \Gamma^k$. All input strings in the set S_y will map to the output string y.

By analogy with $p_1, \ldots, p_c$, let p_y denote the probability of the word $y = s_1 \cdots s_k$ in the output process. Since the symbols of y are chosen independently, p_y is the product of the probabilities of the individual symbols. It is therefore of the form $p_y = a_y d^{-k}$, where $a_y = a_{s_1} \cdots a_{s_k}$ is an integer.

Let $m_y = \lfloor \log_d a_y \rfloor$ and let $a_y = \sum_{j=0}^{m_y} a_{yj} d^j$ be the d-ary expansion of a_y. We will choose a set of $\sum_{y \in \Gamma^k} \sum_{j=0}^{m_y} a_{yj}$ prefix-incomparable codewords and assign them to the S_y so that each S_y contains a_{yj} codewords of length $k - j$ for each $0 \leq j \leq m_y$. This is possible by the Kraft inequality (see [5, Theorem 5.2.1] or [1, Theorem 1.6]); we need only verify that $\sum_{y \in \Gamma^k} \sum_{j=0}^{m_y} a_{yj} d^{-(k-j)} \leq 1$. In fact, equality holds:

$$\sum_{j=0}^{m_y} a_{yj} d^{-(k-j)} = a_y d^{-k} = p_y, \quad \text{so} \sum_{y \in \Gamma^k} \sum_{j=0}^{m_y} a_{yj} d^{-(k-j)} = \sum_{y \in \Gamma^k} p_y = 1. \tag{10}$$

Since the d symbols of the input process are distributed uniformly, the probability that the input stream begins with a given string of length n is d^{-n}. So

$$\Pr(y \preceq \delta^{\omega}{}_k(*, x)) = \Pr(\exists x \in S_y : x \preceq x) = \sum_{x \in S_y} d^{-|x|} = \sum_{j=0}^{m_y} a_{yj} d^{-(k-j)}$$

is p_y as required, and D_k is indeed a reduction. Moreover, by (10), the probability that a prefix is in some S_y is 1, so the code is exhaustive.

To analyze the efficiency of the simulation, we will use the following lemma.

Lemma 6. *Let the d-ary expansion of a be $\sum_{i=0}^{m} a_i d^i$, where $m = \lfloor \log_d a \rfloor$. Then*

$$\left(\log_d a - \frac{2d-1}{d-1}\right) a < \left(m - \frac{d}{d-1}\right) a < \sum_{i=0}^{m} i a_i d^i \leq ma.$$

Proof. Please see [11]. □

Observe now that

$$\sum_{x\in S_y} d^{-|x|}\cdot|x| = \sum_{j=0}^{m_y} a_{yj}d^{j-k}(k-j) = kp_y - \sum_{j=0}^{m_y} ja_{yj}d^{j-k} = kp_y - d^{-k}\sum_{j=0}^{m_y} ja_{yj}d^j$$

$$\overset{6}{<} kp_y - d^{-k}\left(\log_d a_y - \frac{2d-1}{d-1}\right)a_y = kp_y - \left(\log_d (p_y d^k) - \frac{2d-1}{d-1}\right)a_y d^{-k}$$

$$= kp_y - \left(\log_d p_y + k - \frac{2d-1}{d-1}\right)p_y = \frac{2d-1}{d-1}p_y - p_y\log_d p_y.$$

Thus the expected number of input symbols consumed is

$$\sum_{y\in\Gamma^k}\sum_{x\in S_y} d^{-|x|}\cdot|x| < \sum_{y\in\Gamma^k}\left(\frac{2d-1}{d-1}p_y - p_y\log_d p_y\right) = \frac{2d-1}{d-1} + \frac{kH(p_1,\ldots,p_c)}{\log d}.$$

and as $H(U_d) = \log d$, the expected consumption of entropy is at most

$$H(U_d)\cdot\left(\frac{2d-1}{d-1} + \frac{kH(p_1,\ldots,p_c)}{\log d}\right) = \log d\cdot\frac{2d-1}{d-1} + kH(p_1,\ldots,p_c).$$

The number of output symbols is k, so the production of entropy is $kH(p_1,\ldots,p_c)$. Thus the efficiency is at least

$$\frac{kH(p_1,\ldots,p_c)}{kH(p_1,\ldots,p_c) + \log d\cdot\frac{2d-1}{d-1}} = \frac{1}{1+\Theta(k^{-1})} = 1-\Theta(k^{-1}).$$

4.3 Uniform ⇒ Arbitrary

Now suppose the target distribution is over an alphabet $\Gamma = \{1,\ldots,c\}$ with arbitrary real probabilities $p_1^*,\ldots,p_c^*$. We exhibit a family of restart protocols D_k that map the uniform distribution over a d-symbol alphabet Σ to the distribution $\{p_i^*\}$ with efficiency $1-\Theta(k^{-1})$. Moreover, if the p_i^* and basic arithmetic operations are computable, then so is D_k. We assume that $d > 1/\min_i p_i^*$, so by the pigeonhole principle, $d > c$. If we want to convert a uniform distribution over fewer symbols to $\{p_i^*\}$, we can treat groups of k subsequent symbols as a single symbol from a d^k-sized alphabet.

Unlike the other protocols we have seen so far, these protocols require infinitely many states in general. This follows from a cardinality argument: there are only countably many reductions specified by finite-state protocols, but uncountably many probability distributions on c symbols.

We will use the set of real probability distributions $\{p_y\}_{y\in\Gamma^k}$ on k-symbol output strings as our state space. As the initial state, we use the extension of the target distribution onto k-symbol strings $\{p_y^*\}$ with $p_{s_1\cdots s_k}^* = p_{s_1}^*\cdots p_{s_k}^*$.

The construction of the protocol D_k at each state $\{p_y\}_{y\in\Gamma^k}$ closely follows the one for rational target distributions presented in Sect. 4.2. Since we can no longer assume that the probabilities p_y are of the form $a_y d^{-k}$ for some integer a_y, we will instead use the greatest a_y such that $q_y := a_y d^{-k} \le p_y$, namely $a_y = \lfloor p_y d^k \rfloor$. We have $p_y - q_y > d^{-k}$. Of course, these may no longer sum to 1, and so we also define a **residual probability** $rd^{-k} = 1 - \sum_{y \in \Gamma^k} q_y < (c/d)^k$.

As in Sect. 4.2, we construct sets of prefix-incomparable codewords S_y for each k-symbol output word y based on the d-ary expansion of a_y, with the aim that the probability of encountering a codeword in S_y is exactly $a_y d^{-k}$. If the protocol encounters a codeword in S_y, it outputs y and restarts.

We also construct a set S_r based on the d-ary expansion of the residual r. If a codeword in S_r is encountered, then we output nothing, and instead transition to a different state to run the protocol again with the **residual distribution** $\{p'_y\}_{y \in \Gamma^k}$, where p'_y is the probability we lost when rounding down earlier:

$$p'_y = \frac{p_y - q_y}{\sum_{y \in \Gamma^k}(p_y - q_y)} = \frac{p_y - q_y}{rd^{-k}}.$$

The correctness of the protocol follows because each additional generation of states acts contractively on the distribution of output symbols, with a unique fixpoint at the true distribution. Roughly speaking, suppose the protocol we execute at the state $\{p'_y\}$ has an error within ε, i.e. the probability that it will output the string y is bounded by $p'_y \pm \varepsilon$. As in Sect. 4.2, at state $\{p_y\}$, we encounter a string in S_y and output y with probability q_y. With probability rd^{-k}, we encounter a string in S_r and pass to the state $\{p'_y\}$, where we output y with probability bounded by $p'_y \pm \varepsilon$. Hence the total probability of emitting y is bounded by

$$q_y + rd^{-k}\left(\frac{p_y - q_y}{rd^{-k}} \pm \varepsilon\right) = p_y \pm rd^{-k}\varepsilon.$$

In particular, the error at $\{p_y\}$ is at most $(c/d)^k\varepsilon$.

Lemma 7. *If $\{q_y\}_{y \in \Gamma^k}$ is such that $0 < p^*_y - q_y < 1/d^k$ for all y, then*

$$\left| -\sum_{y \in \Gamma^k} q_y \log q_y - kH(p^*_1, \ldots, p^*_c) \right| = O(kC^{-k})$$

for some constant $C > 1$.

Proof. Please see [11]. □

Following the analysis of Sect. 4.2, the expected number of input symbols consumed in the initial state $\{p^*_y\}$ is

$$\begin{aligned}
&\sum_{y \in \Gamma^k} \sum_{x \in S_y} d^{-|x|} \cdot |x| + \sum_{x \in S_r} d^{-|x|} \cdot |x| \\
&< \frac{2d-1}{d-1} + \frac{-\sum_{y \in \Gamma^k} q_y \log q_y - rd^{-k}\log(rd^{-k})}{\log d} \\
&\overset{\text{Lemma 7}}{\le} \frac{2d-1}{d-1} + \frac{kH(p^*_1, \ldots, p^*_c) + O(kC^{-k})}{\log d} + O(1). \qquad (11)
\end{aligned}$$

At any state, we emit nothing and pass to a residual distribution with probability $rd^{-k} < c^k/d^k$. Since we know nothing about the structure of the residual distribution in relation to the original distribution $\{p_y^*\}$, the bound (11) does not apply for the expected number of input symbols consumed at these other states. However, we have a naive bound of

$$\sum_{y\in\Gamma^k}\sum_{x\in S_y} d^{-|x|}\cdot|x| + \sum_{x\in S_r} d^{-|x|}\cdot|x|$$
$$< \frac{2d-1}{d-1} + \frac{-\sum_{y\in\Gamma^k} q_y\log q_y - rd^{-k}\log(rd^{-k})}{\log d}$$
$$\le \frac{2d-1}{d-1} + \frac{kH(U_c)}{\log d} + O(1) = \Theta(k)$$

since the uniform distribution maximizes entropy over all possible distributions $\{q_y\}$, and this is sufficient for our purposes: by the geometric sum formula, the expected number of additional states we will traverse without emitting anything is just $c^k/(d^k - c^k) = \Theta(D^{-k})$, where $D = d/c > 1$ by assumption. Hence, in the initial state, we expect to only consume $\Theta(kD^{-k})$ symbols for some constant D while dealing with residual distributions. We conclude that the total expected number of input symbols consumed to produce k output symbols, hence the latency, is at most

$$\frac{2d-1}{d-1} + \frac{kH(p_1^*,\ldots,p_c^*)}{\log d} + \Theta(kE^{-k})$$

for a constant $E > 1$, so as in Sect. 4.2, the efficiency is at least $1 - \Theta(k^{-1})$.

There is still one issue to resolve if we wish to construct a serial protocol with kth component D_k. The observant reader will have noticed that, as D_k is not finite-state, its consumption is not uniformly bounded by some m_k, as required by Lemma 5. However, the computation of one epoch of D_k consists of a series of stages, and the consumption at each stage is uniformly bounded by $m_k = k\log c/\log d + 2$. In each stage, if digits are produced, the epoch halts, otherwise the computation proceeds to the next stage. Each stage, when started in its start state, consumes at most m_k digits and produces exactly k digits with probability at least $1 - (c/d)^k$ and produces no digits with probability at most $(c/d)^k$. The next lemma shows that this is enough to derive the conclusion of Lemma 5.

Lemma 8. *Let m_k be a uniform bound on the consumption in each stage of one epoch of D_k, as defined in the preceding paragraph. If the m_k satisfy condition* (4)*, then the variances* (5) *vanish in the limit.*

Proof. Please see [11]. □

4.4 Arbitrary ⇒ Uniform with $\Theta(\log K/k)$ Loss

In this section describe a family of restart protocols B_k for transforming an arbitrary d-ary distribution with real probabilities $p_1,\ldots,p_d$ to a c-ary uniform distribution with efficiency $H(p_1,\ldots,p_d)/\log c - \Theta(\log k/k)$. The remainder of this section is omitted due to space constraints; please see [11].

4.5 $(\frac{1}{r},\frac{r-1}{r}) \Rightarrow (r-1)$-Uniform with $\Theta(k^{-1})$ Loss

Let $r \in \mathbb{N}$, $r \geq 2$. In this section we show that a coin with bias $1/r$ can generate an $(r-1)$-ary uniform distribution with $\Theta(k^{-1})$ loss of efficiency. This improves the result of the previous section in this special case. The remainder of this section is omitted due to space constraints; please see [11].

5 Conclusion

We have introduced a coalgebraic model for constructing and reasoning about state-based protocols that implement entropy-conserving reductions between random processes. We have provided basic tools that allow efficient protocols to be constructed in a compositional way and analyzed in terms of the tradeoff between latency and loss of entropy. We have illustrated the use of the model in various reductions.

An intriguing open problem is to improve the loss of the protocol of Sect. 4.4 to $\Theta(1/k)$. Partial progress has been made in Sect. 4.5, but we were not able to generalize this approach.

5.1 Discussion: The Case for Coalgebra

What are the benefits of a coalgebraic view? Many constructions in the information theory literature are expressed in terms of trees; e.g. [3,10]. Here we have defined protocols as coalgebras (S,δ), where $\delta : S \times \Sigma \to S \times \Gamma^*$, a form of Mealy automata. These are not trees in general. However, the class admits a final coalgebra $D : (\Gamma^*)^{\Sigma^+} \times \Sigma \to (\Gamma^*)^{\Sigma^+} \times \Gamma^*$, where

$$D(f,a) = (f@a, f(a)) \qquad\qquad f@a(x) = f(ax),\ a \in \Sigma,\ x \in \Sigma^+.$$

Here the extension to streams $D^\omega : (\Gamma^*)^{\Sigma^+} \times \Sigma^\omega \rightharpoonup \Gamma^\omega$ takes the simpler form

$$D^\omega(f,a\alpha) = f(a) \cdot D^\omega(f@a,\alpha).$$

A state $f : \Sigma^+ \to \Gamma^*$ can be viewed as a labeled tree with nodes Σ^* and edge labels Γ^*. The nodes xa are the children of x for $x \in \Sigma^*$ and $a \in \Sigma$. The label on the edge (x,xa) is $f(xa)$. The tree $f@x$ is the subtree rooted at $x \in \Sigma^*$, where $f@x(y) = f(xy)$. For any coalgebra (S,δ), there is a unique coalgebra morphism $h : (S,\delta) \to ((\Gamma^*)^{\Sigma^+}, D)$ defined coinductively by

$$(h(s)@a, h(s)(a)) = \text{let } (t,z) = \delta(s,a) \text{ in } (h(t),z),$$

where $s \in S$ and $a \in \Sigma$. The coalgebraic view allows arbitrary protocols to inherit structure from the final coalgebra under h^{-1}, thereby providing a mechanism for transferring results on trees, such as entropy rate, to results on state transition systems.

There are other advantages as well. In this paper we have considered only **homogeneous** measures on Σ^ω and Γ^ω, those induced by Bernoulli processes in which the probabilistic choices are independent and identically distributed, for finite Σ and Γ. However, the coalgebraic definitions of protocol and reduction make sense even if Σ and Γ are countably infinite and even if the measures are non-homogeneous.

We have observed that a fixed measure μ on Σ induces a unique homogeneous measure, also called μ, on Σ^ω. But in the final coalgebra, we can go the other direction: For an arbitrary probability measure μ on Σ^ω and state $f : \Sigma^+ \to \Gamma^*$, there is a unique assignment of transition probabilities on Σ^+ compatible with μ, namely the conditional probability

$$f(xa) = \frac{\mu(\{\alpha \mid xa \prec \alpha\})}{\mu(\{\alpha \mid x \prec \alpha\})},$$

or 0 if the denominator is 0. This determines the probabilistic behavior of the final coalgebra as a protocol starting in state f when the input stream is distributed as μ. This behavior would also be reflected in any protocol (S, δ) starting in any state $s \in h^{-1}(f)$ under the same measure on input streams, thus providing a semantics for (S, δ) even under non-homogeneous conditions.

In addition, as in Lemma 3(iii), any measure μ on Σ^ω induces a push-forward measure $\mu \circ (D^\omega)^{-1}$ on Γ^ω. This gives a notion of reduction even in the non-homogeneous case. Thus we can lift the entire theory to Mealy automata that operate probabilistically relative to an arbitrary measure μ on Σ^ω. These are essentially discrete Markov transition systems with observations in Γ^*.

Even more generally, one can envision a continuous-space setting in which the state set S and alphabets Σ and Γ need not be discrete. The appropriate generalization would give reductions between discrete-time, continuous-space Markov transition systems as defined for example in [6,15].

As should be apparent, in this paper we have only scratched the surface of this theory, and there is much left to be done.

Acknowledgments. Thanks to Joel Ouaknine and Aaron Wagner for valuable discussions. Thanks to the Bellairs Research Institute of McGill University for providing a wonderful research environment. Research was supported by NSF grants CCF1637532, IIS-1703846, and IIS-1718108, AFOSR grant FA9550-12-1-0040, ARO grant W911NF-17-1-0592, and a grant from the Open Philanthropy project.

References

1. Adamek, J.: Foundations of Coding. Wiley, Hoboken (1991)
2. Blum, M.: Independent unbiased coin flips from a correlated biased source: a finite state Markov chain. Combinatorica **6**(2), 97–108 (1986)

3. Böcherer, G., Ali Amjad, R.: Informational divergence and entropy rate on rooted trees with probabilities. In: Proceedings of the IEEE International Symposium on Information Theory (June 2014)
4. Chung, K.L.: A Course in Probability Theory, 2nd edn. Academic Press, Cambridge (1974)
5. Cover, T.M., Thomas, J.A.: Elements of Information Theory. Wiley-Interscience, Hoboken (1991)
6. Doberkat, E.-E.: Stochastic Relations: Foundations for Markov Transition Systems. Studies in Informatics. Chapman Hall, London (2007)
7. Dodis, Y., Elbaz, A., Oliveira, R., Raz, R.: Improved randomness extraction from two independent sources. In: Jansen, K., Khanna, S., Rolim, J.D.P., Ron, D. (eds.) APPROX/RANDOM -2004. LNCS, vol. 3122, pp. 334–344. Springer, Heidelberg (2004). https://doi.org/10.1007/978-3-540-27821-4_30
8. Elias, P.: The efficient construction of an unbiased random sequence. Ann. Math. Stat. **43**(3), 865–870 (1992)
9. Feller, W.: An Introduction to Probability Theory and Its Applications, vol. 1, 2nd edn. Wiley, Hoboken (1971)
10. Hirschler, T., Woess, W.: Comparing entropy rates on finite and infinite rooted trees with length functions. IEEE Trans. Inf. Theory (2017)
11. Kozen, D., Soloviev, M.: Coalgebraic tools for randomness-conserving protocols. Technical report, Cornell, July 2018. https://arxiv.org/abs/1807.02735
12. Nisan, N., Ta-shma, A.: Extracting randomness: a survey and new constructions. J. Comput. Syst. Sci. **58**, 148–173 (1999)
13. Pae, S., Loui, M.C.: Optimal random number generation from a biased coin. In: Proceedings of the 16th ACM-SIAM Symposium on Discrete Algorithms, Vancouver, Canada, pp. 1079–1088, January 2005
14. Pae, S., Loui, M.C.: Randomizing functions: Simulation of discrete probability distribution using a source of unknown distribution. Trans. Inf. Theory **52**(11), 4965–4976 (2006)
15. Panangaden, P.: Labelled Markov Processes. Imperial College Press, London (2009)
16. Peres, Y.: Iterating von Neumann's procedure for extracting random bits. Ann. Stat. **20**(1), 590–597 (1992)
17. Peres, Y., Mossel, E., Hillar, C.: New coins from old: computing with unknown bias. Combinatorica **25**(6), 707–724 (2005)
18. Srinivasan, A., Zuckerman, D.: Computing with very weak random sources. SIAM J. Comput. **28**, 264–275 (1999)
19. Ta-shma, A.: On extracting randomness from weak random sources. In: Proceedings of the 28th ACM Symposium Theory of Computing, pp. 276–285 (1996)

Applications and Tools

Algebraic Solution of Weighted Minimax Single-Facility Constrained Location Problems

Nikolai Krivulin(✉)

St. Petersburg State University, St. Petersburg 199034, Russia
nkk@math.spbu.ru
http://www.math.spbu.ru/user/krivulin/

Abstract. We consider location problems to find the optimal sites of placement of a new facility, which minimize the maximum weighted Chebyshev or rectilinear distance to existing facilities under constraints on the feasible location domain. We examine a Chebyshev location problem in multidimensional space to represent and solve the problem in the framework of tropical (idempotent) algebra, which deals with the theory and applications of semirings and semifields with idempotent addition. The solution approach involves formulating the problem as a tropical optimization problem, introducing a parameter that represents the minimum value in the problem, and reducing the problem to a system of parametrized inequalities. The necessary and sufficient conditions for the existence of a solution to the system serve to evaluate the minimum, whereas all corresponding solutions of the system present a complete solution of the optimization problem. With this approach, we obtain a direct, exact solution represented in a compact closed form, which is appropriate for further analysis and straightforward computations with polynomial time complexity. The solution of the Chebyshev problem is then used to solve a location problem with rectilinear distance in the two-dimensional plane. The obtained solutions extend previous results on the Chebyshev and rectilinear location problems without weights.

Keywords: Tropical mathematics · Idempotent semifield
Constrained optimization problem · Single-facility location problem

1 Introduction

Location problems present an important research domain in optimization, which dates back to the XVII century and originates in the influential works of P. Fermat, E. Torricelli, J. J. Sylvester, J. Steiner and A. Weber. Many results achieved in this domain are recognized as notable contributions to various fields, such as operations research, computer science and engineering.

The work was supported in part by the Russian Foundation for Basic Research (grant number 18-010-00723).

J. Desharnais et al. (Eds.): RAMiCS 2018, LNCS 11194, pp. 317–332, 2018.
https://doi.org/10.1007/978-3-030-02149-8_19

To solve location problems, which are formulated in different settings, a variety of analytical approaches and computational techniques exists, including methods of linear and mixed-integer linear programming, methods of discrete, combinatorial and graph optimization [7,9,17,25,28]. Another approach, which finds increasing application in solving some classes of optimization problems, is to use models and methods of tropical mathematics.

Tropical (idempotent) mathematics deals with the theory and applications of semirings and semifields with idempotent addition (see, e.g., [12–14,26]). It includes tropical optimization as a research area concerned with optimization problems that are formulated and solved in the framework of tropical mathematics. In many cases, tropical optimization problems can be solved directly in closed form under general assumptions, whereas other problems have only algorithmic solutions based on iterative numerical procedures. For a brief overview of tropical optimization problems, one can consult, e.g., [22].

As a solution framework, tropical mathematics is used in [4,5] to handle one-dimensional minimax location problems on graphs. A similar algebraic approach based on the theory of max-separable functions is implemented in [15,16,29–31] to solve constrained minimax location problems. Further examples include the solutions, given in [18–21,24] in terms of idempotent algebra, to unconstrained and constrained minimax single-facility location problems with Chebyshev and rectilinear distances.

In this paper, we consider location problems to find the optimal sites of placement of a new facility, which minimize the maximum weighted Chebyshev or rectilinear distance to existing facilities under constraints on the feasible location domain. For any two vectors $\boldsymbol{r} = (r_1, \ldots, r_n)^T$ and $\boldsymbol{s} = (s_1, \ldots, s_n)^T$ in the real space $\mathbb{R}^n$, the Chebyshev distance (maximum or l_∞-metric) is given by

$$d_\infty(\boldsymbol{r}, \boldsymbol{s}) = \max_{1 \leq i \leq n} |r_i - s_i| = \max_{1 \leq i \leq n} \max\{r_i - s_i, s_i - r_i\}. \tag{1}$$

The rectilinear distance (Manhattan, rectangular, taxi-cab, city-block or l_1-metric) is calculated as

$$d_1(\boldsymbol{r}, \boldsymbol{s}) = \sum_{1 \leq i \leq n} |r_i - s_i| = \sum_{1 \leq i \leq n} \max\{r_i - s_i, s_i - r_i\}. \tag{2}$$

Suppose that we are given m points $\boldsymbol{r}_i = (r_{1i}, \ldots, r_{ni})^T \in \mathbb{R}^n$, positive reals w_i (weights), c_i (upper bounds), and reals h_i (addends) for all $i = 1, \ldots, m$. We need to locate a new point $\boldsymbol{x} = (x_1, \ldots, x_n)^T$ in a feasible location domain $S \subset \mathbb{R}^n$ to minimize the maximum distance, in the sense of a metric d, from $\boldsymbol{x}$ to existing points, under upper bound constraints on these distances. The problem is formulated in the form

$$\begin{aligned} \min \quad & \max_{1 \leq i \leq m} (w_i d(\boldsymbol{x}, \boldsymbol{r}_i) + h_i); \\ \text{s. t.} \quad & d(\boldsymbol{x}, \boldsymbol{r}_i) \leq c_i, \quad i = 1, \ldots, m; \\ & \boldsymbol{x} \in S. \end{aligned} \tag{3}$$

We examine the problem with Chebyshev and rectilinear distances under different settings of the dimension n and of the feasible location area S. In the case of Chebyshev distance, we retain the general setting of a real space of arbitrary dimension n. Given real numbers g_{ij}, p_i and q_i such that $p_i \leq q_i$, for all $i, j = 1, \ldots, n$, the location area is described by the set

$$S = \{(x_1, \ldots, x_n)^T |\ g_{ij} + x_j \leq x_i,\ p_i \leq x_i \leq q_i,\ 1 \leq i, j \leq n\}, \tag{4}$$

and takes the form of the intersection, if it exists, of the half-spaces defined by the inequalities $g_{ij} + x_j \leq x_i$ and of the hyper-rectangle defined by $p_i \leq x_i \leq q_i$.

In the rectilinear case, we consider a more specific two-dimensional problem defined on the real plane as follows. Given real numbers p_1, p_2, q_1, q_2, a and b such that $p_1 \leq q_1$, $p_2 \leq q_2$ and $a \leq b$, the location area is given by the set

$$S = \{(x_1, x_2)^T |\ p_1 - x_1 \leq x_2 \leq q_1 - x_1,\ p_2 + x_2 \leq x_1 \leq q_2 + x_2,\ a \leq x_2 \leq b\}, \tag{5}$$

which presents the intersection of the tilted rectangle defined by the inequalities $p_1 - x_1 \leq x_2 \leq q_1 - x_1$ and $p_2 + x_2 \leq x_1 \leq q_2 + x_2$, and the horizontal strip area given by the inequality $a \leq x_2 \leq b$.

The above-described problems and their special cases are examined in many works, which offer various solutions to the problems. First note that these problems can be formulated as linear programs, and then solved using an appropriate linear programming computational procedure such as the simplex or Karmarkar algorithm. This approach, however, provides a numerical solution, if it exists, rather than a direct, complete solution in an exact analytical form.

For the unconstrained problems with rectilinear distance and equal weights, direct explicit solutions are obtained in [8,10] using geometric arguments. A solution for the weighted problem with rectilinear distance is given in [6], which involves decomposition into independent one-dimensional subproblems solved by reducing to equivalent network flow problems. In [18–21,24], an approach based on idempotent algebra is applied to solve unweighted unconstrained and constrained location problems. Further results on both unweighted and weighted location problems can be found in the survey papers [1–3,11,27], as well as in the books [7,9,17,25,28].

In this paper, we represent and examine the location problems in the framework of tropical (idempotent) algebra. We start with the solution of a location problem with Chebyshev distance in multidimensional space. The solution approach follows the analytical technique developed in [20,21,23,24], which involves formulating the problem as a tropical optimization problem, introducing a parameter that represents the minimum value in the problem, and reducing the problem to a system of parametrized inequalities. The necessary and sufficient conditions for the existence of a solution to the system serve to evaluate the minimum, whereas all corresponding solutions of the system present a complete solution of the optimization problem. With this approach, we obtain a direct, exact solution represented in a compact closed form, which is appropriate for further analysis and straightforward computations with polynomial time complexity. The solution of the Chebyshev problem is then used to solve a location problem with rectilinear distance in the two-dimensional plane.

The proposed solutions extend and further develop previous results in [21,24] on the location problems without weights (positive equal weighted problems). These new solutions, which are given in an explicit form, can serve to supplement and complement existing methods, and be of particular interest when the application of known algorithmic solutions, for one reason or other, appears to be impractical or impossible.

2 Elements of Tropical Algebra

In this section, we present a brief introduction to tropical (idempotent) algebra to provide a formal analytical framework for the solution of the location problems in the sequel. For further details on the theory and applications of tropical mathematics, one can refer, for example, to recent works [12–14,26].

An idempotent semifield is an algebraic system $(\mathbb{X}, \oplus, \otimes, \mathbb{0}, \mathbb{1})$, where $\mathbb{X}$ is a nonempty set that has distinct elements $\mathbb{0}$ (zero) and $\mathbb{1}$ (one), and is equipped with binary operations $\oplus$ (addition) and $\otimes$ (multiplication) such that $(\mathbb{X}, \oplus, \mathbb{0})$ is a commutative idempotent monoid, $(\mathbb{X} \setminus \{\mathbb{0}\}, \otimes, \mathbb{1})$ is an Abelian group, and $\otimes$ distributes over $\oplus$.

In the semifield, addition is idempotent, which means that $x \oplus x = x$ for all $x \in \mathbb{X}$, and induces a partial order by the rule: $x \leq y$ if and only if $x \oplus y = y$. This order is assumed to constitute a total order on $\mathbb{X}$. With respect to this order, the operations $\oplus$ and $\otimes$ are monotone, which implies that the inequality $x \leq y$ results in $x \oplus z \leq y \oplus z$ and $x \otimes z \leq y \otimes z$. Furthermore, the inequalities $x \leq x \oplus y$ and $y \leq x \oplus y$ hold for all $x, y \in \mathbb{X}$. Finally, the inequality $x \oplus y \leq z$ is equivalent to the system of inequalities $x \leq z$ and $y \leq z$.

Multiplication is invertible, which provides each $x \neq \mathbb{0}$ with its inverse x^{-1} such that $x \otimes x^{-1} = \mathbb{1}$. Inversion is antitone to turn the inequality $x \leq y$, where $x, y \neq \mathbb{0}$, into $x^{-1} \geq y^{-1}$. In what follows, the multiplication sign $\otimes$ is, as usual, omitted to save writing.

The integer powers are used in the standard way to indicate iterated products: $x^0 = \mathbb{1}$, $x^p = x x^{p-1}$, $x^{-p} = (x^{-1})^p$ and $\mathbb{0}^p = \mathbb{0}$ for all $x \in \mathbb{X}$ and integer $p > 0$. Furthermore, the equation $x^p = a$ has the unique solution $x = a^{1/p}$ for each $a \in \mathbb{X}$ and integer $p > 0$, which allows the powers to have rational exponents. Moreover, it is assumed that the power notation can be further extended to real exponents (e.g., by the usual extra limiting process) to have the real powers and the power rules well defined. Exponentiation is monotone, which means that the inequality $x \leq y$ yields $x^p \leq y^p$ if $p > 0$, and $x^p \geq y^p$ if $p < 0$.

An analogue of the binomial identity holds in the form $(a \oplus b)^r = a^r \oplus b^r$ for any $a, b \in \mathbb{X}$ and nonnegative real r.

As an example, we consider the real semifield $(\mathbb{R} \cup \{-\infty\}, \max, +, -\infty, 0)$, also known as the $(\max, +)$-algebra, where $\oplus = \max$, $\otimes = +$, $\mathbb{0} = -\infty$ and $\mathbb{1} = 0$. In this semifield, the power x^y coincides with the usual arithmetic product xy, and the inverse x^{-1} with the opposite number $-x$. The order induced by the idempotent addition corresponds to the natural linear order on $\mathbb{R}$.

The algebra of vectors and matrices over idempotent semifields is introduced in the ordinary way. The vector (matrix) operations follow the conventional rules, where the operations $\oplus$ and $\otimes$ are used instead of arithmetic addition and multiplication. In the following, all vectors are considered column vectors unless otherwise specified. A vector that has all elements equal to $\mathbb{0}$ is the zero vector. A vector without zero elements is called regular.

A square matrix that has all entries equal to $\mathbb{1}$ on the diagonal, and to $\mathbb{0}$ everywhere else, is the identity matrix denoted by $\boldsymbol{I}$. For any square matrix $\boldsymbol{A}$ and positive integer p, the power notation indicates iterated matrix products $\boldsymbol{A}^0 = \boldsymbol{I}$, $\boldsymbol{A}^p = \boldsymbol{A}\boldsymbol{A}^{p-1}$. For any $(n \times n)$-matrix $\boldsymbol{A} = (a_{ij})$, the trace is given by

$$\operatorname{tr} A = a_{11} \oplus \cdots \oplus a_{nn}.$$

The properties of the scalar operations $\oplus$ and $\otimes$ with respect to the order relation $\leq$ are readily extended to the vector (matrix) operations, where the inequalities are considered componentwise. For any nonzero vector $\boldsymbol{x} = (x_i)$, the multiplicative conjugate transpose is a row vector $\boldsymbol{x}^- = (x_i^-)$ with elements $x_i^- = x_i^{-1}$ if $x_i \neq \mathbb{0}$, and $x_i^- = \mathbb{0}$ otherwise. For any regular vectors $\boldsymbol{x}$ and $\boldsymbol{y}$ such that $\boldsymbol{x} \leq \boldsymbol{y}$, the conjugate transposition yields $\boldsymbol{x}^- \geq \boldsymbol{y}^-$.

We conclude the overview with two results of tropical linear algebra. First suppose that, given an $(m \times n)$-matrix $\boldsymbol{A}$ and m-vector $\boldsymbol{d}$, we need to find all n-vectors $\boldsymbol{x}$ that are solutions of the inequality

$$\boldsymbol{A}\boldsymbol{x} \leq \boldsymbol{d}. \tag{6}$$

Lemma 1. *Let $\boldsymbol{A}$ be a matrix with regular columns, and $\boldsymbol{d}$ a regular vector. Then, all solutions of Inequality* (6) *are given by $\boldsymbol{x} \leq (\boldsymbol{d}^-\boldsymbol{A})^-$.*

Furthermore, given an $(n \times n)$-matrix $\boldsymbol{A}$ and n-vector $\boldsymbol{b}$, we consider the problem to find all regular n-vectors $\boldsymbol{x}$ that satisfy the inequality

$$\boldsymbol{A}\boldsymbol{x} \oplus \boldsymbol{b} \leq \boldsymbol{x}. \tag{7}$$

To describe a solution to the problem, we introduce a function that maps any $(n \times n)$-matrix $\boldsymbol{A}$ onto the scalar

$$\operatorname{Tr}(\boldsymbol{A}) = \operatorname{tr} \boldsymbol{A} \oplus \cdots \oplus \operatorname{tr} \boldsymbol{A}^n.$$

Provided that $\operatorname{Tr}(\boldsymbol{A}) \leq \mathbb{1}$, the asterate operator (the Kleene star) transforms the matrix $\boldsymbol{A}$ into the matrix

$$\boldsymbol{A}^* = \boldsymbol{A} \oplus \cdots \oplus \boldsymbol{A}^n.$$

The next statement presents a solution proposed in [23] to Inequality (7).

Theorem 1. *For any matrix $\boldsymbol{A}$ and vector $\boldsymbol{b}$, the following statements hold:*

1. *If $\operatorname{Tr}(\boldsymbol{A}) \leq \mathbb{1}$, then all regular solutions of Inequality* (7) *are given by $\boldsymbol{x} = \boldsymbol{A}^*\boldsymbol{u}$ for any vector $\boldsymbol{u} \geq \boldsymbol{b}$.*
2. *If $\operatorname{Tr}(\boldsymbol{A}) > \mathbb{1}$, then there are no regular solutions.*

Below, we represent the location problems under study in terms of idempotent algebra, and obtain direct, complete solutions to the problems.

3 Location with Chebyshev Distance

We start with a solution of the location problem defined on the n-dimensional vector space with Chebyshev metric (1). In the framework of $(\max, +)$-algebra, the Chebyshev distance between vectors $\boldsymbol{r} = (r_i)$ and $\boldsymbol{s} = (s_i)$ in $\mathbb{R}^n$ is given by

$$d_\infty(\boldsymbol{r}, \boldsymbol{s}) = \bigoplus_{1 \leq i \leq n} (s_i^{-1} r_i \oplus r_i^{-1} s_i) = \boldsymbol{s}^- \boldsymbol{r} \oplus \boldsymbol{r}^- \boldsymbol{s}.$$

The objective function in Problem (3) takes the form

$$\bigoplus_{1 \leq i \leq m} h_i (\boldsymbol{r}_i^- \boldsymbol{x} \oplus \boldsymbol{x}^- \boldsymbol{r}_i)^{w_i}.$$

The feasible location area, which is defined by (4), becomes

$$S = \{(x_1, \ldots, x_n)^T |\ g_{ij} x_j \leq x_i,\ p_i \leq x_i \leq q_i,\ 1 \leq i, j \leq n\}.$$

With the matrix and vector notation

$$\boldsymbol{G} = \begin{pmatrix} g_{11} & \cdots & g_{1n} \\ \vdots & \ddots & \vdots \\ g_{n1} & \cdots & g_{nn} \end{pmatrix}, \qquad \boldsymbol{p} = \begin{pmatrix} p_1 \\ \vdots \\ p_n \end{pmatrix}, \qquad \boldsymbol{q} = \begin{pmatrix} q_1 \\ \vdots \\ q_n \end{pmatrix},$$

we describe the location area in vector form through the system of inequalities

$$\boldsymbol{G}\boldsymbol{x} \leq \boldsymbol{x}, \qquad \boldsymbol{p} \leq \boldsymbol{x} \leq \boldsymbol{q}.$$

After substitution of the Chebyshev metric and vector description of the location area in terms of $(\max, +)$-algebra into Problem (3), we formulate the problem as follows:

$$\begin{aligned} \min \quad & \bigoplus_{1 \leq i \leq m} h_i (\boldsymbol{r}_i^- \boldsymbol{x} \oplus \boldsymbol{x}^- \boldsymbol{r}_i)^{w_i}; \\ \text{s. t.} \quad & \boldsymbol{r}_i^- \boldsymbol{x} \oplus \boldsymbol{x}^- \boldsymbol{r}_i \leq c_i, \quad i = 1, \ldots, m; \\ & \boldsymbol{G}\boldsymbol{x} \leq \boldsymbol{x}, \quad \boldsymbol{p} \leq \boldsymbol{x} \leq \boldsymbol{q}. \end{aligned} \tag{8}$$

Note that we assume all data involved in the formulation of Problem (3) to be real numbers, and thus none of them is equal to the tropical zero $\mathbb{0} = -\infty$. Specifically, both the known vectors $\boldsymbol{r}_i$ for all $i = 1, \ldots, m$, and the unknown vector $\boldsymbol{x}$ are considered regular.

To solve the problem obtained, we first introduce an additional parameter to represent the minimum value of the objective function, and then reduce the problem to a parameterized system of inequalities. Subsequently, we use existence conditions for solutions of the system to evaluate the value of the parameter. Finally, all solutions of the system, which correspond to this value, serve as a complete solution to the initial optimization problem.

Let us denote the minimum value of the objective function by θ. Then, all solutions of the problem must satisfy the equation

$$\bigoplus_{1\leq i\leq m} h_i(\boldsymbol{r}_i^{-}\boldsymbol{x}\oplus\boldsymbol{x}^{-}\boldsymbol{r}_i)^{w_i}=\theta.$$

Since we assume θ to be the minimum of the objective function, the set of solutions remains unchanged after replacing the equation by the inequality

$$\bigoplus_{1\leq i\leq m} h_i(\boldsymbol{r}_i^{-}\boldsymbol{x}\oplus\boldsymbol{x}^{-}\boldsymbol{r}_i)^{w_i}\leq\theta.$$

Using the extremal property of idempotent addition, we replace the last inequality by an equivalent system of inequalities to describe all solutions of Problem (8) as follows:

$$\begin{aligned} h_i(\boldsymbol{r}_i^{-}\boldsymbol{x}\oplus\boldsymbol{x}^{-}\boldsymbol{r}_i)^{w_i} &\leq\theta,\\ \boldsymbol{r}_i^{-}\boldsymbol{x}\oplus\boldsymbol{x}^{-}\boldsymbol{r}_i &\leq c_i,\quad i=1,\ldots,m;\\ \boldsymbol{G}\boldsymbol{x} &\leq\boldsymbol{x},\\ \boldsymbol{p}\leq\boldsymbol{x} &\leq\boldsymbol{q}. \end{aligned}\tag{9}$$

We use the tropical analogue of the binomial identity to replace the inequality $h_i(\boldsymbol{r}_i^{-}\boldsymbol{x}\oplus\boldsymbol{x}^{-}\boldsymbol{r}_i)^{w_i}\leq\theta$ by the inequalities $h_i(\boldsymbol{r}_i^{-}\boldsymbol{x})^{w_i}\leq\theta$ and $h_i(\boldsymbol{x}^{-}\boldsymbol{r}_i)^{w_i}\leq\theta$. Since exponentiation is monotone, these inequalities can be further rewritten by the usual power rules as $h_i^{1/w_i}\boldsymbol{r}_i^{-}\boldsymbol{x}\leq\theta^{1/w_i}$ and $h_i^{1/w_i}\boldsymbol{x}^{-}\boldsymbol{r}_i\leq\theta^{1/w_i}$, and then represented as the inequalities $\boldsymbol{r}_i^{-}\boldsymbol{x}\leq\theta^{1/w_i}h_i^{-1/w_i}$ and $\boldsymbol{x}^{-}\boldsymbol{r}_i\leq\theta^{1/w_i}h_i^{-1/w_i}$. Application of Lemma 1 to solve the first inequality with respect to $\boldsymbol{x}$ yields $\boldsymbol{x}\leq\theta^{1/w_i}h_i^{-1/w_i}\boldsymbol{r}_i$.

Furthermore, we again use Lemma 1 to solve the second inequality with respect to $\boldsymbol{r}_i$, and then multiply both sides of the result by $\theta^{-1/w_i}h_i^{1/w_i}$ to obtain the inequality $\theta^{-1/w_i}h_i^{1/w_i}\boldsymbol{r}_i\leq\boldsymbol{x}$. Finally, we combine the results into the double inequality $\theta^{-1/w_i}h_i^{1/w_i}\boldsymbol{r}_i\leq\boldsymbol{x}\leq\theta^{1/w_i}h_i^{-1/w_i}\boldsymbol{r}_i$.

In the same way, we replace the inequality $\boldsymbol{r}_i^{-}\boldsymbol{x}\oplus\boldsymbol{x}^{-}\boldsymbol{r}_i\leq c_i$ by the inequalities $\boldsymbol{x}\leq c_i\boldsymbol{r}_i$ and $c_i^{-1}\boldsymbol{r}_i\leq\boldsymbol{x}$, and then represent them as $c_i^{-1}\boldsymbol{r}_i\leq\boldsymbol{x}\leq c_i\boldsymbol{r}_i$.

We now rewrite System (9) in the form

$$\begin{aligned} \theta^{-1/w_i}h_i^{1/w_i}\boldsymbol{r}_i\leq\boldsymbol{x} &\leq\theta^{1/w_i}h_i^{-1/w_i}\boldsymbol{r}_i,\\ c_i^{-1}\boldsymbol{r}_i\leq\boldsymbol{x} &\leq c_i\boldsymbol{r}_i,\quad i=1,\ldots,m;\\ \boldsymbol{G}\boldsymbol{x} &\leq\boldsymbol{x},\\ \boldsymbol{p}\leq\boldsymbol{x} &\leq\boldsymbol{q}. \end{aligned}$$

Furthermore, we combine the left inequalities for all $i=1,\ldots,m$ into one, which provides a lower bound for $\boldsymbol{x}$. Next, we represent the right inequalities as $\boldsymbol{x}^{-}\geq\theta^{-1/w_i}h_i^{1/w_i}\boldsymbol{r}_i^{-}$ and $\boldsymbol{x}^{-}\geq c_i^{-1}\boldsymbol{r}_i^{-}$. Summing up these inequalities and conjugate-transposing the result yield an upper bound.

After adding the remaining inequalities, we have a parameterized description of all solutions in the form of the double inequality

$$
\begin{aligned}
\boldsymbol{G}\boldsymbol{x} \oplus \bigoplus_{1\leq i\leq m} (\theta^{-1/w_i} h_i^{1/w_i} \oplus c_i^{-1})\boldsymbol{r}_i \oplus \boldsymbol{p} \leq \boldsymbol{x} \\
\leq \left(\bigoplus_{1\leq i\leq m} (\theta^{-1/w_i} h_i^{1/w_i} \oplus c_i^{-1})\boldsymbol{r}_i^{-} \oplus \boldsymbol{q}^{-} \right)^{-}.
\end{aligned}
$$

First, we assume that $\mathrm{Tr}(\boldsymbol{G}) \leq \mathbb{1}$ and apply Theorem 1 to solve the left inequality and obtain a solution represented using a vector of parameters $\boldsymbol{u}$ as follows:

$$
\boldsymbol{x} = \boldsymbol{G}^{*}\boldsymbol{u}, \qquad \boldsymbol{u} \geq \bigoplus_{1\leq i\leq m} (\theta^{-1/w_i} h_i^{1/w_i} \oplus c_i^{-1})\boldsymbol{r}_i \oplus \boldsymbol{p}.
$$

Next, we substitute $\boldsymbol{x}$ by $\boldsymbol{G}^{*}\boldsymbol{u}$ into the right inequality, and then apply Lemma 1 to solve the obtained inequality for $\boldsymbol{u}$ in the form

$$
\boldsymbol{u} \leq \left(\bigoplus_{1\leq i\leq m} ((\theta^{-1/w_i} h_i^{1/w_i} \oplus c_i^{-1})\boldsymbol{r}_i^{-} \oplus \boldsymbol{q}^{-})\boldsymbol{G}^{*} \right)^{-}.
$$

The set of parameter vectors $\boldsymbol{u}$ defined by the obtained inequalities is nonempty if and only if the following condition holds:

$$
\bigoplus_{1\leq i\leq m} (\theta^{-1/w_i} h_i^{1/w_i} \oplus c_i^{-1})\boldsymbol{r}_i \oplus \boldsymbol{p} \leq \left(\bigoplus_{1\leq i\leq m} ((\theta^{-1/w_i} h_i^{1/w_i} \oplus c_i^{-1})\boldsymbol{r}_i^{-} \oplus \boldsymbol{q}^{-})\boldsymbol{G}^{*} \right)^{-}.
$$

We now use the last inequality to evaluate the parameter θ. Multiplying both sides of the inequality by the conjugate transpose of the right-hand side yields the equivalent inequality

$$
\bigoplus_{1\leq i,j\leq m} ((\theta^{-1/w_i} h_i^{1/w_i} \oplus c_i^{-1})\boldsymbol{r}_i^{-} \oplus \boldsymbol{q}^{-})\boldsymbol{G}^{*}((\theta^{-1/w_j} h_j^{1/w_j} \oplus c_j^{-1})\boldsymbol{r}_j \oplus \boldsymbol{p}) \leq \mathbb{1},
$$

which we further break down into the inequalities

$$
\begin{aligned}
&((\theta^{-1/w_i} h_i^{1/w_i} \oplus c_i^{-1})\boldsymbol{r}_i^{-} \oplus \boldsymbol{q}^{-})\boldsymbol{G}^{*} \\
&\qquad \otimes ((\theta^{-1/w_j} h_j^{1/w_j} \oplus c_j^{-1})\boldsymbol{r}_j \oplus \boldsymbol{p}) \leq \mathbb{1}, \quad i,j = 1,\ldots,m.
\end{aligned}
$$

After multiplying the terms on the left-hand side, we replace each inequality by four inequalities to write

$$
\begin{aligned}
\theta^{-1/w_i-1/w_j} h_i^{1/w_i} h_j^{1/w_j} \boldsymbol{r}_i^{-}\boldsymbol{G}^{*}\boldsymbol{r}_j &\leq \mathbb{1}, \\
\theta^{-1/w_i} h_i^{1/w_i} \boldsymbol{r}_i^{-}\boldsymbol{G}^{*}(c_j^{-1}\boldsymbol{r}_j \oplus \boldsymbol{p}) &\leq \mathbb{1}, \\
\theta^{-1/w_j} h_j^{1/w_j} (c_i^{-1}\boldsymbol{r}_i^{-} \oplus \boldsymbol{q}^{-})\boldsymbol{G}^{*}\boldsymbol{r}_j &\leq \mathbb{1}, \\
(c_i^{-1}\boldsymbol{r}_i^{-} \oplus \boldsymbol{q}^{-})\boldsymbol{G}^{*}(c_j^{-1}\boldsymbol{r}_j \oplus \boldsymbol{p}) &\leq \mathbb{1}, \quad i,j = 1,\ldots,m.
\end{aligned}
$$

The solution of the first three inequalities with respect to θ yields the result

$$
\begin{aligned}
(h_i^{1/w_i} h_j^{1/w_j} \boldsymbol{r}_i^- \boldsymbol{G}^* \boldsymbol{r}_j)^{\frac{w_i w_j}{w_i + w_j}} &\leq \theta, \\
h_i(\boldsymbol{r}_i^- \boldsymbol{G}^*(c_j^{-1} \boldsymbol{r}_j \oplus \boldsymbol{p}))^{w_i} &\leq \theta, \\
h_j((c_i^{-1} \boldsymbol{r}_i^- \oplus \boldsymbol{q}^-) \boldsymbol{G}^* \boldsymbol{r}_j)^{w_j} &\leq \theta, \\
(c_i^{-1} \boldsymbol{r}_i^- \oplus \boldsymbol{q}^-) \boldsymbol{G}^*(c_j^{-1} \boldsymbol{r}_j \oplus \boldsymbol{p}) &\leq \mathbb{1}, \quad i, j = 1, \ldots, m.
\end{aligned}
$$

By combining the inequalities for each i and j, we obtain the system

$$
\begin{aligned}
\theta \geq \bigoplus_{1 \leq i,j \leq m} \Bigg(&(h_i^{1/w_i} h_j^{1/w_j} \boldsymbol{r}_i^- \boldsymbol{G}^* \boldsymbol{r}_j)^{\frac{w_i w_j}{w_i + w_j}} \oplus h_i(\boldsymbol{r}_i^- \boldsymbol{G}^*(c_j^{-1} \boldsymbol{r}_j \oplus \boldsymbol{p}))^{w_i} \\
&\oplus h_j((c_i^{-1} \boldsymbol{r}_i^- \oplus \boldsymbol{q}^-) \boldsymbol{G}^* \boldsymbol{r}_j)^{w_j} \Bigg), \\
\bigoplus_{1 \leq i,j \leq m} &(c_i^{-1} \boldsymbol{r}_i^- \oplus \boldsymbol{q}^-) \boldsymbol{G}^*(c_j^{-1} \boldsymbol{r}_j \oplus \boldsymbol{p}) \leq \mathbb{1}.
\end{aligned}
$$

Consider the first inequality, which gives the lower bound for the parameter θ. Since θ is assumed to represent the minimum in the problem, we set it to be equal to the right-hand side of the inequality.

As one can see, the second inequality serves as necessary and sufficient conditions for the constraints of the problem to be consistent.

We now summarize the result obtained in the following statement.

Theorem 2. *Suppose that the following conditions hold:*

1. $\mathrm{Tr}(\boldsymbol{G}) \leq \mathbb{1}$;
2. $(c_i^{-1} \boldsymbol{r}_i^- \oplus \boldsymbol{q}^-) \boldsymbol{G}^*(c_j^{-1} \boldsymbol{r}_j \oplus \boldsymbol{p}) \leq \mathbb{1}$ *for all* $i, j = 1, \ldots, m$.

Then, the minimum value in Problem (8) *is equal to*

$$
\begin{aligned}
\theta = \bigoplus_{1 \leq i,j \leq m} \Bigg(&(h_i^{1/w_i} h_j^{1/w_j} \boldsymbol{r}_i^- \boldsymbol{G}^* \boldsymbol{r}_j)^{\frac{w_i w_j}{w_i + w_j}} \oplus h_i(\boldsymbol{r}_i^- \boldsymbol{G}^*(c_j^{-1} \boldsymbol{r}_j \oplus \boldsymbol{p}))^{w_i} \\
&\oplus h_j((c_i^{-1} \boldsymbol{r}_i^- \oplus \boldsymbol{q}^-) \boldsymbol{G}^* \boldsymbol{r}_j)^{w_j} \Bigg), \qquad (10)
\end{aligned}
$$

and all solutions are given by

$$
\boldsymbol{x} = \boldsymbol{G}^* \boldsymbol{u},
$$

where the vector $\boldsymbol{u}$ *satisfies the condition*

$$
\begin{aligned}
&\bigoplus_{1 \leq i \leq m} (\theta^{-1/w_i} h_i^{1/w_i} \oplus c_i^{-1}) \boldsymbol{r}_i \oplus \boldsymbol{p} \leq \boldsymbol{u} \\
&\qquad \leq \left(\bigoplus_{1 \leq i \leq m} ((\theta^{-1/w_i} h_i^{1/w_i} \oplus c_i^{-1}) \boldsymbol{r}_i^- \oplus \boldsymbol{q}^-) \boldsymbol{G}^* \right)^- .
\end{aligned}
$$

It is not difficult to see that the solution given by Theorem 2 has a polynomial time complexity in the number of points m and the dimension of space n. Clearly, the most computationally demanding part of the solution is the calculation of the parameter θ according to (10). The evaluation of θ requires calculating the Kleene star matrix $\boldsymbol{G}^*$ with the computational time, which is at most $O(n^4)$, when computed by direct matrix multiplications. Given the matrix $\boldsymbol{G}^*$, each of three terms in the big brackets on the right-hand side of (10) takes time of $O(n^2)$, and thus the overall time to compute θ is no more than $O(m^2n^2)$.

Note that, in the $(\max,+)$-algebra setting, Problem (8) can be solved as a linear program using a polynomial time iterative procedure such as the Karmarkar algorithm. However, this approach can offer a numerical solution rather than a complete, direct solution in an analytical form like that provided by Theorem 2.

Finally, we represent the result of Theorem 2 in terms of conventional algebra. For the matrix $\boldsymbol{G}$, we denote the entries of the matrix $\boldsymbol{G}^*$ as g_{ij}^* and note that

$$g_{ij}^* = \begin{cases} \gamma_{ij}, & \text{if } i \neq j, \\ \max\{\gamma_{ij}, 0\}, & \text{if } i = j; \end{cases} \qquad \gamma_{ij} = \max_{1 \leq k \leq n-1} \max_{\substack{1 \leq i_1, \ldots, i_{k-1} \leq n \\ i_0 = i,\ i_k = j}} (g_{i_0 i_1} + \cdots + g_{i_{k-1} i_k}).$$

With the identity $\max(a,b) = -\min(-a,-b)$ used to save writing, we arrive at the next corollary.

Corollary 1. *Suppose that the following conditions hold:*

1. $\max\limits_{\substack{1 \leq i_1, \ldots, i_{k-1} \leq n \\ i_0 = i_k = i}} (g_{i_0 i_1} + \cdots + g_{i_{k-1} i_k}) \leq 0$ *for all* $i, k = 1, \ldots, n$;
2. $g_{kl}^* + \max\limits_{1 \leq j \leq m} \max\{-c_j + r_{lj}, p_l\} \leq \min\limits_{1 \leq i \leq m} \max\{c_i + r_{ki}, q_k\}$ *for all* $k, l = 1, \ldots, n$.

Then, the minimum value in Problem (8) *is equal to*

$$\theta = \max_{1 \leq i,j \leq m} \max_{1 \leq k,l \leq n} \max \left\{ \frac{w_j h_i}{w_i + w_j} + \frac{w_i h_j}{w_i + w_j} + \frac{w_i w_j}{w_i + w_j}(g_{kl}^* - r_{ki} + r_{lj}), \right.$$
$$h_i + w_i(g_{kl}^* - r_{ki} + r_{lj} - c_j, g_{kl}^* - r_{ki} + p_k),$$
$$\left. h_j + w_j(g_{kl}^* + r_{lj} - r_{ki} - c_i, g_{kl}^* + r_{lj} - q_k) \right\},$$

and all solutions $\boldsymbol{x} = (x_k)$ *are given by*

$$x_k = \max_{1 \leq j \leq n} (g_{kj}^* + u_j), \qquad k = 1, \ldots, n;$$

where the numbers u_j *for each* $j = 1, \ldots, n$ *satisfy the condition*

$$\max_{1 \leq i \leq m} \max\{r_{ji} + (h_i - \theta)/w_i, r_{ji} - c_i, p_j\} \leq u_j$$
$$\leq \min_{1 \leq i \leq m} \min_{1 \leq l \leq n} (\min\{r_{li} + (\theta - h_i)/w_i, r_{li} + c_i, q_l\} - g_{lj}^*).$$

4 Location with Rectilinear Distance

We now turn to the solution of a location problem defined on the plane with rectilinear distance. To solve the problem, we extend and further develop the technique, which is proposed in [24] to solve unweighted two-dimensional rectilinear location problems. The technique involves the representation of the problem in the form of a tropical optimization problem, following by the change of variables, which reduces the optimization problem to the problem in the form of (8). Note that this technique can hardly provide solutions to the rectilinear location problems in three and more dimensions, which require different solution methods.

First, we represent the rectilinear distance between two-dimensional vectors $\boldsymbol{r} = (r_1, r_2)^T$ and $\boldsymbol{s} = (s_1, s_2)^T$ in terms of (max, +)-algebra and write

$$d_1(\boldsymbol{r}, \boldsymbol{s}) = (s_1^{-1} r_1 \oplus r_1^{-1} s_1)(s_2^{-1} r_2 \oplus r_2^{-1} s_2).$$

Furthermore, we describe the feasible location area given by (5) as follows

$$S = \{(x_1, x_2)^T |\ p_1 x_1^{-1} \leq x_2 \leq q_1 x_1^{-1},\ p_2 x_2 \leq x_1 \leq q_2 x_2,\ a \leq x_2 \leq b\}.$$

Problem (3) now takes the form

$$\begin{aligned} \min \quad & \bigoplus_{1 \leq i \leq m} h_i((r_{1i}^{-1} x_1 \oplus x_1^{-1} r_{1i})(r_{2i}^{-1} x_2 \oplus x_2^{-1} r_{2i}))^{w_i}; \\ \text{s. t.} \quad & (r_{1i}^{-1} x_1 \oplus x_1^{-1} r_{1i})(r_{2i}^{-1} x_2 \oplus x_2^{-1} r_{2i}) \leq c_i, \quad i = 1, \ldots, m; \\ & p_1 x_1^{-1} \leq x_2 \leq q_1 x_1^{-1}, \quad p_2 x_2 \leq x_1 \leq q_2 x_2, \quad a \leq x_2 \leq b. \end{aligned} \tag{11}$$

As before, we assume all parameters and vectors, involved in the problem formulation, to have non-zero values in the sense of (max, +)-algebra.

The solution of Problem (11) is based on changing variables to reduce it to Problem (8), and thus to take advantage of the above-obtained results. Note that this transformation from (11) to (8) reflects the well-known relationship between the solutions of location problems on the plane with rectilinear and Chebyshev distances.

To solve Problem (11), we first introduce new vectors

$$\boldsymbol{y} = \begin{pmatrix} y_1 \\ y_2 \end{pmatrix}, \qquad \boldsymbol{s}_i = \begin{pmatrix} s_{1i} \\ s_{2i} \end{pmatrix}, \qquad i = 1, \ldots, m,$$

with their elements given by the conditions

$$y_1 = x_1 x_2, \quad y_2 = x_1 x_2^{-1}, \quad s_{1i} = r_{1i} r_{2i}, \quad s_{2i} = r_{1i} r_{2i}^{-1}.$$

Clearly, the elements of the vector $\boldsymbol{x} = (x_1, x_2)^T$ are related with those of $\boldsymbol{y}$ by the equalities

$$x_1 = y_1^{1/2} y_2^{1/2}, \qquad x_2 = y_1^{1/2} y_2^{-1/2}.$$

With the new notation, for all $i = 1, \ldots, m$, we can write

$$(r_{1i}^{-1}x_1 \oplus x_1^{-1}r_{1i})(r_{2i}^{-1}x_2 \oplus x_2^{-1}r_{2i}) = s_{1i}^{-1}y_1 \oplus s_{2i}^{-1}y_2 \oplus s_{2i}y_2^{-1} \oplus s_{1i}y_1^{-1} = \boldsymbol{s}_i^{-}\boldsymbol{y} \oplus \boldsymbol{y}^{-}\boldsymbol{s}_i,$$

It remains to rewrite the constraints, which determine the feasible location area S. The first inequality $p_1 x_1^{-1} \leq x_2 \leq q_1 x_1^{-1}$ is equivalent to $p_1 \leq x_1 x_2 \leq q_1$, which can be written as $p_1 \leq y_1 \leq q_1$. In the same way, we represent the second inequality $p_2 x_2 \leq x_1 \leq q_2 x_2$ as $p_2 \leq x_1 x_2^{-1} \leq q_2$, and then as $p_2 \leq y_2 \leq q_2$.

We take the last inequality $a \leq x_2 \leq b$, and put its left-hand side in the equivalent form $a^2 x_1 x_2^{-1} \leq x_1 x_2$, which can be expressed as $a^2 y_2 \leq y_1$. The right-hand side takes the form $x_1 x_2 \leq b^2 x_1 x_2^{-1}$, and then becomes $b^{-2} y_1 \leq y_2$.

With the vector and matrix notation

$$\boldsymbol{p} = \begin{pmatrix} p_1 \\ p_2 \end{pmatrix}, \qquad \boldsymbol{q} = \begin{pmatrix} q_1 \\ q_2 \end{pmatrix}, \qquad \boldsymbol{G} = \begin{pmatrix} \mathbb{0} & a^2 \\ b^{-2} & \mathbb{0} \end{pmatrix},$$

where $\mathbb{0} = -\infty$, we express the constraints in vector form as

$$\boldsymbol{p} \leq \boldsymbol{y} \leq \boldsymbol{q}, \qquad \boldsymbol{G}\boldsymbol{y} \leq \boldsymbol{y}.$$

Finally, we have Problem (11) formulated in terms of (max, +)-algebra as follows

$$\begin{aligned} \min \quad & \bigoplus_{1 \leq i \leq m} h_i(\boldsymbol{s}_i^{-}\boldsymbol{y} \oplus \boldsymbol{y}^{-}\boldsymbol{s}_i)^{w_i}; \\ \text{s. t.} \quad & \boldsymbol{s}_i^{-}\boldsymbol{y} \oplus \boldsymbol{y}^{-}\boldsymbol{s}_i \leq c_i, \quad i = 1, \ldots, m; \\ & \boldsymbol{G}\boldsymbol{y} \leq \boldsymbol{y}, \quad \boldsymbol{p} \leq \boldsymbol{y} \leq \boldsymbol{q}. \end{aligned} \tag{12}$$

Since the problem obtained takes the form of (8), we apply Theorem 2 to derive a complete solution to Problem (12) given by the next statement.

Theorem 3. *Let* $\boldsymbol{s}_i = (s_{1i}, s_{2i})^T$, *where* $s_{1i} = r_{1i} r_{2i}$ *and* $s_{2i} = r_{1i} r_{2i}^{-1}$, *and suppose that* $(c_i^{-1}\boldsymbol{s}_i^{-} \oplus \boldsymbol{q}^{-})\boldsymbol{G}^{*}(c_j^{-1}\boldsymbol{s}_j \oplus \boldsymbol{p}) \leq \mathbb{1}$ *for all* $i, j = 1, \ldots, m$.

Then, the minimum value in Problem (12) *is equal to*

$$\theta = \bigoplus_{1 \leq i,j \leq m} \Bigg((h_i^{1/w_i} h_j^{1/w_j} \boldsymbol{s}_i^{-}\boldsymbol{G}^{*}\boldsymbol{s}_j)^{\frac{w_i w_j}{w_i + w_j}} \oplus h_i(\boldsymbol{s}_i^{-}\boldsymbol{G}^{*}(c_j^{-1}\boldsymbol{s}_j \oplus \boldsymbol{p}))^{w_i} \\ \oplus h_j((c_i^{-1}\boldsymbol{s}_i^{-} \oplus \boldsymbol{q}^{-})\boldsymbol{G}^{*}\boldsymbol{s}_j)^{w_j} \Bigg),$$

and all solution vectors $\boldsymbol{x} = (x_1, x_2)^T$ *have the elements*

$$x_1 = y_1^{1/2} y_2^{1/2}, \qquad x_2 = y_1^{1/2} y_2^{-1/2},$$

defined by the elements of vectors $\boldsymbol{y} = (y_1, y_2)^T$, *which are given by*

$$\boldsymbol{y} = \boldsymbol{G}^* \boldsymbol{u},$$

where the vector $\boldsymbol{u}$ *satisfies the condition*

$$\bigoplus_{1 \leq i \leq m} (\theta^{-1/w_i} h_i^{1/w_i} \oplus c_i^{-1}) \boldsymbol{s}_i \oplus \boldsymbol{p} \leq \boldsymbol{u}$$

$$\leq \left(\bigoplus_{1 \leq i \leq m} ((\theta^{-1/w_i} h_i^{1/w_i} \oplus c_i^{-1}) \boldsymbol{s}_i^- \oplus \boldsymbol{q}^-) \boldsymbol{G}^* \right)^- .$$

In terms of ordinary arithmetic operations, the result reads as follows.

Corollary 2. *Let* $s_{1i} = r_{1i} + r_{2i}$ *and* $s_{2i} = r_{1i} - r_{2i}$, *and suppose that the following conditions hold:*

$$\max_{1 \leq j \leq m} \max\{-c_j + s_{1j}, p_1, -c_j + s_{2j} + 2a, p_2 + 2a\} \leq \min_{1 \leq i \leq m} \min\{c_i + s_{1i}, q_1\},$$
$$\max_{1 \leq j \leq m} \max\{-c_j + s_{1j} - 2b, p_1 - 2b, -c_j + s_{2j}, p_2\} \leq \min_{1 \leq i \leq m} \min\{c_i + s_{2i}, q_2\}.$$

Then, the minimum value in Problem (12) *is equal to*

$$\begin{aligned}
\theta = \max_{1 \leq i,j \leq m} \max \Bigg\{ & \frac{w_j h_i}{w_i + w_j} + \frac{w_i h_j}{w_i + w_j} \\
& + \frac{w_i w_j}{w_i + w_j} \max\{-s_{1i} + s_{1j}, 2a - s_{1i} + s_{2j}, -2b - s_{2i} + s_{1j}, -s_{2i} + s_{2j}\}, \\
& h_i + w_i \max\{-s_{1i} + s_{1j} - c_j, -s_{1i} + p_1, 2a - s_{1i} + s_{2j} - c_j, 2a - s_{1i} + p_1, \\
& \qquad - 2b - s_{2i} + s_{1j} - c_j, -2b - s_{2i} + p_2, -s_{2i} + s_{2j} - c_j, -s_{2i} + p_2\}, \\
& h_j + w_j \max\{s_{1j} - s_{1i} - c_i, s_{1j} - q_1, 2a + s_{2j} - s_{1i} - c_i, 2a + s_{2j} - q_1, \\
& \qquad - 2b + s_{1j} - s_{2i} - c_i, -2b + s_{1j} - q_2, s_{2j} - s_{2i} - c_i, s_{2j} - q_2\} \Bigg\},
\end{aligned}$$

and all solutions $\boldsymbol{x} = (x_1, x_2)^T$ *are given by*

$$x_1 = (y_1 + y_2)/2, \qquad x_2 = (y_1 - y_2)/2,$$

where y_1 *and* y_2 *are calculated as*

$$y_1 = \max\{u_1, u_2 + 2a\}, \qquad y_2 = \max\{u_1 - 2b, u_2\},$$

with the numbers u_1 and u_2 defined by the conditions

$$\max_{1\leq i\leq m}\left\{\frac{h_i-\theta}{w_i}+s_{1i},-c_i+s_{1i},p_1\right\}\leq u_1$$

$$\leq\min_{1\leq i\leq m}\left\{\frac{\theta-h_i}{w_i}+s_{1i},c_i+s_{1i},q_1,\frac{\theta-h_i}{w_i}+s_{2i}-2a,c_i+s_{2i}-2a,q_2-2a\right\},$$

$$\max_{1\leq i\leq m}\left\{\frac{h_i-\theta}{w_i}+s_{2i},-c_i+s_{2i},p_2\right\}\leq u_2$$

$$\leq\min_{1\leq i\leq m}\left\{\frac{\theta-h_i}{w_i}+s_{1i}+2b,c_i+s_{1i}+2b,q_1+2b,\frac{\theta-h_i}{w_i}+s_{2i},c_i+s_{2i},q_2\right\}.$$

5 Conclusions

The paper has examined minimax single-facility location problems in the n-dimensional vector space with Chebyshev distance and in the two-dimensional plane with rectilinear distance. The feasible location areas are given by the intersection of half-spaces defined by a set of inequality constraints.

We have started with the multidimensional problem with Chebyshev distance and general constraints. To handle the problem, we first represented it in terms of (max, +)-algebra as a tropical optimization problem. The solution approach was implemented, which introduces an additional parameter to represent the optimal value of the objective function, and then uses properties of the operations in (max, +)-algebra to reduce the optimization problem to the solution of a set of parameterized inequalities. The existence conditions for solutions of the system serve to evaluate the parameter, whereas all solutions of the system are taken as a complete solution of the optimization problem.

Using this approach, we have derived a new exact, complete solution to the multidimensional location problem with Chebyshev distance in terms of tropical mathematics, and represented the solution in the standard form. The results obtained were extended to examine the two-dimensional problem with rectilinear distance, and to provide new solutions in both tropical and conventional algebra settings. The solutions are given in a closed form, suitable for further analytical study and direct computations with low polynomial complexity in terms of both the dimension of the location space and the number of given points.

Possible lines of further research include the development of algebraic methods to solve rectilinear location problems in the three-dimensional space, as well as to solve Chebyshev and rectilinear problems with new types of constraints.

References

1. Brandeau, M.L., Chiu, S.S.: An overview of representative problems in location research. Manage. Sci. **35**(6), 645–674 (1989). https://doi.org/10.1287/mnsc.35.6.645
2. Brimberg, J., Wesolowsky, G.O.: Optimizing facility location with Euclidean and rectilinear distances. In: Floudas, C.A., Pardalos, P.M. (eds.) Encyclopedia of optimization, pp. 2869–2873. Springer, Boston (2009). https://doi.org/10.1007/978-0-387-74759-0_491
3. Chhajed, D., Francis, R.L., Lowe, T.J.: Facility location. In: Gass, S.I., Fu, M.C. (eds.) Encyclopedia of Operations Research and Management Science, pp. 546–549. Springer, Heidelberg (2013). https://doi.org/10.1007/978-1-4419-1153-7_327
4. Cuninghame-Green, R.A.: Minimax algebra and applications. Fuzzy Sets Syst. **41**(3), 251–267 (1991). https://doi.org/10.1016/0165-0114(91)90130-I
5. Cuninghame-Green, R.A.: Minimax algebra and applications. In: Hawkes, P.W. (ed.) Advances in Imaging and Electron Physics, Advances in Imaging and Electron Physics, vol. 90, pp. 1–121. Academic Press, San Diego (1994). https://doi.org/10.1016/S1076-5670(08)70083-1
6. Dearing, P.M.: On some minimax location problems using rectilinear distance. Ph.D. thesis, University of Florida, Gainesville, FL (1972)
7. Eiselt, H.A., Marianov, V. (eds.): Foundations of Location Analysis. International Series in Operations Research and Management Science, vol. 155. Springer, New York (2011). https://doi.org/10.1007/978-1-4419-7572-0
8. Elzinga, J., Hearn, D.W.: Geometrical solutions for some minimax location problems. Transport. Sci. **6**(4), 379–394 (1972). https://doi.org/10.1287/trsc.6.4.379
9. Farahani, R.Z., Hekmatfar, M. (eds.): Facility Location. Contributions to Management Science. Physica-Verlag, Heidelberg (2009). https://doi.org/10.1007/978-3-7908-2151-2
10. Francis, R.L.: A geometrical solution procedure for a rectilinear distance minimax location problem. AIIE Trans. **4**(4), 328–332 (1972). https://doi.org/10.1080/05695557208974870
11. Francis, R.L., McGinnis, L.F., White, J.A.: Locational analysis. European J. Oper. Res. **12**(3), 220–252 (1983). https://doi.org/10.1016/0377-2217(83)90194-7
12. Golan, J.S.: Semirings and Affine Equations Over Them, Mathematics and Its Applications, vol. 556. Kluwer Acad. Publ., Dordrecht (2003). https://doi.org/10.1007/978-94-017-0383-3
13. Gondran, M., Minoux, M.: Graphs, Dioids and Semirings, Operations Research/Computer Science Interfaces, vol. 41. Springer, New York (2008). https://doi.org/10.1007/978-0-387-75450-5
14. Heidergott, B., Olsder, G.J., van der Woude, J.: Max Plus at Work. Princeton Series in Applied Mathematics. Princeton Univ. Press, Princeton (2006)
15. Hudec, O., Zimmermann, K.: Biobjective center - balance graph location model. Optimization **45**(1–4), 107–115 (1999). https://doi.org/10.1080/02331939908844429
16. Hudec, O., Zimmermann, K.: A service points location problem with min-max distance optimality criterion. Acta Univ. Carolin. Math. Phys. **34**(1), 105–112 (1993)
17. Klamroth, K.: Single-Facility Location Problems with Barriers. Springer Series in Operations Research and Financial Engineering. Springer, New York (2002). https://doi.org/10.1007/b98843

18. Krivulin, N.K.: An extremal property of the eigenvalue for irreducible matrices in idempotent algebra and an algebraic solution to a Rawls location problem. Vestnik St. Petersburg Univ. Math. **44**(4), 272–281 (2011). https://doi.org/10.3103/S1063454111040078
19. Krivulin, N.K., Plotnikov, P.V.: On an algebraic solution of the Rawls location problem in the plane with rectilinear metric. Vestnik St. Petersburg Univ. Math. **48**(2), 75–81 (2015). https://doi.org/10.3103/S1063454115020065
20. Krivulin, N.K., Plotnikov, P.V.: Using tropical optimization to solve minimax location problems with a rectilinear metric on the line. Vestnik St. Petersburg Univ. Math. **49**(4), 340–349 (2016). https://doi.org/10.3103/S1063454115040081
21. Krivulin, N.: Complete solution of a constrained tropical optimization problem with application to location analysis. In: Höfner, P., Jipsen, P., Kahl, W., Müller, M.E. (eds.) RAMICS 2014. LNCS, vol. 8428, pp. 362–378. Springer, Cham (2014). https://doi.org/10.1007/978-3-319-06251-8_22
22. Krivulin, N.: Tropical optimization problems. In: Petrosyan, L.A., Romanovsky, J.V., kay Yeung, D.W. (eds.) Advances in Economics and Optimization: Collected Scientific Studies Dedicated to the Memory of L. V. Kantorovich, Economic Issues, Problems and Perspectives, pp. 195–214. Nova Sci. Publ., New York (2014)
23. Krivulin, N.: A multidimensional tropical optimization problem with nonlinear objective function and linear constraints. Optimization **64**(5), 1107–1129 (2015). https://doi.org/10.1080/02331934.2013.840624
24. Krivulin, N.: Using tropical optimization to solve constrained minimax singlefacility location problems with rectilinear distance. Comput. Manag. Sci. **14**(4), 493–518 (2017). https://doi.org/10.1007/s10287-017-0289-2
25. Laporte, G., Nickel, S., da Gama, F.S. (eds.): Location Science. Springer, Cham (2015). https://doi.org/10.1007/978-3-319-13111-5
26. McEneaney, W.M.: Max-Plus Methods for Nonlinear Control and Estimation. Systems and Control: Foundations and Applications. Birkhäuser. Boston (2006). https://doi.org/10.1007/0-8176-4453-9
27. ReVelle, C.S., Eiselt, H.A.: Location analysis: a synthesis and survey. European J. Oper. Res. **165**(1), 1–19 (2005). https://doi.org/10.1016/j.ejor.2003.11.032
28. Sule, D.R.: Logistics of Facility Location and Allocation. Marcel Dekker, New York (2001)
29. Tharwat, A., Zimmermann, K.: One class of separable optimization problems: solution method, application. Optimization **59**(5), 619–625 (2010). https://doi.org/10.1080/02331930801954698
30. Zimmermann, K.: Optimization problems with unimodal functions in max-separable constraints. Optimization **24**(1–2), 31–41 (1992). https://doi.org/10.1080/02331939208843777
31. Zimmermann, K.: Min-max emergency service location problems with additional conditions. In: Oettli, W., Pallaschke, D. (eds.) Advances in Optimization, Lecture Notes in Economics and Mathematical Systems, vol. 382, pp. 504–512. Springer, Berlin (1992). https://doi.org/10.1007/978-3-642-51682-5_33

A Set Solver for Finite Set Relation Algebra

Maximiliano Cristiá[1(✉)] and Gianfranco Rossi[2]

[1] Universidad Nacional de Rosario and CIFASIS, Rosario, Argentina
cristia@cifasis-conicet.gov.ar
[2] Università di Parma, Parma, Italy
gianfranco.rossi@unipr.it

Abstract. {*log*} ('setlog') is a constraint logic programming framework endowed with a decision procedure for a fragment of the language of finite sets and binary relations. As such, it can decide the satisfiability of a large class of formulas involving classic set and relational operators such as union, domain, and composition. In this paper we first extend {*log*} with the converse and identity operators in such a way that now the language of finite (set) relation algebra is fully supported. As a second improvement, Cartesian products are implemented as first-class entities which renders the solver more efficient. Finally we demonstrate the applicability of {*log*} by conducting an extensive empirical evaluation over 1,300 problems, most of them extracted from the TPTP library.

1 Introduction

In the nineties Gianfranco Rossi and his colleagues developed CLP($\mathcal{SET}$), a constraint logic programming (CLP) language endowed with a decision procedure for an important fragment of the theory of *finite* sets [14]. CLP($\mathcal{SET}$) was implemented by a freely available tool called {*log*} ('setlog') [28]. Later on Cristiá and Rossi [10] extended CLP($\mathcal{SET}$) to incorporate *finite* binary relations as first-class entities. This extension was also added to {*log*}. Formulas in this framework are quantifier-free first-order formulas admitting most of the classical, first-order relational operators such as domain, composition, relational image, domain restriction, etc. (for instance most of Z [31] operators are supported). Three remarkable properties of this framework are: (a) it can solve formulas without binding finite domains to free set variables; (b) it computes a finite representation of all the solutions of the given formula; and (c) sets and binary relations can be freely combined.

However, our framework is still unable to (efficiently) reason about the set relation algebra of finite relations because: (i) the converse operator is not supported; (ii) the identity relation is supported but is not part of the core language; and (iii) reasoning about Cartesian products is too inefficient. Hence, in this paper we present another version of our CLP language, called CLP($\mathcal{RA}$), where all these issues are addressed and implemented in {*log*}. Besides, the satisfiability

J. Desharnais et al. (Eds.): RAMiCS 2018, LNCS 11194, pp. 333–349, 2018.
https://doi.org/10.1007/978-3-030-02149-8_20

solver of CLP($\mathcal{RA}$) is proved to be sound w.r.t. the intended interpretation, and the implementation of CLP($\mathcal{RA}$) as part of $\{log\}$ is briefly shown. Finally, since the set relation algebra of finite relations is not decidable [2], the applicability of the implementation of CLP($\mathcal{RA}$) in $\{log\}$ is empirically assessed. To this end we gather more than 1,300 problems (many taken from the TPTP library [32]) involving all the operators supported by the language.

The structure of the paper is as follows. Section 2 introduces the constraint language of CLP($\mathcal{RA}$), as well as some of its general properties. The satisfiability solver of CLP($\mathcal{RA}$) is presented in Sect. 3. Soundness of the solver is discussed in Sect. 4. Section 5 reports on the empirical evaluation of the $\{log\}$ tool. Sections 6 and 7 present some related work and our conclusions, respectively. More technical and detailed information can be found in an extended version of this paper [12] and in a technical document [9], both of them available on-line.

2 The Constraint Language $\mathcal{L}_{\mathcal{RA}}$

One of the key elements of a CLP scheme is its constraint language. We call $\mathcal{L}_{\mathcal{RA}}$ the constraint language of CLP($\mathcal{RA}$). This section describes the syntax, semantics and expressiveness of $\mathcal{L}_{\mathcal{RA}}$ and some general features of CLP($\mathcal{RA}$).

2.1 Syntax

The syntax of the language is defined primarily by giving the signature upon which terms and formulas are built.

Definition 1 (Signature). *The signature $\Sigma_{\mathcal{RA}}$ of $\mathcal{L}_{\mathcal{RA}}$ is a triple $\langle \mathcal{F}, \Pi, \mathcal{V} \rangle$ where: $\mathcal{F}$ is the set of function symbols partitioned as $\mathcal{F} \mathrel{\widehat{=}} \mathcal{F}_S \cup \mathcal{F}_X$ where $\mathcal{F}_S \mathrel{\widehat{=}} \{\emptyset, \{\cdot \sqcup \cdot\}, \cdot \times \cdot\}$ and $\mathcal{F}_X$ is a set of uninterpreted constant and function symbols, including at least the binary function symbol $(\cdot,\cdot)$; Π is the set of predicate symbols, partitioned as $\Pi \mathrel{\widehat{=}} \Pi_S \cup \Pi_R \cup \Pi_T$ where $\Pi_S \mathrel{\widehat{=}} \{=, \neq, \in, \notin, un, \parallel\}$, $\Pi_R \mathrel{\widehat{=}} \{id, comp, inv\}$ and $\Pi_T \mathrel{\widehat{=}} \{set, nset, rel, nrel, pair, npair\}$; and $\mathcal{V}$ is a denumerable set of variables partitioned as $\mathcal{V} \mathrel{\widehat{=}} \mathcal{V}_S \cup \mathcal{V}_O$.* □

Intuitively, $\emptyset$ represents the empty set; $\{x \sqcup A\}$, called *extensional set term*, represents the set $\{x\} \cup A$; $A \times B$ represents the Cartesian product between sets A and B; and (x, y) represents an ordered pair. The first parameter of an extensional set term is called *element part* and the second is called *set part*. Note that all set terms denote only *finite*, though *unbounded*, sets.

$\mathcal{L}_{\mathcal{RA}}$ defines two sorts, Set and X. For notational convenience, the synonym $\mathsf{U} \mathrel{\widehat{=}} \mathsf{Set} \cup \mathsf{X}$ is also defined. If a symbol h has sort X we write $h : \mathsf{X}$.

Definition 2 (Sorts of function symbols). *The sorts of the symbols defined in $\mathcal{F}$ are as follows: $\emptyset : \mathsf{Set}$; $\{\cdot \sqcup \cdot\} : \mathsf{U} \times \mathsf{Set} \to \mathsf{Set}$; $\cdot \times \cdot : \mathsf{Set} \times \mathsf{Set} \to \mathsf{Set}$; $(\cdot,\cdot) : \mathsf{U} \times \mathsf{U} \to \mathsf{X}$; $s : \mathsf{X}^{n_s} \to \mathsf{X}$, if $s \in \mathcal{F}_X$ for some $n_s \geq 0$; $v : \mathsf{Set}$, if $v \in \mathcal{V}_S$; and $v : \mathsf{X}$, if $v \in \mathcal{V}_O$.* □

The set of admissible $\mathcal{L}_{\mathcal{RA}}$ terms is defined as follows.

Definition 3 ($\mathcal{RA}$-terms). *The set of $\mathcal{RA}$*-terms, *denoted by $\mathcal{T}_{\mathcal{RA}}$, is the minimal subset of the set of $\Sigma_{\mathcal{RA}}$-terms generated by the following grammar respecting the sorts as given in Definition 2 (T represents any non-variable $\mathcal{F}_{\mathcal{X}}$-term and V represents any variable in $\mathcal{V}$):*

$$\mathcal{T}^0_{\mathcal{RA}} ::= T \mid V \mid '(' \ \mathcal{T}^0_{\mathcal{RA}} \ ',' \ \mathcal{T}^0_{\mathcal{RA}} \ ')' \ \mid Set$$
$$Set ::= '\emptyset' \ \mid V \mid '\{' \ \mathcal{T}^0_{\mathcal{RA}} \ '\sqcup' \ Set \ '\}' \mid Set \ '\times' \ Set \qquad \square$$

In view of the intended interpretation, $\Sigma_{\mathcal{RA}}$-terms of sort Set are called *set terms*. Observe that one can write terms representing sets which are nested at any level. Hereafter, we will use the notation $\{t_1, t_2, \ldots, t_n \sqcup t\}$ as a shorthand for $\{t_1 \sqcup \{t_2 \sqcup \cdots \{t_n \sqcup t\} \cdots\}\}$ and the notation $\{t_1, t_2, \ldots, t_n\}$ as a shorthand for $\{t_1, t_2, \ldots, t_n \sqcup \emptyset\}$. For example, the set term $\{a \sqcup \{(b, c) \sqcup \emptyset\}\}$ is written more simply as $\{a, (b, c)\}$.

Symbols $=$, $\neq$, $\in$, $\notin$ and $\|$ are infix; all the other symbols in Π are prefix.

Definition 4 (Sorts of predicate symbols). *The sorts of the predicate symbols defined in Π are as follows: $=, \neq$: $\mathsf{U} \times \mathsf{U}$; $\in, \notin$: $\mathsf{U} \times \mathsf{Set}$; $un, comp$: $\mathsf{Set} \times \mathsf{Set} \times \mathsf{Set}$; $\|, id, inv$: $\mathsf{Set} \times \mathsf{Set}$; $rel, nrel$: Set; and $pair, npair, set, nset$: U.* $\square$

The sets of admissible $\mathcal{L}_{\mathcal{RA}}$ constraints and formulas are defined as follows.

Definition 5 ($\mathcal{RA}$-constraints and formulas). *A $\mathcal{RA}$*-constraint *is any atomic $\Sigma_{\mathcal{RA}}$-predicate respecting the sorts as given in Definition 4. The set of $\mathcal{RA}$-constraints is denoted by $\mathcal{C}_{\mathcal{RA}}$. The set of $\mathcal{RA}$-formulas, denoted by $\Phi_{\mathcal{RA}}$, is given by the following grammar:*

$$\Phi_{\mathcal{RA}} ::= true | \mathcal{C}_{\mathcal{RA}} | \Phi_{\mathcal{RA}} \wedge \Phi_{\mathcal{RA}} | \Phi_{\mathcal{RA}} \vee \Phi_{\mathcal{RA}}$$

where $\mathcal{C}_{\mathcal{RA}}$ represents any element belonging to the set of $\mathcal{RA}$-constraints. $\square$

2.2 Semantics, Sort Constraints, Defined Constraints and Negation

Symbols in $\Sigma_{\mathcal{RA}}$ are interpreted according to the interpretation structure $\mathcal{R} \mathrel{\widehat{=}} \langle D, (\cdot)^{\mathcal{R}} \rangle$, where D and $(\cdot)^{\mathcal{R}}$ are now defined rather informally; the formal definitions can be found in [12, Appendix A].

Definition 6 (Interpretation domain). *The interpretation domain D is partitioned as $D \mathrel{\widehat{=}} D_{\mathsf{Set}} \cup D_{\mathsf{X}}$ where: D_{Set} is the collection of all hereditarily finite hybrid sets built from elements in D; and D_{X} is a collection of non-set objects (i.e.,* ur-elements*).* $\square$

Definition 7 (Interpretation function). *The interpretation function $(\cdot)^{\mathcal{R}}$ for symbols in $\Sigma_{\mathcal{RA}}$ is defined as follows. Each sort $\mathsf{S} \in \{\mathsf{Set}, \mathsf{X}\}$ is mapped to the domain D_{S}. The symbols $(\cdot, \cdot)$, $\emptyset$, $=$ and $\in$ are interpreted as in their intuitive meaning. Then, $\{x \sqcup A\}$ is interpreted as $\{x\} \cup A$; $un(A, B, C)$ as $C = A \cup B$; $A \parallel B$ as $A \cap B = \emptyset$; $set(x)$ as x is a set; $rel(R)$ as R is a set of ordered pairs; $pair(x)$*

as x is of form $(\cdot,\cdot)$; $comp(R,S,T)$ as $T = R \odot S$ (i.e., relational composition); $inv(R,S)$ as $S = R^{\smile}$; and $id(A,R)$ as $R = \mathrm{id}\, A$ (i.e., R is the identity relation over set A). The interpretation of a symbol π' in $\{\neq, \notin, nrel, nset, npair\}$ is $\neg\pi$ for the corresponding symbol in $\{=, \in, rel, set, pair\}$. □

In particular, observe that equality between two set terms is interpreted as the equality in D_{Set}, that is as set equality between hereditarily finite hybrid sets [16]. Equality between ordered pairs follows the standard definition.

Sort Constraints. $\mathcal{L}_{\mathcal{RA}}$ does not provide variable declarations. On the other hand, we would like that, for instance, in $comp(R,S,T)$ all three arguments are binary relations. For this purpose, $\mathcal{L}_{\mathcal{RA}}$ provides the predicate symbols in Π_T, called *sort constraints*. So, for $comp(R,S,T)$ we can conjoin $rel(R) \wedge rel(S) \wedge rel(T)$ to the formula which forces all arguments to be sets of ordered pairs. If, say, R happens to contain a non-ordered pair element, the solver should return *false* because $rel(R)$ will not hold. Furthermore, a solver for $\mathcal{L}_{\mathcal{RA}}$ can (as ours does) automatically conjoin sort constraints to any given formula for every symbol in $\Pi_S \cup \Pi_R$ [10,14]. Then, although $\mathcal{L}_{\mathcal{RA}}$ allows for sets where ordered pairs can coexist with other kinds of elements, *comp*, *id* and *inv* only work when their arguments are as expected.

Defined Constraints and Negation. $\mathcal{L}_{\mathcal{RA}}$ can be extended to support other set and relational operators by defining suitable $\mathcal{L}_{\mathcal{RA}}$-formulas. For example $A \cap B = C$ can be defined as follows [14]:

$$inters(A,B,C) \mathrel{\widehat{=}} un(C,N_1,A) \wedge un(C,N_2,B) \wedge N_1 \parallel N_2$$

where N_i are fresh variables (i.e., implicitly existentially quantified). This means that whenever a formula contains an *inters*-constraint it is replaced by its definition, thus becoming a $\mathcal{L}_{\mathcal{RA}}$ formula. We say that *inters* is a *defined constraint.*

The *relative complementation* operator is also a defined constraint:

$$cpmt(A,B,U) \mathrel{\widehat{=}} un(A,B,U) \wedge A \parallel B \tag{1}$$

that is, B is the complement of A w.r.t. U. We will use *cpmt* in Sect. 2.3.

Several operators can be introduced as defined constraints, in particular the negation of *un* (*nun*), $\parallel$ ($\nparallel$) and *comp* (*ncomp*) [10,14], as well as the negation of *inv* (*ninv*) and *id* (*nid*) (see [12, Appendix B]). For example, $\neg A \cup B = C$ is introduced as:

$$nun(A,B,C) \mathrel{\widehat{=}} (n \in C \wedge n \notin A \wedge n \notin B) \vee (n \in A \wedge n \notin C) \vee (n \in B \wedge n \notin C) \tag{2}$$

The availability of $\neq$, $\notin$, *nrel*, *nset*, *npair* and the introduction of defined constraints for the negation of the other constraints in Π allows us not to introduce explicit negation in $\mathcal{L}_{\mathcal{RA}}$.

2.3 Expressiveness

$\mathcal{L}_{\mathcal{RA}}$ turns out to be at least as expressive as the language of the set relation algebra of finite relations over a countable universe. To prove this claim we show that there exists a natural mapping from the operators of relation algebra (RA) onto the constraints defined in $\mathcal{L}_{\mathcal{RA}}$.

As defined by Maddux [23], a relation algebra is a structure $\langle A, +, ^-, \odot, ^\smile, \mathbf{1}\rangle$ where A is a set called universe of discourse, $+$ is set union, $^-$ is set complementation, $\odot$ is relational composition, $^\smile$ is relational converse, and $\mathbf{1}$ is the identity relation. All these elements are defined on $A \times A$. All formulas in relation algebra are propositional combinations of equalities between binary relations formed by arbitrary combinations of the operators.

A mapping from formulas of a relation algebra $\langle A, +, ^-, \odot, ^\smile, \mathbf{1}\rangle$, where A is a finite set, to $\mathcal{L}_{\mathcal{RA}}$-formulas is defined as follows: A is a $\mathcal{L}_{\mathcal{RA}}$ (finite) set; $A \times A$ is trivially mapped to a $\times$-term, i.e. it is another $\mathcal{L}_{\mathcal{RA}}$ (finite) set; $+$, $\odot$ and $^\smile$ are mapped to *un*, *comp* and *inv*, respectively; $^-$ is mapped to *cpmt* (i.e. formula (1)) in such a way that $S = R^-$ holds if and only if $cpmt(R, S, A \times A)$ holds; identity is mapped to $id(A, R)$ because in this case $R = \mathbf{1}$; equality between relations is simply set equality; and relation algebra formulas are mapped to $\mathcal{RA}$-formulas, i.e., conjunctions, disjunctions and negations of $\mathcal{L}_{\mathcal{RA}}$ constraints. Note that this mapping implies that the complement of a finite binary relation is also a finite binary relation as $A \times A$ is a finite set.

Moreover, $\mathcal{L}_{\mathcal{RA}}$ allows for the definition of the so-called heterogeneous relation algebras [30].

Finally, it is worth noting that in $\mathcal{L}_{\mathcal{RA}}$ it is possible to introduce Cartesian products as a defined constraint. That is, we can define $cp(A, B, C)$ as the $\mathcal{L}_{\mathcal{RA}}$ formula which holds if and only if $A \times B = C$:

$$\begin{aligned} cp(A,B,C) \mathrel{\widehat{=}} {} & dom(N_1, A) \wedge ran(N_1, N_3) \wedge dom(N_2, B) \wedge ran(N_2, N_4) \\ & \wedge N_3 \subseteq \{n\} \wedge N_4 \subseteq \{n\} \wedge inv(N_2, N_5) \wedge comp(N_1, N_5, C) \end{aligned} \tag{3}$$

where N_i and n are new variables, $dom(R, X)$ stands for X is the domain of R, $ran(R, Y)$ stands for Y is the range of R, and $\subseteq$ is the subset relation. In turn, *dom*, *ran* and $\subseteq$ can be introduced as other defined constraints (see [12, Appendix A]).

However, our empirical studies show (see Sect. 5) that introducing Cartesian products with formula (3) makes the solver to incur in unacceptable computing times. Hence, we consider $\times$ as a set constructor that is specially processed by the solver.

2.4 Features of CLP($\mathcal{RA}$)

CLP($\mathcal{RA}$) inherits some basic features and properties from the original CLP language for sets and binary relations [10,14]. In this section we briefly recall some of those properties. The first property is that binary relations and $\times$-terms are just sets of ordered pairs. As such they can be freely combined with extensional sets, and set operators can take relations and ×-terms as arguments. For example, the following is an admissible formula of CLP($\mathcal{RA}$):

$$un(A, B, \{(1, 1), (h, 3) \sqcup C \times D\}) \wedge id(E, A) \wedge inv(B, B) \wedge 1 \notin E \qquad (4)$$

where $A, \ldots, E$ are variables (of sort set), and h is a variable of any sort.

Extensional sets (i.e. $\{\cdot \sqcup \cdot\}$) implement the two basic properties of sets, that is: absorption (e.g., $\{a, a\} = \{a\}$) and commutativity on the left (e.g., $\{a, b \sqcup A\} = \{b, a \sqcup A\}$) [14]. Furthermore, set equality between extensional sets is governed by set unification [15]. Extensional sets are always finite and untyped and they are allowed as set elements (i.e., sets can be nested).

Extensional sets (and thus binary relations) can be *partially specified*, as shown in formula (4). This means that both the element and set parts can be variables. Note that the cardinality of a set S whose set part is a variable is not specified (i.e., S is an unbounded finite set as it can have any finite number of elements). This property comes hand in hand with the ability of $\{log\}$ to decide the satisfiability of formulas without binding a finite domain to each set variable.

As far as constraint solving is concerned, the satisfiability solver of CLP($\mathcal{RA}$) (see Sect. 3) has the ability to produce a finite representation of all the (possibly infinitely many) solutions of a given formula, in the form of a finite disjunction of $\mathcal{L}_{\mathcal{RA}}$ formulas. In this context, a 'solution' is an assignment of values to all the free variables of the formula. For example, the relevant part of the first such disjunct (i.e. solution) of (4) is:

$$A = \{(3, 3) \sqcup N_3\}, B = \{(1, 1) \sqcup N_2\}, h = 3, E = \{3 \sqcup N_1\}$$
$$\textsf{Constraint: } 3 \notin C, un(N_2, N_3, C \times D), id(N_1, N_3), inv(N_2, N_2), \ldots$$

where N_i are fresh variables. That is, each such formula is composed by a (possibly empty) conjunction of equalities between variables and terms and a (possibly empty) conjunction of constraints. The conjunction of constraints is guaranteed to be trivially satisfiable. In other words, whenever CLP($\mathcal{RA}$) terminates, it either produces a proof of unsatisfiability or a finite representation of all countermodels.

The ability to compute the finite representation of all the solutions of a formula comes at a price. Indeed, for some unsatisfiable formulas, computing times may be prohibitive as CLP($\mathcal{RA}$) needs to both compute and discard all their solutions. On the positive side, users interested in more abstract solutions to a problem or in more than one solution, may find CLP($\mathcal{RA}$) useful.

3 A Satisfiability Solver for $\mathcal{L}_{\mathcal{RA}}$

The abstract satisfiability solver for $\mathcal{L}_{\mathcal{RA}}$, called $SAT_{\mathcal{RA}}$, has exactly the same design of the solvers for other versions of our CLP language [10,14]. Hence, $SAT_{\mathcal{RA}}$ is a rewriting system that applies specialized rewriting procedures to the current formula and returns the modified formula. Each rewriting procedure applies a few non-deterministic rewrite rules which reduce the syntactic complexity of the constraints of one kind. When no rewrite rule applies to the formula, the rewriting procedure terminates immediately and the formula remains unchanged. This process is executed until a fixpoint is reached—i.e., the formula cannot be simplified any further. The formulas returned at this point are said to be *irreducible*. Then, the irreducible atomic formulas are returned as part of the computed answer.

$SAT_{\mathcal{RA}}$ uses the same rewriting procedures as CLP($\mathcal{SET}$) for constraints based on symbols in Π_S when their arguments do not include Cartesian product terms; these rewriting procedures can be found in [14]. In particular, $SAT_{\mathcal{RA}}$ exploits *set unification* [16] to deal with equalities between set terms. Similarly, $SAT_{\mathcal{RA}}$ uses the rewriting procedure of [10] for *comp* constraints when their arguments do not include Cartesian product terms. Therefore, in this section we show: (a) the rewriting procedures for the converse and identity constraints (i.e. *inv* and *id*), which were not provided in [10]; and (b) how the rewriting procedures for the other symbols in Π are extended to accommodate Cartesian products. However, due to space restrictions, some of the new rewrite rules are presented in the on-line documents [12, Appendix B] and [9].

The rewrite rules are given as $P \longrightarrow \Phi$ where P is a $\mathcal{RA}$-constraint, and Φ is a $\mathcal{RA}$-formula; if Φ has more than one disjunct then the rule is non-deterministic. Variable names n and N (possibly with sub and superscripts) are used to denote fresh variables. These variables are implicitly existentially quantified. $X \neq \emptyset$ means that term X does not denote the empty set. Furthermore, we distinguish between variable and non-variable products as follows.

Definition 8. *A Cartesian product $X \times Y$ is said to be a* variable product *if and only if either X or Y are variables or variable products, and neither X nor Y are the empty set; otherwise it is called* non-variable product.

Converse and Identity. Figure 1 shows the main rewrite rules for the *inv* constraint, including those dealing with Cartesian products. A literal of the form $inv(R, S)$, where both arguments are variables, is irreducible. Therefore, the recursive application of a rule such as (7) will eventually reach a fixpoint because S will eventually be either a variable (in which case $inv(N, S)$ is irreducible) or the empty set (in which case rule (5) applies and the recursion terminates). $SAT_{\mathcal{RA}}$ also implements the negation of *inv* (i.e., *ninv*) as a defined constraint and sort inference rules for *inv* and *ninv* (see [9]).

Figure 2 lists the main rewrite rules for *id* constraints including those dealing with Cartesian products. Literals like $id(A, R)$, where A and R are variables, are irreducible. The negation of *id* (i.e., *nid*) and sort inference rules for *id* and *nid* are also implemented (see [9]).

Equality, Membership and Union. In Fig. 3 we can see the main rewrite rules for predicates of the form $t = u$, when a product is involved. For example, rule (15) takes care of the case when a product is equal to the empty set and

If R, S, X, Y : Set; x, y : U then:

$$inv(R, \emptyset) \to R = \emptyset \tag{5}$$
$$inv(\emptyset, R) \to R = \emptyset \tag{6}$$
$$inv(R, \{(y,x) \sqcup S\}) \to R = \{(x,y) \sqcup N\} \wedge inv(N, S) \tag{7}$$
$$inv(\{(x,y) \sqcup R\}, S) \to S = \{(y,x) \sqcup N\} \wedge inv(R, N) \tag{8}$$
$$inv(X \times Y, S) \to S = Y \times X \tag{9}$$
$$inv(R, X \times Y) \to R = Y \times X \tag{10}$$

Fig. 1. Rewrite rules for *inv*-constraints

If R, A, X, Y : Set; x, y, a, b : U, then:

$$id(\{x \sqcup A\}, R) \to R = \{(x,x) \sqcup N\} \wedge id(A, N) \tag{11}$$
$$id(A, \{(a,b) \sqcup R\}) \to a = b \wedge A = \{a \sqcup N\} \wedge id(N, R) \tag{12}$$
$$id(\{x \sqcup X\} \times \{y \sqcup Y\}, R) \to R = \{((x,y),(x,y)) \sqcup N_1\} \wedge id(N_2, N_1) \wedge un(\{x\} \times Y, X \times \{y \sqcup Y\}, N_2) \tag{13}$$
$$id(A, X \times Y) \to (A = \emptyset \wedge (X = \emptyset \vee Y = \emptyset)) \vee (A = \{n\} \wedge X = A \wedge Y = A) \tag{14}$$

Fig. 2. Rewrite rules for *id*-constraints

rule (16) the equality between two products. As can be noted, the most complex case is the equality between a product and an extensional set, i.e., rule (17). This rule relies on the rules for extensional sets inherited from CLP($\mathcal{SET}$), as well as on the rules defined for the *un* operator (see rule (21) below). In particular, rule (17) uses a pattern that appears in several other rules. That is, $SAT_{\mathcal{RA}}$ first instantiates the two sets participating in the product to extensional sets and then it exploits the following general result:

$$(\{x\} \cup X) \times (\{y\} \cup Y) = \{(x,y)\} \cup (\{x\} \times Y) \cup (X \times (\{y\} \cup Y))$$

Finally, constraints of the form $X \times Y = Z$, where Z is a variable, remain irreducible (see Sect. 4.1).

The rules to deal with constraints of the form $t \in u$ and $t \notin u$, where u is a product, are given in Fig. 4. They simply follow the definition of set membership.

The rewrite rules for *un* dealing with Cartesian products are given by three cases that are selected as follows.

i. If t_i is a variable or a variable product for all $i \in \{1,2,3\}$, then $un(t_1, t_2, t_3)$ is irreducible.
ii. If at least one t_i ($i = 1,2,3$) is a non-variable product, define the following two functions:

If W, X, Y, Z : Set; z, z_i : X, then:

$$X \times Y = \emptyset \rightarrow X = \emptyset \vee Y = \emptyset \tag{15}$$

$$\begin{aligned} W \times X = Y \times Z \rightarrow \\ (W = Y \wedge X = Z \wedge W \times X \neq \emptyset \wedge Y \times Z \neq \emptyset) \\ \vee (W \times X = \emptyset \wedge Y \times Z = \emptyset) \end{aligned} \tag{16}$$

$$\begin{aligned} X \times Y = \{z \sqcup Z\} \rightarrow X = \{n_1 \sqcup N_1\} \wedge Y = \{n_2 \sqcup N_2\} \wedge z = (n_1, n_2) \\ \wedge\, un(\{n_1\} \times N_2, N_1 \times \{n_2 \sqcup N_2\}, Z) \end{aligned} \tag{17}$$

$$X \times Y \text{ is var. prod.: } X \times Y = \{z_1, \ldots, z_k \sqcup X \times Y\} \rightarrow \bigwedge_{i=1}^{k} z_i \in X \times Y \tag{18}$$

Fig. 3. Rewrite rules for equality

If $X, Y \in$ Set; $z \in$ X, then:

$$z \in X \times Y \longrightarrow z = (n_1, n_2) \wedge n_1 \in X \wedge n_2 \in Y \tag{19}$$

$$z \notin X \times Y \longrightarrow z = (n_1, n_2) \wedge (n_1 \notin X \vee n_2 \notin Y) \tag{20}$$

Fig. 4. Rewrite rules for set membership and its negation

- The first function, $\mathcal{T}$, takes set terms and returns set terms:

$$\mathcal{T}(t) = \begin{cases} \{(x, y) \sqcup N\} & \text{if } t \equiv \{x \sqcup X\} \times \{y \sqcup Y\} \\ \emptyset & \text{if } t \equiv \emptyset \times Y \text{ or } t \equiv X \times \emptyset \\ t & \text{otherwise} \end{cases}$$

- The second function, $\mathcal{U}$, takes set terms and returns literals:

$$\mathcal{U}(t) = \begin{cases} un(\{x\} \times Y, X \times \{y \sqcup Y\}, N_t) & \\ \qquad \text{if } t \equiv \{x \sqcup X\} \times \{y \sqcup Y\} \wedge X \neq \emptyset \wedge Y \neq \emptyset & \\ N_t = X \times \{y\} & \text{if } t \equiv \{x \sqcup X\} \times \{y \sqcup \emptyset\} \wedge X \neq \emptyset \\ N_t = \{x\} \times Y & \text{if } t \equiv \{x \sqcup \emptyset\} \times \{y \sqcup Y\} \wedge Y \neq \emptyset \\ N_t = \emptyset & \text{if } t \equiv \{x \sqcup \emptyset\} \times \{y \sqcup \emptyset\} \\ true & \text{otherwise} \end{cases}$$

Then, in this case, the rewrite rule for *un* is:

$$un(t_1, t_2, t_3) \longrightarrow \mathcal{U}(t_1) \wedge \mathcal{U}(t_2) \wedge \mathcal{U}(t_3) \wedge un(\mathcal{T}(t_1), \mathcal{T}(t_2), \mathcal{T}(t_3)) \tag{21}$$

where we assume that equalities of the form $N_t = s$ are propagated into the *un* constraint after it has been processed. In this way, the rules already available in CLP($\mathcal{SET}$) for *un* constraints can be easily applied as the constraint includes only variables and extensional sets. The intuition behind rule (21) is to transform the union between non-variable products into the union between extensional sets where, in each iteration, one element is pulled out from the products and moved into the extensional sets.

iii. If at least one t_i is an extensional set and the others are either variables or variable products then apply the CLP($\mathcal{SET}$) rules for *un* taking variable products as variables.

Implementation. All the rewrite rules dealing with *inv*, *id* and Cartesian products are implemented as part of $\{log\}$. Using $\{log\}$ one can check satisfiability of $\mathcal{L}_{\mathcal{RA}}$ formulas either in an interactive way, or by calling it as a Prolog library. The following are two sample formulas that $\{log\}$ can deal with.

Example 1. Given the $\mathcal{L}_{\mathcal{RA}}$ formula: $dom(A, X) \wedge ran(A, Y) \wedge A \not\subseteq X \times Y$, the solution returned by $\{log\}$ is no, i.e., the formula is unsatisfiable (and then its negation, $dom(A, X) \wedge ran(A, Y) \implies A \subseteq X \times Y$, is a theorem). □

Example 2. Given the $\mathcal{L}_{\mathcal{RA}}$ formula: $(I, J) \in X \times Y$, $\{log\}$ returns $X = \{I \sqcup N_1\} \wedge Y = \{J \sqcup N_2\} \wedge set(N_1) \wedge set(N_2)$, which is a finite representation of all its (infinitely many) solutions—note that there is no finite domain for I and J. □

4 Soundness of $SAT_{\mathcal{RA}}$

In this section we prove that $SAT_{\mathcal{RA}}$ returns correct answers for $\mathcal{L}_{\mathcal{RA}}$ formulas. Detailed proofs can be found in an extended version of the paper [12, Appendix D].

4.1 Satisfiability of Solved Form

As stated in the previous section, once there are no more rewriting procedures to apply to the current formula, $SAT_{\mathcal{RA}}$ finishes and the last formula is returned as part of the answer. The following definition precisely characterizes the form of the formulas returned by $SAT_{\mathcal{RA}}$.

Definition 9 (Solved form). *Let Φ be a $\mathcal{RA}$-formula and let X and X_i be variables or variable products and t a term. Take 'distinct variables' as a shorthand for 'distinct variables or distinct variable products'. A literal ϕ of Φ is in* solved form *if it has one of the following forms:*

(i) $X = t$ *and neither* t *nor* $\Phi \setminus \{\phi\}$ *contain* X*;*
(ii) $X \neq t$ *and* X *occurs neither in* t *nor as an argument of any predicate* $p(\dots)$, $p \in \{un, id, inv, comp\}$, *in* Φ*;*
(iii) $t \notin X$ *and* X *does not occur in* t*;*
(iv) $un(X_1, X_2, X_3)$, *where* X_1 *and* X_2 *are distinct variables;*
(v) $X_1 \parallel X_2$, *where* X_1 *and* X_2 *are distinct variables;*

(vi) $id(X_1, X_2)$, *where* X_1 *and* X_2 *are distinct variables,* X_2 *is not a product;*
(vii) $inv(X_1, X_2)$;
(viii) $comp(X_1, t, X_2)$, $comp(t, X_1, X_2)$, $comp(X_1, t, \emptyset)$ *and* $comp(t, X_1, \emptyset)$, *where* t *is not the empty set;*
(ix) $set(X)$ *and* $rel(X)$, *and no constraint* $nset(X)$ *is in* Φ;
(x) $npair(X)$ *and* $nset(X)$.

A $\mathcal{RA}$*-formula* Φ *is in solved form if it is true or if all its literals are in solved form.* □

The fact that these are all the constraints returned once $SAT_{\mathcal{RA}}$ finishes can be checked in [9]. In fact, by rule inspection, it is immediate to see that all literals in solved form are dealt as irreducible, that is they are never rewritten by any rule (hence they are returned as part of the computed answer); on the other hand, no other literals are dealt with as irreducible as all cases involving literals not in solved form are "covered" by some of the rewrite rules used by $SAT_{\mathcal{RA}}$.

Besides, the solved form literals allow trivial verification of satisfiability as stated by the following theorem.

Theorem 1 (Satisfiability of solved form). *Any* $\mathcal{L}_{\mathcal{RA}}$ *formula in solved form is satisfiable w.r.t. the intended interpretation structure.*

Termination of $SAT_{\mathcal{RA}}$ cannot be guaranteed since $\mathcal{L}_{\mathcal{RA}}$ is at least as expressive as the set relation algebra whose relations are finite sets, and this theory has been proved to be not decidable [2, Theorem 2.1]. Therefore, $SAT_{\mathcal{RA}}$ applied to a $\mathcal{RA}$-formula returns either *false* or a (finite) collection of $\mathcal{RA}$-formulas in solved form; or it does not terminate. According to Theorem 1, a $\mathcal{L}_{\mathcal{RA}}$ formula in solved form is always satisfiable. Moreover, as we will see in the next subsection, the disjunction of the formulas in solved form generated by $SAT_{\mathcal{RA}}$ preserves the set of solutions of the original formula Φ.

4.2 Equisatisfiability

The following theorem ensures that, after termination, the rewriting process implemented by $SAT_{\mathcal{RA}}$ preserves the set of solutions of the input formula.

Theorem 2 (Equisatisfiability). *Let* Φ *be a* $\mathcal{RA}$*-formula and let* $\{\phi_i\}_{i=1}^n$ *be the collection of* $\mathcal{RA}$*-formulas returned by* $SAT_{\mathcal{RA}}(\Phi)$. *Then* $\bigvee_{i=1}^n \phi_i$ *is equisatisfiable to* Φ, *that is, every possible solution*[1] *of* Φ *is a solution of one of* $\{\phi_i\}_{i=1}^n$ *and, vice versa, every solution of one of these formulas is a solution for* Φ.

Thanks to Theorems 1 and 2 we can conclude that, given a $\mathcal{RA}$-formula Φ, if $SAT_{\mathcal{RA}}(\Phi)$ terminates, then Φ is satisfiable with respect to the intended interpretation structure if and only if there is a non-deterministic choice in $SAT_{\mathcal{RA}}(\Phi)$

[1] More precisely, each solution of Φ expanded to the variables occurring in ϕ_i but not in Φ, so as to account for the possible fresh variables introduced into ϕ_i.

that returns a $\mathcal{RA}$-formula in solved form—i.e., different from *false*. Although termination cannot be guaranteed, the empirical evaluation reported in Sect. 5 shows that the implementation of $SAT_{\mathcal{RA}}$ in $\{log\}$ works in practice.

Remark 1. It is important to note that $\{log\}$ does not require users to set a maximum size for the search space. Thus when $\{log\}$ answers *false* for a formula it means the formula is unsatisfiable *for any finite binary relation.* In this sense, $\{log\}$ works as an automated theorem prover or a specialized SMT solver. Furthermore, if a finite domain is provided for each relation variable, termination of $SAT_{\mathcal{RA}}$ can be proved. Conversely, with tools operating on reduced solution spaces, if one finds that a formula is unsatisfiable in a reduced space, it does not mean the formula is indeed unsatisfiable. Furthermore, in general, if the solution space is reduced, some solution may be lost, while $\{log\}$, if it terminates, will always return a finite representation of *all* the solutions for a given formula.

5 Empirical Evaluation

In this section we present the empirical assessment conducted in order to provide evidence that $\{log\}$ works in practice. The assessment not only measures the effectiveness of the results presented in this paper, but the whole tool in general. This empirical evaluation complements another one carried out by the authors elsewhere [10]. The evaluation consists in running $\{log\}$ over 1,334 problems, of which 391 require Cartesian products (i.e., almost one third of the total), measuring how many of them are solved in less than 10 s. If timeouts are not set, the tool might run for ever or it might exhaust the computer memory. Most of the formulas are the negation of theorems and thus the expected answer is *false*.

The experiments were performed on a Latitude E7470 (06DC) with a 4 core Intel(R) Core^{TM} i7-6600U CPU at 2.60 GHz with 8 Gb of main memory, running Linux Ubuntu 16.04.3 (xenial) 64-bit with kernel 4.4.0-97-generic. $\{log\}$ 4.9.5-10b over SWI-Prolog (multi-threaded, 64 bits, version 7.6.0-rc2) was used during the experiments. The full data set can be downloaded from https://www.dropbox.com/s/oevhb43apovzmm0/experimentsCP.tar.gz?dl=0.

The empirical evaluation is based on five problem collections totaling 1,334 problems[2], see Table 1. The first three collections are taken from the TPTP library [32] and, as can be seen in Table 1, they comprise problems on Boolean algebra, RA and set theory. Collection ZMT corresponds to the theorems of Chaps. 10 and 11 of the Z mathematical toolkit [29] which encode many facts on binary relations and partial functions. Finally, RA is a gathering of problems from: (a) one of the earliest Tarski's works on RA [33] which include 32 results; (b) 20 problems proposed by Höfner and Struth [20]; and (c) some problems of our own including, in particular, the 10 axioms of RA as stated by Givant [17]. A more detailed account of each collection can be found in [12, Appendix C].

[2] There is some overlapping between these collections that is difficult to avoid.

Table 1. Summary of the empirical evaluation

PROBLEM SET	PROBLEMS	UNSAT	SAT	SOLVED	PERCENTAGE	AVG TIME
TPTP.BOO	142	142	0	139	98%	1.55 s
TPTP.REL	72	70	2	70	97%	0.72 s
TPTP.SET	735	731	4	704	96%	0.17 s
ZMT	282	169	113	282	100%	0.22 s
RA	103	93	10	101	98%	0.14 s
TOTALS OR AVERAGES	1,334	1,205	129	1,296	97%	0.31 s

Table 1 shows the number of problems in each collection, the number of (un)satisfiable problems, the number and percentage of answers given by $\{log\}$, and the average computing time spent by the tool on each collection. Average times do not count problems ending in timeout. As can be seen, $\{log\}$ performs well in all five problem collections solving in average 97% of them in 0.31 s on average. In particular, the results shown in TPTP.REL, TPTP.SET, ZMT and RA provide evidence that, in spite of dealing with undecidable problems (i.e., the set relation algebra of finite relations), the implementation in $\{log\}$ of the rewriting system $SAT_{\mathcal{RA}}$ performs well in practice. Last, but not least, the fact that $\{log\}$ is able to solve problems in three different fields constitutes another piece of evidence of its usefulness.

Besides, we ran all the experiments with a version of $\{log\}$ where Cartesian products become defined constraints given by formula (3) (i.e., they are not first-class entities as proposed in this paper). In this case, the tool solves only 76% of the 391 problems requiring at least one Cartesian product; instead, when products are a primitive facility the tool can solve 95% of these 391 problems. Therefore, it is sensible to have Cartesian products as first-class entities while the rewriting procedures are extended accordingly.

6 Related Work

$\mathcal{L}_{\mathcal{RA}}$ can encode problems of specification languages such as Z [31] and B [1] and RA, with the obvious restriction of working only with *finite* sets and relations. As in Z and B, $\mathcal{L}_{\mathcal{RA}}$ seamlessly integrates sets and relations by considering relations as sets of ordered pairs. On the other hand, while RA usually considers homogeneous relations, $\mathcal{L}_{\mathcal{RA}}$ allows for heterogeneous algebras. In $\mathcal{L}_{\mathcal{RA}}$, the complementation operator is not a primitive element of the language as it is in RA. Likewise, there is no universal set in $\mathcal{L}_{\mathcal{RA}}$ but users can use Cartesian products to force relations to be subsets of some set.

Many tools, from interactive theorem provers to SMT solvers, can deal with set and relational formulas. Usually, theorem provers encode set theory in predicate calculus and cannot produce counterexamples (i.e., they cannot generate solutions for satisfiable formulas). On the other hand, SMT solvers usually encode set theory in other theories such as arrays, lists, bit vectors, etc. (e.g. [13,27]), although since recently CVC4 incorporates some native support for sets and binary relations [25]. An interesting tool for proving set-theoretic problems is Why3 [6]. In the end, Why3 invokes external SMT solvers, but it also does a lot of simplifications on its own. Atelier B is a powerful theorem prover for the B notation. As such, it can automatically discharge many proof obligations appearing when B models are analyzed. As Why3, it uses its own reasoning capabilities and calls external tools such as Alt-Ergo [8]. In general, all these tools support both finite and infinite sets and binary relations. The counterexamples produced by SMT solvers are concrete solutions to the input formula. Conversely, $\{log\}$ can deal only with finite sets but it can produce a finite representation of all the solutions of satisfiable formulas. These solutions are more abstract than those returned by SMT solvers as they contain variables. However, a 'point' can be easily obtained, roughly, by substituting all set variables by the empty set.

Twenty problems of the problem set named RA in Table 1 are taken from [20]. Höfner and Struth assess Prover9 and Mace4 [24] on these (and other) relational problems. Prover9 is an automated theorem prover for first-order and equational logic, while Mace4 is a counterexample generator. Note that $\{log\}$ does both tasks with the same solving technique.

Braibant and Pous [7] developed a decision procedure for Kleene algebras in the Coq proof assistant as a proof tactic. Besides the obvious differences with our approach, in Coq finite sets are not first-class entities and are strongly typed. Guttmann and others developed a repository of Tarski-Kleene algebras in the Isabelle/HOL proof system [18]. Armstrong et al. [3] use some of these algebras for program analysis and verification.

Some other tools that natively deal with sets and binary relations are ProB, Alloy and RelView. ProB [22] is a constraint solver for the B notation. It is very good at finding "solutions and counter-examples, but in general it cannot be used to prove formulas using variables with infinite types" [19]. Alloy [21] needs a maximum size for all the base sorts, then it builds all the possible relations in that space (as a Boolean formula) and finally calls a SAT solver to find a solution for the formula. It has been proved to be very efficient in large search spaces as SAT solvers have seen tremendous improvements in recent years. RelView [4] is a tool able to work with relations in the form of matrices or graphs. It implements all the standard operators of RA. In general RelView cannot prove whether a given formula holds or not. For instance, in order to prove properties of relational programs, Berghammer et al. [5] combine RelView with Prover9.

We think that some of these tools may benefit from $SAT_{\mathcal{RA}}$ as it can complement them with regards to solving set theory and relation algebra problems.

7 Concluding Remarks

We have extended a language and a solver for finite binary relations and sets with the converse operator, the identity relation and with a new set term representing Cartesian products. In this way, sets, binary relations and Cartesian products can be freely combined as sets. The new language covers most of the set and relational operators available in specification languages such as Z and B as well as those of RA. These extensions have been implemented in $\{log\}$. We have conducted an extensive empirical evaluation providing evidence that $\{log\}$ works in practice. Then, we believe that CLP($\mathcal{RA}$) should be of interest to the RA and automated reasoning communities.

As part of future work we plan to extend the empirical evaluation to the restricted intensional sets recently added to the framework [11]. Intensional sets further enlarge the expressiveness of the language as they introduce a restricted form of universal quantifiers.

Acknowledgements. Part of the work of M. Cristiá is supported by ANPCyT's grant PICT-2014-2200.

References

1. Abrial, J.R.: The B-book: Assigning Programs to Meanings. Cambridge University Press, New York (1996)
2. Andréka, H., Givant, S.R., Németi, I.: Decision Problems for Equational Theories of Relation Algebras, vol. 604. American Mathematical Society, Providence (1997)
3. Armstrong, A., Struth, G., Weber, T.: Program analysis and verification based on Kleene algebra in Isabelle/HOL. In: Blazy, S., Paulin-Mohring, C., Pichardie, D. (eds.) ITP 2013. LNCS, vol. 7998, pp. 197–212. Springer, Heidelberg (2013). https://doi.org/10.1007/978-3-642-39634-2_16
4. Berghammer, R.: Relview. http://www.informatik.uni-kiel.de/~progsys/relview/
5. Berghammer, R., Höfner, P., Stucke, I.: Automated verification of relational while-programs. In: Höfner, P., Jipsen, P., Kahl, W., Müller, M.E. (eds.) RAMiCS 2014. LNCS, vol. 8428, pp. 173–190. Springer, Cham (2014). https://doi.org/10.1007/978-3-319-06251-8_11
6. Bobot, F., Filliâtre, J.C., Marché, C., Paskevich, A.: Why3: Shepherd your herd of provers. In: Boogie 2011: First International Workshop on Intermediate Verification Languages, Wrocław, Poland, August 2011. http://proval.lri.fr/submissions/boogie11.pdf
7. Braibant, T., Pous, D.: Deciding Kleene algebras in Coq. Log. Methods Comput. **8**(1), 1–42 (2012). https://doi.org/10.2168/LMCS-8(1:16)2012
8. Conchon, S., Iguernlala, M.: Increasing proofs automation rate of Atelier-B thanks to Alt-Ergo. In: Lecomte, T., Pinger, R., Romanovsky, A. (eds.) RSSRail 2016. LNCS, vol. 9707, pp. 243–253. Springer, Cham (2016). https://doi.org/10.1007/978-3-319-33951-1_18
9. Cristiá, M., Rossi, G.: Rewrite rules for a solver for sets, binary relations and partial functions. http://people.dmi.unipr.it/gianfranco.rossi/SETLOG/calculus.pdf
10. Cristiá, M., Rossi, G.: A decision procedure for sets, binary relations and partial functions. In: Chaudhuri, S., Farzan, A. (eds.) CAV 2016. LNCS, vol. 9779, pp. 179–198. Springer, Cham (2016). https://doi.org/10.1007/978-3-319-41528-4_10
11. Cristiá, M., Rossi, G.: A decision procedure for restricted intensional sets. In: de Moura [26], pp. 185–201. https://doi.org/10.1007/978-3-319-63046-5_12
12. Cristiá, M., Rossi, G.: A set solver for finite relation algebra - extended version. Technical report (2018). https://www.researchgate.net/publication/320555347_A_Set_Solver_for_Finite_Relation_Algebra

13. Deters, M., Reynolds, A., King, T., Barrett, C.W., Tinelli, C.: A tour of CVC4: how it works, and how to use it. In: Formal Methods in Computer-Aided Design, FMCAD 2014, Lausanne, Switzerland, 21–24 October 2014, p. 7. IEEE (2014). https://doi.org/10.1109/FMCAD.2014.6987586
14. Dovier, A., Piazza, C., Pontelli, E., Rossi, G.: Sets and constraint logic programming. ACM Trans. Program. Lang. Syst. **22**(5), 861–931 (2000)
15. Dovier, A., Policriti, A., Rossi, G.: A uniform axiomatic view of lists, multisets, and sets, and the relevant unification algorithms. Fundam. Inform. **36**(2–3), 201–234 (1998). https://doi.org/10.3233/FI-1998-36235
16. Dovier, A., Pontelli, E., Rossi, G.: Set unification. Theory Pract. Log. Prog. **6**(6), 645–701 (2006). https://doi.org/10.1017/S1471068406002730
17. Givant, S.: The calculus of relations as a foundation for mathematics. J. Autom. Reasoning **37**(4), 277–322 (2006). https://doi.org/10.1007/s10817-006-9062-x
18. Guttmann, W., Struth, G., Weber, T.: A repository for Tarski-Kleene algebras. In: Höfner, P., McIver, A., Struth, G. (eds.) Proceedings of the First Workshop on Automated Theory Engineering, Wrocław, Poland, 31 July 2011. CEUR Workshop Proceedings, vol. 760, pp. 30–39. CEUR-WS.org (2011). http://ceur-ws.org/Vol-760/paper5.pdf
19. Heinrich-Heine-Universität Düsseldorf: The ProB animator and model checker. https://www3.hhu.de/stups/prob/index.php/Main_Page
20. Höfner, P., Struth, G.: On automating the calculus of relations. In: Armando, A., Baumgartner, P., Dowek, G. (eds.) IJCAR 2008. LNCS (LNAI), vol. 5195, pp. 50–66. Springer, Heidelberg (2008). https://doi.org/10.1007/978-3-540-71070-7_5
21. Jackson, D.: Alloy: a logical modelling language. In: Bert, D., Bowen, J.P., King, S., Waldén, M. (eds.) ZB 2003. LNCS, vol. 2651, pp. 1–1. Springer, Heidelberg (2003). https://doi.org/10.1007/3-540-44880-2_1
22. Leuschel, M., Butler, M.: ProB: a model checker for B. In: Araki, K., Gnesi, S., Mandrioli, D. (eds.) FME 2003. LNCS, vol. 2805, pp. 855–874. Springer, Heidelberg (2003). https://doi.org/10.1007/978-3-540-45236-2_46
23. Maddux, R.: Relation Algebras, vol. 13. Elsevier (2006). https://books.google.it/books?id=fjFH1WvPG9AC
24. McCune, W.: Prover9 and Mace4 (2005–2010). http://www.cs.unm.edu/~mccune/prover9/
25. Meng, B., Reynolds, A., Tinelli, C., Barrett, C.W.: Relational constraint solving in SMT. In: de Moura [26], pp. 148–165. https://doi.org/10.1007/978-3-319-63046-5_10
26. de Moura, L. (ed.): Automated Deduction - CADE 26 - 26th International Conference on Automated Deduction, Gothenburg, Sweden, 6–11 August 2017, Proceedings. LNCS, vol. 10395. Springer, Heidelberg (2017). https://doi.org/10.1007/978-3-319-63046-5
27. de Moura, L., Bjørner, N.: Z3: an efficient SMT solver. In: Ramakrishnan, C.R., Rehof, J. (eds.) TACAS 2008. LNCS, vol. 4963, pp. 337–340. Springer, Heidelberg (2008). https://doi.org/10.1007/978-3-540-78800-3_24
28. Rossi, G.: {*log*} (2008). http://people.dmi.unipr.it/gianfranco.rossi/setlog.Home.html
29. Saaltink, M.: The Z/EVES mathematical toolkit version 2.2 for Z/EVES version 1.5. Technical report, ORA Canada (1997)
30. Schmidt, G., Hattensperger, C., Winter, M.: Heterogeneous Relation Algebra, pp. 39–53. Springer, Vienna (1997). https://doi.org/10.1007/978-3-7091-6510-2_3

31. Spivey, J.M.: The Z Notation: A Reference Manual. Prentice Hall International (UK) Ltd., Hertfordshire (1992)
32. Sutcliffe, G.: The TPTP problem library and associated infrastructure: the FOF and CNF Parts, v3.5.0. J. Autom. Reasoning **43**(4), 337–362 (2009)
33. Tarski, A.: On the calculus of relations. J. Symb. Log. **6**(3), 73–89 (1941). https://doi.org/10.2307/2268577

On the Computational Complexity
of Non-dictatorial Aggregation

Lefteris Kirousis[1], Phokion G. Kolaitis[2], and John Livieratos[1]([✉])

[1] Department of Mathematics,
National and Kapodistrian University of Athens, Athens, Greece
{lkirousis,jlivier89}@math.uoa.gr
[2] Computer Science Department,
UC Santa Cruz and IBM Research - Almaden, Santa Cruz, USA
kolaitis@cs.ucsc.edu

Abstract. We investigate when non-dictatorial aggregation is possible from an algorithmic perspective, where non-dictatorial aggregation means that the votes cast by the members of a society can be aggregated in such a way that the collective outcome is not simply the choices made by a single member of the society. We consider the setting in which the members of a society take a position on a fixed collection of issues, where for each issue several different alternatives are possible, but the combination of choices must belong to a given set X of allowable voting patterns. Such a set X is called a possibility domain if there is an aggregator that is non-dictatorial, operates separately on each issue, and returns values among those cast by the society on each issue. We design a polynomial-time algorithm that decides, given a set X of voting patterns, whether or not X is a possibility domain. Furthermore, if X is a possibility domain, then the algorithm constructs in polynomial time such a non-dictatorial aggregator for X. We also design a polynomial-time algorithm that decides whether X is a uniform possibility domain, that is, whether X admits an aggregator that is non-dictatorial even when restricted to any two positions for each issue. As in the case of possibility domains, the algorithm also constructs in polynomial time a uniform non-dictatorial aggregator, if one exists.

1 Introduction

The study of vote aggregation has occupied a central place in social choice theory. A broad framework for carrying out this study is as follows. There is a fixed collection of issues on each of which every member of a society takes a *position*, that is, for each issue, a member of the society can choose between a number of alternatives. However, not every combination of choices is allowed, which means that the vector of the choices made by a member of the society must belong to a given set X of allowable voting patterns, called *feasible evaluations*. The goal is to investigate properties of *aggregators*, which are functions that take as input the votes cast by the members of the society and return as output a feasible

© Springer Nature Switzerland AG 2018
J. Desharnais et al. (Eds.): RAMiCS 2018, LNCS 11194, pp. 350–365, 2018.
https://doi.org/10.1007/978-3-030-02149-8_21

evaluation that represents the collective position of the society on each of the issues at hand. A concrete key problem studied in this framework is to determine whether or not a *non-dictatorial* aggregator exists, i.e., whether or not it is possible to aggregate votes in such a way that individual members of the society do not impose their voting preferences on the society. A set X of feasible evaluations is called a *possibility domain* if it admits a non-dictatorial aggregator; otherwise, X is called an *impossibility domain*. This framework is broad enough to account for several well-studied cases of vote aggregation, including the case of *preference* aggregation for which Arrow [1] established his celebrated impossibility theorem and the case of *judgment* aggregation [9].

The investigation of the existence of non-dictatorial aggregators is typically carried out under two assumptions: (a) the aggregators are *independent of irrelevant alternatives* (IIA); and (b) the aggregators are *conservative* (also known as *supportive*). The IIA assumption means that the aggregator is an issue-by-issue aggregator, so that an IIA aggregator on m issues can be identified with an m-tuple $(f_1, \ldots, f_m)$ of functions aggregating the votes on each issue. The conservativeness (or supportiveness) assumption means that, for every issue, the position returned by the aggregator is one of the positions held by the members of the society on that issue.

By now, there is a body of research on identifying criteria that characterize when a given set X of feasible evaluations is a possibility domain. The first such criterion was established by Dokow and Holzman [7] in the Boolean framework, where, for each issue, there are exactly two alternatives (say, 0 and 1) for the voters to choose from. Specifically, Dokow and Holzman [7] showed that a set $X \subseteq \{0,1\}^m$ is a possibility domain if and only if X is affine or X is not totally blocked. Informally, the notion of *total blockedness*, which was first introduced in [13], asserts that any position on any issue can be inferred from any position on any issue. As regards the non-Boolean framework (where, for some issues, there may be more than two alternatives), Dokow and Holzman [8] extended the notion of total blockedness and used it to give a sufficient condition for a set X of feasible evaluations to be a possibility domain. Szegedy and Xu [16] used tools from universal algebra to characterize when a totally blocked set X of feasible evaluations is a possibility domain. A consequence of these results is that a set X of feasible evaluations is a possibility domain if and only if X admits a binary non-dictatorial aggregator or a ternary non-dictatorial aggregator; in other words, non-dictatorial aggregation is possible for a society of some size if and only if it is possible for a society with just two members or with just three members. This line of work was pursued further by Kirousis et al. [10], who characterized possibility domains in terms of the existence of binary non-dictatorial aggregators or ternary non-dictatorial aggregators of a particular form.

The aforementioned investigations have characterized possibility domains (in both the Boolean and the non-Boolean frameworks) in terms of *structural* conditions. Our goal is to investigate possibility domains using the *algorithmic lens* and, in particular, to study the following algorithmic problem: given a set X of feasible evaluations, determine whether or not X is a possibility domain. Szegedy and Xu [16, Theorem 36] give algorithms for this problem, but these algorithms have very high running time; in fact, they run in exponential time in the number of issues and in the number of positions over each issue, even when confined to the Boolean framework.

We design a polynomial-time algorithm that, given a set X of feasible evaluations (be it in the Boolean or the non-Boolean framework), determines whether or not X is a possibility domain. Furthermore, if X is a possibility domain, then the algorithm produces a binary non-dictatorial or a ternary non-dictatorial aggregator for X. Along the way, we also show that there is a polynomial-time algorithm for determining, given X, whether or not it is totally blocked.

After this, we turn our attention to *uniform possibility domains*, which were introduced in [10] and form a proper subclass of the class of possibility domains. Intuitively, uniform possibility domains are sets of feasible evaluations that admit an aggregator that is non-dictatorial even when restricted to any two positions for each issue. In [10], a tight connection was established between uniform possibility domains and constraint satisfaction by showing that multi-sorted conservative constraint satisfaction problems on uniform possibility domains are tractable, whereas such constraint satisfaction problems defined on all other domains are NP-complete.

Here, using Carbonnel's result in [5], we give a polynomial-time algorithm for the following decision problem: given a set X of feasible evaluations (be it in the Boolean or the non-Boolean framework), determine whether or not X is a uniform possibility domain; moreover, if X is a uniform possibility domain, then the algorithm produces a suitable uniform non-dictatorial aggregator for X.

The results reported here contribute to the developing field of computational social choice and pave the way for further exploration of algorithmic aspects of vote aggregation.

2 Preliminaries and Earlier Work

In this section, we formally describe the framework we will work on and the necessary tools in order to obtain our results. In Subsect. 2.1 we consider possibility domains both in the Boolean and non-Boolean case, whereas in Subsect. 2.2 we turn our attention to uniform possibility domains.

2.1 Possibility Domains

Let $I = \{1, \ldots, m\}$ be a set of issues. Assume that the *possible position values* of an individual (member of a society) for issue j are given by the finite set A_j, where $j = 1, \ldots, m$.

We assume that each set A_j has cardinality at least 2. If $|A_j| = 2$ for all $j \in \{1, \ldots, m\}$, we say that we are in the *binary* or *Boolean* framework; otherwise we say that we are in the *non-binary* or *non-Boolean* framework.

An *evaluation* is an element of $\prod_{j=1}^{m} A_j$. Let $X \subseteq \prod_{j=1}^{m} A_j$ be a set of *permissible* or *feasible* evaluations. To avoid degenerate cases, we assume that for each $j = 1, \ldots, m$, the j-th projection X_j of X is equal to A_j. Note that this assumption in no way implies that $X = \prod_{j=1}^{m} X_j$.

Let $n \geq 2$ represent the number of individuals. We view the elements of $x \in X^n$ as $n \times m$ matrices that represent the choices of all individuals over every issue. The element x_j^i of such a matrix x will be the choice of the i-th individual over the j-th issue, for $i = 1, \ldots, n$ and $j = 1, \ldots, m$. The i-th row x^i will represent the choices of the i-th individual over every issue, $i = 1, \ldots, n$, and the j-th column x_j the choices of every individual over the j-th issue, $j = 1, \ldots, m$.

To *aggregate* a set of n feasible evaluations, we use m-tuples $\bar{f} = (f_1, \ldots, f_m)$ of functions, where $f_j : A_j^n \to A_j$, $j = 1, \ldots, m$. Such a m-tuple $\bar{f}$ of functions will be called an (n-ary) *aggregator* for X if the following two conditions hold for all $x \in X^n$:

1. $(f_1(x_1), \ldots, f_m(x_m)) \in X$ and
2. $\bar{f}$ is *conservative*, that is, $f_j(x_j) \in \{x_j^1, \ldots, x_j^n\}$, for all $j \in \{1, \ldots, m\}$.

An aggregator $\bar{f} = (f_1, \ldots, f_m)$ is called *dictatorial on* X if there is a number $d \in \{1, \ldots, n\}$ such that $(f_1, \ldots, f_m) = (\mathrm{pr}_d^n, \ldots, \mathrm{pr}_d^n)$, where pr_d^n is the n-ary projection on the d-th coordinate; otherwise, $\bar{f}$ is called *non-dictatorial on* X. We say that X *has a non-dictatorial aggregator* if, for some $n \geq 2$, there is a n-ary non-dictatorial aggregator on X.

A set X of feasible evaluations is a *possibility domain* if it has a non-dictatorial aggregator. Otherwise, it is an *impossibility domain.* A possibility domain is, by definition, one where aggregation is possible for societies of some cardinality, namely, the arity of a non-dictatorial aggregator.

The notion of an aggregator is akin to, but different from, the notion of a polymorphism – a fundamental notion in universal algebra (see Szendrei [17]). Intuitively, a polymorphism is a single-sorted aggregator.

Let A be a finite non-empty set. A *constraint language* over A is a finite set Γ of relations of finite arities.

Let R be an m-ary relation on A. A function $f : A^n \to A$ is a *polymorphism* of R if the following condition holds:

$$\text{if } x^1, \ldots, x^n \in R, \text{ then } (f(x_1), \ldots, f(x_m)) \in R,$$

where $x^i = (x_1^i, \ldots, x_m^i) \in R$, $i = 1, \ldots, n$ and $x_j = (x_j^1, \ldots, x_j^n)$, $j = 1, \ldots, m$. In this case, we also say that R *is closed under* f or that f *preserves* R. Finally, we say that f is a *polymorphism of a constraint language* Γ if f preserves every relation $R \in \Gamma$.

A function $f : A^n \to A$ is *conservative* if, for all $a_1, \ldots, a_n \in A$, we have that $f(a_1, \ldots, a_n) \in \{a_1, \ldots, a_n\}$. Clearly, if $f : A^n \to A$ is a conservative polymorphism of an m-ary relation R on A, then the m-tuple $\bar{f} = (f, \ldots, f)$ is an n-ary aggregator for R.

We say that a ternary operation $f : A^3 \to A$ on an arbitrary set A is a *majority* operation if for all x and y in A,

$$f(x,x,y) = f(x,y,x) = f(y,x,x) = x;$$

we say that f is a *minority* operation if for all x and y in A,

$$f(x,x,y) = f(x,y,x) = f(y,x,x) = y.$$

We also say that a set X of feasible evaluations *admits a majority (respectively, minority) aggregator* if it admits a ternary aggregator every component of which is a majority (respectively, minority) operation. Clearly, X admits a majority aggregator if and only if there is a ternary aggregator $\bar{f} = (f_1, \ldots, f_m)$ for X such that, for all $j = 1, \ldots, m$ and for all two-element subsets $B_j \subseteq X_j$, we have that $f_j \restriction B_j = \text{maj}$, where

$$\text{maj}(x,y,z) = \begin{cases} x & \text{if } x = y \text{ or } x = z, \\ y & \text{if } y = z. \end{cases}$$

Similarly, X admits a minority aggregator if and only if there is a ternary aggregator $\bar{f} = (f_1, \ldots, f_m)$ for X such that, for all $j = 1, \ldots, m$ and for all two-element subsets $B_j \subseteq X_j$, we have that $f_j \restriction B_j = \oplus$, where

$$\oplus(x,y,z) = \begin{cases} z & \text{if } x = y, \\ x & \text{if } y = z, \\ y & \text{if } x = z. \end{cases}$$

In the Boolean framework, a set $X \subseteq \{0,1\}^m$ admits a majority aggregator if and only if the majority operation maj on $\{0,1\}^3$ is a polymorphism of X. Moreover, it is known that this happens precisely when X is a bijunctive logical relation, i.e., X is the set of satisfying assignments of a 2CNF-formula. A set $X \subseteq \{0,1\}^m$ admits a minority aggregator if and only if the minority operation $\oplus$ on $\{0,1\}^3$ is a polymorphism of X. Moreover, it is known that this happens precisely when X is an affine logical relation, i.e., X is the set of solutions of a system of linear equations over the two-element field (see Schaefer [14]).

Example 1. Consider the sets X_1 and X_2 below.

(i) The set $X_1 = \{0,1\}^3 \setminus \{(1,0,1), (0,0,1), (0,0,0)\}$ is bijunctive, since it is the set of satisfying assignments of the 2CNF-formula $(x \vee y) \wedge (y \vee \neg z)$.
(ii) The set $X_2 = \{(0,0,1), (0,1,0), (1,0,0), (1,1,1)\}$ is affine, since it is the set of solutions of the equation $x + y + z = 1$ over the two-element field.
(iii) Both sets X_1 and X_2 are possibility domains, since X_1 admits a majority aggregator and X_2 admits a minority aggregator.

The next two theorems characterize possibility domains in the Boolean and the non-Boolean framework. They are the stepping stones towards showing that the following decision problem is solvable in polynomial time: given a set X of feasible evaluations, is X a possibility domain?

Theorem A (Dokow and Holzman [7]). *Let $X \subseteq \{0,1\}^m$ be a set of feasible evaluations. The following two statements are equivalent.*

1. *X is a possibility domain.*
2. *X is affine or X admits a binary non-dictatorial aggregator.*

Theorem B (Kirousis et al. [10]). *Let X be a set of feasible evaluations. The following two statements are equivalent.*

1. *X is a possibility domain.*
2. *X admits a binary non-dictatorial aggregator, or a majority aggregator, or a minority aggregator.*

We illustrate the two preceding theorems with several examples.

Example 2. Let $X_3 = \{(1,0,0),(0,1,0),(0,0,1)\} \subseteq \{0,1\}^3$ be the set of Boolean triples that contain exactly one 1. By Theorem A, the set X_3 is an impossibility domain, since it is not affine ($\oplus((1,0,0),(0,1,0),(0,0,1)) = (0,0,0) \notin X_3$) and it does not have a binary non-dictatorial aggregator. For the latter, one has to check each of the 62 possible 3-tuples of conservative binary functions over $\{0,1\}$, which is a fairly tedious but straightforward task.

Example 3. The following two sets X_4 and X_5 are possibility domains.

(i) Let $X_4 = \{(0,1,2),(1,2,0),(2,0,1),(0,0,0)\}$. This set has been studied in [7, Example 4]. Let $\bar{f} = (f_1, f_2, f_3)$ be such that, for each $j = 1,2,3$:

$$f_j(x,y,z) = \begin{cases} \mathrm{maj}(x,y,z) & \text{if } |\{x,y,z\}| \le 2, \\ 0 & \text{else.} \end{cases}$$

Clearly, $\bar{f}$ is a majority operation. To see that $\bar{f}$ is an aggregator for X_4, we need to check that $\bar{f}(a,b,c) \in X_4$, only when a,b,c are *pairwise distinct* vectors of X_4. In this case, $\bar{f}(a,b,c) = (0,0,0) \in X_4$, since the input of each f_j contains either two zeros or three pairwise distinct elements.

(ii) Let $X_5 = X_3 \times X_3$, where X_3 is as in Example 2. It is straightforward to check that $(pr_1^2, pr_1^2, pr_1^2, pr_2^2, pr_2^2, pr_2^2)$ is a non-dictatorial aggregator for X_5. In fact, a stronger fact holds: if Y and Z are arbitrary sets, then their Cartesian product $Y \times Z$ is a possibility domain, since it admits non-dictatorial aggregators of any arity $n \ge 2$, defined as the d-th projection pr_d^n on coordinates from Y and as the d'-th projection $\mathrm{pr}_{d'}^n$ on coordinates from Z, where $1 \le d, d' \le n$ and $d \ne d'$.

2.2 Uniform Possibility Domains

We consider a subclass of possibility domains, introduced in [10], called uniform possibility domains.

Let $\bar{f} = (f_1, \ldots, f_m)$ be an n-ary aggregator for X. We say that $\bar{f}$ is a *uniform non-dictatorial* aggregator for X (of arity n) if, for all $j \in \{1, \ldots, m\}$ and for every two-element subset $B_j \subseteq X_j$, it holds that

$$f_j \restriction_{B_j} \neq pr_d^n,$$

for all $d \in \{1, \ldots, n\}$. We say that a set X is a *uniform possibility domain* if it has a uniform non-dictatorial aggregator.

The aforementioned sets X_1, X_2 and X_4 are uniform possibility domains, as X_1 and X_4 admit a majority aggregator, while X_2 admits a minority aggregator. Clearly, if X is a uniform possibility domain, then X is also a possibility domain. The converse, however, is not true. Indeed, suppose that X is a Cartesian product $X = Y \times Z$, where $Y \subseteq \prod_{j=1}^{l} A_j$ and $Z \subseteq \prod_{j=l+1}^{m} A_j$, with $1 \leq l < m$. If Y or Z is an impossibility domain, then X is not a uniform possibility domain, although it is a possibility domain, since, as seen earlier, every Cartesian product of two sets is a possibility domain. It is also clear that if Y and Z are uniform possibility domains, then so is their Cartesian product $Y \times Z$. In particular, the Cartesian product $X_1 \times X_2$ is a uniform possibility domain.

The next result characterizes uniform possibility domains. It is the stepping stone towards showing that the following decision problem is solvable in polynomial time: given a set X of feasible evaluations, is X a uniform possibility domain? To state this result, we first need to give a definition.

We say that $f : A^n \to A$ is a *weak near-unanimity operation* [11] if, for all $x, y \in A$, we have that

$$f(y, x, x, \ldots, x) = f(x, y, x, \ldots, x) = \ldots = f(x, x, x, \ldots, y).$$

In particular, a ternary weak near-unanimity operation is a function $f : A^3 \to A$ such that for all $x, y \in A$, we have that

$$f(y, x, x) = f(x, y, x) = f(x, x, y).$$

Thus, the notion of a ternary weak near-unanimity operation is a common generalization of the notions of a majority operation and a minority operation.

As with the majority/minority aggregators, we say that X *admits a ternary weak near-unanimity aggregator* $\bar{f} = (f_1, \ldots, f_m)$, if it admits a ternary aggregator every component of which is a weak near-unanimity operation, i.e. for all $j = 1, \ldots, m$ and for all $x, y \in X_j$, we have that $f_j(y, x, x) = f_j(x, y, x) = f_j(x, x, y)$.

Theorem C (Kirousis et al. [10]**).** *Let X be a set of feasible evaluations. The following two statements are equivalent.*

1. *X is a uniform possibility domain.*
2. *X admits a ternary weak near-unanimity aggregator.*

3 Results

In this section, we show that there are polynomial-time algorithms for telling, given a set X of feasible evaluations, whether or not X is a possibility domain and whether or not X is a uniform possibility domain.

3.1 Tractability of Possibility Domains

Theorems A and B provide necessary and sufficient conditions for a set X to be a possibility domain in the Boolean framework and in the non-Boolean framework, respectively. Admitting a binary non-dictatorial aggregator is a condition that appears in both of these characterizations. Our first result asserts that this condition can be checked in polynomial time.

Theorem 1. *There is a polynomial-time algorithm for solving the following problem: given a set X of feasible evaluations, determine whether or not X admits a binary non-dictatorial aggregator and, if it does, produce one.*

Proof. We will show that the existence of a binary non-dictatorial aggregator on X is tightly related to connectivity properties of a certain directed graph H_X defined next.

If $X \subseteq \prod_{j=1}^{m} A_j$ is a set of feasible evaluations, then H_X is the following directed graph:

- The vertices of H_X are the pairs of *distinct* elements $u, u' \in X_j$, for $j \in \{1, \ldots, m\}$. Each such vertex will usually be denoted by uu'_j. When the coordinate j is understood from the context, we will often be dropping the subscript j, thus denoting such a vertex by uu'.
  Also, if $u \in X_j$, for some $j \in \{1, \ldots, m\}$, we will often use the notation u_j to indicate that u is an element of X_j.
- Two vertices uu'_k and vv'_l, where $k \neq l$, are connected by a directed edge from uu'_k to vv'_l, denoted by $uu'_k \to vv'_l$, if there are a total evaluation $z \in X$ that extends the partial evaluation (u_k, v_l) and a total evaluation $z' \in X$ that extends the partial evaluation (u'_k, v'_l), such that there is no total evaluation $y \in X$ that extends (u_k, v'_l), and has the property that $y_i = z_i$ or $y_i = z'_i$, for every $i \in \{1, \ldots, m\}$.

For vertices uu'_k, vv'_l, corresponding to issues k, l (that need not be distinct), we write $uu'_k \to\to vv'_l$ to denote the existence of a directed path from uu'_k to vv'_l. In the next example, we describe explicitly the graph H_X for several different sets X of feasible voting patters. Recall that a *directed* graph G is *strongly connected* if for every pair of vertices (u, v) of G, there is a (directed) path from u to v.

Example 4. Recall the sets $X_2 = \{(0,0,1), (0,1,0), (1,0,0), (1,1,1)\}$ and $X_3 = \{(0,0,1), (0,1,0), (1,0,0)\}$ of Examples 1 and 2.

Both H_{X_2} and H_{X_3} have six vertices, namely 01_j and 10_j, for $j = 1, 2, 3$. In the figures below, we use undirected edges between two vertices uu'_k and vv'_l to denote the existence of both $uu'_k \to vv'_l$ and $vv'_l \to uu'_k$.

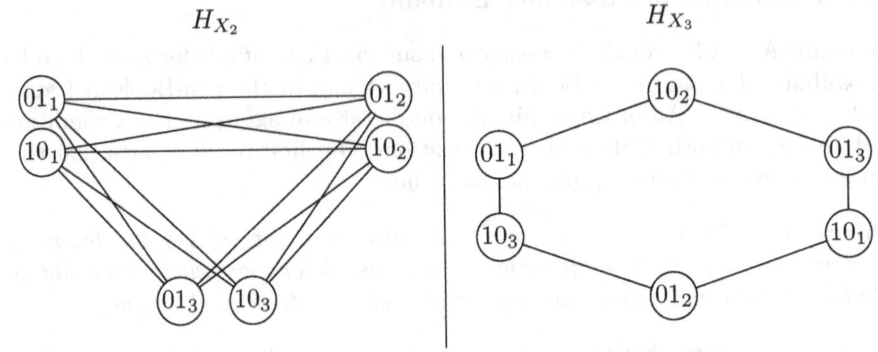

Consider 01_1, 01_2 of H_{X_2}. Since the partial vectors $(0,0)$ and $(1,1)$ extend to $(0,0,1)$ and $(1,1,1)$, respectively, we need to check if there is a vector in X_2 extending $(0,1)$, but whose third coordinate is 1. Since $(0,1,1) \notin X_2$, we have that H_{X_2} contains both edges $01_1 \to 01_2$ and $01_2 \to 01_1$. Now, since the partial vectors $(0,1)$ and $(1,0)$ extend to $(0,1,0)$ and $(1,0,0)$, respectively, and since neither $(0,0,0)$ nor $(1,1,0)$ are in X_2, we have that $01_1 \leftrightarrow 10_2$. By the above and because of the symmetric structure of X_2, it is easy to see that every two vertices uu'_i and vv'_j of H_{X_2} are connected if and only if $i \neq j$.

For X_3, observe that, since *no* partial vector containing two "1"'s, in any two positions, extends to an element of X_3, there are no edges between the vertices 01_i, 01_j and 10_i, 10_j, for any $i, j \in \{1,2,3\}$, $i \neq j$. In the same way as with H_{X_2}, we get that H_{X_3} is a cycle.

There are two observations to be made, concerning H_{X_2} and H_{X_3}. First, they are both strongly connected graphs. Also, neither X_2 nor X_3 admit binary non-dictatorial aggregators (X_2 admits only a minority aggregator and X_3 is an impossibility domain, as shown in Example 2).

Finally, consider $X_6 := \{(0,1),(1,0)\}$. The graph H_{X_6} has four vertices, 01_1, 10_1, 01_2 and 10_2, and it is easy to see that H_{X_6} has only the following edges:

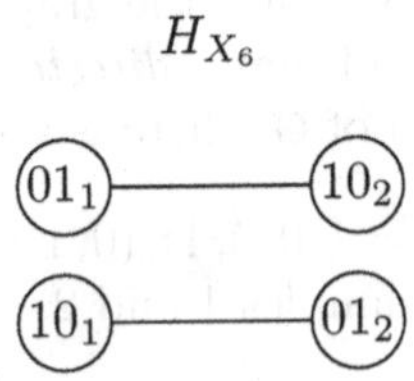

Observe that X_6 is not strongly connected (it is not even connected) and that in contrast to the sets X_2 and X_3, the set X_6 admits two binary non-dictatorial aggregators, namely, $(\wedge, \vee)$ and $(\vee, \wedge)$. In Lemma 2, we establish a tight connection between strong connectedness and the existence of binary non-dictatorial aggregators.

We now state and prove two lemmas about the graph H_X.

Lemma 1. *Assume that* $\bar{f} = (f_1, \ldots, f_m)$ *is a binary aggregator on* X.

1. *If* $uu'_k \to vv'_l$ *and* $f_k(u, u') = u$, *then* $f_l(v, v') = v$.
2. *If* $uu'_k \to\to vv'_l$ *and* $f_k(u, u') = u$, *then* $f_l(v, v') = v$.

Proof. The first part of the lemma follows from the definitions and the fact that $\bar{f}$ is conservative. Indeed, if $uu'_k \to vv'_l$, then there are a total evaluation $z = (z_1, \ldots, z_m) \in X$ that extends (u_k, v_l) (i.e., $z_k = u$ and $z_l = v$) and a total evaluation $z' = (z'_1, \ldots, z'_m) \in X$ that extends (u'_k, v'_l) (i.e., $z'_k = u'$ and $z'_l = v'$), such that there is no total evaluation in X that extends (u_k, v'_l) and agrees with z or with z' on every coordinate. Consider the total evaluation $(f_1(z_1, z'_1), \ldots, f_m(z_m, z'_m))$, which is in X because $\bar{f}$ is an aggregator on X. Since each f_j is conservative, we must have that $f_j(z_j, z'_j) \in \{z_j, z'_j\}$, for every j, hence $f_l(z_l, z'_l) = f_l(v, v') \in \{v, v'\}$. Consequently, if $f_k(u, u') = u$, then we must have $f_l(v, v') = v$, else $(f_1(z_1, z'_1), \ldots, f_m(z_m, z'_m))$ extends (u_k, v'_l) and agrees with z or with z' on every coordinate.

The second part of the lemma follows from the first part by induction.

Lemma 2. X *admits a binary non-dictatorial aggregator if and only if the directed graph* H_X *is not strongly connected.*

Before delving into the proof, consider the graphs of Example 4. Using the fact that the graphs H_{X_2} and H_{X_3} are strongly connected and also using the second item of Lemma 1, it is easy to see that X_2 and X_3 admit no binary non-dictatorial aggregator; indeed, let $\bar{f} = (f_1, f_2, f_3)$ be a binary aggregator of either of these two sets and suppose that $f_1(0, 1) = 0$. Since in both graphs H_{X_2} and H_{X_3}, there are paths from 01_1 to *every* other vertex, it follows that $f_j = pr_1^2$, $j = 1, 2, 3$. If $f_1(0, 1) = 1$, we get that $f_j = pr_2^2$, $j = 1, 2, 3$, in the same way.

In contrast, consider H_{X_6} and let $\bar{g} = (g_1, g_2)$ be a pair of binary functions with $g_1(0, 1) = 0$. For $\bar{g}$ to be an aggregator, Lemma 1 forces us to set $g_2(1, 0) = 1$. But now, by setting $g_1(1, 0) = 0$, and thus $g_2(0, 1) = 1$, we get that $(g_1, g_2) = (\wedge, \vee)$ is a non-dictatorial binary aggregator for X_6.

Proof. We first show that if X admits a binary non-dictatorial aggregator, then H_X is not strongly connected. In the contrapositive form, we show that if H_X is strongly connected, then X admits no binary non-dictatorial aggregator. This is an easy consequence of the preceding Lemma 1. Indeed, assume that H_X is strongly connected and let $\bar{f} = (f_1, \ldots, f_m)$ be a binary aggregator on X. Take two distinct elements x and x' of X_1. Since $\bar{f}$ is conservative, we have that $f_1(x, x') \in \{x, x'\}$. Assume first that $f_1(x, x') = x$. We claim that $f_j = \mathrm{pr}_1^2$, for every $j \in \{1, \ldots, m\}$. To see this, let y and y' be two distinct elements of X_j, for some $j \in \{1, \ldots, m\}$. Since H_X is strongly connected, we have that $xx'_1 \to\to yy'_j$. Since also $f_1(x, x') = x$, Lemma 1 implies that $f_j(y, y') = y = \mathrm{pr}_1^2(y, y')$ and so $f_j = \mathrm{pr}_1^2$. Next, assume that $f_1(x, x') = x'$. We claim that $f_j = \mathrm{pr}_2^2$, for every $j \in \{1, \ldots, m\}$. To see this, let y and y' be two distinct elements of X_j, for some $j \in \{1, \ldots, m\}$. Since H_X is strongly connected, we have that $yy'_j \to\to xx'_1$, hence, if $f_j(y, y') = y$, then, Lemma 1, implies that $f_1(x, x') = x$, which is a contradiction because $x \neq x'$. Thus, $f_j(y, y') = y'$ and so $f_j = \mathrm{pr}_2^2$.

For the converse, assume that H_X is not strongly connected and let uu'_k, vv'_l be two vertices of H_X such that there is no path from uu'_k to vv'_l in H_X, i.e., it is not true that $uu'_k \to\!\!\to vv'_l$. Let V_1, V_2 be a partition of the vertex set such that $uu'_k \in V_1, vv'_l \in V_2$, and there is no edge from a vertex in V_1 to a vertex in V_2. We will now define a binary aggregator $\bar{f} = (f_1, \ldots, f_m)$ and prove that it is non-dictatorial.

Given $z, z' \in X$, we set $f_j(z_j, z'_j) = z_j$ if $zz'_j \in V_1$, and we set $f_j(z_j, z'_j) = z'_j$ if $zz'_j \in V_2$, for $j \in \{1, \ldots, m\}$. Since $uu'_k \in V_1$, we have that $f_k \neq \mathrm{pr}_2^2$; similarly, since $vv'_l \in V_2$, we have that $f_l \neq \mathrm{pr}_1^2$. Consequently, $\bar{f}$ is not a dictatorial function on X. Thus, what remains to be proved is that if $z, z,' \in X$, then $\bar{f}(z, z') \in X$. For this, we will show that if $\bar{f}(z, z') \notin X$, then there is an edge from an element of V_1 to an element of V_2, which is a contradiction.

Assume that $q = \bar{f}(z, z') \notin X$. Let K be a minimal subset of $\{1, \ldots, m\}$ such that $q \restriction K$ cannot be extended to a total evaluation w in X that agrees with z or with z' on $\{1, \ldots, m\} \setminus K$ (i.e., if $j \in \{1, \ldots, m\} \setminus K$, then $w_j = z_j$ or $w_j = z'_j$). Since z' is in X, it does not extend $q \restriction K$, hence there is a number $s \in K$ such that $q_s = f_s(z_s, z'_s) = z_s \neq z'_s$. It follows that the vertex zz'_s is in V_1. Similarly, since z is in X, it does not extend $q \restriction K$, hence there is a number $t \in K$ such that $q_t = f_t(z_t, z'_t) = z'_t \neq z_t$. It follows that the vertex zz'_t is in V_2. Consequently, there is no edge from zz'_s to zz'_t in H_X. We will arrive at a contradiction by showing that $zz'_s \to zz'_t$, i.e., there is an edge zz'_s to zz'_t in H_X. Consider the set $K \setminus \{t\}$. By the minimality of K, there is a total evaluation w in X that extends $q \restriction K \setminus \{t\}$ and agrees with z or with z' outside $K \setminus \{t\}$. In particular, we have that $w_s = q_s = z_s$ and $w_t = z_t$. Similarly, by considering the set $K \setminus \{s\}$, we find that there is a total evaluation w' in X that extends $q \restriction K \setminus \{s\}$ and agrees with z or with z' outside $K \setminus \{s\}$. In particular, we have that $w'_s = z'_s$ and $w_t = q_t = z'_t$. Note that w and w' agree on $K \setminus \{s, t\}$. Since $q \restriction K$ does not extend to a total evaluation that agrees with z or with z' outside K, we conclude that there is no total evaluation y in X that extends (z_s, z'_t) and agrees with w or with w' on every coordinate. Consequently, $zz'_s \to zz'_t$, thus we have arrived at a contradiction.

We are now ready to complete the proof of Theorem 1. Given a set X of feasible evaluations, the graph H_X can be constructed in time bounded by a polynomial in the size $|X|$ of X (in fact, in time $O(|X|^5)$. There are well-known polynomial-time algorithms for testing if a graph is strongly connected and, in case it is not, producing the strongly connected components of the graph (e.g., Kosaraju's algorithm presented in Sharir [15] and Tarjan's algorithm [18]). Consequently, by Lemma 2, there is a polynomial-time algorithm for determining whether or not a given set X admits a binary non-dictatorial aggregator. Moreover, if X admits such an aggregator, then one can be constructed in polynomial-time from the strongly connected components of H_X via the construction in the proof of Lemma 2.

As mentioned in the introduction, the existence of a binary non-dictatorial aggregator on X is closely related to the total blockedness of X.

Theorem D (Kirousis et al. [10]). *Let X be a set of feasible evaluations. The following two statements are equivalent.*

1. *X is totally blocked.*
2. *X admits no binary non-dictatorial aggregator.*

The next corollary follows from Theorems 1 and D.

Corollary 1. *There is a polynomial-time algorithm for the following decision problem: given a set $X \subseteq \{0,1\}^m$ of feasible evaluations, is X totally blocked?*

Furthermore, by combining Theorem 1 with Theorem A, we obtain the following result.

Theorem 2. *There is a polynomial-time algorithm for the following decision problem: given a set $X \subseteq \{0,1\}^m$ of feasible evaluations in the Boolean framework, determine whether or not it is a possibility domain.*

We now turn to the problem of detecting possibility domains in the non-Boolean framework.

Theorem 3. *There is a polynomial-time algorithm for solving the following problem: given a set X of feasible evaluations, determine whether or not X is a possibility domain and, if it is, produce a binary non-dictatorial aggregator for X or a ternary non-dictatorial aggregator for X.*

Proof. It is straightforward to check that, by Theorems B and 1, it suffices to show that there is a polynomial-time algorithm that, given X, detects whether or not X admits a majority aggregator or a minority aggregator, and, if it does, produces such an aggregator.

Let X be a set of feasible evaluations, where $I = \{1, \dots, m\}$ is the set of issues and A_j, $j = 1, \dots, m$, are the sets of the position values. We define the *disjoint union* A of the sets of position values as

$$A = \bigsqcup_{j=1}^{m} A_j = \bigcup_{j=1}^{m}\{(x, j) \mid x \in A_j\}.$$

We also set

$$\tilde{X} = \{((x_1, 1), \dots, (x_m, m)) \mid (x_1, \dots, x_m) \in X\} \subseteq A^m.$$

We will show that we can go back-and-forth between conservative majority or minority polymorphisms for $\tilde{X}$ and majority or minority aggregators for X.

Let $f : A^n \to A$ be a conservative polymorphism for $\tilde{X}$. We define the m-tuple $\bar{f} = (f_1, \dots, f_m)$ of n-ary functions $f_1, \dots, f_m$ as follows: if $x_j^1, \dots, x_j^n \in A_j$, for $j \in \{1, \dots, m\}$, then we set $f_j(x_j^1, \dots, x_j^n) = y_j$, where y_j is such that $f((x_j^1, j), \dots, (x_j^n, j)) = (y_j, j)$. Such a y_j exists and is one of the x_j^i's because

f is conservative, and hence $f((x_j^1, j), \ldots, (x_j^n, j)) \in \{(x_j^1, j), \ldots, (x_j^n, j)\}$. It is easy to see that $\bar{f}$ is an aggregator for X. Moreover, if f is a majority or a minority operation on $\tilde{X}$, then $\bar{f}$ is a majority or a minority aggregator on X.

Next, let $\bar{f} = (f_1, \ldots, f_m)$ be a majority or a minority aggregator for X. We define a ternary function $f : A^3 \to A$ as follows. Let $(x, j), (y, k), (z, l)$ be three elements of A.

- If $j = k = l$, then we set $f((x, j), (y, k), (z, l)) = (f_j(x, y, z), j)$.
- If j, k, l are *not* all equal, then if at least two of $(x, j), (y, k), (z, l)$ are equal to each other, we set

  $$f((x, j), (y, k), (z, l)) = \text{maj}((x, j), (y, k), (z, l)),$$

  if $\bar{f}$ is a majority aggregator on X, and we set

  $$f((x, j), (y, k), (z, l)) = \oplus((x, j), (y, k), (z, l)),$$

  if $\bar{f}$ is a minority aggregator on X;
- otherwise, we set $f((x, j), (y, k), (z, l)) = (x, j)$.

It is easy to see that if $\bar{f}$ is a majority or a minority aggregator for X, then f is a conservative majority or a conservative minority polymorphism on $\tilde{X}$. It follows that X admits a majority or a minority aggregator if and only if $\tilde{X}$ is closed under a conservative majority or minority polymorphism. Bessiere et al. [2] and Carbonnel [6] design polynomial-time algorithms that detect if a given constraint language Γ has a conservative majority or a conservative minority polymorphism, respectively, and, when it has, compute such a polymorphism. Here, we apply these results to $\Gamma = \{\tilde{X}\}$.

3.2 Tractability of Uniform Possibility Domains

Recall that a constraint language is a finite set Γ of relations of finite arities over a finite non-empty set A. The *conservative constraint satisfaction problem for Γ*, denoted by c-CSP(Γ) is the constraint satisfaction problem for the constraint language $\overline{\Gamma}$ that consists of the relations in Γ and, in addition, all unary relations on A. Bulatov [3,4] established a dichotomy theorem for the computational complexity of c-CSP(Γ): if for every two-element subset B of A, there is a conservative polymorphism f of Γ such that f is binary and $f \restriction B \in \{\wedge, \vee\}$ or f is ternary and $f \in \{\text{maj}, \oplus\}$, then c-CSP($\Gamma$) is solvable in polynomial time; otherwise, c-CSP(Γ) is NP-complete. Carbonnel [5] showed that the boundary of the dichotomy for c-CSP(Γ) can be checked in polynomial time.

Theorem E (Carbonnel [5]). *There is a polynomial-time algorithm for solving the following problem: given a constraint language Γ on a set A, determine whether or not for every two-element subset $B \subseteq A$, there is a conservative polymorphism f of Γ such that f is binary and $f \restriction B \in \{\wedge, \vee\}$ or f is ternary and $f \restriction B \in \{\text{maj}, \oplus\}$. Moreover, if such a polymorphism exists, then the algorithm produces one in polynomial time.*

The final result of this section is about the complexity of detecting uniform possibility domains.

Theorem 4. *There is a polynomial-time algorithm for solving the following problem: given a set X of feasible evaluations, determine whether or not X is a uniform possibility domain and, if it is, produce a ternary weak near-unanimity aggregator for X.*

Proof. In what follows, given a two-element set B, we will arbitrarily identify its elements with 0 and 1. Consider the functions $\wedge^3$ and $\vee^3$ on $\{0,1\}^3$, where $\wedge^3(x,y,z) := (\wedge(\wedge(x,y),z))$ and $\vee^3(x,y,z) := (\vee(\vee(x,y),z))$. It is easy to see that the only ternary, conservative, weak near-unanimity functions on $\{0,1\}$ are $\wedge^3$, $\vee^3$, maj, and $\oplus$. We will also make use of the following lemma, which also gives an alternative formulation of the boundary of the dichotomy for conservative constraint satisfaction.

Lemma 3. *Let Γ be a constraint language on set A. The following two statements are equivalent.*

1. *For every two-element subset $B \subseteq A$, there exists a conservative polymorphism f of Γ (which, in general, depends on B), such that f is binary and $f{\restriction}_B \in \{\wedge, \vee\}$ or f is ternary and $f{\restriction}_B \in \{\text{maj}, \oplus\}$.*
2. *Γ has a ternary, conservative, weak near-unanimity polymorphism.*

Proof. (*Sketch*) ($1 \Rightarrow 2$) Given a two-element subset $B \subseteq A$ and a binary conservative polymorphism f of Γ such that $f{\restriction}_B \in \{\wedge, \vee\}$, define f' to be the ternary operation such that $f'(x,y,z) = f(f(x,y),z)$, for all $x,y,z \in A$. It is easy to see that f' is a conservative polymorphism of Γ as well and also that $f'{\restriction}_B \in \{\wedge^{(3)}, \vee^{(3)}\}$.

The hypothesis and the preceding argument imply that, for each two-element subset $B \subseteq A$, there exists a ternary conservative polymorphism f of Γ (which, in general, depends on B) such that $f{\restriction}_B \in \{\wedge^{(3)}, \vee^{(3)}, \text{maj}, \oplus\}$. For each two-element subset $B \subseteq A$, select such a polymorphism and let $f^1, \dots, f^N$, $N \geq 1$, be an enumeration of all these polymorphisms. Clearly, the restriction of each f^i to its respective two element subset is a weak near-unanimity operation.

Consider the '$\diamond$' operator that takes as input two ternary operations $f, g : A^3 \to A$ and returns as output a ternary operation $f \diamond g$ defined by

$$(f \diamond g)(x,y,z) := f(g(x,y,z), g(y,z,x), g(z,x,y)).$$

If f, g are conservative polymorphisms of Γ, then so is $(f \diamond g)$. Also, if B is a two-element subset of A such that $f{\restriction}_B$ or $g{\restriction}_B$ is a weak near-unanimity operation, then so is $(f \diamond g){\restriction}_B$. Consider now the iterated diamond operation h with

$$h := f^1 \diamond (f^2 \diamond (\dots \diamond (f^{N-1} \diamond f^N) \dots)).$$

By the preceding discussion, h is a conservative polymorphism such that $h{\restriction}_B$ is a weak near-unanimous operation for *every* two-element subset B of A, hence h itself is a weak near-unanimity, conservative, ternary operation of Γ.

($2 \Rightarrow 1$) Let h be a ternary, conservative, weak near-unanimity polymorphism of Γ. Thus, for every two-element subset $B \subseteq A$, we have that $h{\restriction_B} \in \{\wedge^{(3)}, \vee^{(3)}, \text{maj}, \oplus\}$.

If there is a two-element subset $B \subseteq A$ such that $h{\restriction_B} \in \{\wedge^{(3)}, \vee^{(3)}\}$, then consider the binary function g defined by

$$g(x, y) := h(x, x, y) = h(pr_1^2(x, y), pr_1^2(x, y), pr_2^2(x, y)).$$

Obviously, g is a binary conservative polymorphism of Γ; moreover, for every two-element subset $B \subseteq A$, if $h{\restriction_B} \in \{\wedge^{(3)}, \vee^{(3)}\}$, then $g{\restriction_B} \in \{\wedge, \vee\}$.

By Theorem C, a set X of feasible evaluations is a uniform possibility domain if and only if there is a ternary aggregator $\bar{f} = (f_1, \ldots, f_m)$ such that each f_j is a weak near-unanimity operation, i.e., for all $j \in \{1, \ldots, m\}$ and for all $x, y \in X_j$, we have that $f_j(x, y, y) = f_j(y, x, y) = f_j(y, y, x)$. As in the proof of Theorem 3, we can go back-and-forth between X and the set $\tilde{X}$, and verify that X is a uniform possibility domain if and only if $\tilde{X}$ has a ternary, conservative, weak near-unanimity polymorphism. Theorem E and Lemma 3 then imply that the existence of such a polymorphism can be tested in polynomial time, and that such a polymorphism can be produced in polynomial time, if one exists.

4 Concluding Remarks

In this paper, we established the first results concerning the tractability of non-dictatorial aggregation. Specifically, we gave polynomial-time algorithms that take as input a set X of feasible evaluations and determine whether or not X is a possibility domain and a uniform possibility domain, respectively. In these algorithms, the set X of feasible evaluations is given to us explicitly, i.e., X is given by listing all its elements. It is natural to ask how the complexity of these problems may change if X is given implicitly via a succinct representation. Notice that Theorems 1, 3 and 4, directly imply that in this case, the problems of whether X is a possibility domain and of whether it is a uniform possibility domain respectively, are EXPTIME. Furthermore, by Theorem 2, we obtain that in the Boolean framework, the problem of whether X is a possibility domain is in PSPACE. It is an open problem weather there are respective lower bounds for the complexity of the above problems in the succinct case. This scenario merits further investigation, because it occurs frequently in other areas of social choice, including the area of judgment aggregation, where X is identified with the set of satisfying assignments of a Boolean formula (for surveys on judgment aggregation, see [9,12]).

The work reported here assumes that the aggregators are conservative, an assumption that has been used heavily throughout the paper. There is a related, but weaker, notion of an *idempotent* (or *Paretian*) aggregator $\bar{f} = (f_1, \ldots, f_m)$ where each f_j is assumed to be an idempotent function, i.e., for all $x \in X_j$, we have that $f(x, \ldots, x) = x$. Clearly, every conservative aggregator is idempotent. In the Boolean framework, idempotent aggregators are conservative, but, in the non-Boolean framework, this need not hold. It remains an open problem to investigate the computational complexity of the existence of non-dictatorial idempotent aggregators in the non-Boolean framework.

Acknowledgements. The research of Lefteris Kirousis was partially supported by the Special Account for Research Grants of the National and Kapodistrian University of Athens. The work of Phokion G. Kolaitis is partially supported by NSF Grant IIS-1814152.

References

1. Arrow, K.J.: Social Choice and Individual Values. Wiley, New York (1951)
2. Bessiere, C., Carbonnel, C., Hebrard, E., Katsirelos, G., Walsh, T.: Detecting and exploiting subproblem tractability. In: IJCAI, pp. 468–474 (2013)
3. Bulatov, A.A.: A dichotomy theorem for constraint satisfaction problems on a 3-element set. J. ACM (JACM) **53**(1), 66–120 (2006)
4. Bulatov, A.A.: Complexity of conservative constraint satisfaction problems. ACM Trans. Comput. Logic (TOCL) **12**(4), 24 (2011)
5. Carbonnel, C.: The dichotomy for conservative constraint satisfaction is polynomially decidable. In: Rueher, M. (ed.) CP 2016. LNCS, vol. 9892, pp. 130–146. Springer, Cham (2016). https://doi.org/10.1007/978-3-319-44953-1_9
6. Carbonnel, C.: The meta-problem for conservative Mal'tsev constraints. In: Thirtieth AAAI Conference on Artificial Intelligence (AAAI-2016) (2016)
7. Dokow, E., Holzman, R.: Aggregation of binary evaluations. J. Econ. Theory **145**(2), 495–511 (2010)
8. Dokow, E., Holzman, R.: Aggregation of non-binary evaluations. Adv. Appl. Math. **45**(4), 487–504 (2010)
9. Endriss, U.: Judgment aggregation. In: Brandt, F., Conitzer, V., Endriss, U., Lang, J., Procaccia, A.D. (eds.), Handbook of Computational Social Choice, pp. 399–426. Cambridge University Press (2016)
10. Kirousis, L., Kolaitis, P.G., Livieratos, J.: Aggregation of votes with multiple positions on each issue. In: Höfner, P., Pous, D., Struth, G. (eds.) RAMICS 2017. LNCS, vol. 10226, pp. 209–225. Springer, Cham (2017). https://doi.org/10.1007/978-3-319-57418-9_13
11. Larose, B.: Algebra and the complexity of digraph CSPs: a survey. In: Dagstuhl Follow-Ups, vol. 7. Schloss Dagstuhl-Leibniz-Zentrum fuer Informatik (2017)
12. List, C., Puppe, C.: Judgment Aggregation: A Survey (2009)
13. Nehring, K., Puppe, C.: Strategy-proof social choice on single-peaked domains: possibility, impossibility and the space between. University of California at Davis (2002). http://vwl1.ets.kit.edu/puppe.php
14. Schaefer, T.J.: The complexity of satisfiability problems. In: Proceedings of the 10th Annual ACM Symposium on Theory of Computing, pp. 216–226 (1978)
15. Sharir, M.: A strong-connectivity algorithm and its applications in data flow analysis. Comput. Math. Appl. **7**(1), 67–72 (1981)
16. Szegedy, M., Xu, Y.: Impossibility theorems and the universal algebraic toolkit. CoRR, abs/1506.01315 (2015)
17. Szendrei, Á.: Clones in Universal Algebra, vol. 99. Presses de l'Université de Montréal, Montreal (1986)
18. Tarjan, R.E.: Depth-first search and linear graph algorithms. SIAM J. Comput. **1**(2), 146–160 (1972)

Calculational Relation-Algebraic Proofs in the Teaching Tool CalcCheck

Wolfram Kahl(✉)

McMaster University, Hamilton, ON, Canada
kahl@cas.mcmaster.ca

Abstract. The proof checker CalcCheck has been developed for teaching calculational proofs in the style of Gries and Schneider's textbook classic "A Logical Approach to Discrete Math". While originally only AC-rewriting was supported, we now added also support for operators that are only associative, which is essential for convenience in reasoning about composition of (in particular) relations. We demonstrate how the CalcCheck language including this and several other recent improvements supports readable and writable machine-checked proofs of interesting relation-algebraic developments.

1 Introduction

Relation-algebraic proofs have a long tradition of calculational presentation, see for example the classic texts "Relations and Graphs" (R&G) by Schmidt and Ströhlein (1993), Aarts and Backhouse *et al.* (1992), and Bird and de Moor (1997). The calculational approach to proof presentation is also central to the textbook "A Logical Approach to Discrete Math" (LADM) by Gries and Schneider (1993) targeted at beginning students of Computer Science and Software Engineering. To support teaching a course using this textbook, we are continuing to develop the proof checker CalcCheck (Kahl 2011, 2018); the latest instance of that course started using this tool also for basic relation algebra. While the CalcCheck features described in (Kahl 2018) allow the students to check their proofs for quite a significant part of the textbook, several peculiarities of typical relation-algebraic proofs turned out to make CalcCheck proofs quite cumbersome; the most prominent of these are the following:

- Relation composition is associative, but not commutative—proof checking by CalcCheck is based on rewriting, and the original matching algorithm, following Contejean (2004), only dealt with (associative-and-)commutative operators.
- Inclusion chains routinely use deeply-nested monotonicity of operators, while this was only occasionally needed for other applications.
- Cyclic inclusion chains are quite frequently used to prove equalities, and splitting them into separate chains as subproofs for an application of antisymmetry adds overhead that noticeably reduces the elegance of proofs.

J. Desharnais et al. (Eds.): RAMiCS 2018, LNCS 11194, pp. 366–384, 2018.
https://doi.org/10.1007/978-3-030-02149-8_22

- The emphasis on inclusion chains also makes reverse chains a more attractive presentation with much higher frequency than previously encountered, so that explicit invocation of the respective reversion laws became a serious nuisance.
- Concise assumptions with far-reaching consequences (such as assuming that a parameter is an equivalence relation) tend to be invoked directly for their consequences; in CALCCHECK, each such assumption reference originally needed to derive the required consequences *in situ*.

We have now addressed these issues; we will show how this allows us to present relation-algebraic proofs with their full elegance in the calculational style of LADM and in a manner that fits well into a course based on LADM.

We start introducing the CALCCHECK language and our treatment of relation algebra using some simple examples of abstract relation-algebraic calculations in Sect. 2. The quantification features of CALCCHECK are then introduced in Sect. 3 using the example of converting between relation-algebraic and predicate-logic formulations. The overview of the CALCCHECK language in Sect. 4 is intentionally delayed until after those first examples established a general understanding. We then show more substantial abstract relation-algebraic calculations in Sect. 5, before turning to infima of sets of relations in Sect. 6.

Some additional material is available at the CALCCHECK home page at http://CalcCheck.McMaster.ca/.

2 Simple Abstract Relation Algebra

Following LADM, we use the prefix operator ~ for set and relation complement; we use the postfix operator $\breve{}$ for relation converse. We write $\bot$ for empty relations, and $\top$ for universal relations. The infix operator ⨾ used for relation forward-composition[1] has higher precedence than $\cup$ and $\cap$. (All these notational choices could be changed by using different preloaded declarations in CALCCHECK.)

As first example, we show a detailed proof of Proposition 2.4.2(i) of Schmidt and Ströhlein (1993), which however already uses automatic associativity of ⨾ — the following is a verbatim copy of Unicode plain-text CALCCHECK source:

[1] In the font used in this paper to display CALCCHECK source listings, the "top" symbol ⊤ comes out looking like a just slightly smaller version of the letter "T", and "⨾" comes out looking almost like a conventional semicolon (which is however currently not used in CALCCHECK).

```
Theorem "Domain restriction of ;":
    (Q ∩ R ; T) ; S  =  Q ; S ∩ R ; T
Proof:
  Using "Mutual inclusion":
    Subproof for `(Q ∩ R ; T) ; S ⊆ Q ; S ∩ R ; T`:
        (Q ∩ R ; T) ; S
      ⊆⟨ "Sub-distributivity of ; over ∩" ⟩
        Q ; S ∩ R ; T ; S
      ⊆⟨ "Monotonicity of ∩" with "Monotonicity of ;"
          with "Greatest relation" ⟩
        Q ; S ∩ R ; T
    Subproof for `Q ; S ∩ R ; T ⊆ (Q ∩ R ; T) ; S`:
        Q ; S ∩ R ; T
      ⊆⟨ "Modal rule" ⟩
        (Q ∩ R ; T ; S ˘) ; S
      ⊆⟨ "Monotonicity of ;" with "Monotonicity of ∩"
          with "Monotonicity of ;" with "Greatest relation" ⟩
        (Q ∩ R ; T) ; S
```

The CalcCheck language is layout-sensitive; nested theorem and proof structure require additional indentation by two spaces. The proof above is structured into two subproofs by invocation of the theorem "Mutual inclusion"; subproofs proven by direct calculations such as those here do not need their goals to be made explicit via "for '. . . '"; below, we will frequently omit this in such cases. The calculation syntax should be almost self-explaining; the steps are separated by *hints* enclosed in ⟨ ... ⟩ *immediately* preceded by the calculation step operator, which is a binary infix predicate operator such as inclusion "$\subseteq$", equality "$=$", logical equivalence "$\equiv$". Theorem references are either theorem names enclosed in matching double quotes, or theorem numbers in parentheses and not containing spaces. Hints can be structured; the use of "with" we see above connects a reference to an implication (here always a monotonicity law) with a hint resolving the antecedent of that implication.

As mentioned in the introduction, such presentation with explicit reference to "Mutual inclusion" is not very satisfactory; recent improvements to CalcCheck allow us to write the following nicer proof:

```
Theorem "Domain restriction of ;":
    (Q ∩ R ; T) ; S = Q ; S ∩ R ; T
Proof:
    (Q ∩ R ; T) ; S
  ⊆⟨ "Sub-distributivity of ; over ∩" ⟩
    Q ; S ∩ R ; T ; S
  ⊆⟨ Monotonicity with "Greatest relation" ⟩
    Q ; S ∩ R ; T
  ⊆⟨ "Modal rule" ⟩
    (Q ∩ R ; T ; S ˘) ; S
  ⊆⟨ Monotonicity with "Greatest relation" ⟩
    (Q ∩ R ; T) ; S
```

Here, the keyword Monotonicity stands for an arbitrary number of references to monotonicity laws that have been explicitly activated for such use. (CalcCheck also has the corresponding keyword Antitonicity; Gries (1997) argues strongly for the presence of such keywords in order reasoning.) The calculation is now cyclic, and has precisely the same steps as that by Schmidt and Ströhlein (1993) in Proposition 2.4.2(i) except that we simplified the third step slightly by using a modal rule instead of the Dedekind rule. The main difference to the version of Schmidt and Ströhlein is now just the presence of the explanatory hints.

To demonstrate a few more features, we show one way to obtain the Schröder equivalences from the modal rules. Under the name `"Inclusion via ∩ ~"` we have the two theorems `R ⊆ S  ≡  R ∩ ~ S ⊆ ⊥` and `R ⊆ S  ≡  R ∩ ~ S = ⊥` available; both are used to show the first implication of the Schröder equivalences:

```
Lemma "Schröder ⇒": Q ; R ⊆ S  ⇒  Q ˘ ; ~ S  ⊆ ~ R
Proof:
  Assuming `Q ; R ⊆ S`:
    Using "Inclusion via ∩ ~":
      Subproof for `Q ˘ ; ~ S ∩ ~ ~ R ⊆ ⊥`:
          Q ˘ ; ~ S ∩ ~ ~ R
        =⟨ "Self-inverse of ~" ⟩
          Q ˘ ; ~ S ∩ R
        ⊆⟨ "Modal rule", "Self-inverse of ˘" ⟩
          Q ˘ ; (~ S ∩ Q ; R)
        =⟨ Assumption `Q ; R ⊆ S` with "Inclusion via ∩ ~" ⟩
          Q ˘ ; ⊥
        =⟨ "Zero of ;" ⟩
          ⊥
```

We show this implication by "assuming the antecedent", and referring to the assumption within the proof of the consequence. The theorem referred to in "`Using` ..." here is a simple equivalence turning this consequence into the goal of the single subproof here. The "with" in the second-last hint here is different from that seen previously, since one of the `"Inclusion via ∩ ~"` theorems is used to rewrite the assumption into an equivalent equation for use in this step.

From this, we easily obtain the full Schröder equivalences—"Mutual implication" most frequently will be used with two subproofs; here, one side is supplied directly using the first kind of "with", so only one subproof still needs to be supplied:

```
Theorem "Schröder": Q ; R ⊆ S  ≡  Q ˘ ; ~ S  ⊆ ~ R
Proof:
  Using "Mutual implication" with "Schröder ⇒":
    Subproof:
        Q ˘ ; ~ S  ⊆ ~ R
      ⇒⟨ "Schröder ⇒" ⟩
        Q ˘ ˘ ; ~ ~ R  ⊆ ~ ~ S
      ≡⟨ "Self-inverse of ˘", "Self-inverse of ~" ⟩
        Q ; R ⊆ S
```

From this, the dual Schröder equivalence is obtained via a straight-forward calculation using the properties of converse:

```
Theorem "Schröder": Q ; R ⊆ S  ≡  ~ S ; R ˘ ⊆ ~ Q
Proof:
    Q ; R ⊆ S
  ≡( "Isotonicity of ˘", "Converse of ;" )
    R ˘ ; Q ˘ ⊆ S ˘
  ≡( "Schröder" )
    R ˘ ˘ ; ~ (S ˘)  ⊆ ~ (Q ˘)
    ≡( "Converse of ~", "Converse of ;" )
      ((~ S) ; R ˘) ˘ ⊆ (~ Q) ˘
    ≡( "Isotonicity of ˘" )
      ~ S ; R ˘ ⊆ ~ Q
```

3 Relationship Reasoning in Predicate Logic

Practice in conversion between relation-algebraic and predicate-logic formulations can be useful for gaining confidence with the former. The connection between the two is established via reasoning about relationship; LADM uses the notation "$a\ R\ b$" as equivalent to $(a, b) \in R$. Since CALCCHECK uses juxtaposition already for function application, and since for more complex expressions in the position of R the customary infix notation becomes hard to read in any case, we use "tortoise shell brackets" to delimit the relation in our general infix relationship notation, and write "a ⦗ R ⦘ b". The distinction between non-infix relation names such as R and infix operators such as equality and inclusion is made explicit by the fact that the latter are declared with names including underscores indicating mixfix argument positions, such as "_=_" and "_⊆_"; writing "$R \subseteq S$" is just considered short-hand for the function application "_⊆_ R S" (with two arguments R and S).

LADM also supports *conjunctional* operators, which include _=_ and _⊆_; this means that "$Q = R \subseteq S$" is considered as short-hand for the conjunction "$Q = R \wedge R \subseteq S$". CALCCHECK allows also more-than-binary infix operators to be declared as conjunctional; we use this for "_ ⦗ _ ⦘ _", which is just a user-defined mixfix operator. Conjunctionality then allows us to write "a ⦗ R ⦘ b ⦗ S ⦘ c" instead of "a ⦗ R ⦘ $b \wedge b$ ⦗ S ⦘ c".

For quantification, CALCCHECK follows the spirit of LADM, but the (currently hard-wired) concrete syntax is closer to the Z notation (Spivey 1989): The general pattern of quantified expressions is "*bigOp varDecls* | *rangePredicate* • *body*"; the shorter form "*bigOp varDecls* • *body*" is an abbreviation where the range predicate defaults to *true*. The quantifying "big operators" are user-declared; we are working with the following set-up:

```
Big operator for _+_ is ∑
Big operator for _·_ is ∏
Big operator for _∧_ is ∀
Big operator for _∨_ is ∃
Big operator for _∩_ is ⋂
Big operator for _∪_ is ⋃
```

A proof for a universal quantification $(\forall x \bullet P)$ by ∀-introduction is presented in CALCCHECK as the heading "For any 'x':" followed by a proof for P, or, using a different variable and writing "For any 'y':" followed by a proof for $P[x := y]$ (where this substitution does not need to be written explicitly); in the context of LADM this is justified by metatheorem (9.16): "P is a theorem iff $(\forall x \bullet P)$ is a theorem." Since LADM emphasises calculational reasoning also for manipulation of quantifications, ∀-introduction is currently the only quantification inference rule for which we have special syntax. In the element-wise proof of the Dedekind rule below (following R&G, pp. 16–17), "For any" is used to continue after application of the definition of "Relation inclusion":

$$R \subseteq S \quad \equiv \quad (\forall\, x, y \bullet x \,\llparenthesis\, R \,\rrparenthesis\, y \Rightarrow x \,\llparenthesis\, S \,\rrparenthesis\, y)$$

The resulting subproof goal is not shown explicitly—we make use of the fact that in "Using" with a single subgoal, the keyword heading "Subproof:" can be omitted (if the proof below does not require an explicit statement of the goal).

```
Theorem "Dedekind rule": (Q ; R) ∩ S ⊆ (Q ∩ S ; R ˘) ; (R ∩ Q ˘ ; S)
Proof:
  Using "Relation inclusion":
    For any `a`, `c`:
        a ⦇ (Q ; R) ∩ S ⦈ c
      ≡⟨ "Relation intersection", "Relation composition" ⟩
        (∃ b₂ • a ⦇ Q ⦈ b₂ ⦇ R ⦈ c) ∧ a ⦇ S ⦈ c
      ≡⟨ "Distributivity of ∧ over ∃" ⟩
        ∃ b₂ • a ⦇ Q ⦈ b₂ ∧ b₂ ⦇ R ⦈ c ∧ a ⦇ S ⦈ c
      ≡⟨ "Idempotency of ∧", "Relation converse" ⟩
        ∃ b₂ • a ⦇ Q ⦈ b₂ ∧ c ⦇ R ˘ ⦈ b₂ ∧ a ⦇ S ⦈ c
             ∧ b₂ ⦇ Q ˘ ⦈ a ∧ b₂ ⦇ R ⦈ c ∧ a ⦇ S ⦈ c
      ≡⟨ Substitution ⟩
        ∃ b₂ • a ⦇ Q ⦈ b₂ ∧ (a ⦇ S ⦈ c₂ ∧ c₂ ⦇ R ˘ ⦈ b₂)[c₂ ≔ c]
             ∧ b₂ ⦇ R ⦈ c ∧ (b₂ ⦇ Q ˘ ⦈ a₂ ∧ a₂ ⦇ S ⦈ c)[a₂ ≔ a]
      ⇒⟨ Monotonicity with "∃-Introduction" ⟩
        ∃ b₂ • a ⦇ Q ⦈ b₂ ∧ (a ⦇ S ⦈ c₂ ∧ c₂ ⦇ R ˘ ⦈ b₂)[c₂ ≔ c]
             ∧ b₂ ⦇ R ⦈ c ∧ (∃ a₂ • b₂ ⦇ Q ˘ ⦈ a₂ ∧ a₂ ⦇ S ⦈ c)
      ⇒⟨ Monotonicity with "∃-Introduction" ⟩
        ∃ b₂ • a ⦇ Q ⦈ b₂ ∧ (∃ c₂ • a ⦇ S ⦈ c₂ ∧ c₂ ⦇ R ˘ ⦈ b₂)
             ∧ b₂ ⦇ R ⦈ c ∧ (∃ a₂ • b₂ ⦇ Q ˘ ⦈ a₂ ∧ a₂ ⦇ S ⦈ c)
      ≡⟨ "Relation composition" ⟩
        ∃ b₂ • a ⦇ Q ⦈ b₂ ∧ a ⦇ S ; R ˘ ⦈ b₂ ∧ b₂ ⦇ R ⦈ c ∧ b₂ ⦇ Q ˘ ; S ⦈ c
      ≡⟨ "Relation intersection", "Relation composition" ⟩
        ∃ b₂ • a ⦇ Q ∩ S ; R ˘ ⦈ b₂ ∧ b₂ ⦇ R ∩ Q ˘ ; S ⦈ c
      ≡⟨ "Relation composition" ⟩
        a ⦇ (Q ∩ S ; R ˘) ; (R ∩ Q ˘ ; S) ⦈ c
```

The second and third step in the calculation here are done together as one step by Schmidt and Ströhlein; the last three steps directly transliterate the last three steps of Schmidt and Ströhlein; merging both of these groups into single steps is still checked.

The second step uses the theorem "Distributivity of ∧ over ∃"

$$P \wedge (\exists\, x \mid R \bullet Q) \quad \equiv \quad (\exists\, x \mid R \bullet P \wedge Q)$$

which in LADM comes with the *proviso* "¬*occurs*('x', 'P')"—technically, this is a meta-theorem with metavariables P, Q, and R standing for expressions and metavariable x standing for a variable; the proviso means that theoremhood is asserted only for instantiations where P is instantiated with an expression in which the instantiation of x does not occur free. CALCCHECK derives such provisos automatically from the theorem statement, and matching fails where such provisos would be violated.

The keyword Substitution used in the fourth calculation hint above justifies performing substitutions; this reverse Substitution is necessary here as preparation for the "∃-Introduction" step since CALCCHECK does not (yet) attempt second-order matching.

The keyword Monotonicity used in the two implication steps here triggers use not only of monotonicity of $\wedge$, but also of "Body monotonicity of $\exists$":

$$(\forall x \mid R \bullet P_1 \Rightarrow P_2) \quad \Rightarrow \quad ((\exists x \mid R \bullet P_1) \Rightarrow (\exists x \mid R \bullet P_2))$$

The proof above also illustrates that the calculation format as such has to be considered as conjunctional, and there cannot be any restriction to conjunctional operators, since in particular implication chains as above are allowed by the "relaxed proof style" in LADM Chap. 4, which does not consider the implication operator $\Rightarrow$ as conjunctional. Indeed, where LADM introduces the calculation format in Chap. 1 on p. 15, it translates the calculation into a (meta-level) conjunction without invoking conjunctionality of the operator "$=$". Therefore, in calculations where the expressions are of type $\mathbb{B}$, the two operators "$\equiv$" and "$=$" can be used interchangeably as calculation operators; we prefer to use "$\equiv$" since it documents the type $\mathbb{B}$.

The proof of the equivalence of relational and predicate-logical definitions of transitivity by (Schmidt and Ströhlein 1993, p. 28) translates directly into CALCCHECK—we only "saved" two additional (mutually inverse) applications of De Morgan laws:

```
Theorem "Transitivity":
  is-transitive R   ≡   ∀ x • ∀ y • ∀ z •
                          x ⦇ R ⦈ y ⦇ R ⦈ z  ⇒  x ⦇ R ⦈ z
Proof:
    is-transitive R
  ≡⟨ "Definition of transitivity" ⟩
    R ; R ⊆ R
  ≡⟨ "Relation inclusion", "Relation composition" ⟩
    ∀ x • ∀ z • (∃ y • x ⦇ R ⦈ y ⦇ R ⦈ z) ⇒ x ⦇ R ⦈ z
  ≡⟨ (3.59) `p ⇒ q ≡ ¬ p ∨ q`, "Generalised De Morgan" ⟩
    ∀ x • ∀ z • (∀ y • ¬ (x ⦇ R ⦈ y ⦇ R ⦈ z)) ∨ x ⦇ R ⦈ z
  ≡⟨ "Distributivity of ∨ over ∀", (3.59) ⟩
    ∀ x • ∀ z • ∀ y • x ⦇ R ⦈ y ⦇ R ⦈ z ⇒ x ⦇ R ⦈ z
  ≡⟨ "Nesting for ∀", "Dummy list permutation for ∀" ⟩
    ∀ x • ∀ y • ∀ z • x ⦇ R ⦈ y ⦇ R ⦈ z ⇒ x ⦇ R ⦈ z
```

In the third hint above, we juxtaposed two theorem references, one by theorem number "(3.59)", and one by theorem expression—this pattern frequently serves documentation in LADM; in CALCCHECK this is defined as referring to the intersection of the sets of theorems referred to by the individual references (using only expressions as stand-alone theorem references is by default disabled for pedagogical reasons). Following LADM, quantifiers with lists of quantified variables are not considered short-hand for nested quantifiers; therefore we need to invoke "Nesting for ∀" in the last hint above to be able to use dummy list permutation, which happens to be stated without nesting:

```
(∀ x, y | R • P) ≡ (∀ y, x | R • P)
```

Finally we show the conversion of the relation-algebraic definition of symmetric quotients into predicate-logic shape:

```
Axiom "Definition of `syq`":  syq Q R =  (Q ╲ R) ∩ (Q ˘ ╱ R ˘)

Theorem "Symmetric quotient":
    b ( syq Q R ) c  ≡  ∀ a • a ( Q ) b ≡ a ( R ) c
Proof:
    b ( syq Q R ) c
  ≡⟨ "Definition of `syq`"  "Relation intersection" ⟩
    b ( Q ╲ R ) c  ∧  b ( Q ˘ ╱ R ˘ ) c
  ≡⟨ "Relationship via left residual",
      "Relationship via right residual", "Relation converse" ⟩
    (∀ a • a ( Q ) b ⇒ a ( R ) c) ∧ (∀ a • a ( R ) c ⇒ a ( Q ) b)
  ≡⟨ "Distributivity of ∀ over ∧", "Mutual implication" ⟩
    ∀ a • a ( Q ) b ≡ a ( R ) c
```

4 Overview of the CALCCHECK Language

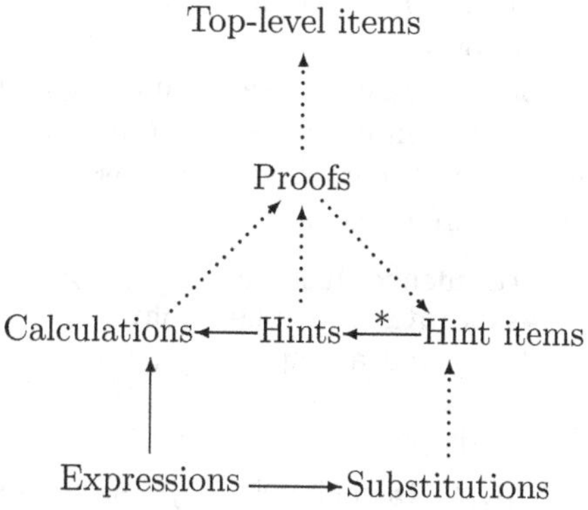

In the diagram to the right, the main syntactic categories of the CALCCHECK language are indicated—full arrows denote the "are immediate constituents of" relation, and dotted arrows the "may be immediate constituents of" relation. A CALCCHECK module currently just consists of a sequence of "top-level items" (TLIs); CALCCHECK notebooks used for teaching are technically module "suffixes" that may not contain module import declarations; these and arbitrary other TLIs may be contained in the preloaded "prefix".

The most prominent kind of TLI is introduced by one of the (completely interchangeable) keywords Theorem, Lemma, Corollary, Proposition, Fact, with the following syntax:

Theorem $\{ThmName \mid ThmNum\}^*$: *expression*
Proof: *proof*

Each theorem can have any number of theorem numbers (in parentheses) or theorem names (in "..."), each of which may also be used in several theorems—this corresponds to the practice in LADM. Lack of a Proof is flagged as an error.

The same syntax as for Theorems is also used for Axioms, except that for an Axiom, presence of a proof is flagged as an error.

Currently, CalcCheck (like LADM) is lacking support for formal definition principles (like inductive datatypes or definition by primitive recursion); new identifiers are introduced used Declarations and may in addition be provided Explanations; they are then defined using axioms, as in the following example:

```
Declaration: _\_ : A ↔ B  →  A ↔ C  → B ↔ C
Explanation: R \ S = "The right-residual of `S` by `R`"
Explanation: R \ S = "Left-division of `S` by `R`"
Explanation: R \ S = "`R` under `S`"
Axiom "Characterisation of right residual" "Characterisation of \":
    X ⊆ R \ S  ≡  R ; X ⊆ S
```

Operator declaration such as this actually need to be preceded by precedence declarations (and possibly declarations specifying association to the left or right) which are then valid for all operators using the same name—this corresponds again to the practice of LADM, which features a single precedence table on its insider cover.

For relation composition and residuals, we fix the same precedence, but higher than that of intersection and union, and to avoid potential confusion we declare the residual operators as non-associating, so that nested applications always require parentheses:

```
Precedence 105 for: _;_ , _/_ , _\_
Associating to the right: _;_
Non-associating: _/_ , _\_
```

TLIs of shape

Activate {associativity | symmetry | monotonicity | ...} property *ThmRef*

are used to enable matching up to associativity and/or commutativity for binary operators, use of the keywords Monotonicity and Antitonicity, and other features. Finally, MarkDown TLIs accommodate documentation.

For expressions, besides the quantification syntax explained in Sect. 3, only the function application syntax by juxtaposition $f\ a$ and the infix function type constructor $\_{\rightarrow}\_$ are currently hard-coded. Mixfix operators can be declared with underscores indicating argument positions; mixfix operators that act as infix, prefix, or postfix operators have to be assigned a single precedence level and associating convention as explained above. (Not all pairs of operators at the same precedence level can be used with mutually associating parentheses conventions; in LADM, $+$ and $-$ can, but in particular $\wedge$ and $\vee$ cannot.)

The hints in calculations are sequences of hint items separated by comma or the keyword and; in nested hints, and has lower precedence than with, and comma has higher precedence. We have already seen most kinds of hints; more explanation can be found in (Kahl 2018). For a quick summary, a hint can be:

- A non-empty sequence of theorem references, denoting the intersection of their theorems
- Assumption \` *expr* \` or Assumption *ThmRef*
- Induction hypothesis $\{$ \` *expr* \` $\}^?$
- Substitution
- Evaluation — on ground terms of built-in operators registered using TLIs
  Register {type | operator} *Identifier* for built-in *BuiltIn*
- hi_0 with hi_1 and ... and hi_n
  — instantiation by substitution, premise discharge, theorem rewriting briefly described above in Sect. 2; for more detail see (Kahl 2018, Sect. 6)
- Monotonicity, Antitonicity — only before with
- $x_1, \ldots, x_n := e_1, \ldots, e_n$ — only after with
- SubProof{ for \` *expr* \` $\}^?$:
  proof

Besides calculations, CalcCheck has the following proof structures, explained in more detail in (Kahl 2018):

- By *Hint* — discharging simple proof obligations,
- Assuming $ThmName^?$ \` *expression* \` : — assuming the antecedent, corresponding to implication introduction.
  Assuming $ThmName^?$ \` *expression* \` and using with *hint* : — see Sect. 5
- By cases: \` $expression_1$ \`, ..., \` $expression_n$ \` — case analysis
- By induction on \` *var* : *type* \`: — proofs by induction
- For any \` *var* : *type* \`: — corresponding to ∀-introduction
- Using *HintItem* : — turning theorems into inference rules

As the examples also in the current paper demonstrate, these proof structures are actually quite sufficient to support and encourage fully formal proofs. (In (Kahl 2018) we show several examples of elegant fully formal proofs where LADM resorts to prose for significant parts of the proof structure.)

5 More Abstract Relation Algebra

Proposition 3.1.6(ii) of R&G has a statement that contains a three-element equation chain, using conjunctionality of =, and proves this via a cyclic inclusion chain. CalcCheck now supports such proofs, too, for arbitrary transitive and antisymmetric "calculation operators", and arbitrary numbers of equalities in the goal.

Even with that, the following proof is still unsatisfactory due to the repetitive effort expanded towards accessing consequences of the assumption that **Θ** is an equivalence relation:

```
Theorem "Equivalence with ∩ and ;":
    is-equivalence Θ  ⇒  (A ; Θ ∩ B) ; Θ  =  A ; Θ ∩ B ; Θ  =  (A ∩ B ; Θ) ; Θ
Proof:
  Assuming "Θ-E" `is-equivalence Θ`:
      (A ; Θ ∩ B) ; Θ
    ⊆⟨ "Sub-distributivity of ; over ∩" ⟩
      A ; Θ ; Θ ∩ B ; Θ
    =⟨ "Definition of idempotency" with "Idempotency from symmetric and transitive"
       with assumption `is-equivalence Θ` with "Definition of equivalence" ⟩
      A ; Θ ∩ B ; Θ
    ⊆⟨ "Modal rule" ⟩
      (A ∩ B ; Θ ; Θ ˘) ; Θ
    =⟨ "Definition of symmetry" with assumption "Θ-E" with "Definition of equivalence" ⟩
      (A ∩ B ; Θ ; Θ) ; Θ
    =⟨ "Definition of idempotency" with "Idempotency from symmetric and transitive"
       with assumption `is-equivalence Θ` with "Definition of equivalence" ⟩
      (A ∩ B ; Θ) ; Θ
    ⊆⟨ "Sub-distributivity of ; over ∩" ⟩
      A ; Θ ∩ B ; Θ ; Θ
    =⟨ "Definition of idempotency" with "Idempotency from symmetric and transitive"
       with assumption `is-equivalence Θ` with "Definition of equivalence" ⟩
      A ; Θ ∩ B ; Θ
    ⊆⟨ "Modal rule" ⟩
      (A ; Θ ; Θ ˘ ∩ B) ; Θ
    =⟨ "Definition of symmetry" with assumption "Θ-E" with "Definition of equivalence" ⟩
      (A ; Θ ; Θ ∩ B) ; Θ
    =⟨ "Definition of idempotency" with "Idempotency from symmetric and transitive"
       with assumption `is-equivalence Θ` with "Definition of equivalence" ⟩
      (A ; Θ ∩ B) ; Θ
```

Since this is a recurring pattern, CalcCheck provides a variant of the Assuming construct where an additional hint is used to derive consequences from the assumption; any of these consequences can be used where that assumption is invoked in the proof body:

```
Theorem “Equivalence with ∩ and _;”:
   is-equivalence Θ  ⇒  (A ; Θ ∩ B) ; Θ  =  A ; Θ ∩ B ; Θ  =  (A ∩ B ; Θ) ; Θ
Proof:
  Assuming `is-equivalence Θ`
    and using with “Definition of equivalence” and “Definition of symmetry”
             and “Idempotency from symmetric and transitive”
                 with “Definition of idempotency”:
      (A ; Θ ∩ B) ; Θ
    ⊆⟨ “Sub-distributivity of ; over ∩” ⟩
      A ; Θ ; Θ ∩ B ; Θ
    =⟨ Assumption `is-equivalence Θ` ⟩
      A ; Θ ∩ B ; Θ
    ⊆⟨ “Modal rule” ⟩
      (A ∩ B ; Θ ; Θ ˘) ; Θ
    =⟨ Assumption `is-equivalence Θ` ⟩
      (A ∩ B ; Θ) ; Θ
    ⊆⟨ “Sub-distributivity of ; over ∩” ⟩
      A ; Θ ∩ B ; Θ ; Θ
    =⟨ Assumption `is-equivalence Θ` ⟩
      A ; Θ ∩ B ; Θ
    ⊆⟨ “Modal rule” ⟩
      (A ; Θ ; Θ ˘ ∩ B) ; Θ
    =⟨ Assumption `is-equivalence Θ` ⟩
      (A ; Θ ∩ B) ; Θ
```

The hint in the “and using with” clause above is still rather verbose, but it captures exactly the consequences that are used in the proof following it, as can be seen from the first proof attempt above. It is useful to introduce special-purpose theorems to be used in such “and using with” clauses; for further equivalence assumptions below we will be using the following:

```
Theorem “Equivalence properties”:
  is-equivalence E
     ⇒  is-reflexive E   ∧  Id ⊆ E    ∧ is-symmetric E   ∧  E ˘ = E
     ∧  is-transitive E  ∧  E ; E ⊆ E  ∧ is-idempotent E  ∧  E ; E = E
```

This theorem is designed to be used when assuming antecedents in the following way:

Assuming `is-equivalence E` and using with “Equivalence properties”:,

In the proof following that, all the conjuncts of the consequence of the theorem “Equivalence properties” are referenced directly by just using the hint item Assumption `is-equivalence E`.

We will use this when porting the calculation for Proposition 3.1.7 in R&G to CALCCHECK, where we prove exactly the same proposition, although the assumption $G^{\smile} ; \Theta ; G \subseteq \Omega$ could actually be weakened to $G^{\smile} ; G \subseteq \Omega$, making Θ superfluous. In comparison with the proof of Schmidt and Ströhlein, our first two steps are a single equality there, and we explicitly mention an identity that Schmidt and Ströhlein only introduce implicitly one step later; apart from that, our steps correspond to theirs. Using the second-last assumption works without further effort since the property that $\supseteq$ is the converse of $\subseteq$ has been activated.

```
Theorem "Congruence property":
  is-equivalence θ ⇒ is-equivalence Ω
   ⇒ G ˘ ; θ ; G ⊆ Ω
   ⇒ F ; Ω ⊆ G ; Ω
   ⇒ F ; T ⊇ G ; T
   ⇒ F ; Ω = G ; Ω
Proof:
  Assuming `is-equivalence θ` and using with "Equivalence properties":
    Assuming `is-equivalence Ω` and using with "Equivalence properties":
      Assuming `G ˘ ; θ ; G ⊆ Ω`, `F ; Ω ⊆ G ; Ω`, `F ; T ⊇ G ; T`:
          G ; Ω
        =⟨ "Inclusion via ∩" with "Greatest relation" ⟩
          G ; (T ∩ Ω)
        ⊆⟨ "Sub-distributivity of ; over ∩" ⟩
          G ; T ∩ G ; Ω
        ⊆⟨ Monotonicity with assumption `F ; T ⊇ G ; T` ⟩
          F ; T ∩ G ; Ω
        ⊆⟨ "Modal rule" ⟩
          F ; (T ∩ F ˘ ; G ; Ω)
        ⊆⟨ Monotonicity with "Weakening for ∩" ⟩
          F ; F ˘ ; G ; Ω
        ⊆⟨ Monotonicity with "Identity of ;" and assumption `is-equivalence Ω` ⟩
          F ; (F ; Ω) ˘ ; G ; Ω
        ⊆⟨ Monotonicity with assumption `F ; Ω ⊆ G ; Ω` ⟩
          F ; (G ; Ω) ˘ ; G ; Ω
        =⟨ "Converse of ;", "Identity of ;", assumption `is-equivalence Ω` ⟩
          F ; Ω ; G ˘ ; Id ; G ; Ω
        ⊆⟨ Monotonicity with assumption `is-equivalence θ` ⟩
          F ; Ω ; G ˘ ; θ ; G ; Ω
        ⊆⟨ Monotonicity with assumption `G ˘ ; θ ; G ⊆ Ω` ⟩
          F ; Ω ; Ω ; Ω
        =⟨ Assumption `is-equivalence Ω` ⟩
          F ; Ω
        ⊆⟨ Assumption `F ; Ω ⊆ G ; Ω` ⟩
          G ; Ω
```

As an example proof for a compound statement, we show Proposition 3.1.8 of R&G. We open the proof using a trick: Since symmetry of ∧ is already activated, referring to "Symmetry of ∧" in "Using" gives us ∧-introduction by allowing us to prove the two conjuncts separately. The hints are kept reasonably short since we added "and using with ..." to the assumptions again. However, here we see the effects of a remaining current weakness of CalcCheck: Inclusions are proved internally by rewriting the whole inclusion to *true*, unlike equations where one side is rewritten to the other side. This has as a consequence that several equational rewrites can be performed in a single calculation step, but it is usually not possible to chain several inclusions (or implications) automatically into a single step.

```
Theorem “Preorder with compatible equivalence”:
    is-preorder R ⇒ is-equivalence θ ⇒ R ; θ ⊆ θ ; R ⇒ R ; θ ∩ θ ; R ˘ ⊆ θ
    ⇒ is-preorder (θ ; R ; θ) ∧ θ ; R ; θ ∩ θ ; R ˘ ; θ = θ
Proof:
  Assuming `is-preorder R` and using with “Preorder properties”:
    Assuming `is-equivalence θ` and using with “Equivalence properties”:
      Assuming `R ; θ ⊆ θ ; R`, `R ; θ ∩ θ ; R ˘ ⊆ θ`:
        Using “Symmetry of ∧”:
          Subproof for `is-preorder (θ ; R ; θ)`:
            Using “Definition of preorder”:
              Subproof for `is-reflexive (θ ; R ; θ)`:
                Using “Definition of reflexivity”:
                  Subproof:
                      Id
                    ⊆⟨ Assumption `is-preorder R` ⟩
                      R
                    =⟨ “Identity of ;” ⟩
                      Id ; R ; Id
                    ⊆⟨ Monotonicity with assumption `is-equivalence θ` ⟩
                      θ ; R ; Id
                    ⊆⟨ Monotonicity with assumption `is-equivalence θ` ⟩
                      θ ; R ; θ
              Subproof for `is-transitive (θ ; R ; θ)`:
                Using “Definition of transitivity”:
                  Subproof:
                      (θ ; R ; θ) ; (θ ; R ; θ)
                    ⊆⟨ Monotonicity with assumption `R ; θ ⊆ θ ; R` ⟩
                      θ ; θ ; R ; θ ; R ; θ
                    ⊆⟨ Monotonicity with assumption `R ; θ ⊆ θ ; R` ⟩
                      θ ; θ ; θ ; R ; R ; θ
                    =⟨ Assumption `is-preorder R` ⟩
                      θ ; θ ; θ ; R ; θ
                    ⊆⟨ Monotonicity with assumption `is-equivalence θ` ⟩
                      θ ; θ ; R ; θ
                    ⊆⟨ Monotonicity with assumption `is-equivalence θ` ⟩
                      θ ; R ; θ
          Subproof for `θ ; R ; θ ∩ θ ; R ˘ ; θ = θ`:
              θ
            =⟨ “Idempotency of ∩”, “Identity of ;”, “Converse of `Id`” ⟩
              θ ; Id ; Id ∩ θ ; Id ˘ ; Id
            ⊆⟨ Monotonicity with assumption `is-preorder R` ⟩
              θ ; R ; Id ∩ θ ; Id ˘ ; Id
            ⊆⟨ Monotonicity with assumption `is-preorder R` ⟩
              θ ; R ; Id ∩ θ ; R ˘ ; Id
            ⊆⟨ Monotonicity with assumption `is-equivalence θ` ⟩
              θ ; R ; θ ∩ θ ; R ˘ ; Id
            ⊆⟨ Monotonicity with assumption `is-equivalence θ` ⟩
              θ ; R ; θ ∩ θ ; R ˘ ; θ
            =⟨ “Equivalence with ∩ and ;_” with assumption `is-equivalence θ` ⟩
              θ ; (R ; θ ∩ θ ; R ˘ ; θ)
            =⟨ “Equivalence with ∩ and _;” with assumption `is-equivalence θ` ⟩
              θ ; (R ; θ ∩ θ ; R ˘) ; θ
            ⊆⟨ Monotonicity with Assumption `R ; θ ∩ θ ; R ˘ ⊆ θ` ⟩
              θ ; θ ; θ
            ⊆⟨ Monotonicity with assumption `is-equivalence θ` ⟩
              θ ; θ
            ⊆⟨ Monotonicity with assumption `is-equivalence θ` ⟩
              θ
```

Among the closure material of R&G, a nice long calculation that translates almost directly can be found for Proposition 3.2.6(ii); in comparison with the proof by Schmidt and Ströhlein, only two applications of "Definition of $^+$" have been moved from the fourth to the fifth step—leaving them both in the fourth leads to time-out with the default settings:

```
Proposition (3.2.6.ii⁺): (R ∪ S) ⁺ = R ⁺ ∪ (R * ; S) ⁺ ; R *
Proof:
    (R ∪ S) ⁺
  =⟨ "Definition of ⁺" ⟩
    (R ∪ S) ; (R ∪ S) *
  =⟨ (3.2.6.ii*) ⟩
    (R ∪ S) ; (R * ; S) * ; R *
  =⟨ "Distributivity of ; over ∪", "* via ⁺" ⟩
    R ; (Id ∪ (R * ; S) ⁺) ; R * ∪ S ; (R * ; S) * ; R *
  =⟨ "Distributivity of ; over ∪", "Identity of ;" ⟩
    R ; R * ∪ R ; (R * ; S) ⁺ ; R * ∪ S ; (R * ; S) * ; R *
  =⟨ "Distributivity of ; over ∪", "Definition of ⁺" ⟩
    R ⁺ ∪ (R ⁺ ; S ∪ S) ; (R * ; S) * ; R *
  =⟨ "Identity of ;", "Distributivity of ; over ∪", "* via ⁺" ⟩
    R ⁺ ∪ (R * ; S) ; (R * ; S) * ; R *
  =⟨ "Definition of ⁺" ⟩
    R ⁺ ∪ (R * ; S) ⁺ ; R *
```

Before showing the proof of a property from Sect. 3.3 of R&G, we first list our versions of the definitions of the relevant bounds functions—we need to make the parameter (order E, respectively strict-order C) explicit, and we prefer to use residuals over nested complements where possible—in the variable names we follow Schmidt and Ströhlein, who motivate these definitions for vectors t:

- max C t selects the maximal elements of t with respect to a strict-order C.
- ubd E t denotes the vector of upper bounds of all elements of t with respect to an order E.
- gre E t selects from t the greatest element (if any) with respect to an order E.

However, the unchanged definitions actually work for unrestricted relations t, and there are also no restrictions implied on C and E:

```
Axiom "Definition of `max`":  max C t  =  t ∩ ~ (C ; t)
Axiom "Definition of `ubd`":  ubd E t  =  E ˘ / t ˘
Axiom "Definition of `gre`":  gre E t  =  t ∩ ubd E t
```

When proving Proposition 3.3.6, they write: "It suffices to show gre(t) ⊆ max(t)". However, to see this, the definitions for gre and max need to be expanded; the overhead this induces in our proof below should be more than compensated by our much more straight-forward proof of gre E R ⊆ max (E ∩ ~ Id) R:

```
Proposition (3.3.6.i.1):
  is-antisymmetric E  ⇒  gre E R = max (E ∩ ~ Id) R ∩ ubd E R
Proof:
  Assuming `is-antisymmetric E`
         and using with “Definition of antisymmetry”:
    Using “Mutual inclusion”:
      Subproof for `gre E R ⊆ max (E ∩ ~ Id) R ∩ ubd E R`:
        Using “Characterisation of ∩”:
          Subproof:
              gre E R
            =⟨ “Definition of `gre`” ⟩
              R ∩ ubd E R
            =⟨ “Definition of `ubd`” ⟩
              R ∩ E ˘ ╱ R ˘
            =⟨ “╱ via ~”, “Self-inverse of ˘” ⟩
              R ∩  ~ (~ (E ˘) ; R)
            ⊆⟨ Antitonicity with subproof for `E ∩ ~ Id ⊆ ~ (E ˘)`:
                       By “Contrapositive of ⊆ with ∩”
                          with assumption `is-antisymmetric E`
              ⟩
              R ∩ ~ ((E ∩ ~ Id) ; R)
            =⟨ “Definition of `max`” ⟩
              max (E ∩ ~ Id) R
          Subproof:
              gre E R
            ⊆⟨ “Definition of `gre`”, “Weakening for ∩” ⟩
              ubd E R
      Subproof:
          max (E ∩ ~ Id) R ∩ ubd E R
        =⟨ “Definition of `max`” ⟩
          R ∩ ~ ((E ∩ ~ Id) ; R) ∩ ubd E R
        ⊆⟨ “Weakening for ∩” ⟩
          R ∩ ubd E R
        =⟨ “Definition of `gre`” ⟩
          gre E R
```

The property used to supply the antecedent of Antitonicity has been wrapped into a subproof only for the purpose of documentation; without that wrapper, this is still checked. Also, the three steps of the last subproof above could be merged into one step, or the whole subproof could be replaced with:

```
Subproof for `max (E ∩ ~ Id) R ∩ ubd E R ⊆ gre E R`:
  By “Definition of `max`”, “Weakening for ∩”, “Definition of `gre`”
```

6 Arbitrary Meets and Joins

In the context of LADM, it is more natural to consider quantification using the meet and join operators instead of inf and sup operators taking sets of relations to relations. Following LADM, such quantifications take the same pattern as universal and existential quantifications in predicate logic; the following is the usual characterisation of arbitrary meets:

```
Axiom “Characterisation of ⋂”:
     Q ⊆ (⋂ x ❙ R • E)   ≡   (∀ x ❙ R • Q ⊆ E)
```

It is straightforward to show that such meets are the universal relation "⊤" exactly when all arguments are ⊤, too:

```
Theorem "Characterisation for `⋂ = T`":
    (⋂ x | R • E) = T  ≡  (∀ x | R • E = T)
Proof:
    (⋂ x | R • E) = T
  ≡⟨ "Inclusion of T" ⟩
    T ⊆ (⋂ x | R • E)
  ≡⟨ "Characterisation of ⋂" ⟩
    ∀ x | R • T ⊆ E
  ≡⟨ "Inclusion of T" ⟩
    ∀ x | R • E = T
```

When using "Instantiation", which is the LADM theorem corresponding to ∀-elimination, usually an explicit Substitution step is necessary in CalcCheck:

```
Theorem "Weakening for ⋂": R[x ≔ e] ⇒ (⋂ x | R • E) ⊆ E[x ≔ e]
Proof:
    (⋂ x | R • E) ⊆ (⋂ x | R • E)   — This is "Reflexivity of ⊆"
  ≡⟨ "Characterisation of ⋂" ⟩
    ∀ x | R • (⋂ x | R • E) ⊆ E
  ⇒⟨ "Instantiation" ⟩
    (R ⇒ (⋂ x | R • E) ⊆ E)[x ≔ e]
  ≡⟨ Substitution ⟩
    R[x ≔ e]  ⇒  (⋂ x | R • E) ⊆ E[x ≔ e]
```

We take the definitions of initial part and progressive finiteness from Definition 6.3.2 of R&G:

```
Axiom "Definition of `iniPart`":
    iniPart B  =  (⋂ x | x ; T = x  ∧  ~ x ⊆ B ; ~ x  •  x)
```

The simpler characterisation of progressive finiteness that follows on p. 121 requires an application of the LADM theorem "Change of dummy" of which we showed a more useful version in Kahl (2018).

```
Theorem "Characterisation of `is-prog-finite`":
    is-prog-finite B  ≡  (∀ y  |  y ; T = y  •  y ⊆ B ; y ⇒ y = ⊥)
Proof:
    is-prog-finite B
  ≡⟨ "Definition of `is-prog-finite`" ⟩
    iniPart B = T
  ≡⟨ "Definition of `iniPart`" ⟩
    (⋂ x | x ; T = x  ∧  ~ x ⊆ B ; ~ x • x) = T
  ≡⟨ "Characterisation for `⋂ = T`" ⟩
    ∀ x | x ; T = x  ∧  ~ x ⊆ B ; ~ x • x = T
  ≡⟨ "Trading for ∀" ⟩
    ∀ x | x ; T = x  •  ~ x ⊆ B ; ~ x ⇒ x = T
  ≡⟨ "Change of dummy " (8.22) with
     subproof for `(∀ x • ∀ y • x = ~ y  ⇒  y = ~ x)`:
         For any `x`, `y`:
           By "Equality with ~", "Reflexivity of ⇒"
  ⟩
    ∀ y | (x ; T = x)[x ≔ ~ y]  •  (~ x ⊆ B ; ~ x ⇒ x = T)[x ≔ ~ y]
  ≡⟨ Substitution ⟩
    ∀ y | ~ y ; T = ~ y  •  ~ ~ y ⊆ B ; ~ ~ y ⇒ ~ y = T
  ≡⟨ "Complement of vector", "Self-inverse of ~" ⟩
    ∀ y | y ; T = y  •  y ⊆ B ; y ⇒ ~ y = T
  ≡⟨ "Equality with ~", "Complement of T" ⟩
    ∀ y | y ; T = y  •  y ⊆ B ; y ⇒ y = ⊥
```

7 Conclusion

We picked mostly longer calculations from the textbook classic "Relations and Graphs" (R&G) by Schmidt and Ströhlein (1993), and in most cases were able to render them in CALCCHECK-checkable form with very similar step numbers and step organisation, to large part due to recent additions to CALCCHECK addressing the previous shortcomings listed in the introduction:

- Monotonicity/Antitonicity now also works for range- and body-monotonicity/-antitonicity of quantifications.
- Cyclic chains in orders can now be used to prove equalities.
- Declared converses of operators are internally translated.
- Assuming the antecedent can now be done with an additional "and using with" clause that gives easy access to consequences of the assumption.

Optically, the main difference still remaining is the fact that full hints are required in CALCCHECK; we posit that for educational purposes this is actually an advantage.

Yet more elegance can easily be achieved by producing more lemmas and giving them nicer and/or shorter names; in most cases we have deliberately shown the result of working in rather "raw" theories.

Some of the shortcomings seen in the examples will probably be addressed by adding new features to CALCCHECK:

- Activation of automatic definition expansion will reduce the cost (in hint space) of fine-granularity conceptual hierarchies realised though definitions building on each other.
- Proving non-equational pre-order steps by rewriting not of the whole step, but of the two sides of the step will make in particular inclusion reasoning much more concise.

Apart from these well-defined extensions, we will also investigate adding facilities for local definitions and for proofs of local facts.

By seamlessly integrating relation algebra with the quantification facilities and predicate-logic theories of LADM, we obtain a setting where large parts of R&G synergistically join LADM for the purpose of teaching discrete mathematics to beginning computer science students.

References

Aarts, C., Backhouse, R.C., Hoogendijk, P., Voermans, E., van der Woude, J.: A relational theory of datatypes. Working document, 387 pp. (1992). http://www.cs.nott.ac.uk/~rcb/papers/abstract.html#book

Bird, R.S., de Moor, O.: Algebra of Programming. International Series in Computer Science, vol. 100. Prentice Hall, Upper Saddle River (1997)

Contejean, E.: A certified AC matching algorithm. In: van Oostrom, V. (ed.) RTA 2004. LNCS, vol. 3091, pp. 70–84. Springer, Heidelberg (2004). https://doi.org/10.1007/978-3-540-25979-4_5

Gries, D.: Foundations for calculational logic. In: Broy, M., Schieder, B. (eds.) Mathematical Methods in Program Development. NATO ASI Series (Series F: Computer and Systems Sciences), pp. 83–126. Springer, Heidelberg (1997). https://doi.org/10.1007/978-3-642-60858-2_16

Gries, D., Schneider, F.B.: A Logical Approach to Discrete Math. Monographs in Computer Science. Springer, Heidelberg (1993). https://doi.org/10.1007/978-1-4757-3837-7

Kahl, W.: The teaching tool CalcCheck a proof-checker for Gries and Schneider's "logical approach to discrete math". In: Jouannaud, J.-P., Shao, Z. (eds.) CPP 2011. LNCS, vol. 7086, pp. 216–230. Springer, Heidelberg (2011). https://doi.org/10.1007/978-3-642-25379-9_17

Kahl, W.: CalcCheck: a proof checker for teaching the "logical approach to discrete math". In: Avigad, J., Mahboubi, A. (eds.) ITP 2018. LNCS, vol. 10895, pp. 324–341. Springer, Cham (2018). https://doi.org/10.1007/978-3-319-94821-8_19

Schmidt, G., Ströhlein, T.: Relations and Graphs. EATCS Monographs on Theoretical Computer Science. Discrete Mathematics for Computer Scientists. Springer, Heidelberg (1993). https://doi.org/10.1007/978-3-642-77968-8

Spivey, J.M.: The Z Notation: A Reference Manual. Prentice Hall International Series in Computer Science. Prentice Hall, Upper Saddle River (1989)

Author Index

Printed in the United States
By Bookmasters